Studies in Fuzziness and Soft Computing

Volume 364

Series editor

Janusz Kacprzyk, Polish Academy of Sciences, Warsaw, Poland
e-mail: kacprzyk@ibspan.waw.pl

The series "Studies in Fuzziness and Soft Computing" contains publications on various topics in the area of soft computing, which include fuzzy sets, rough sets, neural networks, evolutionary computation, probabilistic and evidential reasoning, multi-valued logic, and related fields. The publications within "Studies in Fuzziness and Soft Computing" are primarily monographs and edited volumes. They cover significant recent developments in the field, both of a foundational and applicable character. An important feature of the series is its short publication time and world-wide distribution. This permits a rapid and broad dissemination of research results.

More information about this series at http://www.springer.com/series/2941

Ali Ebrahimnejad · José Luis Verdegay

Fuzzy Sets-Based Methods and Techniques for Modern Analytics

 Springer

Ali Ebrahimnejad
Department of Mathematics
Qaemshahr Branch Islamic Azad University
Qaemshahr
Iran

José Luis Verdegay
Department of Computer Science
and Artificial Intelligence
University of Granada
Granada
Spain

ISSN 1434-9922   ISSN 1860-0808   (electronic)
Studies in Fuzziness and Soft Computing
ISBN 978-3-030-08882-8   ISBN 978-3-319-73903-8   (eBook)
https://doi.org/10.1007/978-3-319-73903-8

Printed on acid-free paper

This Springer imprint is published by the registered company Springer International Publishing AG part of Springer Nature
The registered company address is: Gewerbestrasse 11, 6330 Cham, Switzerland

To Arvin and Daniel

Preface

Analytics is the scientific process of transforming data into information to enable better decisions to be made. Therefore, analytics is fundamentally an interdisciplinary field, which has significant effects on attempts to design its core, its contents, and its boundaries. But this is not a easy task since, among other topics, analytics can consider

- Decision making (multicriteria and multiperson)
- Fusion of information from different sources and natures
- Analysis of data and information
- Management of sophisticated math software
- Problem solving
- Interpreting the meaning of information for others, etc.

On the one hand, analytics involves a great deal of different disciplines. On the other hand, if we review degrees in analytics from universities around the world, it is evident that the courses included therein are integrated into a variety of departments or faculties, such as Computing and Engineering departments, Business departments, or Management schools. Likewise, if we focus on master's degrees, a similar dispersion is observed, with some universities even offering courses within Mathematics or Statistics. This highlights that there is a certain discrepancy as what type of discipline analytics actually is, and there are a variety of aspects involved which come from different academic traditions.

Upon broadening the panorama a little further, it can be seen how analytics is related with three major scientific areas, namely Statistics, Intelligence of Business and Information Systems and Modeling and finally with Optimization. There are a few topics that are common to these areas, including Data Mining, Simulation and Risk, Intelligent Decision Support Systems, and Data Visualization.

Being analytics such an interdisciplinary field, an exhaustive coverage of all the topics that must be known to become an expert in analytics would be an impossible task for one single book. But it would also seem obvious that anyone who seeks to become an expert in analytics has the obligation to know the basic theoretical elements which will support their reports, analysis, and recommendations. Joining these two facts together leads us to the fundamentals that are to be found in the area that they share—Computer Science, Statistics, and Operations Research. In that common place the three pillars of: Design of Algorithms, Descriptive Statistics, and Mathematical Programming must be highlighted.

Now considering the point of view of professional practice, analytics cannot be understood as a purely theoretical subject; on the contrary, it must be principally seen from a practical point of view. This does not mean that some solid and rigorous scientific foundations should be excluded but that they must be oriented to solve real-world problems. In this context, it is necessary to consider the omnipresent inaccuracy in the majority of the information we have. In fact, we usually express measurements, judgments, classifications, etc., with linguistic (verbal) terms which, although they are perfectly understood by humans, are not easily transferred neither to the context of computers, nor to mathematical models based on Boolean logic. Thus, the models that fuzzy logic provides seem to be the most appropriate to find the solutions we are interested in.

The methods and models of Fuzzy Mathematical Programming are in the intersection between Statistics, Mathematical Programming, and Computational Intelligence. Rigorous knowledge of Fuzzy Mathematical Programming is essential for what we call modern analytics, i.e., analytics not exclusively based on conventional (Boolean) models, but in Computational Intelligence (fuzzy) models, which is to say in what we could refer to as human-centric models.

Therefore, in the context of modern analytics, Fuzzy Mathematical Programming plays a fundamental role in dealing with a various range of problems such as those concerning Smart Cities, Intelligent and Multimodal Transport, Renewable Energies, etc., with the right amount of seriousness and scientific rigor. Consequently, this book is dedicated to the study of different Fuzzy Mathematical Programming models and methods. By presenting extensive theoretical information together with practical tools, it represents an outstanding reference guide helping experts to find the best solutions to the real problems they must face in their different work environments in today's Smart Society.

The book is divided into five chapters. In the first, a brief introduction is given of the key concepts of Fuzzy Sets that are most relevant for the models that are considered in the rest of the volume. The following chapter is dedicated to the different types of problems that can be considered in Fuzzy Linear Programming (FLP) as well as to rigorously describe the different solving methods that exist, which, as a whole, constitute the basis of all the subsequent theoretical and practical developments. Following that, the third and fourth chapters cover Linear Programming problems involving fuzzy parameters, by means of focuses based on the Simplex Algorithm and not based on the Simplex Algorithm, respectively. Finally, one of the problems that is most frequently the motive for analytics is considered: the Transport Problem. The relevance today of this type of problem in a variety of areas such as scheduling, production, investment, deciding plant location, inventory control, as well as all those related in some way to Intelligent Transport, means that its inclusion as a demonstration of the practical applicability of FLP problems is more than justified.

Because of this, the preparation of this manuscript has been supported through Projects TIN2014-55024-P and P11-TIC-8001 from the Spanish Ministry of Economy and Competitiveness and the Andalusian Counseling of Economy, Innovation and Science (both including FEDER funds), respectively.

To conclude, we the authors wish to express our special recognition to Prof. Janusz Kacprzyk, who from the very first moment accepted our book proposal and who encouraged us continually throughout the preparation of this volume. We would also like to mention the assistance provided by Mr. Anthony Carlson in reviewing the language used in the manuscript. And needless to say, the authors wish to manifest our sincere gratitude to our respective families for the support, understanding, and patience that they have shown us throughout the time that we have dedicated to preparing this book.

Qaemshahr, Iran   Ali Ebrahimnejad
Granada, Spain   José Luis Verdegay

Contents

Chapter 1
Fuzzy Set Theory

1.1 Introduction

Fuzzy sets as defined by Zadeh [1] provide a useful, effective and operative tool for handling imprecise data. It not only provides a powerful representation of measurement uncertainty, but also gives a meaningful representation of vague concepts expressed in natural language.

When knowledge and information are incomplete, precise mathematics is not sufficient to model a complex system. The probability theory is traditionally used to handle this uncertainty. However, because this approach is suitable to represent some information defined with certain boundaries; the probability theory cannot formulate all the possible problems of the incomplete aspects. The fuzzy set theory has been developed to solve uncertain problems with sharp boundaries. The aim of this chapter is to introduce the main concepts of fuzzy set theory.

Some appropriate references for this chapter are Klir and Yuan [2], Lai and Hwang [3], Dubois and Prade [4], Zimmermann [5] and Lodwick and Thipwiwatpotjana [6].

1.2 Crisp Sets

The fuzzy set theory is inseparable from the classical (crisp) set theory and it is difficult to understand the concepts of fuzzy sets before understanding the concepts of crisp sets. Thus, prior to discussing the concepts of fuzzy set theory, we shall briefly explore the main components of crisp sets.

Definition 1.1 Let X denote the universal set. A set is defined by a function, namely a characteristic function, that declares which elements of the universal set X are members of the set and which are not. Set A in terms of its characteristic function, $\chi_A(x)$, is defined as follows:

© Springer International Publishing AG, part of Springer Nature 2018
A. Ebrahimnejad and J. L. Verdegay, *Fuzzy Sets-Based Methods and Techniques for Modern Analytics*, Studies in Fuzziness and Soft Computing 364,
https://doi.org/10.1007/978-3-319-73903-8_1

$$\chi_A(x) = \begin{cases} 1, & x \in A, \\ 0, & x \notin A, \end{cases} \tag{1.1}$$

This means that the characteristic function maps elements of X to elements of the set $\{0, 1\}$ and expressed formally by $\chi_A : X \to \{0, 1\}$. For each $x \in X$, when $\chi_A(x) = 1$, x is a member of A; when $\chi_A(x) = 0$, x is a non-member of A. Therefore, in this case the set A is completely characterized by the set of ordered pairs $A = \{(x, \chi_A(x)) | x \in X\}$ where the second component of this ordered pair declares the membership or non-membership of the first component to the set A.

Definition 1.2 Set A is called a subset of set B, if every member of set A is also a member of set B and this will be denoted by $A \subseteq B$. Therefore, $A \subseteq B$, if for each $x \in X$, $\chi_A(x) \le \chi_B(x)$.

Definition 1.3 Sets A and B are called equal if $A \subseteq B$ and $B \subseteq A$ this will be denoted by $A = B$. Therefore, $A = B$, if for each $x \in X$, $\chi_A(x) = \chi_B(x)$.

Definition 1.4 The complement of set A is the set containing all the members of the universal set X that are not also members of A and this will be denoted by $\bar{A}$. Therefore,

$$\bar{A} = \{(x, \chi_{\bar{A}}(x)) | x \in X, \chi_{\bar{A}}(x) = 1 - \chi_A(x)\}.$$

Remark 1.1 The complement of the empty set equals the universal set, and the complement of the universal set equals the empty set. That is, $\bar{\varnothing} = X$ and $\bar{X} = \varnothing$.

Definition 1.5 The union of A and B, denoted by $A \cup B$, is the set containing all the elements that belong either to set A alone, to set B alone, or to both sets A and B. Therefore,

$$A \cup B = \{(x, \chi_{A \cup B}(x)) | x \in X, \chi_{A \cup B}(x) = \max\{\chi_A(x), \chi_B(x)\}\}.$$

Definition 1.6 The intersection of A and B, denoted by $A \cap B$, is the set containing all the elements that belong to both sets A and B. Therefore,

$$A \cap B = \{(x, \chi_{A \cap B}(x)) | x \in X, \chi_{A \cap B}(x) = \min\{\chi_A(x), \chi_B(x)\}\}.$$

Remark 1.2 One of the fundamentals in the crisp set is the law of excluded middle $A \cup \bar{A} = X$ and contradiction $A \cap \bar{A} = \varnothing$.

Definition 1.7 The Cartesian product of two sets A and B, denoted by $A \times B$, is the set of all ordered pairs such that the first element of each pair belongs to A and the second element belongs to B. Therefore,

$$A \times B = \{((x, y), \chi_{A \times B}(x, y)) | x \in X, \chi_{A \times B}(x, y) = \min\{\chi_A(x), \chi_B(y)\}\}.$$

In a similar way, the Cartesian product of a family $\{A_1, A_2, \ldots, A_n\}$, denoted by $A_1 \times A_2 \times \cdots \times A_n$, is defined as follows:

$$A_1 \times A_2 \times \cdots \times A_n = \left\{ ((x_1, \ldots, x_n), \chi_{A_1 \times A_2 \times \cdots \times A_n}(x_1, \ldots, x_n)) | x_i \in X_i, i = 1, \ldots, n \right\}$$

where

$$\chi_{A_1 \times A_2 \times \cdots \times A_n}(x_1, \ldots, x_n) = \min\left\{ \chi_{A_i}(x_i), i = 1, 2, \ldots, n \right\}$$

Definition 1.8 A crisp set A on X is called a convex set if $\lambda x_1 + (1 - \lambda)x_2 \in X$ for all $x_1, x_2 \in X$ and all $\lambda \in [0, 1]$. That is,

$$\chi_A(\lambda x_1 + (1 - \lambda)x_2) \geq \min\{\chi_A(x_1), \chi_A(x_2)\}$$

1.3 Basic Concepts of Fuzzy Sets

In this section, we shall introduce some basic concepts of fuzzy sets.

Definition 1.9 The characteristic function χ_A of a crisp set A assigns a value of either one or zero to each individual in the universal set X. This function can be generalized to a function μ such that the values assigned to the element of the universal set X fall within the specified range $[0, 1]$. The assigned value indicates the membership grade of the element in the set under consideration. Larger values denote higher degrees of set membership. Such a function is called a membership function and the set defined by this function is called a fuzzy set. In sum, the membership function of a fuzzy set $\tilde{A}$, denoted by $\mu_{\tilde{A}}$, maps elements of the universal set X into real numbers in $[0, 1]$; i.e., $\mu_{\tilde{A}} : X \rightarrow [0, 1]$. Therefore, in this case the fuzzy set $\tilde{A}$ is completely characterized by the set of ordered pairs $\tilde{A} = \{(x, \mu_{\tilde{A}}(x)) | x \in X\}$ where the second component of this ordered pair declares the membership degree of the first component to the fuzzy set $\tilde{A}$.

Example 1.1 Consider the fuzzy set $\tilde{A}$ as natural numbers that are close to number five. The membership function of the fuzzy set $\tilde{A}$ can be defined as follows:

$$\tilde{A} = \{(1, 0.1), (2, 0.4), (3, 0.6), (4, 0.8), (5, 1), (6, 0.8), (7, 0.6), (8, 0.4), (9, 0.1)\} \tag{1.2}$$

Example 1.2 Consider the fuzzy set $\tilde{B}$ as real numbers that are close to number five. The membership function of the fuzzy set $\tilde{B}$ can be defined as follows:

$$\tilde{B} = \left\{ (x, \mu_{\tilde{B}}(x)) \,\middle|\, \mu_{\tilde{B}}(x) = \frac{1}{1 + (x - 5)^2}, x \in \mathbb{R} \right\} \tag{1.3}$$

Remark 1.3 A fuzzy set $\tilde{A}$ defined on a finite universal set $X = \{x_1, x_2, \ldots, x_n\}$ can be rewritten as follows:

$$\tilde{A} = \frac{\mu_{\tilde{A}}(x_1)}{x_1} + \frac{\mu_{\tilde{A}}(x_2)}{x_2} + \cdots + \frac{\mu_{\tilde{A}}(x_n)}{x_n} = \sum_{i=1}^{n} \frac{\mu_{\tilde{A}}(x_i)}{x_i} \tag{1.4}$$

Example 1.3 Given fuzzy set $\tilde{A}$ in the Eq. (1.2) is written as follows:

$$\tilde{B} = \frac{0.1}{1} + \frac{0.4}{2} + \frac{0.6}{3} + \frac{0.8}{4} + \frac{1}{5} + \frac{0.8}{6} + \frac{0.6}{7} + \frac{0.4}{8} + \frac{0.1}{9} \tag{1.5}$$

Remark 1.4 A fuzzy set $\tilde{A}$ defined on the infinite universal set X can be rewritten as follows:

$$\tilde{A} = \int_{X} \frac{\mu_{\tilde{A}}(x)}{x} \tag{1.6}$$

Example 1.4 The fuzzy set $\tilde{B}$ given in the Eq. (1.3) is written as follows:

$$\tilde{B} = \int_{-\infty}^{\infty} \frac{\frac{1}{1 + (x-5)^2}}{x} \tag{1.7}$$

Example 1.5 Let the membership function of the fuzzy set $\tilde{C}$, the set of real numbers close to number six, be defined as follows (see Fig. 1.1):

$$\mu_{\tilde{C}}(x) = \begin{cases} 0 & x \leq 2 \\ \frac{x-2}{4} & 2 < x \leq 6 \\ \frac{10-x}{4} & 6 < x \leq 10 \\ 0 & x > 10 \end{cases} \tag{1.8}$$

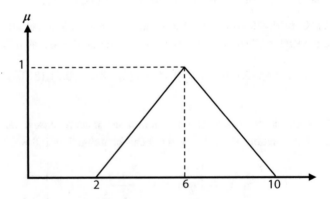

Fig. 1.1 The membership function of fuzzy set $\tilde{C}$

In this case, the fuzzy set $\tilde{C}$ can be written as follows:

$$\tilde{C} = \int_2^6 \frac{\frac{x-2}{4}}{x} + \int_6^{10} \frac{\frac{10-x}{4}}{x} \tag{1.9}$$

Definition 1.10 The support of a fuzzy set $\tilde{A}$, denoted by $supp\,(\tilde{A})$, is a crisp set of the universal set X containing all the elements of X with non-zero membership degree in $\tilde{A}$. That is,

$$supp\,(\tilde{A}) = \{x \in X; \mu_{\tilde{A}}(x) > 0\} \tag{1.10}$$

Example 1.6 The support of fuzzy set $\tilde{A}$, given in Example 1.1, is $supp\,(\tilde{A}) = \{1, 2, 3, 4, 5, 6, 7, 8, 9\}$.

Example 1.7 The support of fuzzy set $\tilde{C}$, given in Example 1.5, is $supp\,(\tilde{C}) = (2, 10)$.

Definition 1.11 The height of a fuzzy set $\tilde{A}$, denoted by $h\,(\tilde{A})$, is the largest membership degree obtained by any element in that set. That is,

$$h\,(\tilde{A}) = \sup\{\mu_{\tilde{A}}(x)|x \in X\} \tag{1.11}$$

Definition 1.12 A fuzzy set $\tilde{A}$ is called normal when $h\,(\tilde{A}) = 1$; it is called subnormal when $h\,(\tilde{A}) < 1$ (Fig. 1.2).

Definition 1.13 A fuzzy set $\tilde{A}$ on X is called a convex set if $\mu_{\tilde{A}}(\lambda x_1 + (1 - \lambda)x_2) \geq \min\{\mu_{\tilde{A}}(x_1), \mu_{\tilde{A}}(x_2)\}$ for all $x_1, x_2 \in X$ and all $\lambda \in [0, 1]$.
Figure 1.3 illustrates a convex fuzzy set and a nonconvex fuzzy set.

Definition 1.14 The core of a fuzzy set $\tilde{A}$, denoted by $core\,(\tilde{A})$, is a subset of X that contains all the elements of X that have membership degrees of one. That is,

$$core\,(\tilde{A}) = \{x \in X|\mu_{\tilde{A}}(x) = 1\} \tag{1.12}$$

Fig. 1.2 Normal and subnormal fuzzy sets

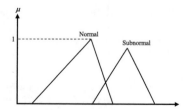

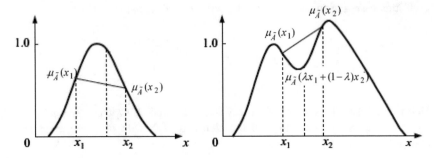

Fig. 1.3 Convex fuzzy set and nonconvex fuzzy set

Example 1.8 For the fuzzy set $\tilde{A}$, given in Example 1.1, $core\,(\tilde{A}) = 5$ and for the fuzzy set $\tilde{C}$, given in Example 1.5, $core\,(\tilde{C}) = 6$.

Definition 1.15 Let $\tilde{A}_1$, $\tilde{A}_2, \ldots, \tilde{A}_n$ be n fuzzy subsets of X_1, $X_2, \ldots, X_n$, respectively. The Cartesian product of $\tilde{A}_1$, $\tilde{A}_2, \ldots, \tilde{A}_n$ is a fuzzy set on $X = X_1 \times X_2 \times \cdots \times X_n$ defined as follows:

$$\tilde{A}_1 \times \tilde{A}_2 \times \cdots \times \tilde{A}_n = \left\{((x_1, \ldots, x_n), \mu_{\tilde{A}_1 \times \tilde{A}_2 \times \cdots \times \tilde{A}_n}(x_1, \ldots, x_n))|x_i \in X_i, i = 1, \ldots, n\right\}$$

where

$$\mu_{\tilde{A}_1 \times \tilde{A}_2 \times \cdots \times \tilde{A}_n}(x_1, \ldots, x_n) = \min\left\{\mu_{\tilde{A}_i}(x_i), i = 1, 2, \ldots, n\right\}$$

Example 1.9 Let $\tilde{A}_1 = \{(-1, 0.5), (0, 1), (1, 0.2)\}$ and $\tilde{A}_2 = \{(2, 0.5), (3, 1)\}$. Then,

$$\tilde{A}_1 \times \tilde{A}_2 = \{((-1, 2), 0.5), ((-1, 3), 0.5), ((0, 2), 0.5), ((0, 3), 1), ((1, 2), 0.2), ((1, 3), 0.2)\}$$

Definition 1.16 (Extension principle [7]): Let f be a mapping form $X = X_1 \times X_2 \times \cdots \times X_n$ to a universe Y, i.e., $f : X \rightarrow Y$, and $\tilde{A} = \tilde{A}_1 \times \tilde{A}_2 \times \cdots \times \tilde{A}_n$ be a fuzzy set on X. Then, $\tilde{B} = f(\tilde{A})$ is a fuzzy set on Y defined as follows:

$$\tilde{B} = f(\tilde{A}_1 \times \tilde{A}_2 \times \cdots \times \tilde{A}_n) = \left\{(y, \mu_{\tilde{B}}(y)), y = f(x_1, x_2, \ldots, x_n), x_i \in X_i, 1 \le i \le n\right\}$$

where

$$\mu_{\tilde{B}}(y) = \begin{cases} \sup\limits_{x\,:\,y = f(x_1, x_2, \ldots, x_n)}\;\min\limits_{1 \le i \le n}\left\{\mu_{\tilde{A}_i}(x)\right\} & f^{-1}(y) \ne \varnothing \\ 0 & f^{-1}(y) = \varnothing \end{cases}$$

Example 1.10 For the fuzzy sets given in Example 1.9, let $f(x_1, x_2) = x_1 x_2$. Then, we have $\tilde{B} = f(\tilde{A}_1, \tilde{A}_2) = \{(-3, 0.5), (-2, 0.5), (0, 1), (2, 0.2), (3, 0.2)\}$.

1.4 Basic Operations of Fuzzy Sets

In this section, we develop the basic operations on crisp sets to fuzzy sets. As the characteristic function has an important role to define operations on crisp sets, the membership function plays a similar role to define operations on fuzzy sets.

Definition 1.17 Fuzzy set $\tilde{A}$ is called a subset of fuzzy set $\tilde{B}$, if for each $x \in X$, $\mu_{\tilde{A}}(x) \leq \mu_{\tilde{B}}(x)$.

Example 1.11 Let $\tilde{A} = \{(0, 0.4), (1, 0.7)\}$ and $\tilde{B} = \{(-1, 0.5), (0, 1), (1, 0.8)\}$. Then, $\tilde{A} \subseteq \tilde{B}$.

Definition 1.18 Two fuzzy sets $\tilde{A}$ and $\tilde{B}$ are called equal if for each $x \in X$, $\mu_{\tilde{A}}(x) = \mu_{\tilde{B}}(x)$.

Definition 1.19 The complement of a fuzzy set $\tilde{A}$, denoted by $\bar{\tilde{A}}$, is defined as follows:

$$\bar{\tilde{A}} = \left\{ (x, \mu_{\bar{\tilde{A}}}(x)) \big| x \in X, \mu_{\bar{\tilde{A}}}(x) = 1 - \mu_{\tilde{A}}(x) \right\}.$$

Example 1.12 The complement of the fuzzy set $\tilde{A}$ given in Example 1.1 is as follows:

$$\bar{\tilde{A}} = \{(1, 0.9), (2, 0.6), (3, 0.4), (4, 0.2), (5, 0), (6, 0.2), (7, 0.4), (8, 0.6), (9, 0.9)\}$$

Example 1.13 The complement of the fuzzy set $\tilde{C}$ given in Example 1.5 is defined as follows (see Fig. 1.4):

$$\mu_{\bar{\tilde{C}}}(x) = \begin{cases} 1 & x \leq 2 \\ \frac{6-x}{4} & 2 < x \leq 6 \\ \frac{x-6}{4} & 6 < x \leq 10 \\ 1 & x > 10 \end{cases} \tag{1.13}$$

Definition 1.20 The union of $\tilde{A}$ and $\tilde{B}$, denoted by $\tilde{A} \tilde{\cup} \tilde{B}$, is defined as follows:

$$\tilde{A} \tilde{\cup} \tilde{B} = \left\{ (x, \mu_{\tilde{A} \cup \tilde{B}}(x)) \big| x \in X, \mu_{\tilde{A} \cup \tilde{B}}(x) = \max\{\mu_{\tilde{A}}(x), \mu_{\tilde{B}}(x)\} \right\}$$

Example 1.14 The union of the fuzzy set $\tilde{C}$ given in Example 1.5 and its complement $\bar{\tilde{C}}$ is as follows (see Fig. 1.5):

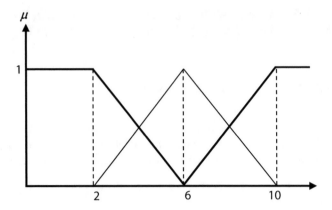

Fig. 1.4 The membership function of the complement of $\tilde{C}$

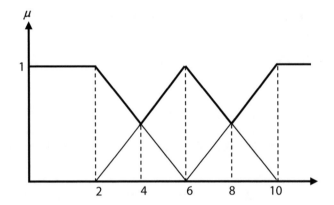

Fig. 1.5 The membership function of the union set

$$\mu_{\tilde{C}\cup\bar{\tilde{C}}}(x) == \begin{cases} 1 & x \leq 2 \\ \frac{6-x}{4} & 2 < x \leq 4 \\ \frac{x-2}{4} & 4 < x \leq 6 \\ \frac{10-x}{4} & 6 < x \leq 8 \\ \frac{x-6}{4} & 8 < x \leq 10 \\ 1 & x > 10 \end{cases} \tag{1.14}$$

Definition 1.21 The intersection of $\tilde{A}$ and $\tilde{B}$, denoted by $\tilde{A} \cap \tilde{B}$, is defined as follows:

$$\tilde{A} \cap \tilde{B} = \left\{ (x, \mu_{\tilde{A}\cap\tilde{B}}(x)) \,\middle|\, x \in X, \mu_{\tilde{A}\cap\tilde{B}}(x) = \min\{\mu_{\tilde{A}}(x), \mu_{\tilde{B}}(x)\} \right\}$$

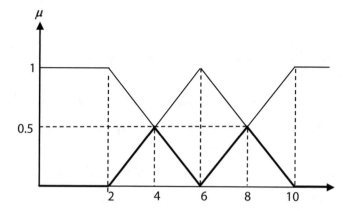

Fig. 1.6 The membership function of the intersection set

Example 1.15 The intersection of the fuzzy set $\tilde{C}$ given in Example 1.5 and its complement $\bar{\tilde{C}}$ is as follows (see Fig. 1.6):

$$\mu_{\tilde{C} \cap \bar{\tilde{C}}}(x) = \begin{cases} 0 & x \leq 2 \\ \frac{x-2}{4} & 2 < x \leq 4 \\ \frac{6-x}{4} & 4 < x \leq 6 \\ \frac{x-6}{4} & 6 < x \leq 8 \\ \frac{10-x}{4} & 8 < x \leq 10 \\ 0 & x > 10 \end{cases} \qquad (1.15)$$

Remark 1.5 The Examples 1.14 and 1.15 verify that the laws of excluded middle and contradiction are violated for the fuzzy sets.

1.5 Alpha-Cut and Its Properties

In this section, we introduce the concept of alpha-cut as one of the important concepts of fuzzy sets which plays a principal role in the relation between fuzzy sets and crisp sets.

Definition 1.22 The $\alpha-$ cut of fuzzy set $\tilde{A}$, denoted by $\left[\tilde{A}\right]_{\alpha}$, is a crisp set that contains all the elements of the universal set X whose membership degree in $\tilde{A}$ are greater than or equal to the value of α. That is,

$$\left[\tilde{A}\right]_{\alpha} = \left\{x \in X \mid \mu_{\tilde{A}}(x) \geq \alpha\right\}$$

Definition 1.23 The set of all levels $\alpha \in [0, 1]$ that represent distinct $\alpha-$ cuts of a given fuzzy set $\tilde{A}$ is called a level set of $\tilde{A}$. That is,

$$\Lambda(\tilde{A}) = \left\{\alpha \,\middle|\, \mu_{\tilde{A}}(x) = \alpha \text{ for some } x \in X\right\}$$

Example 1.16 For the fuzzy set $\tilde{A}$ given in Example 1.1, we have $[\tilde{A}]_{0.5} = \{3, 4, 5, 6, 7\}$.

Example 1.17 For the fuzzy set $\tilde{C}$ given in Example 1.5, we have $[\tilde{C}]_{0.5} = [4, 8]$.

Remark 1.6 For the fuzzy sets $\tilde{A}_1$ and $\tilde{A}_2$ defined on the universal set X, we have $\tilde{A}_1 \subseteq \tilde{A}_2$ if, and only if, $[\tilde{A}_1]_\alpha \subseteq [\tilde{A}_2]_\alpha$ for each $\alpha \in [0, 1]$.

Theorem 1.1 [2] *Let $\tilde{A}$ and $\tilde{B}$ be two fuzzy sets of a universal set X. Then, we have*

   i. $\alpha \leq \beta$ implies $[\tilde{A}]_\beta \subseteq [\tilde{B}]_\alpha$;

  ii. $[\tilde{A} \cup \tilde{B}]_\alpha = [\tilde{A}]_\alpha \cup [\tilde{B}]_\alpha$;

 iii. $[\tilde{A} \cap \tilde{B}]_\alpha = [\tilde{A}]_\alpha \cap [\tilde{B}]_\alpha$.

Proof

(i)   For any $x \in [\tilde{A}]_\beta$, we have $\mu_{\tilde{A}}(x) \geq \beta$ and hence $\mu_{\tilde{A}}(x) \geq \alpha$. This implies that $x \in [\tilde{B}]_\beta$ and, consequently, $[\tilde{A}]_\beta \subseteq [\tilde{B}]_\alpha$.

(ii)   For any $x \in [\tilde{A} \cup \tilde{B}]_\alpha$, we have $\mu_{\tilde{A} \cup \tilde{B}}(x) = \max\{\mu_{\tilde{A}}(x), \mu_{\tilde{B}}(x)\} \geq \alpha$. This means that $\mu_{\tilde{A}}(x) \geq \alpha$ or $\mu_{\tilde{B}}(x) \geq \alpha$. This implies that $x \in [\tilde{A}]_\alpha$ or $x \in [\tilde{B}]_\alpha$, and, hence, $x \in [\tilde{A}]_\alpha \cup [\tilde{B}]_\alpha$. Consequently $[\tilde{A} \cup \tilde{B}]_\alpha \subseteq [\tilde{A}]_\alpha \cup [\tilde{B}]_\alpha$. Conversely, for any $x \in [\tilde{A}]_\alpha \cup [\tilde{B}]_\alpha$, we have $x \in [\tilde{A}]_\alpha$ or $x \in [\tilde{B}]_\alpha$. This means that $\mu_{\tilde{A}}(x) \geq \alpha$ or $\mu_{\tilde{B}}(x) \geq \alpha$. This implies that $\mu_{\tilde{A} \cup \tilde{B}}(x) = \max\{\mu_{\tilde{A}}(x), \mu_{\tilde{B}}(x)\} \geq \alpha$, and hence, $x \in [\tilde{A} \cup \tilde{B}]_\alpha$. Consequently, $[\tilde{A}]_\alpha \cup [\tilde{B}]_\alpha \subseteq [\tilde{A} \cup \tilde{B}]_\alpha$. This concludes that $[\tilde{A} \cup \tilde{B}]_\alpha = [\tilde{A}]_\alpha \cup [\tilde{B}]_\alpha$.

(iii)   For any $x \in [\tilde{A} \cap \tilde{B}]_\alpha$, we have $\mu_{\tilde{A} \cap \tilde{B}}(x) = \min\{\mu_{\tilde{A}}(x), \mu_{\tilde{B}}(x)\} \geq \alpha$. This means that $\mu_{\tilde{A}}(x) \geq \alpha$ and $\mu_{\tilde{B}}(x) \geq \alpha$. This implies that $x \in [\tilde{A}]_\alpha$ and $x \in [\tilde{B}]_\alpha$, and hence, $x \in [\tilde{A}]_\alpha \cap [\tilde{B}]_\alpha$. Consequently $[\tilde{A} \cap \tilde{B}]_\alpha \subseteq [\tilde{A}]_\alpha \cap [\tilde{B}]_\alpha$. Conversely, for any $x \in [\tilde{A}]_\alpha \cap [\tilde{B}]_\alpha$, we have $x \in [\tilde{A}]_\alpha$ and $x \in [\tilde{B}]_\alpha$. This means that $\mu_{\tilde{A}}(x) \geq \alpha$ and $\mu_{\tilde{B}}(x) \geq \alpha$. This implies that $\mu_{\tilde{A} \cap \tilde{B}}(x) = \min\{\mu_{\tilde{A}}(x), \mu_{\tilde{B}}(x)\} \geq \alpha$, and hence, $x \in [\tilde{A} \cap \tilde{B}]_\alpha$. Consequently, $[\tilde{A}]_\alpha \cap [\tilde{B}]_\alpha \subseteq [\tilde{A} \cap \tilde{B}]_\alpha$. This concludes that $[\tilde{A} \cap \tilde{B}]_\alpha = [\tilde{A}]_\alpha \cap [\tilde{B}]_\alpha$.   □

Theorem 1.2 [2] *A fuzzy set $\tilde{A}$ of a universal set X is convex if, and only if, $[\tilde{A}]_\alpha$ is convex for any $\alpha \in (0, 1]$.*

Proof Assume that a fuzzy set $\tilde{A}$ is a convex set in the sense of Definition 1.13 and take $x_1, x_2 \in [\tilde{A}]_\alpha$ for any $\alpha \in (0, 1]$. Then, $\mu_{\tilde{A}}(x_1) \geq \alpha$ and $\mu_{\tilde{A}}(x_2) \geq \alpha$. Hence, according to Definition 1.13, we have $\mu_{\tilde{A}}(\lambda x_1 + (1 - \lambda)x_2) \geq \alpha$ for any $\lambda \in [0, 1]$. This implies that $(\lambda x_1 + (1 - \lambda)x_2) \in [\tilde{A}]_\alpha$ and hence, $[\tilde{A}]_\alpha$ is convex for any

$\alpha \in (0, 1]$. Conversely, assume that $\left[\tilde{A}\right]_{\alpha}$ is convex for any $\alpha \in (0, 1]$ and there exist $x_1', x_2' \in X$ and $\lambda' \in [0, 1]$ such that

$$\mu_{\tilde{A}}(\lambda' x_1' + (1 - \lambda') x_2') < \min\{\mu_{\tilde{A}}(x_1'), \mu_{\tilde{A}}(x_2')\}$$

Let $\alpha' = \min\{\mu_{\tilde{A}}(x_1'), \mu_{\tilde{A}}(x_2')\}$. Then, $x_1', x_2' \in \left[\tilde{A}\right]_{\alpha'}$ and $(\lambda' x_1' + (1 - \lambda') x_2') \notin \left[\tilde{A}\right]_{\alpha'}$ which means $\left[\tilde{A}\right]_{\alpha}$ is not convex and leads to a contradiction. □

Note that using the concept of $\alpha-$ level sets, the relationship between ordinary sets and fuzzy sets can be characterized by the following theorem.

Theorem 1.3 (Decomposition Theorem [2]) *A fuzzy set $\tilde{A}$ can be represented by* $\tilde{A} = \cup_{\alpha \in [0,1]} \alpha [A]_{\alpha}$.

Proof We have

$$\mu_{\cup_{\alpha \in [0,1]} \alpha [\tilde{A}]_{\alpha}}(x) = \sup_{\alpha \in [0,1]} \mu_{\alpha [\tilde{A}]_{\alpha}}(x) = \sup_{\alpha \in [0,1]} \alpha \chi_{[A]_{\alpha}}(x) = \sup_{\alpha \le \mu_{\tilde{A}}(x)} \alpha = \mu_{\tilde{A}}(x)$$

This completes the proof. □

It should be noted that Theorem 1.3 states that a fuzzy set $\tilde{A}$ can be decomposed into a series of $\alpha-$ cuts by which $\tilde{A}$ can be reconstructed. Thus any fuzzy set can be considered as a family of ordinary sets.

1.6 Fuzzy Numbers

In this section, we shall first introduce the concept of a special kind of fuzzy sets that are defined on the set of real numbers known as fuzzy numbers and then present a complete characterization of such fuzzy sets.

Definition 1.24 A fuzzy number is a convex normalized fuzzy set of the real line $\mathbb{R}$, whose membership function is piecewise continuous. We denote the set of fuzzy numbers on $\mathbb{R}$ with $F(\mathbb{R})$.

The following theorem gives a complete characterization of the membership functions of fuzzy numbers [2].

Theorem 1.4 *A fuzzy set $\tilde{A}$ on $\mathbb{R}$ is a fuzzy number if, and only if, there exists a nonempty closed interval $[m, n]$ such that*

$$\mu_{\tilde{A}}(x) = \begin{cases} l(x), & x \le m, \\ 1, & m \le x \le n, \\ r(x), & x \ge n. \end{cases} \tag{1.16}$$

where (i) $l : (-\infty, m) \to [0, 1]$ is monotonic increasing continuous from the right, and $l(x) = 0$ for $x \in (-\infty, w_1)$, $w_1 \le m$; and (ii) $r : (n, \infty) \to [0, 1]$ is monotonic decreasing continuous from the left, and $r(x) = 0$ for $x \in (w_2, \infty)$, $w_2 \ge n$.

Remark 1.7 In the membership function defined in (1.16), if $m = n$, the fuzzy set $\tilde{A}$ is called a fuzzy number and if $m \ne n$ it is called a flat fuzzy number (fuzzy interval).

Remark 1.8 Each real number like a is viewed as a fuzzy number. To do this, let $w_1 = m = n = w_2 = a$, $l(x) = 0$ for $x \in (-\infty, a)$ and $r(x) = 0$ for $x \in (a, \infty)$.

Remark 1.9 Each real interval like $[a, b]$ is viewed as a flat fuzzy number. To do this, let $w_1 = m = a$, $w_2 = n = b$, $l(x) = 0$ for $x \in (-\infty, a)$ and $r(x) = 0$ for $x \in (b, \infty)$.

Example 1.18 The fuzzy set $\tilde{C}$ given in Example 1.5 is a fuzzy number and can be represented in the form of (1.16) as follows: $w_1 = 2$, $m = n = 6$, $w_2 = 10$, and

$$l(x) = \begin{cases} 0, & x < 2 \\ \frac{x-2}{4}, & 2 \le x < 6. \end{cases}, r(x) = \begin{cases} \frac{10-x}{4}, & 6 < x \le 10 \\ 0, & x > 10. \end{cases}$$

Example 1.19 The fuzzy set $\tilde{D}$ with the following membership function is a flat fuzzy number (see Fig. 1.7). This fuzzy set can represent the concept of a middle-aged person.

$$\mu_{\tilde{D}}(x) = \begin{cases} 0, & x \le 20, \\ \frac{x-20}{15}, & 20 < x < 35 \\ 1, & 35 \le x \le 45, \\ \frac{60-x}{15}, & 45 < x < 60, \\ 0, & x \ge 60, \end{cases} \tag{1.17}$$

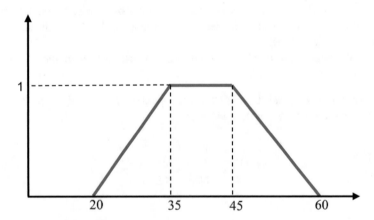

Fig. 1.7 The membership function of the flat fuzzy number $\tilde{D}$

This flat fuzzy number is defined in terms of (1.16) as follows: $w_1 = 20$, $m = 35$, $n = 45$, $w_2 = 60$, and

$$l(x) = \begin{cases} 0, & x < 20 \\ \frac{x-20}{15}, & 20 \leq x < 35. \end{cases}, r(x) = \begin{cases} \frac{60-x}{15}, & 45 < x \leq 60 \\ 0, & x > 60. \end{cases}$$

Definition 1.25 A fuzzy number $\tilde{A}$ is called positive (resp. negative) if its membership function is such that $\mu_{\tilde{A}}(x) = 0$ for all $x < 0$ (resp. $x > 0$).

Example 1.20 The fuzzy sets $\tilde{A}_1$ and $\tilde{A}_2$ with the following membership functions are respectively positive and negative fuzzy numbers. They can be viewed as fuzzy numbers close to 3 and -3, respectively.

$$\mu_{\tilde{A}_1}(x) = \begin{cases} 0, & x < 1, x > 5 \\ \frac{x-1}{2}, & 1 \leq x \leq 3, \\ \frac{5-x}{2}, & 3 < x \leq 5 \end{cases} \quad (1.18)$$

$$\mu_{\tilde{A}_2}(x) = \begin{cases} 0, & x < -5, x > -1 \\ \frac{x+5}{2}, & -5 \leq x \leq -3, \\ \frac{-1-x}{2}, & -3 < x \leq -1. \end{cases} \quad (1.19)$$

Definition 1.26 [4] A function $L : (-\infty, \infty) \rightarrow [0, 1]$ is said to be the reference function of fuzzy numbers if

(i) $L(x) = L(-x) \; \forall x \in (-\infty, \infty)$,
(ii) $L(0) = 1$,
(iii) $L(.)$ is non-increasing and upper semi-continuous on $[0, \infty)$,
(iv) Either $L(1) = 0$ or $\lim_{x \to \infty} L(x) = 0$.

Example 1.21 The commonly used linear reference functions and non-linear reference functions with parameter q, denoted as RF_q, are summarized as follows:

(i) Linear: $\max\{0, 1 - |x|\}$,
(ii) Power: $RF_q = \max\{0, 1 - |x|^q\}$, $q > 0$,
(iii) Exponential Power: $RF_q = e^{-|x|^q}$, $q > 0$,
(iv) Rational: $RF_q = \frac{1}{1+|x|^q}$, $q > 0$.

Definition 1.27 [4] A flat fuzzy number $\tilde{A}$ is called an *LR* flat fuzzy number (see Fig. 1.8), if the membership function $\mu_{\tilde{A}}(x)$ is given by

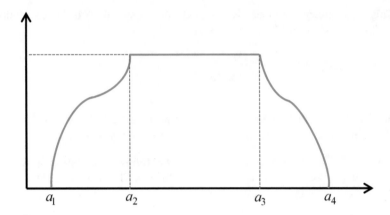

Fig. 1.8 The *LR* flat fuzzy number $\tilde{A} = (a_1, a_2, a_3, a_4)_{LR}$

$$\mu_{\tilde{A}}(x) = \begin{cases} L\left(\frac{a_2-x}{a_2-a_1}\right), & a_1 \leq x \leq a_2, \\ 1, & a_2 \leq x \leq a_3, \\ R\left(\frac{x-a_3}{a_4-a_3}\right), & a_3 \leq x \leq a_4. \end{cases} \tag{1.20}$$

where L and R are the reference functions of $\tilde{A}$. The *LR* flat fuzzy number $\tilde{A}$ can be written as $\tilde{A} = (a_1, a_2, a_3, a_4)_{LR}$. The set of *LR* flat fuzzy numbers on a real line is denoted by $\ell\Re(\mathbb{R})$.

Definition 1.28 A flat fuzzy number $\tilde{A}$ is called a trapezoidal fuzzy number if its membership function is given by

$$\mu_{\tilde{A}}(x) = \begin{cases} \frac{x-a_1}{a_2-a_1}, & a_1 \leq x \leq a_2, \\ 1, & a_2 \leq x \leq a_3, \\ \frac{a_4-x}{a_4-a_3}, & a_3 \leq x \leq a_4. \end{cases} \tag{1.21}$$

The trapezoidal fuzzy number $\tilde{A}$ is denoted by $\tilde{A} = (a_1, a_2, a_3, a_4)$.

Example 1.22 The flat fuzzy number $\tilde{D}$ given in Example 1.19, with the membership function (1.17) is a trapezoidal fuzzy number than can be denoted as $\tilde{D} = (20, 35, 45, 60)$.

Definition 1.29 [4] A fuzzy number $\tilde{A}$ is called an *LR* fuzzy number, if its membership function is given by

$$\mu_{\tilde{A}}(x) = \begin{cases} L\left(\frac{a_2-x}{a_2-a_1}\right), & a_1 \leq x \leq a_2, \\ R\left(\frac{x-a_2}{a_3-a_2}\right), & a_2 \leq x \leq a_3. \end{cases} \tag{1.22}$$

where L and R are the reference functions of $\tilde{A}$.

Definition 1.30 A fuzzy number $\tilde{A}$ is called a triangular fuzzy number if its membership function is given by

$$\mu_{\tilde{A}}(x) = \begin{cases} \frac{x-a_1}{a_2-a_1}, & a_1 \leq x \leq a_2, \\ \frac{a_3-x}{a_3-a_2}, & a_2 \leq x \leq a_3. \end{cases} \tag{1.23}$$

The triangular fuzzy number $\tilde{A}$ is denoted by $\tilde{A} = (a_1, a_2, a_3)$.

Remark 1.10 If $L(1) = 0$ and $R(1) = 0$, the *LR* (flat) fuzzy number $\tilde{A}$ is called a finite (flat) fuzzy number. If $\lim_{x \to \infty} L(x) = 0$ and $\lim_{x \to \infty} R(x) = 0$, it is called an infinite (flat) fuzzy number.

Remark 1.11 The triangular (trapezoidal) fuzzy number $\tilde{A}$ is a special kind of the *LR* (flat) fuzzy number $\tilde{A}$ with $L(x) = R(x) = \max\{0, 1 - |x|\}$.

Definition 1.31 Two *LR* flat fuzzy numbers $\tilde{A} = (a_1, a_2, a_3, a_4)_{LR}$ and $B = (b_1, b_2, b_3, b_4)_{LR}$ are said to be equal, i.e. $\tilde{A} = \tilde{B}$ if, and only if, $a_1 = b_1$, $a_2 = b_2$, $a_3 = b_3$ and $a_4 = b_4$.

Definition 1.32 An *LR* flat fuzzy number $\tilde{A} = (a_1, a_2, a_3, a_4)_{LR}$ is said to be a non-negative *LR* flat fuzzy number if, and only if, $a_1 \geq 0$.

Theorem 1.5 *The $\alpha-$ cut of the LR flat fuzzy number $\tilde{A} = (a_1, a_2, a_3, a_4)_{LR}$ is the closed interval*

$$[\tilde{A}]_\alpha = \left[(\tilde{A})_\alpha^L, (\tilde{A})_\alpha^U\right] = \left[a_2 - (a_2 - a_1)L^{-1}(\alpha), a_3 + (a_4 - a_3)R^{-1}(\alpha)\right] \tag{1.24}$$

Proof According to the definition of $\alpha-$ cut, we should find all $x \in \mathbb{R}$ where $\mu_{\tilde{A}}(x) \geq \alpha$. Regarding the membership function (1.20), we have

$$L\left(\frac{a_2 - x}{a_2 - a_1}\right) \geq \alpha, \ R\left(\frac{x - a_3}{a_4 - a_3}\right) \geq \alpha$$

To find the beginning point of the $\alpha-$ cut, i.e. $(\tilde{A})_\alpha^L$, we have

$$L\left(\frac{a_2 - x}{a_2 - a_1}\right) \geq \alpha \Leftrightarrow \frac{a_2 - x}{a_2 - a_1} \leq L^{-1}(\alpha) \Leftrightarrow x \geq a_2 - (a_2 - a_1)L^{-1}(\alpha)$$

This means that $(\tilde{A})_\alpha^L = a_2 - (a_2 - a_1)L^{-1}(\alpha)$. In a similar way, in order to find the end point of the $\alpha-$ cut, i.e. $(\tilde{A})_\alpha^U$, we have

$$R\left(\frac{x - a_3}{a_4 - a_3}\right) \geq \alpha \Leftrightarrow \frac{x - a_3}{a_4 - a_3} \leq R^{-1}(\alpha) \Leftrightarrow x \leq a_3 + (a_4 - a_3)R^{-1}(\alpha)$$

This means that $(\tilde{A})_\alpha^U = a_3 + (a_4 - a_3)R^{-1}(\alpha)$. This concludes the proof. □

Similarly, we can obtain the $\alpha-$ cut of the *LR* fuzzy number, trapezoidal fuzzy number and triangular fuzzy number. The proofs of the following theorems are analogous to the proof of Theorem 1.5.

Theorem 1.6 *The $\alpha-$ cut of the LR fuzzy number $\tilde{A} = (a_1, a_2, a_3)_{LR}$ is the closed interval*

$$[\tilde{A}]_\alpha = \left[(\tilde{A})_\alpha^L, (\tilde{A})_\alpha^U\right] = \left[a_2 - (a_2 - a_1)L^{-1}(\alpha), a_2 + (a_3 - a_2)R^{-1}(\alpha)\right] \quad (1.25)$$

Theorem 1.7 *The $\alpha-$ cut of the trapezoidal fuzzy number $\tilde{A} = (a_1, a_2, a_3, a_4)$ is the closed interval*

$$[\tilde{A}]_\alpha = \left[(\tilde{A})_\alpha^L, (\tilde{A})_\alpha^U\right] = [a_1 + (a_2 - a_1)\alpha, a_4 - (a_4 - a_3)\alpha] \quad (1.26)$$

Theorem 1.8 *The $\alpha-$ cut of the triangular fuzzy number $\tilde{A} = (a_1, a_2, a_3)$ is the closed interval*

$$[\tilde{A}]_\alpha = \left[(\tilde{A})_\alpha^L, (\tilde{A})_\alpha^U\right] = [a_1 + (a_2 - a_1)\alpha, a_3 - (a_3 - a_2)\alpha] \quad (1.27)$$

1.7 Fuzzy Arithmetic

In this section, we explore two approaches for developing arithmetic operations on fuzzy numbers. The first method is based on $\alpha-$ cuts and the decomposition theorem. The second method is based on the extension principle that extends the arithmetic operations on real numbers to fuzzy numbers.

1.7.1 Fuzzy Arithmetic Based on $\alpha-$ Cuts

We recall that the $\alpha-$ cut of each fuzzy number is a closed interval for every $\alpha \in [0, 1]$. Thus, we define fuzzy arithmetic on fuzzy numbers in terms of their $\alpha-$ cuts which are closed intervals in $\mathbb{R}$.

Definition 1.33 Let $[a_1, a_2]$ and $[b_1, b_2]$ be two closed intervals in $\mathbb{R}$. The four arithmetic operations, addition, subtraction, multiplication and division on these intervals are defined as:

(i)
$$[a_1, a_2] + [b_1, b_2] = [a_1 + b_1, a_2 + b_2],$$

(ii)
$$[a_1, a_2] - [b_1, b_2] = [a_1 - b_2, a_2 - b_1],$$

(iii)
$$[a_1, a_2] \cdot [b_1, b_2] = [\min(a_1 b_1, a_1 b_2, a_2 b_1, a_2 b_2), \max(a_1 b_1, a_1 b_2, a_2 b_1, a_2 b_2)],$$

(iv)
$$[a_1, a_2]/[b_1, b_2] = \left[\min\left(\frac{a_1}{b_1}, \frac{a_1}{b_2}, \frac{a_2}{b_1}, \frac{a_2}{b_2}\right), \max\left(\frac{a_1}{b_1}, \frac{a_1}{b_2}, \frac{a_2}{b_1}, \frac{a_2}{b_2}\right)\right], 0 \notin [b_1, b_2].$$

Remark 1.12 In the case of $a_1, b_1 \geq 0$, multiplication on the intervals $[a_1, a_2]$ and $[b_1, b_2]$ is simplified as follows:

$$[a_1, a_2] \cdot [b_1, b_2] = [a_1 b_1, a_2 b_2] \tag{1.28}$$

Remark 1.13 In the case of $a_1 \geq 0$, $b_1 > 0$, division on the intervals $[a_1, a_2]$ and $[b_1, b_2]$ is simplified as follows:

$$[a_1, a_2]/[b_1, b_2] = \left[\frac{a_1}{b_2}, \frac{a_2}{b_1}\right] \tag{1.29}$$

Definition 1.34 The image of the closed interval $[a_1, a_2]$ is defined as $-[a_1, a_2] = [-a_2, -a_1]$.

Definition 1.35 Let $[a_1, a_2]$ be a closed interval and k be a scaler. The scaler multiplication is defined as follows:

$$k.[a_1, a_2] = \begin{cases} [ka_1, ka_2], & k \geq 0, \\ [ka_2, ka_1], & k < 0. \end{cases} \tag{1.30}$$

Definition 1.36 Let $[a_1, a_2]$ and $[b_1, b_2]$ be two closed intervals in $\mathbb{R}$. The max and min operations on these intervals are defined as:

$$\begin{aligned} \max([a_1, a_2], [b_1, b_2]) &= [\max(a_1, b_1), \max(a_2, b_2)] \\ \min([a_1, a_2], [b_1, b_2]) &= [\min(a_1, b_1), \min(a_2, b_2)] \end{aligned} \tag{1.31}$$

Now, we are in a position to define the fuzzy arithmetic operations on fuzzy numbers based on $\alpha-$ cuts.

Definition 1.37 Let $\tilde{A}$ and $\tilde{B}$ be two fuzzy numbers and let $*$ denote any of the four arithmetic operations. Then, $\tilde{A} * \tilde{B}$ is a fuzzy number that can be defined as follows:

$$\tilde{A} * \tilde{B} = \cup_{\alpha \in [0,1]} \alpha \left[\tilde{A} * \tilde{B}\right]_\alpha \tag{1.32}$$

where

$$\left[\tilde{A} * \tilde{B}\right]_\alpha = \left[\tilde{A}\right]_\alpha * \left[\tilde{B}\right]_\alpha \tag{1.33}$$

Example 1.23 [2] Consider two triangular fuzzy numbers $\tilde{A}_1 = (-1, 1, 3)$ and $\tilde{A}_2 = (1, 3, 5)$. Based on Theorem 1.8, we have $\left[\tilde{A}_1\right]_\alpha = [2\alpha - 1, 3 - 2\alpha]$ and $\left[\tilde{A}_2\right]_\alpha = [2\alpha + 1, 5 - 2\alpha]$.

Regarding the Eq. (1.33) and Definition 1.33 we have

$$\left[\tilde{A}_1 + \tilde{A}_2\right]_\alpha = \left[\tilde{A}_1\right]_\alpha + \left[\tilde{A}_2\right]_\alpha = [2\alpha - 1, 3 - 2\alpha] + [2\alpha + 1, 5 - 2\alpha] = [4\alpha, 8 - 4\alpha]$$

For $\alpha = 1$, we have $\left[\tilde{A}_1 + \tilde{A}_2\right]_\alpha = [4, 4] = 4$. This means that $\mu_{\tilde{A}_1 + \tilde{A}_2}(4) = 1$. Also, for $\alpha = 0$, the support of fuzzy number $\tilde{A}_1 + \tilde{A}_2$ is obtained as $(0, 8)$. The left membership function of $\tilde{A}_1 + \tilde{A}_2$ is obtained by solving the equation $4\alpha = x$ leading to $\alpha = \frac{x}{4}$. Similarly, the right membership function of $\tilde{A}_1 + \tilde{A}_2$ is obtained by solving the equation $8 - 4\alpha = x$. This implies that $\alpha = \frac{8-x}{4}$. This concludes that the membership function of $\tilde{A}_1 + \tilde{A}_2$ is as follows:

$$\mu_{\tilde{A}_1 + \tilde{A}_2}(x) = \begin{cases} 0, & x < 0, x > 8, \\ \frac{x}{4}, & 0 \leq x \leq 4, \\ \frac{8-x}{4}, & 4 < x \leq 8. \end{cases}$$

This implies that $\tilde{A}_1 + \tilde{A}_2 = (0, 4, 8)$ is a triangular fuzzy number.

Similarly, in order to compute the subtraction of these fuzzy numbers, we have:

$$\left[\tilde{A}_1 - \tilde{A}_2\right]_\alpha = \left[\tilde{A}_1\right]_\alpha - \left[\tilde{A}_2\right]_\alpha = [2\alpha - 1, 3 - 2\alpha] - [2\alpha + 1, 5 - 2\alpha]$$
$$= [4\alpha - 6, 2 - 4\alpha]$$

For $\alpha = 1$, we have $\left[\tilde{A}_1 - \tilde{A}_2\right]_\alpha = [-2, -2] = -2$. This means that $\mu_{\tilde{A}_1 - \tilde{A}_2}(-2) = 1$. Also, for $\alpha = 0$, the support of fuzzy number $\tilde{A}_1 - \tilde{A}_2$ is obtained as $(-6, 2)$. The left membership function of $\tilde{A}_1 - \tilde{A}_2$ is obtained by solving the equation $4\alpha - 6 = x$ leading to $\alpha = \frac{x+6}{4}$. Similarly, the right membership function of $\tilde{A}_1 - \tilde{A}_2$ is obtained by solving the equation $2 - 4\alpha = x$. This implies that $\alpha = \frac{2-x}{4}$. This concludes that the membership function of $\tilde{A}_1 - \tilde{A}_2$ is as follows:

$$\mu_{\tilde{A}_1-\tilde{A}_2}(x) = \begin{cases} 0, & x < -6, x > 2, \\ \frac{x+6}{4}, & -6 \le x \le -2, \\ \frac{2-x}{4}, & -2 < x \le 2. \end{cases}$$

This implies that $\tilde{A}_1 - \tilde{A}_2 = (-6, -2, 2)$ is a triangular fuzzy number.

To obtain the multiplication $\tilde{A}_1 \cdot \tilde{A}_2$ using Definition (1.33) and the Eq. (1.33) we have:

$$\begin{aligned} \left[\tilde{A}_1 \cdot \tilde{A}_2\right]_\alpha &= \left[\tilde{A}_1\right]_\alpha \cdot \left[\tilde{A}_2\right]_\alpha = [2\alpha - 1, 3 - 2\alpha] \cdot [2\alpha + 1, 5 - 2\alpha] \\ &= \begin{cases} [-4\alpha^2 + 12\alpha - 5, 4\alpha^2 - 16\alpha + 15], & \alpha \in \left(0, \frac{1}{2}\right), \\ [4\alpha^2 - 1, 4\alpha^2 - 16\alpha + 15], & \alpha \in \left(\frac{1}{2}, 1\right]. \end{cases} \end{aligned} \qquad (1.34)$$

For $\alpha = 1$, from the second relation of (1.34) we have d $\left[\tilde{A}_1 \cdot \tilde{A}_2\right]_\alpha = [3, 3] = 3$. This means that $\mu_{\tilde{A}_1 \cdot \tilde{A}_2}(3) = 1$. Also, for $\alpha = 0$ from the first relation of (1.34), the support of fuzzy number $\tilde{A}_1 \cdot \tilde{A}_2$ is obtained as $(-15, 5)$. Since the end points of the $\alpha-$ cut given in (1.34) for $\alpha \in \left(0, \frac{1}{2}\right)$ and $\alpha \in \left(\frac{1}{2}, 1\right]$ are equal, the right membership function of $\tilde{A}_1 \cdot \tilde{A}_2$ is obtained by solving $4\alpha^2 - 16\alpha + 15 = x$. By solving this quadratic equation, we obtain $\alpha = \frac{4 - \sqrt{1+x}}{2}$. As, the starting points of the $\alpha-$ cut given in (1.34) for $\alpha \in \left(0, \frac{1}{2}\right)$ and $\alpha \in \left(\frac{1}{2}, 1\right]$ are different, the left membership function of $\tilde{A}_1 \cdot \tilde{A}_2$ is a two-piece function. The first piece is obtained by solving the equation $-4\alpha^2 + 12\alpha - 5 = x$ that leads to $\alpha = \frac{3 - \sqrt{4-x}}{2}$. The second one is obtained by solving the equation $4\alpha^2 - 1 = x$ which leads to $\alpha = \frac{\sqrt{1+x}}{2}$. This concludes that the membership function of $\tilde{A}_1 \cdot \tilde{A}_2$ is given as follows:

$$\mu_{\tilde{A}_1 \cdot \tilde{A}_2}(x) = \begin{cases} 0, & x < -5, x > 15, \\ \frac{3 - \sqrt{4-x}}{3}, & -5 \le x \le 0, \\ \frac{\sqrt{1+x}}{2}, & 0 < x \le 3, \\ \frac{4 - \sqrt{1+x}}{2}, & 3 \le x \le 15. \end{cases}$$

Finally, to obtain the division $\tilde{A}_1 / \tilde{A}_2$ using Definition (1.33) and the Eq. (1.33) we have:

$$\left[\tilde{A}_1 / \tilde{A}_2\right]_\alpha = [2\alpha - 1, 3 - 2\alpha] / [2\alpha + 1, 5 - 2\alpha] = \begin{cases} \left[\frac{2\alpha-1}{2\alpha+1}, \frac{3-2\alpha}{2\alpha+1}\right], & \alpha \in \left(0, \frac{1}{2}\right), \\ \left[\frac{2\alpha-1}{5-2\alpha}, \frac{3-2\alpha}{2\alpha+1}\right], & \alpha \in \left(\frac{1}{2}, 1\right]. \end{cases}$$

Similar to the discussion of multiplication of these fuzzy numbers, one can obtain the membership function of $\tilde{A}_1 / \tilde{A}_2$ as follows:

$$\mu_{\tilde{A}_1/\tilde{A}_2}(x) = \begin{cases} 0, & x < -1, x > 3, \\ \frac{x+1}{2-2x}, & -1 \leq x \leq 0, \\ \frac{5x+1}{2x+2}, & 0 < x \leq \frac{1}{3}, \\ \frac{3-x}{2x+2}, & \frac{1}{3} \leq x \leq 3. \end{cases}$$

Remark 1.14 Example 1.23 shows that the multiplication and division of two triangular fuzzy numbers are not always as triangular fuzzy numbers in general.

Theorem 1.9 *Let* $\tilde{A} = (a_1, a_2, a_3, a_4)_{LR}$ *and* $\tilde{B} = (b_1, b_2, b_3, b_4)_{LR}$ *be any LR flat fuzzy numbers and* $\tilde{C} = (c_1, c_2, c_3, c_4)_{RL}$ *be any RL flat fuzzy numbers. Then,*

(i) $\qquad\qquad\qquad \tilde{A} + \tilde{B} = (a_1 + b_1, a_2 + b_2, a_3 + b_3, a_4 + b_4)_{LR},$

(ii) $\qquad\qquad\qquad \tilde{A} - \tilde{C} = (a_1 - c_4, a_2 - c_3, a_3 - c_2, a_4 - c_1)_{LR}.$

Theorem 1.10 *Let* $\tilde{A} = (a_1, a_2, a_3, a_4)$ *and* $\tilde{B} = (b_1, b_2, b_3, b_4)$ *be two trapezoidal fuzzy numbers. Then,*

(i) $\qquad\qquad\qquad \tilde{A} + \tilde{B} = (a_1 + b_1, a_2 + b_2, a_3 + b_3, a_4 + b_4),$

(ii) $\qquad\qquad\qquad \tilde{A} - \tilde{B} = (a_1 - b_4, a_2 - b_3, a_3 - b_2, a_4 - b_1).$

Proof We only prove the proposition (i) and the proof if the proposition (ii) is analogous.

Using Theorem 1.7 and Definition 1.33, we have

$$\left[\tilde{A} + \tilde{B}\right]_\alpha == [(a_1 + b_1) + [(a_2 - a_1) + (b_2 - b_1)]\alpha, (a_4 + b_4) - [(a_4 - a_3) + (b_4 - b_3)]\alpha]$$

For $\alpha = 1$, we have $\left[\tilde{A} + \tilde{B}\right]_\alpha = [a_2 + b_2, a_3 + b_3]$. This means that $\mu_{\tilde{A}_1 + \tilde{A}_2}(x) = 1$ for all $x \in [a_2 + b_2, a_3 + b_3]$. Also, for $\alpha = 0$, the support of fuzzy number $\tilde{A} + \tilde{B}$ is obtained as $(a_1 + b_1, a_4 + b_4)$. The left membership function of $\tilde{A} + \tilde{B}$ is obtained by solving the equation $(a_1 + b_1) + [(a_2 - a_1) + (b_2 - b_1)]\alpha = x$ leading to $\alpha = \frac{x - (a_1 + b_1)}{(a_2 - a_1) + (b_2 - b_1)}$. Similarly, the right membership function of $\tilde{A} + \tilde{B}$ is obtained by solving the equation $(a_4 + b_4) - [(a_4 - a_3) + (b_4 - b_3)]\alpha = x$. This implies that $\alpha = \frac{(a_4 + b_4) - x}{(a_4 - a_3) + (b_4 - b_3)}$. This concludes that the membership function of $\tilde{A} + \tilde{B}$ is as follows:

$$\mu_{\tilde{A}_1 + \tilde{A}_2}(x) = \begin{cases} 0, & x < a_1 + b_1, \\ \frac{x - (a_1 + b_1)}{(a_2 - a_1) + (b_2 - b_1)}, & a_1 + b_1 \leq x < a_2 + b_2, \\ 1, & a_2 + b_2 \leq x \leq a_3 + b_3. \\ \frac{(a_4 + b_4) - x}{(a_4 - a_3) + (b_4 - b_3)}, & a_3 + b_3 < x \leq a_4 + b_4, \\ 0, & x > a_4 + b_4. \end{cases}$$

This implies that $\tilde{A} + \tilde{B} = (a_1 + b_1, a_2 + b_2, a_3 + b_3, a_4 + b_4)$. □

Theorem 1.11 *Let* $\tilde{A} = (a_1, a_2, a_3)$ *and* $\tilde{B} = (b_1, b_2, b_3)$ *be two triangular fuzzy numbers. Then,*

(i) $$\tilde{A} + \tilde{B} = (a_1 + b_1, a_2 + b_2, a_3 + b_3),$$

(ii) $$\tilde{A} - \tilde{B} = (a_1 - b_3, a_2 - b_2, a_3 - b_1).$$

Proof Analogous to the proof of Theorem 1.10. □

It should be noted here that approximate formulas [4] exist for the product of *LR* flat fuzzy numbers in which the spreads $a_2 - a_1$, $a_4 - a_3$, $b_2 - b_1$ and $b_4 - b_3$ are smaller as compared to the mean values a_2, a_3, b_2 and b_3. This is the subject of the following theorem in the case of non-negative fuzzy numbers.

Theorem 1.12 *Let* $\tilde{A} = (a_1, a_2, a_3, a_4)_{LR}$ *and* $\tilde{B} = (b_1, b_2, b_3, b_4)_{LR}$ *be two non-negative LR flat fuzzy numbers. Then,*

$$\tilde{A} \otimes \tilde{B} = (a_1 b_1, a_2 b_2, a_3 b_3, a_4 b_4)_{LR}$$

Theorem 1.13 *Let* $\tilde{A} = (a_1, a_2, a_3, a_4)_{LR}$ *be any LR flat fuzzy numbers and k be a scaler. Then,*

$$k\tilde{A} = \begin{cases} (ka_1, ka_2, ka_3, ka_4)_{LR}, & k \geq 0, \\ (ka_4, ka_3, ka_2, ka_1)_{RL}, & k < 0. \end{cases}$$

1.7.2 Fuzzy Arithmetic Based on Extension Principle

In this subsection, we develop to the second method, which is based on the extension principle, to define the arithmetic operations on fuzzy numbers.

Let $\tilde{A}$ and $\tilde{B}$ be two fuzzy numbers and let $*$ denote any of the four arithmetic operations. Then, $\tilde{A} * \tilde{B}$ is a fuzzy number with a membership function which for all $z \in \mathbb{R}$ is defined as follows:

$$\mu_{\tilde{A}*\tilde{B}}(z) = \sup_{z = x*y} \min\{\mu_{\tilde{A}}(x), \mu_{\tilde{B}}(y)\} \tag{1.35}$$

More specifically, for all $z \in \mathbb{R}$ we define:

$$\mu_{\tilde{A}+\tilde{B}}(z) = \sup_{z = x+y} \min\{\mu_{\tilde{A}}(x), \mu_{\tilde{B}}(y)\} \tag{1.36}$$

$$\mu_{\tilde{A}-\tilde{B}}(z) = \sup_{z = x-y} \min\{\mu_{\tilde{A}}(x), \mu_{\tilde{B}}(y)\} \tag{1.37}$$

$$\mu_{\tilde{A}\cdot\tilde{B}}(z) = \sup_{z=xy} \min\left\{\mu_{\tilde{A}}(x), \mu_{\tilde{B}}(y)\right\} \tag{1.38}$$

$$\mu_{\tilde{A}/\tilde{B}}(z) = \sup_{z=\frac{x}{y}} \min\left\{\mu_{\tilde{A}}(x), \mu_{\tilde{B}}(y)\right\} \tag{1.39}$$

1.8 Fuzzy Ranking Methods

Ranking fuzzy numbers plays a very important role in decision-making, optimization, and other usages. The last few decades have seen a large number of approaches investigated for ranking fuzzy numbers. In this section, we describe three ranking approaches that will be needed in the rest of this book.

1.8.1 Ranking Function Approach

One convenient approach for ranking fuzzy numbers is based on the concept of comparison of fuzzy numbers by use of ranking functions. An effective approach for ordering the elements of $F(\mathbb{R})$ is to define a ranking function $\Re : F(\mathbb{R}) \to \mathbb{R}$ which maps each fuzzy number into the real line, where a natural order exists. A ranking function $\Re$ is said to be a linear ranking function if $\Re(k\tilde{A}+\tilde{B}) = k\,\Re(\tilde{A}) + \Re(\tilde{B})$ for any $k \in \mathbb{R}$ and any fuzzy numbers $\tilde{A}$ and $\tilde{B}$. Now, according to this approach, we have $\tilde{A} \prec \tilde{B}$ when $\Re(\tilde{A}) < \Re(\tilde{B})$.

A special version of the above linear ranking function was first proposed by Yager [8] as follows:

$$Y(\tilde{A}) = \frac{1}{2} \int_0^1 \left[(\tilde{A})_\alpha^L + (\tilde{A})_\alpha^U \right] d\alpha \tag{1.40}$$

Remark 1.15 For a trapezoidal fuzzy number $\tilde{A} = (a_1, a_2, a_3, a_4)$, the value of Yager's index is reduced to

$$Y(\tilde{A}) = \frac{a_1 + a_2 + a_3 + a_4}{4} \tag{1.41}$$

Remark 1.16 For the *LR* flat fuzzy number $\tilde{A} = (a_1, a_2, a_3, a_4)_{LR}$, if $L(x) = \max\{0, 1 - |x|\}$ and $R(x) = \max\left\{0, 1 - |x|^2\right\}$, then the value of Yager's index is reduced to

$$Y(\tilde{A}) = \frac{3a_1 + 3a_2 + 2a_3 + 4a_4}{12} \qquad (1.42)$$

Remark 1.17 For the *LR* flat fuzzy number $\tilde{A} = (a_1, a_2, a_3, a_4)_{LR}$, if $L(x) = \max\{0, 1 - |x|^2\}$ and $R(x) = \max\{0, 1 - |x|\}$, then the value of Yager's index is reduced to

$$Y(\tilde{A}) = \frac{4a_1 + 2a_2 + 3a_3 + 3a_4}{12} \qquad (1.43)$$

Example 1.24 Consider two trapezoidal fuzzy numbers $\tilde{A} = (10, 22, 32, 45)$ and $\tilde{B} = (32, 45, 55, 67)$. Based on Remark 1.15, we have $Y(\tilde{A}) = 27.25$ and $Y(\tilde{A}) = 49.75$. This concludes that $\tilde{A} \prec \tilde{B}$.

1.8.2 Lattice of Fuzzy Numbers Approach

In this subsection we introduce the lattice of fuzzy numbers $\langle F(\mathbb{R}), MIN, MAX \rangle$ by means of which fuzzy numbers can be ordered partially in a natural way.

Definition 1.38 For any two fuzzy numbers $\tilde{A}$ and $\tilde{B}$, the operations *MIN* and *MAX* are defined as follows:

$$\mu_{MIN(\tilde{A}, \tilde{B})}(z) = \sup_{z = \min(x, y)} \min\left[\mu_{\tilde{A}}(x), \mu_{\tilde{B}}(y)\right] \qquad (1.44)$$

$$\mu_{MAX(\tilde{A}, \tilde{B})}(z) = \sup_{z = \max(x, y)} \min\left[\mu_{\tilde{A}}(x), \mu_{\tilde{B}}(y)\right] \qquad (1.45)$$

Remark 1.18 For two fuzzy numbers $\tilde{A}$ and $\tilde{B}$, $MIN(\tilde{A}, \tilde{B})$ and $MAX(\tilde{A}, \tilde{B})$ are fuzzy numbers.

Example 1.25 Consider two triangular fuzzy numbers $\tilde{A} = (-2, 1, 4)$ and $\tilde{B} = (1, 2, 3)$. The membership functions of fuzzy numbers $MIN(\tilde{A}, \tilde{B})$ and $MAX(\tilde{A}, \tilde{B})$ are given as follows with regard to (1.44) and (1.45) (see Figs. 1.9 and 1.10):

$$\mu_{MIN(\tilde{A}, \tilde{B})}(z) = \begin{cases} 0, & x < -2, x > 3, \\ \frac{x+2}{3}, & -2 \le x \le 1, \\ \frac{4-x}{3}, & 1 \le x \le \frac{5}{2}, \\ 3 - x, & 1 \le x \le \frac{5}{2}. \end{cases}$$

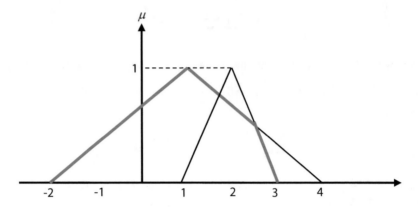

Fig. 1.9 The membership function $MIN(\tilde{A}, \tilde{B})$

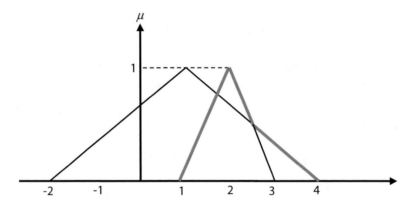

Fig. 1.10 The membership function $MAX(\tilde{A}, \tilde{B})$

$$\mu_{MAX(\tilde{A},\tilde{B})}(x) = \begin{cases} 0, & x<1, x>4, \\ x-1, & 1\leq x\leq 2, \\ 3-x, & 2\leq x\leq \frac{5}{2}, \\ \frac{4-x}{3}, & \frac{5}{2}\leq x\leq 4. \end{cases}$$

Theorem 1.14 [2] *Let MIN and MAX be binary operations on the set of all fuzzy numbers $F(\mathbb{R})$ defined in Definition 1.38. Then, the triple $\langle F(\mathbb{R}), MIN, MAX\rangle$ is a distributive lattice.*

Remark 1.19 The lattice $\langle F(\mathbb{R}), MIN, MAX\rangle$ can also be expressed as the pair $\langle F(\mathbb{R}), \preceq\rangle$, where $\preceq$ is a partial ordering defined as:

$$\tilde{A} \preceq \tilde{B} \Leftrightarrow MIN(\tilde{A}, \tilde{B}) = \tilde{A} \text{ or, alternatively,}$$

$$\tilde{A} \preceq \tilde{B} \Leftrightarrow MAX(\tilde{A}, \tilde{B}) = \tilde{B}.$$

Remark 1.20 Regarding Definition 1.36, the partial ordering defined in Remark 1.19 in terms of $\alpha-$ cuts is reduced to:

$$\tilde{A} \preceq \tilde{B} \Leftrightarrow \min\left[\left[\tilde{A}\right]_\alpha, \left[\tilde{B}\right]_\alpha\right] = \left[\tilde{A}\right]_\alpha \text{ or, alternatively,}$$

$$\tilde{A} \preceq \tilde{B} \Leftrightarrow \max\left[\left[\tilde{A}\right]_\alpha, \left[\tilde{B}\right]_\alpha\right] = \left[\tilde{B}\right]_\alpha.$$

Remark 1.20 means that

$$\tilde{A} \preceq \tilde{B} \Leftrightarrow \left[\tilde{A}\right]_\alpha \leq \left[\tilde{B}\right]_\alpha.$$

Remark 1.21 For any two trapezoidal fuzzy numbers $\tilde{A} = (a_1, a_2, a_3, a_4)$ and $\tilde{B} = (b_1, b_2, b_3, b_4)$, $MAX(\tilde{A}, \tilde{B}) = \tilde{B}$ or $\tilde{A} \preceq \tilde{B}$ if,   and   only   if, $a_1 \leq b_1, a_2 \leq b_2, a_3 \leq b_3$ and $a_4 \leq b_4$.

Example 1.25 Consider two trapezoidal fuzzy numbers $\tilde{A} = (10, 22, 32, 45)$ and $\tilde{B} = (32, 45, 55, 67)$. Based on Remark 1.21, we have $\tilde{A} \preceq \tilde{B}$.

1.8.3   Possibility Theory Approach

Dubois and Prade [4] developed a set of four ranking indexes within the possibility theory framework.

Definition 1.39 For any two fuzzy numbers $\tilde{A}$ and $\tilde{B}$, the inequality relations are defined as follows:

$$Pos(\tilde{A} \preceq \tilde{B}) = \sup_{x \leq y} \min\left[\mu_{\tilde{A}}(x), \mu_{\tilde{B}}(y)\right] \tag{1.46}$$

$$Nec(\tilde{A} \prec \tilde{B}) = \inf_{x \geq y} \max\left[1 - \mu_{\tilde{A}}(x), 1 - \mu_{\tilde{B}}(y)\right] \tag{1.47}$$

According to Definition 1.39, the inequality relation between two fuzzy numbers $\tilde{A}$ and $\tilde{B}$ is defined as follows:

$$\tilde{A} \preceq \tilde{B} \Leftrightarrow Pos(\tilde{A} \preceq \tilde{B}) \geq Pos(\tilde{B} \preceq \tilde{A}) \tag{1.48}$$

Theorem 1.15 *For any two fuzzy numbers $\tilde{A}$ and $\tilde{B}$, we have*

$$Pos\left(\tilde{A} \preceq \tilde{B}\right) \geq \alpha \Leftrightarrow \left(\tilde{A}\right)_{\alpha}^{L} \leq \left(\tilde{B}\right)_{\alpha}^{U} \tag{1.49}$$

$$Nec\left(\tilde{A} \prec \tilde{B}\right) \geq \alpha \Leftrightarrow \left(\tilde{A}\right)_{1-\alpha}^{U} \leq \left(\tilde{B}\right)_{1-\alpha}^{L} \tag{1.50}$$

We may write $Pos\left(\tilde{A} \preceq \tilde{B}\right) = \tilde{A} \preceq^{Pos} \tilde{B}$ *and* $Nec\left(\tilde{A} \prec \tilde{B}\right) = \tilde{A} \prec^{Nec} \tilde{B}.$

Theorem 1.16 [9] *For any two trapezoidal fuzzy numbers* $\tilde{A} = (a_1, a_2, a_3, a_4)$ *and* $\tilde{B} = (b_1, b_2, b_3, b_4)$, *we have*

$$Pos\left(\tilde{A} \preceq \tilde{B}\right) = \begin{cases} 1, & b_3 \geq a_2, \\ \frac{b_4 - a_1}{(b_4 - b_3) + (a_2 - a_1)}, & 0 < a_2 - b_3 \leq (b_4 - b_3) + (a_2 - a_1), \\ 0, & a_2 - b_3 > (b_4 - b_3) + (a_2 - a_1). \end{cases} \tag{1.51}$$

$$Nec\left(\tilde{A} \prec \tilde{B}\right) = \begin{cases} 1, & b_2 - a_3 \geq (b_2 - b_1) + (a_4 - a_3), \\ \frac{b_2 - a_3}{(b_2 - b_1) + (a_4 - a_3)}, & 0 \leq b_2 - a_3 < (b_2 - b_1) + (a_4 - a_3), \\ 0, & b_2 < a_3. \end{cases} \tag{1.52}$$

Theorem 1.17 [9] *Let* $\tilde{A} = (a_1, a_2, a_3, a_4)$ *and* $\tilde{B} = (b_1, b_2, b_3, b_4)$ *be two trapezoidal fuzzy numbers* $\tilde{A} = (a_1, a_2, a_3, a_4)$ *and* $\tilde{B} = (b_1, b_2, b_3, b_4)$, *and* $\alpha \in (0, 1]$. *Then*

$$Pos\left(\tilde{A} \preceq \tilde{B}\right) \geq \alpha \Leftrightarrow b_3 - a_2 \geq (\alpha - 1)[(b_4 - b_3) + (a_2 - a_1)]$$

Proof If $\alpha = 1$ then from $Pos\left(\tilde{A} \preceq \tilde{B}\right) \geq 1$ and the Eq. (1.51) one can obtain $b_3 \geq a_2$ and vice versa. If $0 < \alpha < 1$ then $b_3 < a_2$ and $b_3 + (b_4 - b_3) > a_2 - (a_2 - a_1)$. In this case from the Eq. (1.51) $Pos\left(\tilde{A} \preceq \tilde{B}\right) \geq \alpha$ if, and only if, $\frac{b_4 - a_1}{(b_4 - b_3) + (a_2 - a_1)} \geq \alpha$, that is $b_3 - a_2 \geq (\alpha - 1)[(b_4 - b_3) + (a_2 - a_1)]$. □

Theorem 1.18 [9] *Let* $\tilde{A} = (a_1, a_2, a_3, a_4)$ *and* $\tilde{B} = (b_1, b_2, b_3, b_4)$ *be two trapezoidal fuzzy numbers* $\tilde{A} = (a_1, a_2, a_3, a_4)$ *and* $\tilde{B} = (b_1, b_2, b_3, b_4)$, *and* $\alpha \in (0, 1]$. *Then*

$$Nec\left(\tilde{A} \prec \tilde{B}\right) \geq \alpha \Leftrightarrow b_2 - a_3 \geq \alpha[(b_2 - b_1) + (a_4 - a_3)]$$

Proof If $\alpha = 1$ then from $Nec\left(\tilde{A} \prec \tilde{B}\right) \geq 1$ and the Eq. (1.52) one can obtain $b_2 - a_3 \geq (b_2 - b_1) + (a_4 - a_3)$ and vice versa. If $0 < \alpha < 1$ then $b_2 > a_3$ and $b_2 - a_3 > (b_2 - b_1) + (a_4 - a_3)$. In this case from the Eq. (1.52) $Nec\left(\tilde{A} \prec \tilde{B}\right) \geq \alpha$ if, and only if, $\frac{b_2 - a_3}{(b_2 - b_1) + (a_4 - a_3)} \geq \alpha$, that is $b_2 - a_3 \geq \alpha[(b_2 - b_1) + (a_4 - a_3)]$. □

Example 1.26 Consider two trapezoidal fuzzy numbers $\tilde{A} = (10, 22, 32, 45)$ and $\tilde{B} = (32, 45, 55, 67)$. Based on Theorem 1.16 and the Eq. (1.51), we have $Pos\left(\tilde{A} \preceq \tilde{B}\right) = 1$ and $Pos\left(\tilde{B} \preceq \tilde{A}\right) = 0.5$. Now, from relation (1.48), we conclude that $\tilde{A} \preceq \tilde{B}$.

1.9 Conclusions

In this chapter we have provided basic definitions and results on fuzzy sets and related topics. We have also given a brief discussion on fuzzy numbers and fuzzy arithmetic. Since the ranking of fuzzy numbers is an important component of the decision-making process, three approaches on this topic have been presented and which will be used in the following chapters.

References

1. Zadeh, L.A.: Fuzzy sets. Inf. Con. **8**(3), 338–353 (1965)
2. Klir, G.J., Yuan, B.: Fuzzy Sets and Fuzzy Logic. Theory and Applications. Prentice-Hall, PTR, Englewood Cliffs (1995)
3. Lai, Y.J., Hwang, C.L.: Fuzzy mathematical programming. Springer, Berlin (1992)
4. Dubois, D., Prade, H.: Fuzzy Sets and Systems Theory and Applications. Academic New York, (1980)
5. Zimmermann, H.J.: Fuzzy Sets, Decision Making and Expert Systems. Kluwer Academic Publishers, Boston (1987)
6. Lodwick, W.A. Thipwiwatpotjana, P.: Flexible and generalized uncertainty optimization. Theory and methods. Studies in Computational Intelligence, vol. 696. *Springer, Cham,* (2017)
7. Bellman, R.E., Zadeh, L.A.: Decision making in a fuzzy environment. Manage. Sci. **17**(4), 141–164 (1970)
8. Yager, R.R.: A procedure for ordering fuzzy subsets of the unit interval. Inf. Sci. **24**, 143–161 (1981)
9. Li, H., Gong, Z.: Fuzzy linear programming with possibility and necessity relation. In: Fuzzy Information and Engineering. Advances in Intelligent and Soft Computing, pp. 305–311. Springer-Verlag, Berlin-Heidelberg (2010)

Chapter 2
Fuzzy Linear Programming

2.1 Introduction

Linear programming (LP) is a mathematical technique for the optimal allocation of scarce resources to several competing activities on the basis of given criteria of optimality. Mathematically, the general LP problem may be formulated as follows [1]:

$$\max z = \sum_{j=1}^{n} c_j x_j$$
$$s.t. \quad \sum_{j=1}^{n} a_{ij} x_j \leq b_i, \quad i = 1, 2, \ldots, m, \tag{2.1}$$
$$x_j \geq 0, \qquad j = 1, 2, \ldots, n.$$

The function $z = \sum_{j=1}^{n} c_j x_j$ to be maximized is called an objective function. The vector $b = (b_1, b_2, \ldots, b_m)$ is called the right-hand-side vector or resources vector. The vector $x = (x_1, x_2, \ldots, x_n)$ is called the decision variables vector. The vector $c = (c_1, c_2, \ldots, c_n)$ is called the cost coefficients or the coefficients of decision variables in the objective function and matrix $A = (a_{ij})_{m \times n}$ is called the constraints matrix or the matrix of technical coefficients. Using this notation, the LP problem (2.1) can be formulated in a vector-matrix form as follows:

$$\max z = cx$$
$$s.t. \quad Ax \leq b, \tag{2.2}$$
$$x \geq 0.$$

In conventional LP problems, it is assumed that the data have precise values. This means that the elements of the problem are crisp numbers, inequality "$\leq$" is defined in the crisp sense and "max" is a strict imperative. However, the observed

© Springer International Publishing AG, part of Springer Nature 2018
A. Ebrahimnejad and J. L. Verdegay, *Fuzzy Sets-Based Methods and Techniques for Modern Analytics*, Studies in Fuzziness and Soft Computing 364,
https://doi.org/10.1007/978-3-319-73903-8_2

values of the data in real-life problems are often imprecise because of incomplete or non-obtainable information. In such situations, fuzzy sets theory is an ideal approach to handle imprecise data in LP by generalizing the notion of membership in a set and this leads to the concept of fuzzy LP problems.

Fuzzy LP (FLP) problems allow working with imprecise data and constraints, leading to more realistic models. They have often been used for solving a wide variety of problems in sciences and engineering [2–8]. Fuzzy mathematical programming has been researched by a number of authors. One of the earliest works on fuzzy mathematics programming problems was presented by Tanaka et al. [9] based on the fuzzy decision framework of Bellman and Zadeh [10]. Since Tanaka et al. [11] there have been a number of fuzzy LP models. In the literature, fuzzy LP has been classified into different categories, depending on how imprecise parameters are modeled by subjective preference-based membership functions or possibility distributions. Indeed, Lai and Hwang [12] for the first time, classified LP problems in a fuzzy environment into two general categories: FLP problems and LP problems with fuzzy parameters. The fuzzy input data in FLP problems is modeled by subjective preference-based membership functions. The imprecise data in LP problems with fuzzy parameters is modeled by subjective or objective possibility distributions. In this chapter, we first sub-classify FLP problems into several groups and then discuss the solution techniques for solving each group. LP problems with fuzzy parameters will be explored in Chaps. 3 and 4.

Based on possible combinations of the fuzziness of the constraints matrix, resources vector, the cost coefficients and the objective function, Lai and Hwang [12] classified FLP problems into the following five general groups:

- FLP of type-I: The FLP problems with a fuzzy resources vector or fuzzy inequality constraints belong to this group.
- FLP of type-II: The FLP problems with a fuzzy resources vector (fuzzy inequality constraints) and a fuzzy objective function belong to this group.
- FLP of type-III: The FLP problems with fuzzy cost coefficients belong to this group.
- FLP of type-IV: The FLP problems with fuzzy cost coefficients, a fuzzy constraints matrix and a fuzzy resources vector (fuzzy inequality constraints) belong to this group.
- FLP of type-V: The FLP problems with a fuzzy objective function, a fuzzy constraints matrix and a fuzzy resources vector (fuzzy inequality constraints) belong to this group.

Zimmermann [13] has classified FLP problems into two categories: symmetrical problems and non-symmetrical problems. In a symmetrical problem there is no difference between the weight of the objectives and constraints whilst in non-symmetrical problems, the objectives and constraints are not equally important and have different weights. Luhandjula [14] indicated important milestones in the development of fuzzy optimization emphasizing the macro-level view and highlighting Tanaka and Asai [15] contribution. Verdegay [16] described the main developments, results and solution methods achieved in this topic, from a personal point of view. A comprehensive literature review on FLPs is presented in the next section.

2.2   A Brief Literature Review on FLP Problems

An overview of the papers which have studied FLPs and discussed solution methodologies can be summarized as follows: Tanaka and Asai [15] proposed a possibilistic LP formulation where the coefficients of the decision variables were crisp while the decision variables were fuzzy numbers. Verdegay [17] proved that the optimal solution of an FLP of type-I can be found by the use of solving an equivalent crisp parametric LP problem assuming that the objective function is crisp. An approach for solving an FLP problem with a triangular membership function was presented by Chanas and Kulej [18]. On the other hand, Werners [19] suggested that the objective function should be fuzzy because of fuzzy inequality constraints and computed the lower and upper bounds of the optimal values by solving two crisp LP problems. Then, using the max-min operator of Bellman and Zadeh [10], he proposed a non-symmetric model for finding the solution of the FLP problem of type-I that satisfies the constraints and objective with the maximum degree. But the solution obtained by the max-min operator may not be efficient since the situation in the model proposed by Werners [19] has multiple optimal solutions. To overcome this shortcoming, Guu and Wu [20] proposed a two-phase approach for solving the FLP problem of type-I that not only pursues the highest membership degree in the objective, but also pursues a better utilization of each constrained resource.

In the FLPs of type-II not only a goal and its corresponding tolerance are considered for the objective function, but also a tolerance for each fuzzy constraint is known. Zimmermann [13] proposed a symmetric approach for solving this kind of FLPs. In the approach proposed by Zimmerman it is assumed that the goal and its corresponding tolerance of the objective function are given initially. According to this approach, the optimal solution of such a problem can be found by solving a crisp LP problem. Lai and Hwang [21] proposed a modified method of Zimmermann's approach for solving the FLP problem of type-II by assuming that only the aspiration level of the fuzzy objective is given, but its tolerance is not determined by the decision maker. On the other hand, Chanas [22] proposed a symmetric approach for solving the FLPs of type-II. In the approach proposed by Chanas [22], the goal and its corresponding tolerance are determined by solving two crisp LP problems. Then, based on the obtained results and the optimal solution of a crisp parametric LP problem, the optimal solution of the FLP problem under consideration is determined. Safi et al. [23] proved that when this LP problem proposed by Zimmermann [13] has multiple optimal solutions, it may not always present the best solution to the decision maker. Moreover, Zimmerman's approach may not distinguish the case that the FLP problem under consideration is unbounded. An algorithm for eliminating these difficulties was proposed by Safi et al. [23]. Furthermore, in all the above-mentioned approaches for solving the FLP of type-II there is no distinction between the objective and constraints function because they are all aggregated together via the 'min' operator. To overcome this difficulty, Chandra and Aggarwal [24] presented a new formulation for solving this kind of fuzzy problems in the framework of a bi-objective optimization problem.

In the FLPs of type-III it is assumed that the coefficient of decision variables in the objective function cannot be precisely determined. For solving this kind of fuzzy problems, two approaches have been proposed by Verdegay [25, 26]. In the first approach, an equivalent cost parametric LP problem is constructed for finding the optimal solution. For developing the second approach, Verdegay [26] proved that for a given FLP of type-III there always exists an FLP of type-II with the same solution. Hence, he proposed that the optimal solution of an FLP of type-III can be found by solving the dual problem of an FLP problem of type-II problem.

In the FLPs of type-IV it is assumed that all the coefficients of an LP are fuzzy. In contrast to Chanas' approach [22] which does not provide any continuous trade-off between the degrees of violation constraints, Carlsson and Korhonen [27] provided an efficient approach to find the optimal solution of FLPs of type-IV. Gasimov and Yenilmez [28] proposed the modified subgradient method and the fuzzy decisive set method for solving the FLPs of type-IV by use of the symmetric method of Bellman and Zadeh [10] for a defuzzification of this kind of fuzzy problems. Farhadinia [29] suggested a revised formula for the membership function of fuzzy constraints involved in the FLP of type-IV. According to this approach, the number of required iterations to obtain the desired solution is reduced; the maximum satisfaction degree of the fuzzy decision set reaches a more accurate optimum; and the sequence which evaluates how much constraints are violated is controlled by a smaller upper bound.

In the FLPs of type-V it is assumed that the membership function of the constraints matrix and the resources vector are linear. Based on the concept of non-symmetrical models, Lai and Hwang [21] proposed an approach to establish the membership function of the fuzzy objective function. To obtain the optimal solution of the FLP of type-V, they first defuzzified the fuzzy constraints matrix and the fuzzy resources vector according to Carlsson and Korhonen's approach. Then, they considered the Zimmermann's and Werners' membership functions for situations in which the decision maker may like to provide his/her goal and its tolerance or not. An open-source R package to deal with fuzzy constraints, fuzzy costs and fuzzy coefficients in LP can be found in [30].

2.3 Decision Making Under a Fuzzy Environment

The solution approaches for solving FLP problems are based on Bellman and Zadeh's concept for decision making under fuzzy conditions [31]. Bellman and Zadeh [10] by introducing three basic concepts, namely fuzzy objective, fuzzy constraint, and fuzzy decision, explored the application of these concepts to decision-making processes under fuzziness. Let us now introduce the conceptual framework for decision making in a fuzzy environment.

For the fuzzy objective, the main aim is not really to maximize or minimize the objective function; rather it is to achieve some aspiration level which might not be even definable crisply. Let X be a given set of possible alternatives which contains the solution of a decision-making problem under consideration. A fuzzy objective $\tilde{G}$ is a fuzzy set on X characterized by its membership function $\mu_{\tilde{G}} : X \rightarrow [0, 1]$.

For the fuzzy inequality (fuzzy constraint), the "$\leq$" sign might not be meant in the strict mathematical sense but subjectively determined violations may be acceptable. A fuzzy constraint $\tilde{C}$ is a fuzzy set on X characterized by its membership function $\mu_{\tilde{C}} : X \rightarrow [0, 1]$.

A fuzzy decision $\tilde{D}$, resulting from the confluence of fuzzy objectives and fuzzy constraints, is a fuzzy set in the space of alternatives and may be defined by a max-min operator. The membership function of this set orders the alternatives according to the criterion of the degree of simultaneous satisfaction of fuzzy constraints and of fuzzy objective achievement. To be more explicit, the fuzzy decision of Bellman and Zadeh is the fuzzy set $\tilde{D}$ on X defined as $\tilde{D} = \tilde{G} \cap \tilde{C}$ and is characterized by its membership function $\mu_{\tilde{D}}(x) = \min\{\mu_{\tilde{G}}(x), \mu_{\tilde{C}}(x)\}$. More generally, the fuzzy decision $\tilde{D}$, the resulting fuzzy objective $\tilde{G}_j (j = 1, 2, \ldots, n)$ and the fuzzy constraints $\tilde{C}_i (i = 1, 2, \ldots, m)$ are defined by $\tilde{D} = \left(\bigcap_{j=1}^{n} \tilde{G}_j \right) \cap \left(\bigcap_{i=1}^{m} \tilde{C}_i \right)$ with the membership function $\mu_{\tilde{D}}(x) = \min_{j,i}\{\mu_{\tilde{G}_j}(x), \mu_{\tilde{C}_i}(x)\}$. In this case, the corresponding maximizing decision is defined as follows:

$$\mu_{\tilde{D}}(x^*) = \max_{x \in X} \mu_{\tilde{D}}(x) = \max_{x \in X} \min_{j,i}\{\mu_{\tilde{G}_j}(x), \mu_{\tilde{C}_i}(x)\} \tag{2.3}$$

Another possibility could be to choose an $\alpha (0 < \alpha < 1)$ and determine all alternatives $x^* \in X$ for which $\mu_{\tilde{D}}(x^*) \geq \alpha$.

It is significant to realize here that in the fuzzy decision defined by Bellman and Zadeh [10], the fuzzy objectives and the fuzzy constraints have equal importance and there is no longer a difference between them. In this case, Bellman and Zadeh [10] also suggested the convex fuzzy decision, defined as the weighted sum of the fuzzy objectives and the fuzzy constraints:

$$\bar{\mu}_{\tilde{D}}(x) = \sum_{j=1}^{n} u_j \mu_{\tilde{G}_j}(x) + \sum_{i=1}^{m} w_i \mu_{\tilde{C}_i}(x)$$

$$\sum_{j=1}^{n} u_j + \sum_{i=1}^{m} w_i = 1, u_j \geq 0, w_i \geq 0, \quad \forall i, j.$$

Fig. 2.1 Fuzzy decision and
convex fuzzy decision

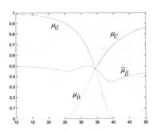

where the weighting coefficients reflect the relative importance among the fuzzy
objectives and constraints.

In this case, the corresponding maximizing convex fuzzy decision is defined as
follows

$$\bar{\mu}_{\tilde{D}}(x^*) = \max_{x \in X} \bar{\mu}_{\tilde{D}}(x) = \max_{x \in X} \left[\sum_{j=1}^{n} u_j \mu_{\tilde{G}_j}(x) + \sum_{i=1}^{m} w_i \mu_{\tilde{C}_i}(x) \right] \qquad (2.4)$$

Example 2.1 Consider a fuzzy decision problem in which we have to find a real
number $x \in X = [0, 50]$ which is in much larger than 25 (fuzzy objective $\tilde{G}$) and is
substantially smaller than 40 (fuzzy constraint $\tilde{C}$). For this problem, the goal and
the constraint can be expressed in terms of their membership functions as follows

$$\mu_{\tilde{G}}(x) = \begin{cases} 0, & x < 25, \\ 1 - \dfrac{1}{1 + \frac{1}{100(x-25)^2}}, & x \geq 25 \end{cases}$$

$$\mu_{\tilde{C}}(x) = \begin{cases} 0, & x > 40, \\ \dfrac{1}{1 + \frac{x}{(x-40)^2}}, & x \leq 40. \end{cases}$$

The following diagram depicts this fuzzy decision-making problem and identifies the optimal solutions with regard to the problems (2.3) and (2.4) (Fig. 2.1).

The optimal solution of this problem with regard to the problem (2.3) is $x^* =
34.452$ with $\mu_{\tilde{D}}(x^*) = 0.4719$. Moreover, the optimal solution of this problem with
regard to the problem (2.4) for $u = w = 0.5$ is $x^* = 32.036$ with $\bar{\mu}_{\tilde{D}}(x^*) = 0.4978$.

2.4   FLP Problems of Type-I

The general form of the FLP problems with fuzzy resources can be formulated as
follows [12]:

$$\max z = \sum_{j=1}^{n} c_j x_j$$

$$s.t. \quad g_i(x) = \sum_{j=1}^{n} a_{ij} x_j \preceq b_i, \quad i = 1, 2, \ldots, m, \tag{2.5}$$

$$x_j \geq 0, \qquad \qquad j = 1, 2, \ldots, n.$$

In model (2.5), $\preceq$ is called "fuzzy less than or equal to" and it is assumed that the tolerance p_i for each constraint is given. This means that the decision maker can accept a violation of each constraint up to degree p_i. In this case, constraints $g_i(x) \preceq b_i$, are equivalent to $g_i(x) \leq b_i + \theta p_i, (i = 1, 2, \ldots, m)$, where $\theta \in [0, 1]$.

The problem (2.5) can be equivalently considered as the following fuzzy inequality constraints [32]:

$$\max z = \sum_{j=1}^{n} c_j x_j$$

$$s.t. \quad \sum_{j=1}^{n} a_{ij} x_j \leq \tilde{b}_i, \quad i = 1, 2, \ldots, m, \tag{2.6}$$

$$x_j \geq 0, \qquad j = 1, 2, \ldots, n.$$

In model (2.6) $\tilde{b}_i$ is a fuzzy number with the following membership function:

$$\mu_{\tilde{b}_i}(x) = \begin{cases} 1, & x < b_i, \\ 1 - \frac{x - b_i}{p_i}, & b_i \leq x \leq b_i + p_i, \\ 0, & x > b_i + p_i. \end{cases} \tag{2.7}$$

In this section, we shall explore three approaches for solving the FLP problem of type-I (2.5).

2.4.1 Verdegay's Approach

Verdegay [17] proved that the FLP problem (2.5) is equivalent to a crisp parametric LP problem when the membership functions of the fuzzy constraints are continuous and non-increasing functions. According to this non-symmetric approach, the membership functions of the fuzzy inequality constraints of problem (2.5) can be modeled as follows (see Fig. 2.2):

Fig. 2.2 The membership functions of the fuzzy inequality constraint

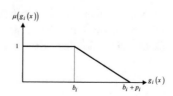

$$\mu_i(g_i(x)) = \begin{cases} 1, & g_i(x) < b_i, \\ 1 - \frac{g_i(x) - b_i}{p_i}, & b_i \le g_i(x) \le b_i + p_i, \\ 0, & g_i(x) > b_i + p_i. \end{cases} \tag{2.8}$$

In this case, the membership function of all constraints of the problem (2.5) according to the Bellman and Zadeh operator is given by

$$\mu(g(x)) = \min\{\mu_1(g_1(x)), \mu_2(g_2(x)), \dots, \mu_m(g_m(x))\} \tag{2.9}$$

Assuming $\alpha = \min\{\mu_1(g_1(x)), \mu_2(g_2(x)), \dots, \mu_m(g_m(x))\}$, the FLP problem of type-I (2.5) is equivalent to:

$$\begin{aligned} \max z &= \sum_{j=1}^{n} c_j x_j \\ s.t. \quad \mu_i(g_i(x)) &\ge \alpha, \qquad i = 1, 2, \dots, m, \\ \alpha &\in [0, 1], x_j \ge 0, \quad j = 1, 2, \dots, n. \end{aligned} \tag{2.10}$$

Now, by substituting the membership functions (2.8) into the problem (2.10), the following crisp parametric LP problem is obtained:

$$\begin{aligned} \max z &= \sum_{j=1}^{n} c_j x_j \\ s.t. \quad g_i(x) &= \sum_{j=1}^{n} a_{ij} x_j \le b_i + (1 - \alpha) p_i, \quad i = 1, 2, \dots, m, \\ \alpha &\in [0, 1], x_j \ge 0, \qquad\qquad j = 1, 2, \dots, n. \end{aligned} \tag{2.11}$$

It should be noted that for each $\alpha \in [0, 1]$, an optimal solution is obtained. This indicates that the solution with α grade of membership function is actually fuzzy.

Example 2.2 [12] Consider the following FLP problem of type-I:

$$\begin{aligned} \max z &= 4x_1 + 5x_2 + 9x_3 + 11x_4 \\ s.t. \quad g_1(x) &= x_1 + x_2 + x_3 + x_4 \precsim 15, \\ g_2(x) &= 7x_1 + 5x_2 + 3x_3 + 2x_4 \le 120, \\ g_3(x) &= 3x_1 + 5x_2 + 10x_3 + 15x_4 \precsim 100, \\ x_1, x_2, x_3, x_4 &\ge 0. \end{aligned} \tag{2.12}$$

Assume that the first and third constraints are imprecise and their maximum tolerances are 3 and 20, respectively, i.e. $p_1 = 3$ and $p_3 = 20$. Then, according to (2.11), the following crisp parametric LP problem is solved:

$$\max z = 4x_1 + 5x_2 + 9x_3 + 11x_4$$
$$s.t. \quad g_1(x) = x_1 + x_2 + x_3 + x_4 \leq 15 + 3(1 - \alpha),$$
$$g_2(x) = 7x_1 + 5x_2 + 3x_3 + 2x_4 \leq 120, \tag{2.13}$$
$$g_3(x) = 3x_1 + 5x_2 + 10x_3 + 15x_4 \leq 100 + 20(1 - \alpha),$$
$$x_1, x_2, x_3, x_4 \geq 0.$$

By using the parametric technique, the optimal solution and the optimal objective value are obtained as follows:

$$x^* = \left(\frac{50}{7} + \frac{10}{7}(1 - \alpha), 0, \frac{55}{7} + \frac{11}{7}(1 - \alpha), 0 \right),$$
$$z^* = \frac{834}{7} - \frac{139}{7}\alpha. \tag{2.14}$$

2.4.2 Werners's Approach

In the symmetric method proposed by Werners [19], it is assumed that the objective function of the problem (2.5) should be fuzzy because of fuzzy inequality constraints. For doing this, he calculated the lower and upper bounds of the optimal values by solving the standard LP problems (2.15) and (2.16), respectively, as follows:

$$z^l = \max z = \sum_{j=1}^{n} c_j x_j$$
$$s.t. \quad \sum_{j=1}^{n} a_{ij} x_j \leq b_i, \quad i = 1, 2, \ldots, m, \tag{2.15}$$
$$x_j \geq 0, \quad j = 1, 2, \ldots, n.$$

$$z^u = \max z = \sum_{j=1}^{n} c_j x_j$$
$$s.t. \quad \sum_{j=1}^{n} a_{ij} x_j \leq b_i + p_i, \quad i = 1, 2, \ldots, m, \tag{2.16}$$
$$x_j \geq 0, \quad j = 1, 2, \ldots, n.$$

In this case, the membership function of the objective function, $\mu_0(z)$, is defined as follows (see Fig. 2.3):

Fig. 2.3 The membership functions of the fuzzy objective function

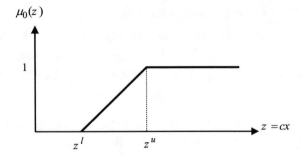

$$\mu_0(z) = \begin{cases} 1, & z > z^u, \\ 1 - \frac{z^u - z}{z^u - z^l}, & z^l \le z \le z^u, \\ 0, & z < z^l. \end{cases} \tag{2.17}$$

Now, the problem (2.5) can be solved by solving the following problem using the max-min operator of Bellman and Zadeh [10], a problem of finding a solution that satisfies the constraints and objective with the maximum degree:

$$\begin{aligned} \max z &= \alpha \\ s.t. \quad \mu_0(z) &\ge \alpha, \\ \mu_i(g_i(x)) &\ge \alpha, i = 1, 2, \ldots, m, \\ \alpha &\in [0, 1], x_j \ge 0, j = 1, 2, \ldots, n. \end{aligned} \tag{2.18}$$

By substituting the membership functions of the constraints given in (2.8) and the membership function of the objective function given in (2.17) into the problem (2.18), the following crisp LP problem is obtained:

$$\begin{aligned} \max \alpha \\ s.t. \quad \sum_{j=1}^{n} c_j x_j &\ge z^u - (1 - \alpha)(z^u - z^l), \\ g_i(x) = \sum_{j=1}^{n} a_{ij} x_j &\le b_i + (1 - \alpha)p_i, \quad i = 1, 2, \ldots, m, \\ \alpha &\in [0, 1], x_j \ge 0, \quad j = 1, 2, \ldots, n. \end{aligned} \tag{2.19}$$

Example 2.3 [33] Consider the following FLP problem of type-I:

$$\begin{aligned} \max z &= 0.4x_1 + 0.3x_2 \\ s.t. \quad g_1(x) &= x_1 + x_2 \le \widetilde{400}, \\ g_2(x) &= 2x_1 + x_2 \le \widetilde{500}, \\ x_1, x_2 &\ge 0. \end{aligned} \tag{2.20}$$

where the membership functions of the fuzzy inequality constraints are defined by:

$$\mu_1(x) = \begin{cases} 1, & g_1(x) < 400, \\ 1 - \frac{g_1(x)-400}{100}, & 400 \le g_1(x) \le 500, \\ 0, & g_1(x) > 500. \end{cases} \tag{2.21}$$

$$\mu_2(x) = \begin{cases} 1, & g_2(x) < 500, \\ 1 - \frac{g_2(x)-500}{100}, & 500 \le g_2(x) \le 600, \\ 0, & g_2(x) > 600. \end{cases} \tag{2.22}$$

Then, the lower and upper bounds of the objective function are calculated by solving the following two crisp LP problems (2.23) and (2.24), respectively:

$$\begin{aligned} z^l = \max z &= 0.4x_1 + 0.3x_2 \\ s.t. \quad g_1(x) &= x_1 + x_2 \le 400, \\ g_2(x) &= 2x_1 + x_2 \le 500, \\ x_1, x_2 &\ge 0. \end{aligned} \tag{2.23}$$

$$\begin{aligned} z^u = \max z &= 0.4x_1 + 0.3x_2 \\ s.t. \quad g_1(x) &= x_1 + x_2 \le 500, \\ g_2(x) &= 2x_1 + x_2 \le 600, \\ x_1, x_2 &\ge 0. \end{aligned} \tag{2.24}$$

Solving the two problems (2.23) and (2.24) give $z^l = 130$ and $z^u = 160$, respectively. Then, the membership function μ_0 of the objective function is defined as follows:

$$\mu_0(z) = \begin{cases} 1, & z > 160, \\ 1 - \frac{160-z}{30}, & 130 \le z \le 160, \\ 0, & z < 130. \end{cases} \tag{2.25}$$

Then, according to the problem (2.19), the FLP problem of type-I (2.20) becomes:

$$\begin{aligned} \max \alpha \\ s.t. \quad 0.4x_1 + 0.3x_2 &\ge 160 - 30(1 - \alpha), \\ g_1(x) = x_1 + x_2 &\le 400 + 100(1 - \alpha), \\ g_2(x) = 2x_1 + x_2 &\le 500 + 100(1 - \alpha), \\ \alpha \in [0, 1], \ x_1, x_2 &\ge 0. \end{aligned} \tag{2.26}$$

The optimal solution of the problem (2.26) is $x^* = (x_1^*, x_2^*) = (100, 350)$ with $\alpha^* = 0.5$. The optimal value of the objective function is then calculated by

$$z^* = 0.4x_1^* + 0.3x_2^* = 145.$$

2.4.3   Guu and Wu's Approach

Guu and Wu [20] proposed a two-phase approach for solving the FLP problem of type-I (2.5) to fix the situation where the max-min operator is not efficient. The two-phase method consists of the max-min operator as its first phase. In the second phase, a solution is yielded which is at least "better" than the solution obtained by the max-min operator. The two-phase method not only pursues the highest membership degree in the objective, but also pursues a better utilization of each constrained resource.

According to this approach, for solving the FLP problem of type-I (2.5), in the first phase, the problem (2.18) or (2.19) is solved. If the optimal solution of the problem (2.18) or (2.19) is unique, then this optimal solution is a fuzzy-efficient compromise solution for the problem (2.5). On the other hand, if the problem (2.18) or (2.19) has multiple optimal solutions, then the solution obtained by the max-min operator may not be efficient. In this case, assuming (x^*, α^*) to be an optimal solution of the problem (2.18) or (2.19), in the second phase, the following LP problem is solved:

$$
\begin{aligned}
\max \ & \sum_{i=0}^{m} \alpha_i \\
s.t. \quad & \mu_0(z^* = cx^*) \leq \alpha_0, \\
& \mu_i(g_i(x^*)) \leq \alpha_i, \quad i = 1, 2, \ldots, m, \\
& \mu_0(z = cx) \geq \alpha_0, \\
& \mu_i(g_i(x)) \geq \alpha_i, \quad i = 1, 2, \ldots, m, \\
& \alpha_i \in [0, 1], i = 0, 1, \ldots, m, \\
& x_j \geq 0, \quad j = 1, 2, \ldots, n.
\end{aligned}
\tag{2.27}
$$

The problem (2.27) is equivalent to the problem (2.28):

$$
\begin{aligned}
\max \ & \sum_{i=0}^{m} \alpha_i \\
s.t. \quad & \sum_{j=1}^{n} c_j x_j \geq z^u - (1 - \alpha_0)(z^u - z^l),
\end{aligned}
$$

$$g_i(x) = \sum_{j=1}^{n} a_{ij}x_j \le b_i + (1 - \alpha_i)p_i, \quad i = 1, 2, \ldots, m,$$

$$\sum_{j=1}^{n} c_j x_j^* \le z^u - (1 - \alpha_0)(z^u - z^l),$$

$$g_i(x^*) = \sum_{j=1}^{n} a_{ij}x_j^* \ge b_i + (1 - \alpha_i)p_i, \quad i = 1, 2, \ldots, m, \tag{2.28}$$

$$\alpha_i \in [0, 1], i = 0, 1, \ldots, m,$$

$$x_j \ge 0, \quad j = 1, 2, \ldots, n.$$

Assume $(x^{**}, \alpha_0^{**}, \alpha_1^{**}, \ldots, \alpha_m^{**})$ to be an optimal solution of the problem (2.27) or (2.28). In this case, this solution is a fuzzy efficient solution of the FLP problem of type-I (2.5).

Theorem 2.1 [20] *The optimal solution* x^{**} *of the problem* (2.27) *is a fuzzy efficient (compromise) solution of the problem* (2.5).

Proof Suppose, on the contrary, that x^{**} is not a fuzzy efficient (compromise) solution of the problem (2.5). Then a solution $\bar{x}$ exists such that $\mu_i(x^{**}) \le \mu_i(\bar{x})$ for all $i = 0, 1, \ldots, m$ and $\mu_k(x^{**}) < \mu_k(\bar{x})$ for some k.

Since $(x^{**}, \alpha_0^{**}, \alpha_1^{**}, \ldots, \alpha_m^{**})$ is an optimal solution of the problem (2.27), and the coefficient in the objective function of the problem (2.27) is positive, we have $\alpha_i^{**} = \mu_i(x^{**})$ for all $i = 0, 1, \ldots, m$. Now let $\bar{\alpha}_i = \mu_i(\bar{x})$ for all $i = 0, 1, \ldots, m$. In this case, we obtain a feasible solution $(\bar{x}, \bar{\alpha}_0, \bar{\alpha}_1, \ldots, \bar{\alpha}_m)$ for the problem (2.27) such that

$$\sum_{i=0}^{m} \alpha_i^{**} = \sum_{i=0}^{m} \mu_i(x^{**}) < \sum_{i=0}^{m} \mu_i(\bar{x}) = \bar{\alpha}_k + \sum_{i=0,i\neq k}^{m} \bar{\alpha}_i$$

However, this implies that $(x^{**}, \alpha_0^{**}, \alpha_1^{**}, \ldots, \alpha_m^{**})$ is not an optimal solution of the problem (2.27), which is a contradiction.□

Now, we solve the following example to illustrate the two-phase approach proposed by Guu and Wu [20] for solving the FLP problem of type-I.

Example 2.4 [20] Consider the following FLP problem of type-I:

$$\max z = 4x_1 + 5x_2 + 9x_3 + 11x_4$$

$$\text{s.t.} \quad g_1(x) = x_1 + x_2 + x_3 + x_4 \le \widetilde{15},$$

$$g_2(x) = 7x_1 + 5x_2 + 3x_3 + 2x_4 \le \widetilde{80}, \tag{2.29}$$

$$g_3(x) = 3x_1 + 5x_2 + 10x_3 + 15x_4 \le \widetilde{100},$$

$$x_1, x_2, x_3, x_4 \ge 0.$$

where the membership functions of the fuzzy inequality constraints are defined by:

$$\mu_1(x) = \begin{cases} 1, & g_1(x) < 15, \\ 1 - \frac{g_1(x)-15}{5}, & 15 \leq g_1(x) \leq 20, \\ 0, & g_1(x) > 20. \end{cases} \tag{2.30}$$

$$\mu_2(x) = \begin{cases} 1, & g_2(x) < 80, \\ 1 - \frac{g_2(x)-80}{40}, & 80 \leq g_2(x) \leq 120, \\ 0, & g_2(x) > 120. \end{cases} \tag{2.31}$$

$$\mu_3(x) = \begin{cases} 1, & g_3(x) < 100, \\ 1 - \frac{g_3(x)-100}{30}, & 100 \leq g_3(x) \leq 130, \\ 0, & g_3(x) > 130. \end{cases} \tag{2.32}$$

Then, the lower and upper bounds of the objective function are calculated by solving the following two crisp LP problems (2.33) and (2.34), respectively:

$$z^L = \max z = 4x_1 + 5x_2 + 9x_3 + 11x_4$$
$$s.t. \quad g_1(x) = x_1 + x_2 + x_3 + x_4 \leq 15,$$
$$g_2(x) = 7x_1 + 5x_2 + 3x_3 + 2x_4 \leq 80, \tag{2.33}$$
$$g_3(x) = 3x_1 + 5x_2 + 10x_3 + 15x_4 \leq 100,$$
$$x_1, x_2, x_3, x_4 \geq 0.$$

$$z^u = \max z = 4x_1 + 5x_2 + 9x_3 + 11x_4$$
$$s.t. \quad g_1(x) = x_1 + x_2 + x_3 + x_4 \leq 20,$$
$$g_2(x) = 7x_1 + 5x_2 + 3x_3 + 2x_4 \leq 120, \tag{2.34}$$
$$g_3(x) = 3x_1 + 5x_2 + 10x_3 + 15x_4 \leq 130,$$
$$x_1, x_2, x_3, x_4 \geq 0.$$

Solving the two problems (2.33) and (2.34) gives $z^l = 99.29$ and $z^u = 130$, respectively. Then, the membership function μ_0 of the objective function is defined as follows:

$$\mu_0(z) = \begin{cases} 1, & z > 130, \\ 1 - \frac{130-z}{30.71}, & 99.29 \leq z \leq 130, \\ 0, & z < 99.29. \end{cases} \tag{2.35}$$

Then, according to the problem (2.19), we solve the following LP problem as the first phase problem of the two-phase method for solving the FLP problem of type-I (2.29):

$$\max \alpha$$

$$
\begin{aligned}
s.t. \quad & 4x_1 + 5x_2 + 9x_3 + 11x_4 \geq 130 - 30.71(1 - \alpha), \\
& g_1(x) = x_1 + x_2 + x_3 + x_4 \leq 15 + 5(1 - \alpha), \\
& g_2(x) = 7x_1 + 5x_2 + 3x_3 + 2x_4 \leq 80 + 40(1 - \alpha), \\
& g_3(x) = 3x_1 + 5x_2 + 10x_3 + 15x_4 \leq 100 + 30(1 - \alpha), \\
& \alpha \in [0, 1], x_1, x_2, x_3, x_4 \geq 0.
\end{aligned}
\tag{2.36}
$$

The optimal solution of the problem (2.36) is $x^* = (x_1^*, x_2^*, x_3^*, x_4^*) = (8.57, 0, 8.93, 0)$ with $\alpha^* = 0.5$. By substituting this optimal solution in the objective function and in the constraints of the problem (2.29), we obtain $z(x^*) = z^* = 114.65, g_1(x^*) = 17.5, g_2(x^*) = 86.78$ and $g_3(x^*) = 115.01$. Thus, according to the Eqs. (2.30), (2.31), (2.32) and (2.35), we obtain $\mu_0(x^*) = \mu_1(x^*) = \mu_3(x^*) = 0.5$ and $\mu_2(x^*) = 0.83$.

Now, according to the problem (2.27), we solve the following LP problem as the second-phase problem of the two-phase method for solving the FLP problem of type-I (2.29):

$$\max \alpha_0 + \alpha_1 + \alpha_2 + \alpha_3$$

$$
\begin{aligned}
s.t. \quad & 4x_1 + 5x_2 + 9x_3 + 11x_4 \geq 130 - 30.71(1 - \alpha_0), \\
& x_1 + x_2 + x_3 + x_4 \leq 15 + 5(1 - \alpha_1), \\
& 7x_1 + 5x_2 + 3x_3 + 2x_4 \leq 80 + 40(1 - \alpha_2), \\
& x_1 + 5x_2 + 10x_3 + 15x_4 \leq 100 + 30(1 - \alpha_3), \\
& 0.5 \leq \alpha_0 \leq 1, 0.5 \leq \alpha_1 \leq 1, 0.83 \leq \alpha_2 \leq 1, 0.5 \leq \alpha_3 \leq 1, \\
& x_1, x_2, x_3, x_4 \geq 0.
\end{aligned}
\tag{2.37}
$$

The optimal solution of the problem (2.37) is $x^{**} = (x_1^{**}, x_2^{**}, x_3^{**}, x_4^{**}) = (4.05, 5.65, 7.8, 0)$. By substituting this optimal solution in the objective function and in the constraints of the problem (2.29), we obtain $z(x^{**}) = z^{**} = 114.65, g_1(x^{**}) = 17.5, g_2(x^{**}) = 80$ and $g_3(x^{**}) = 115.01$. Thus, according to the Eqs. (2.30), (2.31), (2.32) and (2.35), we obtain $\mu_0(x^{**}) = \mu_1(x^{**}) = \mu_3(x^{**}) = 0.5$ and $\mu_2(x^{**}) = 1$. This means that x^{**} not only achieves the optimal objective value but also attains a higher degree in $\mu_2(x)$. More precisely, the two-phase production decision x^{**} utilizes 80 units of second resource, while the max-min operator's solution x^* requires 86.78 units.

2.5 FLP Problems of Type-II

The general form of FLP problems with a fuzzy resources vector and a fuzzy objective can be formulated as the following two equivalent problems [12]:

$$\tilde{\max} \, z = \sum_{j=1}^{n} c_j x_j$$

$$s.t. \quad g_i(x) = \sum_{j=1}^{n} a_{ij} x_j \tilde{\preceq} b_i, \quad i = 1, 2, \ldots, m, \tag{2.38}$$

$$x_j \geq 0, \qquad j = 1, 2, \ldots, n.$$

$$\tilde{\max} \, z = \sum_{j=1}^{n} c_j x_j$$

$$s.t. \quad \sum_{j=1}^{n} a_{ij} x_j \leq \tilde{b}_i, \quad i = 1, 2, \ldots, m, \tag{2.39}$$

$$x_j \geq 0, \qquad j = 1, 2, \ldots, n.$$

In this section, we shall explore five approaches for solving the FLP problem of type-II (2.38).

2.5.1 Zimmermann's Approach

Zimmermann [34] proposed a symmetric approach for solving the FLP problem of type-II (2.38). According to this approach, the membership functions $\mu_i (i = 1, 2, \ldots, m)$ of the fuzzy inequality constraints are given by (2.8) and also, an aspiration level z_0 and its corresponding tolerance p_0 for the objective function are given initially by decision maker. In this case, the problem (2.38) can be described as follows:

$$Find \quad x$$

$$s.t. \quad z = \sum_{j=1}^{n} c_j x_j \succeq z_0$$

$$g_i(x) = \sum_{j=1}^{n} a_{ij} x_j \preceq b_i, \quad i = 1, 2, \ldots, m, \tag{2.40}$$

$$x_j \geq 0, \quad j = 1, 2, \ldots, n.$$

Note that p_0 is the permissible tolerance for the objective function, the membership function μ_0 of the fuzzy objective is given as follows (see Fig. 2.4):

$$\mu_0(z) = \begin{cases} 1, & z > z_0, \\ 1 - \frac{z_0 - z}{p_0}, & z_0 - p_0 \leq z \leq z_0, \\ 0, & z < z_0 - p_0. \end{cases} \tag{2.41}$$

Using the min operator as an aggregator for the fuzzy objective function and fuzzy constraints, we obtain $\mu_D(x) = \min\{\mu_0(z), \mu_1(g_1(x)), \ldots, \mu_m(g_m(x))\}$ as the

Fig. 2.4 The membership functions of the fuzzy objective constraint $z \succeq z_0$

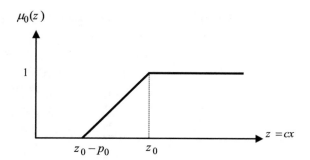

membership function of the decision space. Zimmermann [34] then used the max-min operator of Bellman and Zadeh [10] for solving the problem (2.40). Thus, the optimal solution can be obtained by

$$\max \mu_D(x) = \min\{\mu_0(z), \mu_1(g_1(x)), \ldots, \mu_m(g_m(x))\} \tag{2.42}$$

Let $\alpha = \mu_D(x) = \min\{\mu_0(z), \mu_1(g_1(x)), \ldots, \mu_m(g_m(x))\}$. In this case, to solve the fuzzy system of inequalities (2.40) corresponding to the problem (2.38), the following crisp LP problem is solved:

$$
\begin{aligned}
\max \quad & \alpha \\
\text{s.t.} \quad & \mu_0(z) \geq \alpha, \\
& \mu_i(g_i(x)) \geq \alpha, \quad i = 1, 2, \ldots, m, \\
& \alpha \in [0, 1], x_j \geq 0, \quad j = 1, 2, \ldots, n.
\end{aligned}
\tag{2.43}
$$

By substituting the membership functions of the fuzzy constraints given in (2.8) and the membership function of the objective function given in (2.41) into the problem (2.43), the following crisp LP problem is obtained:

$$
\begin{aligned}
\max \quad & \alpha \\
\text{s.t.} \quad & \sum_{j=1}^{n} c_j x_j \geq z_0 - (1 - \alpha) p_0, \\
& \sum_{j=1}^{n} a_{ij} x_j \leq b_i + (1 - \alpha) p_i, \quad i = 1, 2, \ldots, m, \\
& \alpha \in [0, 1], x_j \geq 0, \quad j = 1, 2, \ldots, n.
\end{aligned}
\tag{2.44}
$$

It should be pointed out that if (x^*, α^*) is an optimal solution of the crisp LP problem (2.44), then it is said to be an optimal solution of the FLP problem (2.38) with α^* as the degree up to which the aspiration level z_0 of the decision maker is satisfied.

Example 2.5 [12] Consider the following FLP problem of type-II:

$$\tilde{\max} \, z = x_1 + x_2$$
$$\text{s.t.} \quad g_1(x) = -x_1 + 3x_2 \precsim 21,$$
$$g_2(x) = x_1 + 3x_2 \precsim 27,$$
$$g_3(x) = 4x_1 + 3x_2 \precsim 45, \tag{2.45}$$
$$g_4(x) = 3x_1 + x_2 \le 30,$$
$$x_1, x_2 \ge 0.$$

Assume that $z_0 = 14.5, p_0 = 2, p_1 = 3, p_2 = 6$, and $p_3 = 6$. To obtain the optimal solution of the problem (2.45), we have to solve the following crisp LP problem with regard to the problem (2.44):

$$\max \alpha$$
$$\text{s.t.} \quad z = x_1 + x_2 \ge 14.5 - 2(1 - \alpha)$$
$$g_1(x) = -x_1 + 3x_2 \le 21 + 3(1 - \alpha),$$
$$g_2(x) = x_1 + 3x_2 \le 27 + 6(1 - \alpha),$$
$$g_3(x) = 4x_1 + 3x_2 \le 45 + 6(1 - \alpha), \tag{2.46}$$
$$g_4(x) = 3x_1 + x_2 \le 30,$$
$$0 \le \alpha \le 1,$$
$$x_1, x_2 \ge 0.$$

The optimal solution of the crisp LP problem (2.46) is $(x_1^*, x_2^*, \alpha^*) = (6, 7.75, 0.625)$ with the optimal objective function value $z^* = 13.75$. Then, $(x_1^*, x_2^*) = (6, 7.75)$ is the optimal solution of the FLP problem of type-II (2.45).

It is worth noting that in the Zimmermann FLP model an aspiration level of the objective function and its maximum tolerance should be given initially by the decision maker. Since in real-world application; the decision maker has little information about the data of the problem under consideration, it is difficult for him/ her to establish the mentioned data. Thus, the membership function of the fuzzy objective and the solution derived based on this approach are questionable. Moreover, if the given tolerance is too large, the Zimmermann FLP model is infeasible. At the same time, if the aspiration level of the objective function is too large, then the membership function has no meaning. In this case the solution obtained from this approach is questionable. To overcome this shortcoming, the aspiration level of the objective function and its maximum tolerance can be provided by Werners's approach by solving the models (2.15) and (2.17) as $z_0 = z^u$ and $p_0 = z^u - z^l$. However, if $z^l \le z_0 \le z^u$ and $z_0 - p_0 \ge z^l$, the decision maker often considers the membership function of Zimmermann's approach as a more acceptable one.

2.5.2 Lai and Hwang's Approach

Lai and Hwang [21] proposed a modified method of Zimmermann's approach for solving the FLP problem of type-II (2.38) when only the aspiration level z_0 of the fuzzy objective is given, but its tolerance p_0 is not given by the decision maker. In this case, the tolerance p_0 should obviously be between 0 and $z_0 - z^l$ where z^l is given by solving the LP problem (2.15). Now, for each $p_0 \in [0, z_0 - z^l]$ the membership function of the fuzzy objective is obtained as the Eq. (2.41) and the problem (2.44) is solved with each given p_0. Then, the obtained solution for each given p_0 is presented to the decision maker. In this case, a satisfying solution is chosen by the decision maker and the solution procedure is terminated.

Example 2.6 [12] Consider the following FLP problem of type-II with $z_0 = 111.57$ $p_1 = 5$, $p_2 = 40$, and $p_3 = 30$.

$$\tilde{\max}\, z = 4x_1 + 5x_2 + 9x_3 + 11x_4$$

$$s.t. \quad g_1(x) = x_1 + x_2 + x_3 + x_4 \tilde{\leq} 15,$$

$$g_2(x) = 7x_1 + 5x_2 + 3x_3 + 2x_4 \tilde{\leq} 80, \tag{2.47}$$

$$g_3(x) = 3x_1 + 5x_2 + 10x_3 + 15x_4 \tilde{\leq} 100,$$

$$x_1, x_2, x_3, x_4 \geq 0.$$

The membership functions of the fuzzy inequality constraints are defined as (2.30), (2.31) and (2.32).

The membership function μ_0 of the fuzzy objective is given by

$$\mu_0(z) = \begin{cases} 1, & z > 111.57, \\ 1 - \frac{111.57 - z}{p_0}, & 111.57 - p_0 \leq z \leq 111.57, \\ 0, & z < 111.57 - p_0. \end{cases} \tag{2.48}$$

Solving the LP problem (2.33) gives $z^l = 99.29$. Thus, the reasonable range of p_0 is $[0, z_0 - z^l] = [0, 12.28]$. We choose five possible values of p_0 as $0, 3, 6, 9, 12, 12.28$. For each p_0, the membership function μ_0 of the fuzzy objective is given as the Eq. (2.48). For example, for $p_0 = 9$, we have:

$$\mu_0(z) = \begin{cases} 1, & z > 111.57, \\ 1 - \frac{111.57 - z}{9}, & 102.57 \leq z \leq 111.57, \\ 0, & z < 102.57. \end{cases} \tag{2.49}$$

In this case, the problem (2.44) is solved with each given p_0. For example, for $p_0 = 9$, the following problem is solved:

Table 2.1 The solutions of the problem (2.47)

p_0	α^*	x^*	z^*
0	0.400	(8.286, 0, 8.714, 0)	111.570
3	0.364	(8.184, 0, 8.638, 0)	110.478
6	0.335	(8.099, 0, 8.574, 0)	109.562
9	0.309	(8.027, 0, 8.520, 0)	108.788
12.28	0.286	(7.959, 0, 8.469, 0)	108.057

$$\max \alpha$$
$$\begin{aligned}
s.t. \quad & 4x_1 + 5x_2 + 9x_3 + 11x_4 \geq 111.57 - 9(1 - \alpha), \\
& g_1(x) = x_1 + x_2 + x_3 + x_4 \leq 15 + 5(1 - \alpha), \\
& g_2(x) = 7x_1 + 5x_2 + 3x_3 + 2x_4 \leq 80 + 40(1 - \alpha), \\
& g_3(x) = 3x_1 + 5x_2 + 10x_3 + 15x_4 \leq 100 + 30(1 - \alpha), \\
& \alpha \in [0, 1], x_1, x_2, x_3, x_4 \geq 0.
\end{aligned} \tag{2.50}$$

The optimal solution of the problem (2.49) is $x^* = (x_1^*, x_2^*, x_3^*, x_4^*) = (8.027, 0, 8.52, 0)$ with $\alpha^* = 0.691$. By substituting this optimal solution in the objective we obtain $z(x^*) = z^* = 108.788$. In a similar way, the optimal solution of the problem (2.47) with $z_0 = 111.57$ for each of the above-mentioned p_0 is provided in Table 2.1. The decision maker finally indicates that $p_0 = 10$ will be perfect. Thus, the following LP problem is solved:

$$\max \alpha$$
$$\begin{aligned}
s.t. \quad & 4x_1 + 5x_2 + 9x_3 + 11x_4 \geq 111.57 - 10(1 - \alpha), \\
& g_1(x) = x_1 + x_2 + x_3 + x_4 \leq 15 + 5(1 - \alpha), \\
& g_2(x) = 7x_1 + 5x_2 + 3x_3 + 2x_4 \leq 80 + 40(1 - \alpha), \\
& g_3(x) = 3x_1 + 5x_2 + 10x_3 + 15x_4 \leq 100 + 30(1 - \alpha), \\
& \alpha \in [0, 1], x_1, x_2, x_3, x_4 \geq 0.
\end{aligned} \tag{2.51}$$

The optimal solution of the problem (2.51) is $x^* = (x_1^*, x_2^*, x_3^*, x_4^*) = (8.01, 0, 8.5, 0)$ with $\alpha^* = 0.7$. By substituting this optimal solution in the objective we obtain $z(x^*) = z^* = 108.54$.

2.5.3 Chanas's Approach

Chanas [22] proposed a non-symmetric model for solving the FLP problem of type-II (2.38). According to this approach, the decision maker cannot initially provide an aspiration level z_0 and its corresponding tolerance p_0 for the objective function due to a lack of knowledge about the fuzzy feasible space. Chanas [22] therefore proposed first to solve the following problem and then present the results to the decision maker to estimate the aspiration level and the tolerance of the objective function:

$$\max z = \sum_{j=1}^{n} c_j x_j$$

$$s.t. \quad g_i(x) = \sum_{j=1}^{n} a_{ij} x_j \tilde{\leq} b_i, \quad i = 1, 2, \ldots, m, \tag{2.52}$$

$$x_j \geq 0, \qquad\qquad j = 1, 2, \ldots, n.$$

In this problem, the membership functions of the fuzzy inequality constraints are considered as the Eq. (2.8). Therefore, following Verdegay's approach [17] and assuming $\theta = 1 - \alpha$ the problem (2.52) becomes:

$$\max z = \sum_{j=1}^{n} c_j x_j$$

$$s.t. \quad g_i(x) = \sum_{j=1}^{n} a_{ij} x_j \leq b_i + \theta p_i, \quad i = 1, 2, \ldots, m, \tag{2.53}$$

$$\theta \in [0, 1], x_j \geq 0, \quad j = 1, 2, \ldots, n.$$

The optimal solution of the parametric LP problem (2.53) can be found by means of the parametric techniques. For a given θ, assume that $x^*(\theta)$ and $z^*(\theta)$ are, respectively an optimal solution and the corresponding optimal objective value of the parametric LP problem (2.53). Then, the following constraints hold:

$$\mu_i(g_i(x^*(\theta))) \geq 1 - \theta, \quad i = 1, 2, \ldots, m, \tag{2.54}$$

On the other hand, in the case of $p_i > 0$, for every non-zero basic solution, there is at least one i such that $\mu_i(g_i(x^*(\theta))) = 1 - \theta$. This follows that the common degree of the constraints satisfaction is $\mu_c(g(x^*(\theta))) = \min_{1 \leq i \leq m} \mu_i(g_i(x^*(\theta))) = 1 - \theta$. This means that for every θ, a solution can be obtained which satisfies all the constraints with degree $1 - \theta$. Now, such a solution is presented to the decision maker to determine the aspiration level z_0 of the objective function and its tolerance p_0. Therefore, the membership function μ_0 of the objective function becomes:

$$\mu_0(z^*(\theta)) = \begin{cases} 1, & z^*(\theta) > z_0, \\ 1 - \frac{z_0 - z^*(\theta)}{p_0}, & z_0 - p_0 \leq z^*(\theta) \leq z_0, \\ 0, & z^*(\theta) < z_0 - p_0. \end{cases} \tag{2.55}$$

In this case, as depicted in Fig. 2.5, the final optimal solution $x^*(\theta^*)$ of the FLP problem of type-II (2.38) will exist at:

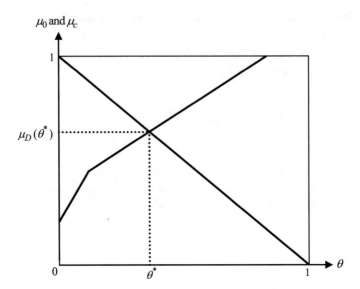

Fig. 2.5 The intersection of μ_0 and μ_D

$$\mu_D(\theta^*) = \max\{\mu_D(\theta)\} = \mu_c(g(x^*(\theta))) = \max\{\min\{\mu_o(\theta), \mu_c(\theta)\}\}$$

This approach can be summarized as follows:

Step 1: Solve the parametric LP problem (2.53) to determine $x^*(\theta)$ and $z^*(\theta)$.
Step 2: Determine an appropriate value θ to obtain z_0 and choose p_0.
Step 3: Obtain the membership function $\mu_0(z^*(\theta))$ as (2.55).
Step 4: Set $\mu_c = \mu_0 = 1 - \theta$ to find θ^*.
Step 5: The optimal solution of the FLP problem (2.28) is $x^*(\theta^*)$ with the optimal objective value $z^*(\theta^*)$.

Example 2.7 [12] Again, consider the FLP problem of type-II (2.45) with $p_1 = 3, p_2 = 6$, and $p_3 = 6$.

The solution procedure is illustrated as follows:

Step 1: According to the problem (2.53), the following parametric LP problem should be solved:

$$
\begin{aligned}
\max z &= x_1 + x_2 \\
s.t. \quad g_1(x) &= -x_1 + 3x_2 \le 21 + 3\theta, \\
g_2(x) &= x_1 + 3x_2 \le 27 + 6\theta, \\
g_3(x) &= 4x_1 + 3x_2 \le 45 + 6\theta, \\
g_4(x) &= 3x_1 + x_2 \le 30, \\
\theta &\in [0, 1], \quad x_1, x_2 \ge 0.
\end{aligned}
\tag{2.56}
$$

The optimal solution is $x^*(\theta) = \left(x_1^*(\theta), x_2^*(\theta)\right) = (6, 7 + 2\theta)$ with $z^*(\theta) = 13 + 2\theta$.

Step 2: Assume $z_0 = 14$ and choose $p_0 = 1$.
Step 3: Now, the membership function $\mu_0(z^*(\theta))$ is then given by

$$\mu_0(z^*(\theta)) = \begin{cases} 1, & 13 + 2\theta > 14, \\ 1 - \frac{14 - (13 + 2\theta)}{1} = 2\theta, & 13 \le 13 + 2\theta \le 14, \\ 0, & 13 + 2\theta < 13. \end{cases} \qquad (2.57)$$

Step 4: In order to find θ^*, we set $\mu_c = \mu_0 = 1 - \theta$. We obtain $2\theta = 1 - \theta$ and then $\theta^* = \frac{1}{3}$.
Step 5: The optimal solution is then $x^*\left(\frac{1}{3}\right) = \left(x_1^*\left(\frac{1}{3}\right), x_2^*\left(\frac{1}{3}\right)\right) = \left(6, \frac{23}{3}\right)$ with $z^*\left(\frac{1}{3}\right) = 13 + \frac{2}{3} = 13.667$.

It is worth noting that the membership function of the fuzzy objective suggested by Chanas [22] is constructed by use of the parametric optimal solution as a function of the parameter instead of the general function of decision variables. This approach is not practical because the number of constraints and variables is rather large in real-world application problems.

2.5.4 Safi et al.'s Approach

For solving the FLP problem of type-II (2.38), according to Zimmermanns's approach, it is required to solve the crisp LP problem (2.43) or the problem (2.44). Safi et al. [23] discussed the case that there are alternative optimal solutions (AOS) for the problem (2.43). Since all of the AOS have the same α, they have the same values for the LP problem (2.43). Therefore, unless the value of $z = cx$ is checked for all AOS, it may be that it is not the best solution that is presented to the decision maker. It is possible that, among the AOS, the first solution obtained by the software is presented, whereas the others may have better values for cx. It may actually also be that cycling occurs during the process of finding the AOS, and hence the software cannot obtain all of the AOS. In this case, it may lose the best value for the objective function $z = cx$. Moreover, in Zimmerman's approach, if an optimal solution exists for the corresponding LP, then it is bounded, whereas it is possible that $z = cx$ is unbounded. Hence, when one of the AOS of the corresponding LP is presented to DM, it is clearly not optimal. Therefore, Safi et al. [23] proposed an algorithm for eliminating these difficulties.

Safi et al. [23] illustrated the shortcomings of Zimmermann's approach with the help of the following examples.

Example 2.8 Consider the following FLP problem of type-II with $z_0 = 3$ $p_0 = 1, p_1 = 2, p_2 = 3$, and $p_3 = 3$:

$$\widetilde{\max}\ z = x_1 + x_2$$
$$s.t.\quad g_1(x) = x_1 + 2x_2 \preceq 10,$$
$$g_2(x) = -2x_1 + x_2 \preceq 3,\qquad\qquad (2.58)$$
$$g_3(x) = 2x_1 + x_2 \preceq 12,$$
$$x_1, x_2 \geq 0.$$

According to Zimmermann's approach, discussed in Sect. 2.4.1, to obtain the optimal solution of the problem (2.58), we have to solve the following crisp LP problem with regard to the problem (2.44):

$$\max\ \alpha$$
$$s.t.\quad z = x_1 + x_2 \geq 3 - (1 - \alpha)$$
$$g_1(x) = x_1 + 2x_2 \leq 10 + 2(1 - \alpha),$$
$$g_2(x) = -2x_1 + x_2 \leq 3 + 3(1 - \alpha),\qquad (2.59)$$
$$g_3(x) = 2x_1 + x_2 \leq 12 + 3(1 - \alpha),$$
$$0 \leq \alpha \leq 1,$$
$$x_1, x_2 \geq 0.$$

The alternative optimal solutions of the crisp LP problem (2.59) with $\alpha^* = 1$ are given in Table 2.2. Since the purpose of problem is to obtain the best value for α, one of the AOS is not preferred to the others. This means that we may present one of the AOS to the decision maker as the optimal solution of the problem (2.58), whereas the best value for the objective value z occurs at another AOS. As can be seen from Table 2.2, the optimal solution x_E^* has the best value of the objective function and should be considered as the optimal solution of the FLP problem of type-II (2.58).

Moreover, as mentioned by Safi et al. [23], when solving this problem by WinQSB there was a cycling between x_A^* and x_B^*, and hence only these two solutions, as alternative basic optimal solutions, were obtained. Thus it is not possible to find the other three alternative basic optimal solutions that have better values for the objective function z.

Table 2.2 The alternative optimal solutions	The optimal solution	The objective function
	$x_A^* = (3, 0)$	$z_A^* = 3$
	$x_B^* = (0, 3)$	$z_B^* = 3$
	$x_C^* = (0.8, 4.6)$	$z_C^* = 5.4$
	$x_D^* = (6, 0)$	$z_D^* = 6$
	$x_E^* = (4.6667, 2.6667)$	$z_E^* = 7.3334$

Example 2.9 Consider the following FLP problem of type-II with $z_0 = 7$ $p_0 = 10$, $p_1 = p_2 = p_4 = p_5 = 2$ and $p_3 = p_6 = 6$:

$$
\begin{aligned}
\widetilde{\max}\, z = {} & 2x_1 + 2.2x_2 \\
s.t. \quad g_1(x) = {} & 2x_1 + x_2 \precsim 8, \\
g_2(x) = {} & x_1 + x_2 \succsim 3, \\
g_3(x) = {} & -x_1 + x_2 \precsim 2, \\
g_4(x) = {} & -x_1 + 10x_2 \precsim 10, \\
g_5(x) = {} & 3x_1 + 3x_2 \precsim 6, \\
g_6(x) = {} & x_1 - x_2 \succsim 2, \\
& x_1, x_2 \geq 0.
\end{aligned}
\tag{2.60}
$$

For solving the FLP problem (2.60), we should solve a crisp LP problem with regard to the problem (2.44). The optimal solution of the corresponding LP problem is $x_A^* = (2.25, 0)$ and $x_B^* = (10.5625, 00.6875)$ with $\alpha^* = 0.625$. The objective function values for x_A^* and x_B^* are $z_A^* = 4.5$ and $z_B^* = 4.6375$, respectively. These optimal solutions not only have different objective values, but also have different membership degrees $\mu_z(x_A^*) = 0.75$ and $\mu_z(x_B^*) = 0.7637$.

Example 2.10 Consider the following FLP problem of type-II with $z_0 = 6$ $p_0 = 1$, $p_1 = 2$ and $p_2 = 3$:

$$
\begin{aligned}
\widetilde{\max}\, z = {} & x_1 + x_2 \\
s.t. \quad g_1(x) = {} & 2x_1 - 5x_2 \precsim 10, \\
g_2(x) = {} & 5x_1 - 2x_2 \precsim 30, \\
& x_1, x_2 \geq 0.
\end{aligned}
\tag{2.61}
$$

For solving the FLP problem (2.61), we should solve a crisp LP problem with regard to the problem (2.44). The optimal solutions of the corresponding LP problem are $x_A^* = (5.7143, 0.2857), x_B^* = (6.1905, 0.4762)$ and $x_C^* = (0, 6)$ with $\alpha^* = 1$. The objective function values for x_A^*, x_B^* and x_C^* are $z_A^* = 6, z_B^* = 6.6667$ and $z_C^* = 6$, respectively. But, none of these is the best value for the objective function z. In fact, the coefficients of x_2 in the objective function and the constraints show that the value of the variable x_2 can be increased without any restriction, and hence the objective function z is unbounded. However, Zimmerman's approach does not distinguish this case.

The above examples show that when the crisp LP problem (2.44) corresponding to the FLP problem of type-II (2.38) has alternative optimal solutions, Zimmermanns's approach does not give the best value for the objective function z. To overcome this problem, Safi et al. [23] improved Zimmermanns's approach.

In the case of the problem (2.44) having a unique optimal solution, this will be the optimal solution of the problem (2.38). Otherwise, assume that (x^*, α^*) is an optimal solution of the crisp LP problem (2.44) and solve the following LP problem:

$$\max z = \sum_{j=1}^{n} c_j x_j$$

$$s.t. \quad \sum_{j=1}^{n} c_j x_j \geq z_0 - (1 - \alpha^*)p_0, \tag{2.62}$$

$$\sum_{j=1}^{n} a_{ij} x_j \leq b_i + (1 - \alpha^*)p_i, \quad i = 1, 2, \ldots, m,$$

$$x_j \geq 0, \quad j = 1, 2, \ldots, n.$$

If the problem (2.62) is unbounded, then the problem (2.38) does not have any bounded optimal solution. If the problem (2.62) has the unique optimal solution x^{**}, the best value of the objective function is $z^{**} = \sum_{j=1}^{n} c_j x_j^{**}$. Also, the degree of satisfaction for the main objective function and the degree of satisfaction for the constraint are $\mu_0(x^{**})$ and $\mu_i(x^{**})$, respectively. Finally, if the problem (2.62) has alternative optimal solutions, the following problem is solved:

$$\max \sum_{i=0}^{m} \alpha_i$$

$$s.t. \quad \mu_0(z^{**} = cx^{**}) \leq \alpha_0,$$
$$\mu_i(g_i(x^{**})) \leq \alpha_i, \quad i = 1, 2, \ldots, m,$$
$$\mu_0(z = cx) \geq \alpha_0, \tag{2.63}$$
$$\mu_i(g_i(x)) \geq \alpha_i, \quad i = 1, 2, \ldots, m,$$
$$cx = cx^{**}$$
$$\alpha_i \in [0, 1], i = 0, 1, \ldots, m,$$
$$x_j \geq 0, \quad j = 1, 2, \ldots, n.$$

The problem (2.63) is equivalent to the problem (2.64):

$$\max \sum_{i=0}^{m} \alpha_i$$

$$s.t. \quad \sum_{j=1}^{n} c_j x_j \geq z_0 - (1 - \alpha_0)p_0,$$

$$g_i(x) = \sum_{j=1}^{n} a_{ij} x_j \leq b_i + (1 - \alpha_i)p_i, \quad i = 1, 2, \ldots, m,$$

$$\sum_{j=1}^{n} c_j x_j^{**} \leq z_0 - (1 - \alpha_0)p_0, \tag{2.64}$$

$$g_i(x^{**}) = \sum_{j=1}^{n} a_{ij} x_j^{**} \geq b_i + (1 - \alpha_i)p_i, \quad i = 1, 2, \ldots, m,$$

$$\sum_{j=1}^{n} c_j x_j = \sum_{j=1}^{n} c_j x_j^{**},$$
$$\alpha_i \in [0, 1], i = 0, 1, \ldots, m,$$
$$x_j \geq 0, \quad j = 1, 2, \ldots, n.$$

The optimal solution of the problem (2.64) has the best value for the objective function $z = cx$ and it is a fuzzy efficient solution in the set of alternative optimal solutions.

Example 2.11 Reconsider the FLP problem of type-II (2.58) given in Example 2.8. Since the problem (2.59) has alternative optimal solutions with $\alpha^* = 1$, the following LP problem is solved with regard to the problem (2.62):

$$
\begin{aligned}
\max \ & x_1 + x_2 \\
s.t. \quad & z = x_1 + x_2 \geq 3 \\
& g_1(x) = x_1 + 2x_2 \leq 10, \\
& g_2(x) = -2x_1 + x_2 \leq 3, \\
& g_3(x) = 2x_1 + x_2 \leq 12, \\
& x_1, x_2 \geq 0.
\end{aligned}
\tag{2.65}
$$

This problem has a unique optimal solution as $x^{**} = (4.6667, 2.6667)$ with $z^{**} = 7.3333$. This is the best value for the objective function $z = cx$.

Example 2.12 Reconsider the FLP problem of type-II (2.60) given in Example 2.9. Note that the corresponding LP problem with regard to the problem (2.44) has alternative optimal solutions with $\alpha^* = 0.625$. Solving the corresponding LP problem (2.62) leads to $x^{**} = (1.5625, 0.6875)$ as the unique optimal solution with $\mu_0(x^{**}) = 0.625$ and $z^{**} = 4.6375$.

Example 2.13 Reconsider the FLP problem of type-II (2.61) given in Example 2.10. Note that the corresponding LP problem with regard to the problem (2.44) has alternative optimal solutions with $\alpha^* = 1$. Solving the corresponding LP problem (2.62) leads to an unbounded solution. Thus, the problem (2.61) does not have any bounded optimal solution.

Example 2.14 [23] Consider the following FLP problem of type-II with $z_0 = 130$ $p_0 = 30.7143, p_1 = 5, p_2 = 40$ and $p_3 = 30$:

$$
\begin{aligned}
\widetilde{\max} \ z = \ & 4x_1 + 5x_2 + 9x_3 + 11x_4 \\
s.t. \quad & g_1(x) = x_1 + x_2 + x_3 + x_4 \leq \widetilde{15}, \\
& g_2(x) = 7x_1 + 5x_2 + 3x_3 + 2x_4 \leq \widetilde{80}, \\
& g_3(x) = 3x_1 + 5x_2 + 10x_3 + 15x_4 \leq \widetilde{100}, \\
& x_1, x_2, x_3, x_4 \geq 0.
\end{aligned}
\tag{2.66}
$$

For solving this problem, we first solve the following crisp LP problem with regard to the problem (2.44):

$$\max \alpha$$
$$\text{s.t.} \quad 4x_1 + 5x_2 + 9x_3 + 11x_4 \geq 130 - 30.7143(1 - \alpha),$$
$$g_1(x) = x_1 + x_2 + x_3 + x_4 \leq 15 + 5(1 - \alpha),$$
$$g_2(x) = 7x_1 + 5x_2 + 3x_3 + 2x_4 \leq 80 + 40(1 - \alpha), \qquad (2.67)$$
$$g_3(x) = 3x_1 + 5x_2 + 10x_3 + 15x_4 \leq 100 + 30(1 - \alpha),$$
$$\alpha \in [0, 1], \quad x_1, x_2, x_3, x_4 \geq 0.$$

This problem has alternative optimal solutions with $\alpha^* = 0.5$. Thus, the following LP problem is solved with regard to the problem (2.62):

$$\max z = 4x_1 + 5x_2 + 9x_3 + 11x_4$$
$$\text{s.t.} \quad 4x_1 + 5x_2 + 9x_3 + 11x_4 \geq 114.64285,$$
$$g_1(x) = x_1 + x_2 + x_3 + x_4 \leq 17.5,$$
$$g_2(x) = 7x_1 + 5x_2 + 3x_3 + 2x_4 \leq 100, \qquad (2.68)$$
$$g_3(x) = 3x_1 + 5x_2 + 10x_3 + 15x_4 \leq 115,$$
$$x_1, x_2, x_3, x_4 \geq 0.$$

Solving this problem gives $x^{**} = (8.5714, 0, 8.9286, 0)$ as one of the alternative optimal solutions with $z^{**} = 114.64286$.

Now, according to the problem (2.63), we solve the following LP problem:

$$\max \alpha_0 + \alpha_1 + \alpha_2 + \alpha_3$$
$$\text{s.t.} \quad 4x_1 + 5x_2 + 9x_3 + 11x_4 \geq 130 - 30.71(1 - \alpha_0),$$
$$x_1 + x_2 + x_3 + x_4 \leq 15 + 5(1 - \alpha_1),$$
$$7x_1 + 5x_2 + 3x_3 + 2x_4 \leq 80 + 40(1 - \alpha_2),$$
$$x_1 + 5x_2 + 10x_3 + 15x_4 \leq 100 + 30(1 - \alpha_3), \qquad (2.69)$$
$$0.5 \leq \alpha_0 \leq 1, \quad 0.5 \leq \alpha_1 \leq 1, \quad 0.83 \leq \alpha_2 \leq 1, \quad 0.5 \leq \alpha_3 \leq 1,$$
$$4x_1 + 5x_2 + 9x_3 + 11x_4 = 114.64286$$
$$x_1, x_2, x_3, x_4 \geq 0.$$

The optimal solution of the problem (2.69) is $x^{***} = (4.048, 5.655, 7.798, 0)$ as the fuzzy efficient solution of the problem (2.67).

2.5.5 Chandra and Aggarwal's Approach

In a crisp scenario, there is a clear distinction between the objective and constraints. An optimal solution must firstly be feasible, i.e. it has to satisfy the given constraints. But in a fuzzy scenario, as per Zimmermann's approach, there is no distinction between the objective and constraints function because they are all aggregated together via the 'min' operator. To overcome this difficulty, Chandra and Aggarwal [24] presented an apparently new formulation for solving the FLP problem of type-II (2.38) in the framework of a bi-objective optimization problem. This formulation attempts to trade-off between the two objectives, namely, 'satisfaction of constraints' and 'attainment of the aspiration level of the objective function'.

Chandra and Aggarwal [24] proposed a new two-phase approach for solving the problem (2.38), highly motivated by the conventional two-phase approach for solving the crisp LP problem. In this approach, similar to the crisp case, the Phase-I problem depends only on the constraint functions. Thus, in Phase-I, only membership functions corresponding to constraints are aggregated via the 'min' operator to obtain the collection of 'most appropriate' feasible solutions. Then in Phase-II the membership degree of the fuzzy objective function is maximized over the feasible set identified in Phase-I. This approach seems to be very natural as it respects the distinction between the objective and constraint functions as happens in the crisp scenario. Furthermore, since the end formulation is based on Zimmermann's approach only, it carries all the nice properties associated with it.

We now describe the details of Chandra and Aggarwal's [24] method for solving the FLP problem of type-II (2.38).

The Phase-I problem to obtain the collection of 'most appropriate' feasible solutions is formulated as follows:

$$
\begin{aligned}
\max \quad & \beta \\
\text{s.t.} \quad & \mu_i(g_i(x)) \geq \beta, \qquad i = 1, 2, \ldots, m, \\
& \beta \in [0, 1], x_j \geq 0, \quad j = 1, 2, \ldots, n.
\end{aligned}
\tag{2.70}
$$

The problem (2.70) is equivalent to the following LP problem:

$$
\begin{aligned}
\max \quad & \beta \\
\text{s.t.} \quad & \sum_{j=1}^{n} a_{ij} x_j \leq b_i + (1 - \beta) p_i, \quad i = 1, 2, \ldots, m, \\
& \beta \in [0, 1], x_j \geq 0, \qquad\qquad j = 1, 2, \ldots, n.
\end{aligned}
\tag{2.71}
$$

Let $(\bar{x}, \bar{\beta})$ be an optimal solution of the problem (2.71). Then $\bar{x}$ is called a $\bar{\beta}-$ feasible solution of the FLP problem of type-II (2.38). Denote the set of all $\bar{\beta}-$ feasible solution of the FLP problem of type-II (2.38) by $S(\bar{\beta})$. In this case, each $\bar{x} \in S(\bar{\beta})$ has the highest common degree up to which it meets all the constraints $\sum_{j=1}^{n} a_{ij} x_j \leq b_i (i = 1, 2, \ldots, m)$.

Now, the Phase-II problem to maximize the membership degree of the fuzzy objective function over the feasible set $S(\bar{\beta})$ is formulated as follows:

$$
\begin{aligned}
\max \quad & \alpha \\
\text{s.t.} \quad & \sum_{j=1}^{n} c_j x_j \geq z_0 - (1 - \alpha) p_0 \\
& \sum_{j=1}^{n} a_{ij} x_j \leq b_i + (1 - \bar{\beta}) p_i, \quad i = 1, 2, \ldots, m, \\
& \alpha \in [0, 1], x_j \geq 0, \quad\quad\quad j = 1, 2, \ldots, n.
\end{aligned}
\tag{2.72}
$$

Let $(\hat{x}, \hat{\alpha})$ be the optimal solution of the problem (2.72). Then $\hat{x}$ denoted by $(\hat{\alpha}, \bar{\beta})-$ optimal solution, is called an optimal solution of the FLP problem of type-II (2.38). Here $\hat{\alpha}$ is the degree up to which the aspiration level of the decision maker is met and $\bar{\beta}$ is the common highest degree up to which $\hat{x}$ satisfies the constraints $\sum_{j=1}^{n} a_{ij} x_j \leq b_i (i = 1, 2, \ldots, m)$.

Note that if (x^*, α^*) is an optimal solution of the crisp LP problem (2.44) obtained by Zimmermann's approach and $(\hat{x}, \hat{\alpha})$ is the optimal solution of the problem (2.72), then $\alpha^* \leq \bar{\beta} \leq \max(\hat{\alpha}, \bar{\beta})$. This means that the two-phase approach proposed by Chandra and Aggarwal's method [24] always gives a solution $\hat{x}$ which does a better trade-off between the twin objectives of 'meeting the aspiration level' and 'satisfaction of constraints'.

Theorem 2.2 [24] *Let $(\hat{x}, \hat{\alpha})$ be the optimal solution of the problem (2.72). Then $(\hat{x}, \hat{\alpha}, \bar{\beta})$ is an efficient solution of the following bi-objective optimization problem:*

$$
\begin{aligned}
\max \quad & (\alpha, \beta) \\
\text{s.t.} \quad & \sum_{j=1}^{n} c_j x_j \geq z_0 - (1 - \alpha) p_0 \\
& \sum_{j=1}^{n} a_{ij} x_j \leq b_i + (1 - \beta) p_i, \quad i = 1, 2, \ldots, m, \\
& \alpha, \beta \in [0, 1], x_j \geq 0, \quad\quad\quad j = 1, 2, \ldots, n.
\end{aligned}
\tag{2.73}
$$

Proof Let $(\hat{x}, \hat{\alpha}, \bar{\beta})$ be not an efficient solution of the bi-objective problem (2.73). This implies that a feasible solution $(\bar{\bar{x}}, \bar{\bar{\alpha}}, \bar{\bar{\beta}})$ exists for the problem (2.73) such that $(\bar{\bar{\alpha}} \geq \hat{\alpha}, \bar{\bar{\beta}} > \bar{\beta})$ or $(\bar{\bar{\alpha}} > \hat{\alpha}, \bar{\bar{\beta}} \geq \bar{\beta})$. But with regard to problem (2.71) it is clear that $\bar{\bar{\beta}} \not> \bar{\beta}$ and therefore $\bar{\bar{\beta}} = \bar{\beta}$. Thus the only case to be considered is as $(\bar{\bar{\alpha}} > \hat{\alpha}, \bar{\bar{\beta}} = \bar{\beta})$. But this is obviously not possible as $\hat{\alpha}$ is the optimal value of the problem (2.72). □

Example 2.15 Consider the following FLP problem of type-II with $z_0 = 9.3$ $p_0 = 1.8, p_1 = 4, p_2 = 3, p_3 = 5$ and $p_4 = 6$:

$$\widetilde{\max}\ z = x_1 + x_2$$
$$\begin{aligned}
s.t. \quad & g_1(x) = -x_1 + x_2 \precsim 2, \\
& g_2(x) = -x_1 + 10x_2 \succsim 15, \\
& g_3(x) = x_1 + 3x_2 \precsim 12, \\
& g_4(x) = 3x_1 - x_2 \succsim 20, \\
& x_1, x_2 \geq 0.
\end{aligned} \tag{2.74}$$

For solving the FLP problem (2.74) using Zimmermann's approach, we should solve the following crisp LP problem with regard to the problem (2.44):

$$\max\ \alpha$$
$$\begin{aligned}
s.t. \quad & x_1 + x_2 \geq 9.3 - 1.8(1 - \alpha), \\
& -x_1 + x_2 \leq 2 + 4(1 - \alpha), \\
& -x_1 + 10x_2 \geq 15 - 3(1 - \alpha), \\
& x_1 + 3x_2 \leq 12 + 5(1 - \alpha), \\
& 3x_1 - x_2 \geq 20 - 6(1 - \alpha), \\
& \alpha \in [0, 1], x_1, x_2 \geq 0.
\end{aligned} \tag{2.75}$$

The optimal solution of the crisp LP problem (2.75) is $(x_1^*, x_2^*, \alpha^*) = (6.88, 2.11, 0.754)$. This means that the highest common degree of satisfaction up to which all the constraints and the aspiration level of the decision maker can be satisfied, is $\alpha^* = 0.754$.

Now, we solve the problem (2.74) using the two-phase method proposed by Chandra and Aggarwal [24]. The Phase-I problem is formulated as follows with regard to the problem (2.71):

$$\max\ \beta$$
$$\begin{aligned}
s.t. \quad & -x_1 + x_2 \leq 2 + 4(1 - \beta), \\
& -x_1 + 10x_2 \geq 15 - 3(1 - \beta), \\
& x_1 + 3x_2 \leq 12 + 5(1 - \beta), \\
& 3x_1 - x_2 \geq 20 - 6(1 - \beta), \\
& \beta \in [0, 1], x_1, x_2 \geq 0.
\end{aligned} \tag{2.76}$$

The optimal solution of the crisp LP problem (2.76) is $(\bar{x}_1, \bar{x}_2, \bar{\beta}) = (6.88, 2.11, 0.754)$. This means that the highest common degree of satisfaction up to which all the constraints are satisfied, is $\bar{\beta} = 0.754$. Now, the Phase-II problem is formulated as follows with regard to the problem (2.72):

$$\max \ \alpha$$
$$s.t. \quad x_1 + x_2 \geq 9.3 - 1.8(1 - \alpha),$$
$$-x_1 + x_2 \leq 2 + 4(1 - \bar{\beta}),$$
$$-x_1 + 10x_2 \geq 15 - 3(1 - \bar{\beta}), \qquad (2.77)$$
$$x_1 + 3x_2 \leq 12 + 5(1 - \bar{\beta}),$$
$$3x_1 - x_2 \geq 20 - 6(1 - \bar{\beta}),$$
$$\alpha \in [0, 1], x_1, x_2 \geq 0.$$

The optimal solution of the crisp LP problem (2.77) is $(\hat{x}_1, \hat{x}_2, \hat{\alpha}) = (6.88, 2.11, 0.844)$.

Here note that $\max(\hat{\alpha}, \bar{\beta}) = \max(0.844, 0.754) = 0.844$ is greater than $\alpha^* = 0.754$ as obtained via Zimmermann's approach.

In a fuzzy scenario, the bi-objective optimization problem (2.73) consists of two objectives, namely, the 'satisfaction' of fuzzy constraints and 'attainment' of the appropriate aspiration level of the objective function. Since none of these can be met exactly, any formulation must carry out some sort of trade-off between these two goals. Zimmermann's symmetric formulation [34] is an indifference formulation as it attempts to find the highest common degree up to which these two objectives can be met. The above-mentioned two-phase approach gives more importance to the satisfaction of constraints, as the Phase-I problem (2.71) involves only constraints.

In a similar manner, Chandra and Aggarwal [24] considered other formulations for solving the FLP problem of type-II (2.38) based on the bi-objective optimization problem (2.73).

Assume that the attainment of the aspiration level of the objective function is more important than the satisfaction of constraints. In this case, the Phase-I problem only considering the membership function of the objective function, is formulated as follows:

$$\max \quad \alpha$$
$$s.t. \quad \mu_i(z) \geq \alpha, \qquad i = 1, 2, \ldots, m, \qquad (2.78)$$
$$\alpha \in [0, 1], x_j \geq 0, \quad j = 1, 2, \ldots, n.$$

The problem (2.78) is equivalent to the following LP problem:

$$\max \quad \alpha$$
$$s.t. \quad \sum_{j=1}^{n} c_j x_j \geq z_0 - (1 - \alpha) p_0, \quad i = 1, 2, \ldots, m, \qquad (2.79)$$
$$\alpha \in [0, 1], x_j \geq 0, \qquad\qquad j = 1, 2, \ldots, n.$$

Let $(\bar{x}, \bar{\alpha})$ be an optimal solution of the problem (2.79). Now, the Phase-II problem to maximize the membership degree of the fuzzy constraints is formulated as follows:

$$\max \quad \beta$$
$$s.t. \quad \sum_{j=1}^{n} c_j x_j \geq z_0 - (1 - \bar{\alpha}) p_0$$
$$\sum_{j=1}^{n} a_{ij} x_j \leq b_i + (1 - \beta) p_i, \quad i = 1, 2, \ldots, m, \qquad (2.80)$$
$$\beta \in [0, 1], x_j \geq 0, \qquad\qquad j = 1, 2, \ldots, n.$$

Let $(\hat{x}, \hat{\beta})$ be the optimal solution of the problem (2.80). Then $(\hat{x}, \bar{\alpha}, \hat{\beta})$ is an efficient solution of the problem (2.38).

Example 2.16 Reconsider the FLP problem of type-II (2.74) given in Example 2.14. Now, we solve this problem with regard to the problems (2.79) and (2.80). For this, the Phase-I problem to maximize the membership degree of the fuzzy objective function is formulated as follows with regard to the problem (2.79):

$$\max \quad \alpha$$
$$s.t. \quad x_1 + x_2 \geq 9.3 - 1.8(1 - \alpha), \qquad (2.81)$$
$$\alpha \in [0, 1], x_1, x_2 \geq 0.$$

The optimal solution of the crisp LP problem (2.77) is $(\bar{x}_1, \bar{x}_2, \bar{\alpha}) = (0, 9.3, 1)$. Now, the Phase-II problem to maximize the membership degree of the fuzzy constraints is formulated as follows with regard to the problem (2.80):

$$\max \quad \beta$$
$$s.t. \quad x_1 + x_2 \geq 9.3 - 1.8(1 - \bar{\alpha}),$$
$$-x_1 + x_2 \leq 2 + 4(1 - \beta),$$
$$-x_1 + 10x_2 \geq 15 - 3(1 - \beta), \qquad (2.82)$$
$$x_1 + 3x_2 \leq 12 + 5(1 - \beta),$$
$$3x_1 - x_2 \geq 20 - 6(1 - \beta),$$
$$\beta \in [0, 1], x_1, x_2 \geq 0.$$

The optimal solution of the crisp LP problem (2.82) is $(\hat{x}_1, \hat{x}_2, \hat{\alpha}) = (7.175410, 2.124590, 0.6901639)$. Here again note that $\max(\bar{\alpha}, \hat{\beta}) = \max(1, 0.6901639) = 1$ is greater than $\alpha^* = 0.754$ as obtained via Zimmermann's approach.

In situations where the 'satisfaction' of fuzzy constraints and the 'attainment' of the appropriate aspiration level of the objective function have a different level of importance, Chandra and Aggarwal [24] also proposed another method for solving the FLP problem of type-II (2.38). To do this, they formulated the following LP problem based on the weighted sum approach in solving the multi-objective optimization problem:

$$\max \quad \lambda\alpha + (1 - \lambda)\beta$$

$$s.t. \quad \sum_{j=1}^{n} c_j x_j \geq z_0 - (1 - \alpha)p_0$$

$$\sum_{j=1}^{n} a_{ij} x_j \leq b_i + (1 - \beta)p_i, i = 1, 2, \ldots, m, \tag{2.83}$$

$$\alpha, \beta \in [0, 1], x_j \geq 0, \qquad j = 1, 2, \ldots, n.$$

where $\lambda \in [0, 1]$.

For a given $\bar{\lambda}$, let $(\bar{x}, \bar{\alpha}, \bar{\beta})$ be a solution of the problem (2.83). Then, $\bar{x}$ is called an $\bar{\lambda}-$ efficient solution of the FLP problem of type-II (2.38).

Here note that the decision maker gives more importance to the 'meeting the aspiration level' of the objective function by choosing $\lambda \in (\frac{1}{2}, 1]$ and in a similar way, he/she gives more importance to the 'satisfaction of constraints' by choosing $\lambda \in [0, \frac{1}{2})$. Furthermore, $\lambda = \frac{1}{2}$ represents the neutral or indifference scenario of the decision maker.

Example 2.17 Reconsider the FLP problem of type-II (2.74) given in Example 2.14. Now, we solve this problem based on the weighted sum approach. For this, the problem (2.74) is formulated as follows with regard to the problem (2.83):

$$\max \quad \lambda\alpha + (1 - \lambda)\beta$$

$$s.t. \quad x_1 + x_2 \geq 9.3 - 1.8(1 - \alpha),$$

$$-x_1 + x_2 \leq 2 + 4(1 - \beta),$$

$$-x_1 + 10x_2 \geq 15 - 3(1 - \beta), \tag{2.84}$$

$$x_1 + 3x_2 \leq 12 + 5(1 - \beta),$$

$$3x_1 - x_2 \geq 20 - 6(1 - \beta),$$

$$\alpha, \beta \in [0, 1], x_1, x_2 \geq 0.$$

By solving the problem (2.84) for different values of λ, the fuzzy efficient solutions as given in Table 2.3, are obtained.

Table 2.3 The fuzzy efficient solutions

λ	$\bar{x}_1$	$\bar{x}_2$	$\bar{\alpha}$	$\bar{\beta}$
0.0	6.881423	2.114625	0.0	0.7549407
0.1	6.881423	2.114625	0.8311375	0.7549407
0.2	6.881423	2.114625	1.0	0.7549407
0.3	7.175410	2.114625	1.0	0.6901639
0.4	7.175410	2.114625	1.0	0.6901639
0.5	7.175410	2.114625	1.0	0.6901639
0.6	7.175410	2.114625	1.0	0.6901639
0.7	7.175410	2.114625	1.0	0.6901639
0.8	7.175410	2.114625	1.0	0.6901639
0.9	7.175410	2.114625	1.0	0.6901639
1.0	7.363	1.936	1.0	0.0

2.6 FLP Problems of Type-III

The general form of the FLP problem of type-III with fuzzy cost coefficients can be formulated as follows [12]:

$$
\begin{aligned}
\max \ \tilde{z} &= \sum_{j=1}^{n} \tilde{c}_j x_j \\
\text{s.t.} \quad g_i(x) &= \sum_{j=1}^{n} a_{ij} x_j \leq b_i, \quad i = 1, 2, \ldots, m, \\
x_j &\geq 0, \qquad\qquad\quad j = 1, 2, \ldots, n.
\end{aligned}
\tag{2.85}
$$

In the model (2.85), assume that the membership functions of $\tilde{c}_j (j = 1, 2, \ldots, n)$ are continuous and strictly monotone and given by $\mu_j(c_j) : \mathbb{R} \to [0, 1]$, $(j = 1, 2, \ldots, n)$.

In this section, we shall explore two approaches for solving the FLP problem of type-III (2.85).

2.6.1 Verdegay's First Approach

Verdegay [17] proposed an equivalent LP problem with parametric cost coefficients for solving the FLP problem of type-III (2.85). Here, we describe the details of the proposed approach.

The membership function of the fuzzy objective $\tilde{z}$ is obviously given by $\mu(\tilde{z}) = \inf\{\mu_1(c_1), \mu_2(c_2), \ldots, \mu_n(c_n)\}$. As shown in [17], the fuzzy solution of (2.85) could be found after solving the following problem:

$$
\begin{aligned}
\max z &= \sum_{j=1}^{n} c_j x_j \\
\text{s.t.} \quad \mu(\tilde{z}) &\geq 1 - \alpha \\
g_i(x) &= \sum_{j=1}^{n} a_{ij} x_j \leq b_i, \quad i = 1, 2, \ldots, m, \\
\alpha &\in [0, 1], x_j \geq 0, \qquad j = 1, 2, \ldots, n.
\end{aligned}
\tag{2.86}
$$

Note that it can be very difficult to solve the problem (2.86). Thus, Verdegay [17] proved that the optimal solution of this problem can be found by solving an equivalent problem according to the following theorem.

Theorem 2.3 [17] *Consider the FLP problem of type-III (2.85). If the membership functions of $\tilde{c}_j (j = 1, 2, \ldots, n)$ are continuous and strictly monotone, then the fuzzy solution of the problem (2.85) is given by the optimal solution of the following parametric LP problem:*

$$\max z = \sum_{j=1}^{n} \eta_j(\beta) x_j$$

$$s.t. \quad g_i(x) = \sum_{j=1}^{n} a_{ij} x_j \le b_i, \quad i = 1, 2, \ldots, m,$$

$$\beta \in [0, 1], x_j \ge 0, \qquad j = 1, 2, \ldots, n. \tag{2.87}$$

where $\eta_j(\beta) : \mathbb{R} \to [0, 1], (j = 1, 2, \ldots, n)$.

Proof The fuzzy optimal solution of (2.85) is found by solving the problem (2.86). From $\mu(\tilde{z}) = \inf_{1 \le j \le n} \{\mu_j(c_j)\} \ge 1 - \alpha$, we have $\mu_j(c_j) \ge 1 - \alpha (j = 1, 2, \ldots, n)$. As $\mu_j(c_j)$ is continuous and strictly monotone, μ_j^{-1} exists and $c_j \ge \mu_j^{-1}(1 - \alpha)(j = 1, 2, \ldots, n)$. Thus, the problem (2.86) can be written as follows:

$$\max z = \sum_{j=1}^{n} c_j x_j$$

$$s.t. \quad c_j \ge \mu_j^{-1}(1 - \alpha), \qquad j = 1, 2, \ldots, n$$

$$g_i(x) = \sum_{j=1}^{n} a_{ij} x_j \le b_i, \quad i = 1, 2, \ldots, m,$$

$$\alpha \in [0, 1], x_j \ge 0, \qquad j = 1, 2, \ldots, n. \tag{2.88}$$

The problem (2.88) is equivalent to the following problem:

$$\max z = \sum_{j=1}^{n} c_j x_j$$

$$s.t. \quad c_j = \mu_j^{-1}(1 - \alpha), \qquad j = 1, 2, \ldots, n$$

$$g_i(x) = \sum_{j=1}^{n} a_{ij} x_j \le b_i, \quad i = 1, 2, \ldots, m,$$

$$\alpha \in [0, 1], x_j \ge 0, \qquad j = 1, 2, \ldots, n. \tag{2.89}$$

This means that every optimal solution of the problem (2.89) is also optimal for the problem (2.88). Thus, we should solve the following problem:

$$\max z = \sum_{j=1}^{n} \mu_j^{-1}(1 - \alpha) x_j$$

$$s.t. \quad g_i(x) = \sum_{j=1}^{n} a_{ij} x_j \le b_i, \quad i = 1, 2, \ldots, m,$$

$$\alpha \in [0, 1], x_j \ge 0, \qquad j = 1, 2, \ldots, n. \tag{2.90}$$

The problem (2.90) is matched with the problem (2.87) by taking $\beta = 1 - \alpha$ and $\eta_j(\beta) = \mu_j^{-1}(1 - \alpha)$, $(j = 1, 2, \ldots, n)$. □

Example 2.18 [17] Consider the following FLP problem of type-III:

$$\max z = \tilde{c}_1 x_1 + 75 x_2$$
$$\text{s.t.} \quad g_1(x) = 3x_1 - x_2 \leq 2,$$
$$g_2(x) = x_1 + 2x_2 \leq 3, \tag{2.91}$$
$$x_1, x_2 \geq 0.$$

where the membership function μ_1 for $\tilde{c}_1$ is given by:

$$\mu_1(c_1) = \begin{cases} 1, & c_1 > 115, \\ \frac{(c_1 - 40)^2}{5625}, & 40 \leq c_1 \leq 115, \\ 0, & c_1 < 40. \end{cases}$$

Then, we have $\mu_1^{-1}(t) = 40 + 75\sqrt{t}$. Now, according to the model (2.90) we have to solve the following problem:

$$\max z = \left(40 + 75\sqrt{1 - \alpha}\right)x_1 + 75 x_2$$
$$\text{s.t.} \quad g_1(x) = 3x_1 - x_2 \leq 2, \tag{2.92}$$
$$g_2(x) = x_1 + 2x_2 \leq 3,$$
$$\alpha \in [0, 1], x_1, x_2 \geq 0.$$

By use of the parametric technique, the optimal solution of problem (2.92) is $x^* = (x_1^*, x_2^*) = (1, 1)$ for each $\alpha \in [0, 1]$. The fuzzy set of the objective function values of the objective function is then obtained as follows:

$$\mu(\tilde{z}) = 75 + \frac{(c_1 - 40)^2}{5625}, \quad 40 \leq c_1 \leq 115.$$

2.6.2 Verdegay's Second Approach

Verdegay [25] proved that the dual of an FLP problem with fuzzy cost coefficients in the objective function [FLP problem of type-III (2.85)] is an FLP problem with fuzzy inequality constraints [FLP problem of type-I (2.5)], and vice versa. In fact, he proved that for an FLP problem of type-III (2.85) or of type-I (2.5), another one always exists that is dual and has the same fuzzy solution. Hence, he proposed that the optimal solution of the problem (2.85) can be found by solving the dual problem of the problem (2.85) by using the approaches discussed in Sect. 2.3. Here, we explain the details of the dual approach for solving an FLP problem of type-III (2.85).

Theorem 2.4 [25] *Given an FLP problem of type-III (2.85) or of type-I (2.5), another one always exists that is dual and has the same fuzzy solution.*

Proof Consider the FLP problem of type-I (2.5) where the membership functions of the fuzzy inequality constraints are given as the Eq. (2.8). As discussed in Sect. 2.4.1., the optimal solution of the FLP problem of type-I (2.5) can be found by solving the parametric LP problem (2.11). As this problem is a classical parametric LP problem, its dual is given by

$$
\begin{aligned}
\min \sum_{i=1}^{m} & [b_i + (1 - \alpha)p_i] y_i \\
s.t. \quad g_i(x) &= \sum_{i=1}^{m} y_i a_{ij} \geq c_j, \quad j = 1, 2, \ldots, n, \\
\alpha &\in [0, 1], y_i \geq 0, \qquad i = 1, 2, \ldots, m.
\end{aligned}
\tag{2.93}
$$

Let $\beta = 1 - \alpha$ and $d_i = b_i + (1 - \alpha)p_i$ $(i = 1, 2, \ldots, m)$. In this case, the problem (2.93) is equivalent to:

$$
\begin{aligned}
\min \sum_{i=1}^{m} & d_i y_i \\
s.t. \quad d_i &= [b_i + \beta p_i], \qquad i = 1, 2, \ldots, m \\
g_i(x) &= \sum_{i=1}^{m} y_i a_{ij} \geq c_j, \quad j = 1, 2, \ldots, n, \\
\beta &\in [0, 1], y_i \geq 0, \qquad i = 1, 2, \ldots, m.
\end{aligned}
\tag{2.94}
$$

Now, problem (2.94) can be equivalently rewritten as follows in the sense that any optimal solution of (2.94) is also an optimal solution of (2.95)

$$
\begin{aligned}
\min \sum_{i=1}^{m} & d_i y_i \\
s.t. \quad d_i &\leq [b_i + \beta p_i], i = 1, 2, \ldots, m \\
g_i(x) &= \sum_{i=1}^{m} y_i a_{ij} \geq c_j, \quad j = 1, 2, \ldots, n, \\
\beta &\in [0, 1], y_i \geq 0, \qquad i = 1, 2, \ldots, m.
\end{aligned}
\tag{2.95}
$$

Note that for $i = 1, 2, \ldots, m$, $d_i \leq b_i + \beta p_i$ if and only if $b_i + p_i - d_i \geq (1 - \beta)p_i$, i.e., $1 - \frac{d_i - b_i}{p_i} \geq 1 - \beta$. Hence, the problem (2.95) can be rewritten as follows with regard to the Eq. (2.8):

$$\min \sum_{i=1}^{m} d_i y_i$$

$$\begin{aligned}
s.t. \quad & \mu_i(d_i) \geq 1 - \beta, & i = 1, 2, \ldots, m, \\
& g_i(x) = \sum_{i=1}^{m} y_i a_{ij} \geq c_j, & j = 1, 2, \ldots, n, \\
& \beta \in [0, 1], y_i \geq 0, & i = 1, 2, \ldots, m.
\end{aligned} \tag{2.96}$$

The problem (2.96) is obviously the classical parametric LP problem for the following FLP problem of type-III:

$$\min \sum_{i=1}^{m} \tilde{d}_i y_i$$

$$\begin{aligned}
s.t. \quad & g_i(x) = \sum_{i=1}^{m} y_i a_{ij} \geq c_j, & j = 1, 2, \ldots, n, \\
& \beta \in [0, 1], y_i \geq 0, & i = 1, 2, \ldots, m.
\end{aligned} \tag{2.97}$$

Since, in the optimum, the problems (2.11) and (2.93) have the same parametric solution, the FLP problem of type-III (2.97) has the same fuzzy solution as the FLP problem of type-I (2.5) by taking $\beta = 1 - \alpha$. If we had initially started from an FLP problem of type-III, we would, by the same development, in a parallel way, have come to an FLP problem of type-I with the same fuzzy solution.   □

Example 2.19 Consider the following FLP problem of type-III:

$$\begin{aligned}
\min \quad & z = \tilde{c}_1 x_1 + 120 x_2 + \tilde{c}_3 x_3 \\
s.t. \quad & x_1 + 7x_2 + 3x_3 \geq 4, \\
& x_1 + 5x_2 + 5x_3 \geq 5, \\
& x_1 + 3x_2 + 10x_3 \geq 9, \\
& x_1 + 2x_2 + 15x_3 \geq 11, \\
& x_1, x_2, x_3 \geq 0.
\end{aligned} \tag{2.98}$$

where the membership functions of the fuzzy cost coefficients are defined by:

$$\mu_1(c_1) = \begin{cases} 1, & c_1 < 15, \\ \frac{18 - c_1}{3}, & 15 \leq c_1 \leq 18, \\ 0, & c_1 > 18. \end{cases} \tag{2.99}$$

$$\mu_3(c_3) = \begin{cases} 1, & c_3 < 100, \\ \frac{120 - c_3}{20}, & 100 \leq c_3 \leq 120, \\ 0, & c_3 > 120. \end{cases} \tag{2.100}$$

The optimal solution of the problem (2.98) is obtained by solving the following parametric LP problem with regard to the problem (2.90):

$$\begin{aligned}
\min \quad & (18 - 3\alpha)x_1 + 120x_2 + (120 - 20\alpha)x_3 \\
s.t. \quad & x_1 + 7x_2 + 3x_3 \geq 4, \\
& x_1 + 5x_2 + 5x_3 \geq 5, \\
& x_1 + 3x_2 + 10x_3 \geq 9, \\
& x_1 + 2x_2 + 15x_3 \geq 11, \\
& \alpha \in [0, 1], x_1, x_2, x_3 \geq 0.
\end{aligned} \tag{2.101}$$

By means of the parametric technique, the optimal solution is obtained as follows:

$$(x_1^*, x_2^*, x_3^*) = \left(\frac{13}{7}, 0, \frac{5}{7}\right) \tag{2.102}$$

By substituting the optimal solution (2.102) in the objective function (2.101), we have the objective value as follows:

$$z^* = \frac{834}{7} - \frac{139}{7}\alpha$$

On the other hand, if we solve (2.98) by means of its dual, we should have

$$\begin{aligned}
\max z = \; & 4x_1 + 5x_2 + 9x_3 + 11x_4 \\
s.t. \quad & g_1(x) = x_1 + x_2 + x_3 + x_4 \precsim 15, \\
& g_2(x) = 7x_1 + 5x_2 + 3x_3 + 2x_4 \leq 120, \\
& g_3(x) = 3x_1 + 5x_2 + 10x_3 + 15x_4 \precsim 100, \\
& x_1, x_2, x_3, x_4 \geq 0.
\end{aligned} \tag{2.103}$$

where the membership functions of the fuzzy inequality constraints are given in (2.99) and (2.100).

It is obvious that the problem (2.103) belongs to the FLP problem of type-I and then can be solved by the proposed approaches discussed in Sect. 2.4. This problem is the same as the problem (2.12) of Example 2.2 where its optimal solution and optimal objective value are given in the Eq. (2.14). This confirms that both the problem (2.98) and its dual problem (2.103) have the same fuzzy optimal objective values.

2.7 FLP Problems of Type-IV

The general form of the FLP problem of type-IV with fuzzy cost coefficients, fuzzy constraints matrix and fuzzy resources vector can be formulated as follows [12]:

$$
\max \tilde{z} = \sum_{j=1}^{n} \tilde{c}_j x_j
$$
$$
s.t. \quad g_i(x) = \sum_{j=1}^{n} \tilde{a}_{ij} x_j \le \tilde{b}_i, \quad i = 1, 2, \ldots, m, \tag{2.104}
$$
$$
x_j \ge 0, \qquad\qquad j = 1, 2, \ldots, n.
$$

In the model (2.104), assume that the membership functions of fuzzy coefficients are linear as follows:

$$
\mu_{\tilde{c}_j}(x) = \begin{cases} 1, & x > c_j, \\ \frac{x-(c_j-p_j)}{p_j}, & c_j - p_j \le x \le c_j, \\ 0, & x < (c_j - p_j). \end{cases} \tag{2.105}
$$

$$
\mu_{\tilde{a}_{ij}}(x) = \begin{cases} 1, & x < a_{ij}, \\ \frac{(a_{ij}+p_{ij})-x}{p_{ij}}, & a_{ij} \le x \le a_{ij} + p_{ij}, \\ 0, & x > a_{ij} + p_{ij}. \end{cases} \tag{2.106}
$$

$$
\mu_{\tilde{b}_i}(x) = \begin{cases} 1, & x < b_i, \\ 1 - \frac{x-b_i}{p_i}, & b_i \le x \le b_i + p_i, \\ 0, & x > b_i + p_i. \end{cases} \tag{2.107}
$$

It is noted that the membership functions of fuzzy cost coefficients are for a maximization problem so that the larger the coefficients, the more we are satisfied. On the other hand, the fewer resources that are consumed, the better we feel, and thus the technique coefficients.

In this section, we shall explore two approaches for solving the FLP problem of type-IV involving fuzzy constraints matrix.

2.7.1 Carlsson and Korhonen's Approach

Carlsson and Korhonen [35] have provided an efficient approach to solve the problem (2.104). They argued that Chanas's approach does not consider any continuous trade-off between grades of constraint violation, and proposed a fully trade-off approach. They pointed out that if full trade-off between the constraints matrix, resources vector and cost coefficients is assumed, the optimal solution will always exist at $\mu = \mu_{\tilde{a}_{ij}} = \mu_{\tilde{c}_j} = \mu_{\tilde{b}_i}$, where the membership functions are non-increasing (or non-decreasing) monotonic linear functions. Therefore, they proposed the following auxiliary problem:

$$\max \sum_{j=1}^{n} \mu_{\tilde{c}_j}^{-1}(\mu) x_j$$

$$s.t. \quad \sum_{j=1}^{n} \mu_{\tilde{a}_{ij}}^{-1}(\mu) x_j \le \mu_{\tilde{b}_i}^{-1}(\mu), \quad i = 1, 2, \ldots, m, \tag{2.108}$$

$$\mu \in [0, 1], x_j \ge 0, \qquad j = 1, 2, \ldots, n.$$

The problem (2.108) can be equivalently rewritten as follows by substituting membership functions (2.105), (2.106) and (2.107) into the problem (2.109):

$$\max \sum_{j=1}^{n} [c_j - (1 - \mu) p_j] x_j$$

$$s.t. \quad \sum_{j=1}^{n} [a_{ij} + (1 - \mu) p_{ij}] x_j \le b_i + (1 - \mu) p_i, \quad i = 1, 2, \ldots, m, \tag{2.109}$$

$$\mu \in [0, 1], x_j \ge 0, \qquad\qquad\qquad j = 1, 2, \ldots, n.$$

Notice that the problem (2.109) is a nonlinear programing problem as the objective function and the constraints in this problem contain the cross-product terms μx_j. However, this problem is an LP problem if μ is fixed first. Thus, for each given μ value, one can obtain the optimal solution of the problem (2.109) and then the results are presented to the decision maker for further decision.

Example 2.20 Consider the following FLP problem of type-IV:

$$\max \ z = \tilde{c}_1 x_1 + \tilde{c}_2 x_2$$

$$s.t. \quad \tilde{a}_{11} x_1 + \tilde{a}_{12} x_2 \le \tilde{b}_1, \tag{2.110}$$

$$\tilde{a}_{21} x_1 + \tilde{a}_{22} x_2 \le \tilde{b}_2,$$

$$x_1, x_2 \ge 0.$$

where the membership functions of the fuzzy coefficients are defined by:

$$\mu_{\tilde{c}_1}(x) = \begin{cases} 1, & x > 3, \\ x - 2, & 2 \le x \le 3, \\ 0, & x < 2. \end{cases} \quad \mu_{\tilde{c}_2}(x) = \begin{cases} 1, & x > 5, \\ x - 4, & 4 \le x \le 5, \\ 0, & x < 4. \end{cases} \tag{2.111}$$

$$\mu_{\tilde{a}_{11}}(x) = \begin{cases} 1, & x < 1, \\ 2 - x, & 1 \le x \le 2, \\ 0, & x > 2. \end{cases} \quad \mu_{\tilde{a}_{12}}(x) = \begin{cases} 1, & x < 2, \\ 3 - x, & 2 \le x \le 3, \\ 0, & x > 3. \end{cases}$$

$$\mu_{\tilde{a}_{21}}(x) = \begin{cases} 1, & x < 3, \\ 4 - x, & 3 \le x \le 4, \\ 0, & x > 4. \end{cases} \quad \mu_{\tilde{a}_{22}}(x) = \begin{cases} 1, & x < 4, \\ 5 - x, & 4 \le x \le 5, \\ 0, & x > 5. \end{cases}$$

$$(2.112)$$

$$\mu_{\tilde{b}_1}(x) == \begin{cases} 1, & x < 8, \\ 9 - x, & 8 \le x \le 9, \\ 0, & x > 9. \end{cases} \quad \mu_{\tilde{b}_2}(x) == \begin{cases} 1, & x < 10, \\ 11 - x, & 10 \le x \le 11, \\ 0, & x > 11. \end{cases} \quad (2.113)$$

The optimal solution of the problem (2.110) can be found by solving the following problem with regard to the problem (2.109):

$$\begin{aligned} \max z = {} & [3 - (1 - \mu)]x_1 + [5 - (1 - \mu)]x_2 \\ s.t. \quad & [1 + (1 - \mu)]x_1 + [2 + (1 - \mu)]x_2 \le [8 + (1 - \mu)], \\ & [3 + (1 - \mu)]x_1 + [4 + (1 - \mu)]x_2 \le [10 + (1 - \mu)], \\ & \mu \in [0, 1] \quad x_1, x_2 \ge 0. \end{aligned} \quad (2.114)$$

The optimal solution and the corresponding objective function value are given in Table 2.4 for different values of $\mu \in [0, 1]$.

Carlsson and Korhonen's approach can also be applied to solving problems involving a fuzzy constraints matrix, fuzzy cost coefficients and a fuzzy resources vector; a fuzzy constraints matrix and a fuzzy resources vector; and fuzzy cost coefficients and a fuzzy constraints matrix.

Table 2.4 Solutions of the problem (2.114)

μ	x_1^*	x_2^*	z^*
0.0	0.00	2.20	8.80
0.1	0.00	2.224490	9.120408
0.2	0.00	2.25	9.45
0.3	0.00	2.276596	9.789362
0.4	0.00	2.304348	10.13913
0.5	0.00	2.33333	10.5
0.6	0.00	2.363636	10.87273
0.7	0.00	2.395349	11.25814
0.8	0.00	2.428571	11.65714
0.9	0.00	2.463415	12.07073
1.0	0.00	2.50	12.5

2.7.2 Gasimov and Yenilmez's Approach

Gasimov and Yenilmez [28] considered two kinds of FLP problems of type-IV: LP problems with only a fuzzy constraints matrix and LP problems in which both the resources vector and the constraints matrix are fuzzy numbers. They used the symmetric method of Bellman and Zadeh [10] for a defuzzification of these problems.

First consider the following FLP problem with a fuzzy constraints matrix as a special kind of the FLP problem of type-IV (2.104):

$$
\max z = \sum_{j=1}^{n} c_j x_j
$$
$$
s.t. \quad g_i(x) = \sum_{j=1}^{n} \tilde{a}_{ij} x_j \leq b_i, \quad i = 1, 2, \ldots, m,
$$
$$
x_j \geq 0, \qquad\qquad\qquad j = 1, 2, \ldots, n.
$$

(2.115)

The linear membership function of the fuzzy constraints matrix of the problem (2.115) is given in the Eq. (2.106).

For defuzzification of the problem (2.115), Gasimov and Yenilmez [28] first fuzzified the objective function by calculating the lower and upper bounds of the optimal values. The bounds of the optimal values are obtained by solving the following standard LP problems, assuming they have the finite optimal values:

$$
z_1 = \max z = \sum_{j=1}^{n} c_j x_j
$$
$$
s.t. \quad \sum_{j=1}^{n} a_{ij} x_j \leq b_i, \quad i = 1, 2, \ldots, m,
$$
$$
x_j \geq 0, \qquad\qquad j = 1, 2, \ldots, n.
$$

(2.116)

$$
z_2 = \max z = \sum_{j=1}^{n} c_j x_j
$$
$$
s.t. \quad \sum_{j=1}^{n} (a_{ij} + p_{ij}) x_j \leq b_i, \quad i = 1, 2, \ldots, m,
$$
$$
x_j \geq 0, \qquad\qquad\qquad j = 1, 2, \ldots, n.
$$

(2.117)

Let $z^l = \min\{z_1, z_2\}$ and $z^u = \max\{z_1, z_2\}$. Then, z^l and z^u are called the lower and upper bounds of the optimal values, respectively. In this case, the fuzzy set of optimal value, $\tilde{G}$, is defined as follows:

$$\mu_{\tilde{G}}(x) = \begin{cases} 1, & \sum_{j=1}^{n} c_j x_j > z^u, \\ \dfrac{\sum_{j=1}^{n} c_j x_j - z^l}{z^u - z^l}, & z^l \le \sum_{j=1}^{n} c_j x_j \le z^u, \\ 0, & \sum_{j=1}^{n} c_j x_j < z^l. \end{cases} \tag{2.118}$$

In a similar way, the fuzzy set of the constraint i, $\tilde{C}_i$, is defined as follows:

$$\mu_{\tilde{C}_i}(x) = \begin{cases} 1, & b_i > \sum_{j=1}^{n} (a_{ij} + p_{ij}) x_j, \\ \dfrac{b_i - \sum_{j=1}^{n} a_{ij} x_j}{\sum_{j=1}^{n} p_{ij} x_j}, & \sum_{j=1}^{n} a_{ij} x_j \le b_i \le \sum_{j=1}^{n} (a_{ij} + p_{ij}) x_j, \\ 0, & b_i < \sum_{j=1}^{n} a_{ij} x_j. \end{cases} \tag{2.119}$$

By using the definition of the fuzzy decision proposed by Bellman and Zadeh [10], discussed in Sect. 2.4, the optimal fuzzy decision $\tilde{D}$, is a solution of the problem $\mu_{\tilde{D}}(x^*) = \max_{x \ge 0} \mu_{\tilde{D}}(x) = \max_{x \ge 0} \min \left\{ \mu_{\tilde{G}}(x), \min_{1 \le i \le m} \left\{ \mu_{\tilde{C}_i}(x) \right\} \right\}$.
Consequently, the problem (2.115) becomes the following optimization problem:

$$\max \alpha$$
$$\begin{aligned} s.t. \quad & \mu_{\tilde{G}}(x) \ge \alpha, \\ & \mu_{\tilde{C}_i}(x) \ge \alpha, i = 1, 2, \ldots, m, \\ & \alpha \in [0, 1], x \ge 0. \end{aligned} \tag{2.120}$$

By using (2.118) and (2.119), the problem (2.120) can be written as follows:

$$\max \alpha$$
$$\begin{aligned} s.t. \quad & \alpha(z^u - z^l) - \sum_{j=1}^{n} c_j x_j + z^l \le 0, \\ & \sum_{j=1}^{n} (a_{ij} + p_{ij}\alpha) x_j - b_i \le 0, \qquad i = 1, 2, \ldots, m, \\ & \alpha \in [0, 1], x_j \ge 0, \qquad\qquad j = 1, 2, \ldots, n. \end{aligned} \tag{2.121}$$

Notice that the problem (2.121) is a nonlinear programing problem as the constraints in this problem contain the cross-product terms αx_j. Therefore, the solution of this problem requires the special approach adopted for solving general nonlinear optimization problems. Gasimov and Yenilmez [28] applied the modified subgradient method [36] and the fuzzy decisive set method [37] for solving the problem (2.121). The former method is beyond the scope of this book and thus we explore only the latter method.

The fuzzy decisive set method [37] is based on the idea that, for a fixed value of α, the problem (2.121) is an LP problem. Obtaining the optimal solution α^* to the problem (2.121) is equivalent to determining the maximum value of α so that the feasible set is nonempty. The algorithm of this method for the problem (2.121) is presented below.

Algorithm 2.1: The fuzzy decisive set method

Step 1 Set $\alpha = 1$ and test whether the problem (2.121) is feasible or not using phase-I of the simplex method. If this problem is feasible, set $\alpha = 1$ and stop. Otherwise, set $\alpha^L = 0$, $\alpha^R = 1$ and go to the next step.

Step 2 Let $\alpha = \frac{\alpha^L + \alpha^R}{2}$ and update the values of α^L and α^R. If the problem is feasible for α, set $\alpha^L = \alpha$. If the problem is infeasible for α, set $\alpha^R = \alpha$.

Consequently, for each α; test if the problem (2.121) is feasible or not using phase-I of the simplex method and determine the maximum value α^* satisfying the constraints of the problem (2.121).

Example 2.21 [28] Consider the following FLP problem with the fuzzy constraints matrix:

$$\begin{aligned} \max z &= 2x_1 + 3x_2 \\ s.t. \quad \tilde{a}_{11}x_1 + \tilde{a}_{12}x_2 &\leq 4, \\ \tilde{a}_{21}x_1 + \tilde{a}_{22}x_2 &\leq 6, \\ x_1, x_2 &\geq 0. \end{aligned} \tag{2.122}$$

where the membership functions of the fuzzy coefficients are defined by:

$$\mu_{\tilde{a}_{11}}(x) = \begin{cases} 1, & x<1, \\ 2-x, & 1\leq x\leq 2, \\ 0, & x>2. \end{cases} \quad \mu_{\tilde{a}_{12}}(x) = \begin{cases} 1, & x<2, \\ \frac{5-x}{3}, & 2\leq x\leq 5, \\ 0, & x>5. \end{cases}$$

$$\mu_{\tilde{a}_{21}}(x) = \begin{cases} 1, & x<3, \\ \frac{5-x}{2}, & 3\leq x\leq 5, \\ 0, & x>5. \end{cases} \quad \mu_{\tilde{a}_{22}}(x) = \begin{cases} 1, & x<1, \\ \frac{4-x}{3}, & 1\leq x\leq 4, \\ 0, & x>4. \end{cases} \tag{2.123}$$

To obtain the lower and upper bounds of the optimal values we first solve the following problem with regard to the problems (2.116) and (2.117):

$$z_1 = \max z = 2x_1 + 3x_2$$
$$s.t.\quad x_1 + 2x_2 \leq 4,$$
$$\qquad 3x_1 + x_2 \leq 6,$$
$$\qquad x_1, x_2 \geq 0.$$
$$(2.124)$$

$$z_2 = \max z = 2x_1 + 3x_2$$
$$s.t.\quad 2x_1 + 5x_2 \leq 4,$$
$$\qquad 5x_1 + 4x_2 \leq 6,$$
$$\qquad x_1, x_2 \geq 0.$$
$$(2.125)$$

By solving the problems (2.124) and (2.125), we obtain $z_1 = 6.8$ and $z_2 = 0.06$, respectively. Thus, we have $z^l = \min\{6.8, 3.06\} = 3.06$ and $z^u = \max\{6.8, 3.06\} = 6.8$. Therefore, the FLP problem (2.122) is reduced to the following nonlinear programming problem with regard to the problem (2.121):

$$\max \alpha$$
$$s.t.\quad 3.74\alpha - (2x_1 + 3x_2) + 3.06 \leq 0,$$
$$\qquad (1 + \alpha)x_1 + (2 + 3\alpha)x_2 - 4 \leq 0,$$
$$\qquad (3 + 2\alpha)x_1 + (1 + 3\alpha)x_2 - 6 \leq 0,$$
$$\qquad \alpha \in [0, 1], x_1, x_2 \geq 0.$$
$$(2.126)$$

Now, we solve the problem (2.122) by use of Algorithm 2.1. For $\alpha = 1$, this problem is reduced to the following system:

$$2x_1 + 3x_2 \geq 6.8,$$
$$2x_1 + 5x_2 \leq 4,$$
$$5x_1 + 4x_2 \leq 6,$$
$$x_1, x_2 \geq 0.$$
$$(2.127)$$

By using phase-I of the simplex algorithm we conclude that the system (2.127) is infeasible. Thus, we set $\alpha^L = 0$ and $\alpha^R = 1$. The new value of α is $\alpha = \frac{0+1}{2} = \frac{1}{2}$. For $\alpha = \frac{1}{2}$, the problem (2.126) is reduced to the following system:

$$2x_1 + 3x_2 \geq 6.8,$$
$$2x_1 + 5x_2 \leq 4,$$
$$5x_1 + 4x_2 \leq 6,$$
$$x_1, x_2 \geq 0.$$
$$(2.128)$$

Since the system (2.128) is infeasible, we set $\alpha^L = 0$ and $\alpha^R = \frac{1}{2}$. The new value of α is $\alpha = \frac{1}{4}$. For $\alpha = \frac{1}{4}$, the problem (2.126) is reduced to the following system:

$$2x_1 + 3x_2 \geq 3.9941,$$
$$1.25x_1 + 2.75x_2 \leq 4,$$
$$3.5x_1 + 1.75x_2 \leq 6,$$
$$x_1, x_2 \geq 0.$$

(2.129)

Since the system (2.129) is feasible, we set $\alpha^L = \frac{1}{4}$ and $\alpha^R = \frac{1}{2}$. The new value of α is $\alpha = \frac{3}{8}$. For $\alpha = \frac{3}{8}$, the problem (2.126) is reduced to the following system:

$$2x_1 + 3x_2 \geq 4.4618,$$
$$1.375x_1 + 3.125x_2 \leq 4,$$
$$3.75x_1 + 2.125x_2 \leq 6,$$
$$x_1, x_2 \geq 0.$$

(2.130)

Since the system (2.130) is feasible, we set $\alpha^L = \frac{1}{4}$ and $\alpha^R = \frac{1}{2}$. The new value of α is $\alpha = \frac{3}{8}$. For $\alpha = \frac{3}{8}$, the problem (2.126) is reduced to the following system:

$$2x_1 + 3x_2 \geq 4.4618,$$
$$1.375x_1 + 3.125x_2 \leq 4,$$
$$3.75x_1 + 2.125x_2 \leq 6,$$
$$x_1, x_2 \geq 0.$$

(2.131)

By continuing this process the optimal value of α is obtained at the twenty first iteration of the fuzzy decisive set method (Algorithm 2.1).

It should be note that the optimal value of α is found at the second iteration of the modified subgradient method [36] is approximately equal to the optimal value of α calculated at the twenty first iteration of the fuzzy decisive set method.

Now consider the following FLP problem with a fuzzy constraints matrix and a fuzzy resources vector as a special kind of the FLP problem of type-IV (2.104):

$$\max z = \sum_{j=1}^{n} c_j x_j$$

$$s.t. \quad \sum_{j=1}^{n} \tilde{a}_{ij} x_j \leq \tilde{b}_i, \quad i = 1, 2, \ldots, m,$$

$$x_j \geq 0, \quad j = 1, 2, \ldots, n.$$

(2.132)

The linear membership functions of fuzzy constraints matrix and fuzzy resources vector of the problem (2.132) are given in the Eqs. (2.106) and (2.107), respectively.

For defuzzification of the problem (2.132), Gasimov and Yenilmez [28] first fuzzified the objective function by calculating the bounds of the optimal values. The bounds of the optimal values are obtained by solving the following standard LP problems, assuming they have the finite optimal values:

$$z_1 = \max z = \sum_{j=1}^{n} c_j x_j$$

$$\text{s.t.} \quad \sum_{j=1}^{n} (a_{ij} + p_{ij}) x_j \leq b_i, \quad i = 1, 2, \ldots, m,$$

$$x_j \geq 0, \quad j = 1, 2, \ldots, n. \tag{2.133}$$

$$z_2 = \max z = \sum_{j=1}^{n} c_j x_j$$

$$\text{s.t.} \quad \sum_{j=1}^{n} a_{ij} x_j \leq b_i + p_i, \quad i = 1, 2, \ldots, m,$$

$$x_j \geq 0, \quad j = 1, 2, \ldots, n. \tag{2.134}$$

$$z_3 = \max z = \sum_{j=1}^{n} c_j x_j$$

$$\text{s.t.} \quad \sum_{j=1}^{n} (a_{ij} + p_{ij}) x_j \leq b_i + p_i, \quad i = 1, 2, \ldots, m,$$

$$x_j \geq 0, \quad j = 1, 2, \ldots, n. \tag{2.135}$$

$$z_4 = \max z = \sum_{j=1}^{n} c_j x_j$$

$$\text{s.t.} \quad \sum_{j=1}^{n} a_{ij} x_j \leq b_i, \quad i = 1, 2, \ldots, m,$$

$$x_j \geq 0, \quad j = 1, 2, \ldots, n. \tag{2.136}$$

In this case $z^l = \min\{z_1, z_2, z_3, z_4\}$ and $z^u = \max\{z_1, z_2, z_3, z_4\}$ are the lower and upper bounds of the optimal values, respectively. Now, the fuzzy set of optimal value, $\tilde{G}$, is defined as (2.118).

In addition, the fuzzy set of the constraint i, $\tilde{C}_i$, is defined as follows:

$$\mu_{\tilde{C}_i}(x) = \begin{cases} 1, & b_i > \sum_{j=1}^{n} (a_{ij} + p_{ij}) x_j + p_i, \\ \dfrac{b_i - \sum_{j=1}^{n} a_{ij} x_j}{\sum_{j=1}^{n} p_{ij} x_j + p_i}, & \sum_{j=1}^{n} a_{ij} x_j \leq b_i \leq \sum_{j=1}^{n} (a_{ij} + p_{ij}) x_j + p_i, \\ 0, & b_i < \sum_{j=1}^{n} a_{ij} x_j. \end{cases} \tag{2.137}$$

Then, by using the method of defuzzification as for the problem (2.120), the problem (2.132) is reduced to the following crisp nonlinear programming problem:

$$\max \alpha$$

$$s.t. \quad \alpha(z^u - z^l) - \sum_{j=1}^{n} c_j x_j + z^l \leq 0,$$

$$\sum_{j=1}^{n} (a_{ij} + p_{ij}\alpha)x_j + p_i\alpha - b_i \leq 0, \quad i = 1, 2, \ldots, m,$$

$$\alpha \in [0, 1], x_j \geq 0, \qquad\qquad j = 1, 2, \ldots, n. \tag{2.138}$$

Again, the fuzzy decisive set method (Algorithm 2.1) can be used for solving the problem (2.138).

Example 2.22 [28] Consider the following FLP problem with a fuzzy constraints matrix and a fuzzy resources vector:

$$\max z = x_1 + x_2$$

$$s.t. \quad \tilde{a}_{11}x_1 + \tilde{a}_{12}x_2 \leq \tilde{b}_1,$$

$$\tilde{a}_{21}x_1 + \tilde{a}_{22}x_2 \leq \tilde{b}_1, \tag{2.139}$$

$$x_1, x_2 \geq 0.$$

where the membership functions of the fuzzy coefficients are defined as follows:

$$\mu_{\tilde{a}_{11}}(x) = \begin{cases} 1, & x < 1, \\ 2 - x, & 1 \leq x \leq 2, \\ 0, & x > 2. \end{cases} \quad \mu_{\tilde{a}_{12}}(x) = \begin{cases} 1, & x < 2, \\ 3 - x, & 2 \leq x \leq 3, \\ 0, & x > 3. \end{cases}$$

$$\mu_{\tilde{a}_{21}}(x) = \begin{cases} 1, & x < 2, \\ \frac{4-x}{2}, & 2 \leq x \leq 4, \\ 0, & x > 4. \end{cases} \quad \mu_{\tilde{a}_{22}}(x) = \begin{cases} 1, & x < 3, \\ \frac{5-x}{2}, & 3 \leq x \leq 5, \\ 0, & x > 5. \end{cases} \tag{2.140}$$

$$\mu_{\tilde{b}_1}(x) == \begin{cases} 1, & x < 3, \\ \frac{5-x}{2}, & 3 \leq x \leq 5, \\ 0, & x > 5. \end{cases} \quad \mu_{\tilde{b}_2}(x) == \begin{cases} 1, & x < 4, \\ \frac{7-x}{3}, & 4 \leq x \leq 7, \\ 0, & x > 7. \end{cases} \tag{2.141}$$

To obtain the lower and upper bounds of the optimal values we first solve the following problem with regard to the problems (2.133)–(2.136):

$$z_1 = \max z = x_1 + x_2$$

$$s.t. \quad 2x_1 + 3x_2 \leq 3,$$

$$4x_1 + 5x_2 \leq 4, \tag{2.142}$$

$$x_1, x_2 \geq 0.$$

$$z_2 = \max z = x_1 + x_2$$
$$s.t. \quad x_1 + 2x_2 \le 5,$$
$$2x_1 + 3x_2 \le 7,$$
$$x_1, x_2 \ge 0.$$

(2.143)

$$z_3 = \max z = x_1 + x_2$$
$$s.t. \quad 2x_1 + 3x_2 \le 5,$$
$$4x_1 + 5x_2 \le 7,$$
$$x_1, x_2 \ge 0.$$

(2.144)

$$z_4 = \max z = x_1 + x_2$$
$$s.t. \quad x_1 + 2x_2 \le 3,$$
$$2x_1 + 3x_2 \le 4,$$
$$x_1, x_2 \ge 0.$$

(2.145)

By solving the problems (2.142)–(2.145), the optimal values $z_1 = 6.8, z_2 = 3.5$, $z_3 = 1.75$ and $z_4 = 2$ are obtained. Thus, we have $z^l = \min\{1, 3.5, 2, 1.75\} = 1$ and $z^u = \max\{1, 3.5, 2, 1.75\} = 3.5$. Therefore, the FLP problem (2.139) is reduced to the following nonlinear programming problem with regard to the problem (2.138):

$$\max \alpha$$
$$s.t. \quad 2.5\alpha - (x_1 + x_2) + 1 \le 0,$$
$$(1 + \alpha)x_1 + (2 + \alpha)x_2 + 2\alpha - 3 \le 0,$$
$$(2 + 2\alpha)x_1 + (3 + 2\alpha)x_2 + 3\alpha - 4 \le 0,$$
$$\alpha \in [0, 1], x_1, x_2 \ge 0.$$

(2.146)

Now, we solve the problem (2.146) by use of Algorithm 2.1. For $\alpha = 1$, this problem is reduced to the following system:

$$x_1 + x_2 \ge 3.5,$$
$$2x_1 + 3x_2 \le 1,$$
$$4x_1 + 5x_2 \le 1,$$
$$x_1, x_2 \ge 0.$$

(2.147)

By use of phase-I of the simplex algorithm we conclude that the system (2.147) is infeasible. Thus, we set $\alpha^L = 0$ and $\alpha^R = 1$. The new value of α is $\alpha = \frac{0+1}{2} = \frac{1}{2}$. For $\alpha = \frac{1}{2}$, the problem (2.146) is reduced to the following system:

$$x_1 + x_2 \geq 2.25,$$
$$1.5x_1 + 2.5x_2 \leq 2,$$
$$3x_1 + 4x_2 \leq 2.5,$$
$$x_1, x_2 \geq 0.$$
(2.148)

Since the system (2.148) is infeasible, we set $\alpha^L = 0$ and $\alpha^R = \frac{1}{2}$. The new value of α is $\alpha = \frac{1}{4}$. For $\alpha = \frac{1}{4}$, the problem (2.146) is reduced to the following system:

$$x_1 + x_2 \geq 1.625,$$
$$1.25x_1 + 2.25x_2 \leq 2.5,$$
$$2.5x_1 + 3.5x_2 \leq 3.25,$$
$$x_1, x_2 \geq 0.$$
(2.149)

Since the system (2.149) is infeasible, we set $\alpha^L = 0$ and $\alpha^R = \frac{1}{4}$. The new value of α is $\alpha = \frac{1}{8}$. For $\alpha = \frac{1}{8}$, the problem (2.146) is reduced to the following system:

$$x_1 + x_2 \geq 1.3125,$$
$$1.125x_1 + 2.125x_2 \leq 2.75,$$
$$2.25x_1 + 3.25x_2 \leq 3.625,$$
$$x_1, x_2 \geq 0.$$
(2.150)

Since the system (2.150) is feasible, we set $\alpha^L = \frac{1}{8}$ and $\alpha^R = \frac{1}{4}$. The new value of α is $\alpha = \frac{3}{16}$.

By continuing this process, the optimal value of α is obtained at the twenty fifth iteration of the fuzzy decisive set method (Algorithm 2.1).

It should be noted that the optimal value of α is found at the second iteration of the modified subgradient method [36] and is approximately equal to the optimal value of α calculated at the twenty fifth iteration of the fuzzy decisive set method.

2.7.3 Farhadinia's Approach

For solving the problem (2.132), Farhadinia [29] presented a revised formula for the membership function of fuzzy constraints with respect to the following perfectly acceptable remark in fuzzy contexts.

Remark 2.1 A proper fuzzy membership function should be set 0 if the constraints are strongly violated in the crisp sense, and 1 if they are very well satisfied in the crisp sense. Also, it should increase monotonously from 0 to 1.

As can be seen, the problems (2.133)–(2.136) have the same objective function, and different constraints. By shifting the term $p_i (i = 1, 2, \ldots, m)$ in all the constraints to the left-hand side, one obtains

$$
\begin{aligned}
\sum_{j=1}^{n} (a_{ij} + p_{ij}) x_j &\leq b_i, & i &= 1, 2, \ldots, m, \\
\sum_{j=1}^{n} a_{ij} x_j - p_i &\leq b_i, & i &= 1, 2, \ldots, m, \\
\sum_{j=1}^{n} (a_{ij} + p_{ij}) x_j - p_i &\leq b_i, & i &= 1, 2, \ldots, m, \\
\sum_{j=1}^{n} a_{ij} x_j &\leq b_i, & i &= 1, 2, \ldots, m.
\end{aligned}
\tag{2.151}
$$

For any $x = (x_j)_{1 \times n}$ define,

$$
b_i^{\max}(x) = \max \left\{ \sum_{j=1}^{n} (a_{ij} + p_{ij}) x_j, \sum_{j=1}^{n} a_{ij} x_j - p_i, \sum_{j=1}^{n} (a_{ij} + p_{ij}) x_j - p_i, \sum_{j=1}^{n} a_{ij} x_j \right\}
\tag{2.152}
$$

$$
b_i^{\min}(x) = \max \left\{ \sum_{j=1}^{n} (a_{ij} + p_{ij}) x_j, \sum_{j=1}^{n} a_{ij} x_j - p_i, \sum_{j=1}^{n} (a_{ij} + p_{ij}) x_j - p_i, \sum_{j=1}^{n} a_{ij} x_j \right\}
\tag{2.153}
$$

Hence, by virtue of Remark 2.1 and the Eqs. (2.152) and (2.153), the fuzzy set of the constraint i, $\tilde{C}_i$, should be characterized by:

$$
\mu_{\tilde{C}_i}(x) = \begin{cases}
1, & b_i > b_i^{\max}, \\
\in [0, 1], & b_i^{\min} \leq b_i \leq b_i^{\max}, \\
0, & b_i < b_i^{\min}.
\end{cases}
\tag{2.154}
$$

It can be easily verified that for any $x \geq 0$,

$$
b_i^{\max}(x) = \sum_{j=1}^{n} (a_{ij} + p_{ij}) x_j
\tag{2.155}
$$

$$
b_i^{\min}(x) = \sum_{j=1}^{n} (a_{ij} - p_{ij}) x_j
\tag{2.156}
$$

Consequently, by virtue of $\mu_{\tilde{C}_i}(x)$ defined in (2.154) and the Eqs. (2.155)–(2.156), the revised definition for a fuzzy constraint, $\tilde{C}_i$, may be defined by

$$\mu_{\tilde{C}_i}(x) = \begin{cases} 1, & b_i > \sum_{j=1}^{n}(a_{ij}+p_{ij})x_j+p_i, \\ \dfrac{b_i-\sum_{j=1}^{n}a_{ij}x_j+p_i}{\sum_{j=1}^{n}p_{ij}x_j+p_i}, & \sum_{j=1}^{n}a_{ij}x_j-p_i \le b_i \le \sum_{j=1}^{n}(a_{ij}+p_{ij})x_j+p_i, \quad (2.157) \\ 0, & b_i < \sum_{j=1}^{n}a_{ij}x_j-p_i. \end{cases}$$

Then, by using the defuzzification method as for the problem (2.120), the problem (2.132) is reduced to the following crisp nonlinear programming problem by putting the membership function $\mu_{\tilde{C}_i}(x)$ defined in (2.157) into the problem (2.120):

$$\max \alpha$$
$$s.t. \quad \alpha(z^u-z^l)-\sum_{j=1}^{n}c_j x_j+z^l \le 0,$$
$$\sum_{j=1}^{n}(a_{ij}+p_{ij}\alpha)x_j+p_i\alpha-p_i-b_i \le 0, \quad i=1,2,\dots,m, \qquad (2.158)$$
$$\alpha \in [0,1], x_j \ge 0, \qquad\qquad\qquad\qquad j=1,2,\dots,n.$$

Solving the problem (2.158) proposed by Farhadinia [29] has some advantages with respect to solving the problem (2.138) proposed by Gasimov and Yenilmez [28]: (1) the number of required iterations to obtain the desired solution is reduced; and (2) the maximum satisfaction degree of the fuzzy decision set reaches a more accurate optimum.

Example 2.23 [29] Consider the FLP problem given in the problem (2.139) where the membership functions of the fuzzy matrix constraints are defined (2.140) and of the fuzzy resources vector is given as follows:

$$\mu_{\tilde{b}_1}(x) == \begin{cases} 1, & x<3, \\ \frac{11-x}{8}, & 3\le x\le 11, \\ 0, & x>11. \end{cases} \quad \mu_{\tilde{b}_2}(x) == \begin{cases} 1, & x<4, \\ \frac{14-x}{10}, & 4\le x\le 14, \quad (2.159) \\ 0, & x>14. \end{cases}$$

To obtain the lower and upper bounds of the optimal values we first solve the following problem with regard to the Eqs. (2.155)–(2.156):

$$z^l = \max\ z = x_1+x_2$$
$$s.t. \quad 2x_1+3x_2 \le 3,$$
$$4x_1+5x_2 \le 4, \qquad (2.160)$$
$$x_1, x_2 \ge 0.$$

$$z^u = \max z = x_1 + x_2$$
$$s.t. \quad x_1 + 2x_2 \leq 11,$$
$$2x_1 + 3x_2 \leq 14, \qquad\qquad (2.161)$$
$$x_1, x_2 \geq 0.$$

By solving the problems (2.160) and (2.161), the optimal values $z^l = 1$ and $z^u = 7$ are obtained, respectively. Therefore, the FLP problem (2.139) is reduced to the following nonlinear programming problem with regard to the problem (2.158):

$$\max \alpha$$
$$s.t. \quad 6\alpha - (x_1 + x_2) + 1 \leq 0,$$
$$(1 + \alpha)x_1 + (2 + \alpha)x_2 + 8\alpha - 11 \leq 0, \qquad (2.162)$$
$$(2 + 2\alpha)x_1 + (3 + 2\alpha)x_2 + 10\alpha - 14 \leq 0,$$
$$\alpha \in [0, 1], \quad x_1, x_2 \geq 0.$$

Now, we solve the problem (2.162) by use of Algorithm 2.1. For $\alpha = 1$, this problem is reduced to the following system:

$$x_1 + x_2 \geq 7,$$
$$2x_1 + 3x_2 \leq 3,$$
$$4x_1 + 5x_2 \leq 4, \qquad\qquad (2.163)$$
$$x_1, x_2 \geq 0.$$

By use of phase-I of the simplex algorithm we conclude that the system (2.163) is infeasible. Thus, we set $\alpha^L = 0$ and $\alpha^R = 1$. The new value of α is $\alpha = \frac{0+1}{2} = \frac{1}{2}$. For $\alpha = \frac{1}{2}$, the problem (2.162) is reduced to the following system:

$$x_1 + x_2 \geq 4,$$
$$1.5x_1 + 2.5x_2 \leq 7,$$
$$3x_1 + 4x_2 \leq 9, \qquad\qquad (2.164)$$
$$x_1, x_2 \geq 0.$$

Since the system (2.164) is infeasible, we set $\alpha^L = 0$ and $\alpha^R = \frac{1}{2}$. The new value of α is $\alpha = \frac{1}{4}$. For $\alpha = \frac{1}{4}$, the problem (2.162) is reduced to the following system:

$$x_1 + x_2 \geq 2.5,$$
$$1.25x_1 + 2.25x_2 \leq 9,$$
$$2.5x_1 + 3.5x_2 \leq 11.5, \qquad\qquad (2.165)$$
$$x_1, x_2 \geq 0.$$

Since the system (2.149) is infeasible, we set $\alpha^L = \frac{1}{4}$ and $\alpha^R = \frac{1}{2}$. The new value of α is $\alpha = \frac{3}{8}$. For $\alpha = \frac{3}{8}$, the problem (2.162) is reduced to the following system:

$$
\begin{aligned}
x_1 + x_2 &\geq 3.25, \\
1.375x_1 + 2.375x_2 &\leq 10, \\
2.75x_1 + 3.75x_2 &\leq 6.5, \\
x_1, x_2 &\geq 0.
\end{aligned}
\tag{2.166}
$$

Since the system (2.150) is infeasible, we set $\alpha^L = \frac{1}{4}$ and $\alpha^R = \frac{3}{8}$. The new value of α is $\alpha = \frac{5}{16}$.

By continuing this process the optimal value $\alpha = 0.4143$ is obtained at the sixteen first iteration of the fuzzy decisive set method (Algorithm 2.1).

Note that for solving the FLP problem considered in Example 2.23 by use of Gasimov and Yenilmez's [28] approach, we should solve the following nonlinear problem with regard to the problem (2.138):

$$
\begin{aligned}
\max \ &\alpha \\
s.t. \quad &6\alpha - (x_1 + x_2) + 1 \leq 0, \\
&(1 + \alpha)x_1 + (2 + \alpha)x_2 + 8\alpha - 3 \leq 0, \\
&(2 + 2\alpha)x_1 + (3 + 2\alpha)x_2 + 10\alpha - 4 \leq 0, \\
&\alpha \in [0, 1], x_1, x_2 \geq 0.
\end{aligned}
\tag{2.167}
$$

Solving the problem (2.167) by use of Algorithm 2.1 gives the optimal value $\alpha = 0.0801$ at the twenty-four first iteration of the algorithm.

By comparing the above results we conclude that the number of iterations for solving the FLP problem (2.132) according to Frahadinia's approach [29] is one less than for solving the same problem according to Gasimov and Yenilmez's approach [28], but also the maximum satisfaction degree of the fuzzy decision set, that is, the optimum of the problem (2.162), $\alpha = 0.4143$, has been more improved rather than the counterpart of the problem (2.167), $\alpha = 0.0801$.

2.8 FLP Problems of Type-V

The general form of the FLP problem of type-IV with a fuzzy objective function, a fuzzy constraints matrix and a fuzzy resources vector can be formulated as follows [12]:

$$\tilde{\max} \, z = \sum_{j=1}^{n} c_j x_j$$

$$s.t. \quad g_i(x) = \sum_{j=1}^{n} \tilde{a}_{ij} x_j \leq \tilde{b}_i, \quad i = 1, 2, \ldots, m, \tag{2.168}$$

$$x_j \geq 0, \qquad\qquad\qquad j = 1, 2, \ldots, n.$$

where the membership functions of fuzzy coefficients $\tilde{a}_{ij}$ and $\tilde{b}_i$ are linear as (2.106) and (2.107), respectively.

In this section, we explore Lai and Hwang's [38] approach for solving the FLP problem of type-V (2.168).

Lai and Hwang [38] argued that based on the concept of non-symmetrical models, the membership function of the fuzzy objective function can be established by the analysis of fuzzy constraints and considering the Werners' and Zimmermann's membership functions for this situation.

For solving the FLP problem of type-V (2.168), Lai and Hwang [38] first defuzzified the fuzzy constraints matrix and the fuzzy resources vector according to Carlsson and Korhonen's approach. That is, they first solved the following LP problem:

$$\max z = \sum_{j=1}^{n} c_j x_j$$

$$s.t. \quad \sum_{j=1}^{n} [a_{ij} + (1-\mu)p_{ij}] x_j \leq b_i + (1-\mu)p_i, \quad i = 1, 2, \ldots, m, \tag{2.169}$$

$$\mu \in [0,1], x_j \geq 0, \qquad\qquad j = 1, 2, \ldots, n.$$

Once again, the problem (2.169) is a nonlinear programing problem as the constraints in this problem containing the cross product terms μx_j. Thus, for each given μ value like $\mu = 0.0, 0.1, \ldots, 0.9, 1.0$, one can obtain the optimal solution of the problem (2.169) and the corresponding piece-wise linear (assumed) graph (see Fig. 2.6) between z^* and μ. Thus, Werners' membership function of the objective function (z) can be constructed by the Eq. (2.17) where z^l and z^u are the optimal values of the problem (2.169) for $\mu = 0$ and $\mu = 1$, respectively. The optimal solution z_w^* is at the intersection point of the membership function of the objective function and the assumed linear segment of the relation functions between objective values (z^*) and the degree of satisfaction (μ) as shown in Fig. 2.6.

On the other hand, if the decision maker would like to provide his/her goal (z_0) and its tolerance (p_0), after obtaining the optimal solution of the problem (2.169) for each given $\mu = 0.0, 0.1, \ldots, 0.9, 1.0$, then Zimmermann's membership function is more meaningful. The solution (z_z^*) can be found by solving two simultaneous equations-Zimmermann's membership function and the assumed linear segment of the relation functions between objective values (z^*) and the degree of satisfaction (μ) as is shown in Fig. 2.6.

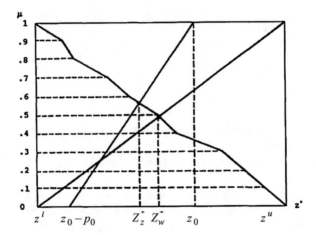

Fig. 2.6 The solution of the problem (2.169)

2.9 Conclusions

The basic techniques for solving FLP problems among various techniques in the literature are Zimmermann's, Werners's, Chanas's, Verdegay's, Carlsson and Korhonen's and Lai and Hwang's approaches where each of them solves one kind of the FLP problems. In this chapter, we have classified the FLP problems that can be solved based on symmetric models, non-symmetric models, parametric models and multi-objective models into five different groups. Then, we explored several approaches for solving the FLP problems belonging to each group. The limitations and advantages of each approach were also pointed out. Moreover, the solution approaches were illustrated with several numerical examples.

References

1. Bazaraa, M.S., Jarvis, J.J., Sherali, H.D.: Linear programming and network flows, 3rd edn. Wiley-Interscience, Wiley, Hoboken (2005)
2. Cadenas, J.M., Pelta, D.A., Pelta, H.R., Verdegay, J.L.: Application of fuzzy optimization to diet problems in Argentinean farms. Eur. J. Oper. Res. **158**, 218–228 (2004)
3. Ebrahimnejad, A.: A new link between output-oriented BCC model with fuzzy data in the present of undesirable outputs and MOLP. Fuzzy Inf. Eng. **3**(2), 113–125 (2011)
4. Ebrahimnejad, A., Karimnejad, K., Alrezaamiri, H.: Particle swarm optimisation algorithm for solving shortest path problems with mixed fuzzy arc weights. Int. J. Appl. Decis. Sci. **8**(2), 203–222 (2015)
5. Ebrahimnejad, A., Tavana, M., Alrezaamiri, A.: A novel artificial bee colony algorithm for shortest path problems with fuzzy arc weights. Measurement **93**, 48–56 (2016)
6. Hatami-Marbini, A., Ebrahimnejad, A., Lozano, S.: Fuzzy efficiency measures in data envelopment analysis using lexicographic multiobjective approach. Comput. Ind. Eng. **105**, 362–376 (2017)

7. Melián, B., Verdegay, J.L.: Fuzzy Optimization Models for the design of WDM networks. IEEE Trans. Fuzzy Syst. **16**(2), 466–476 (2008)
8. Melián, B., Verdegay, J.L.: Using fuzzy numbers in network design optimization problems. IEEE Trans. Fuzzy Syst. **19**(5), 797–806 (2011)
9. Tanaka, H., Okuda, T., Asai, K.: On fuzzy mathematical programming. J. Cybern. **3**(4), 37–46 (1974)
10. Bellman, R.E., Zadeh, L.A.: Decision making in a fuzzy environment. Manage. Sci. **17**(4), 141–164 (1970)
11. Tanaka, H., Ichihashi, H., Asai, K.: A formulation of fuzzy linear programming problems based on comparison of fuzzy numbers. Control Cybern. **13**, 186–194 (1984)
12. Lai, Y.J., Hwang, C.L.: Fuzzy Mathematical Programming. Springer, Berlin (1992)
13. Zimmermann, H.J.: Description and optimization of fuzzy Systems. Int. J. Gen Syst **2**, 209–215 (1976)
14. Luhandjula, M.K.: Fuzzy optimization: milestones and perspectives. Fuzzy Sets Systs. **274**, 4–11 (2015)
15. Tanaka, H., Asai, K.: Fuzzy solution in fuzzy linear programming problems. IEEE Trans. Syst. Man Cybern. **14**(2), 325–328 (1984)
16. Verdegay, J.L.: Progress on fuzzy mathematical programming: a personal perspective. Fuzzy Sets Syst. **281**, 219–226 (2015)
17. Verdegay, J.L.: Fuzzy mathematical programming. In: Gupta, M.M., Sanchez, E. (eds.) Fuzzy Information and Decision Processes. North Holland, Amsterdam (1982)
18. Chanas, S., Kulej, M.: A fuzzy linear programming problem with equality constraints. Control Cybern. **13**(3), 195–201 (1984)
19. Werners, B.: Interactive multiple objective programming subject to flexible constraints. Eur. J. Oper. Res. **31**, 342–349 (1987)
20. Guu, S.-M., Wu, Y.-K.: Two-phase approach for solving the fuzzy linear programming problems. Fuzzy Sets Syst. **107**(2), 191–195 (1999)
21. Lai, Y.J., Hwang, C.L.: Interactive fuzzy linear programming. Fuzzy Sets Syst. **45**(2), 169–183 (1992)
22. Chanas, S.: The use of parametric programming in fuzzy linear programming problems. Fuzzy Sets Syst. **11**, 243–251 (1983)
23. Safi, M.R., Maleki, H.R., Zaeimazad, E.: A note on the Zimmermann method for solving fuzzy linear programming problems. Iran. J. Fuzzy Syst. **4**(2), 31–45 (2007)
24. Chandra, S., Aggarwal, A.: On solving fuzzy linear programming problems: a revisit to Zimmermann's approach. J. Intell. Fuzzy Syst. **279**(5), 2603–2610 (2014)
25. Verdegay, J.L.: Applications of fuzzy optimization in operational research. Control Cybern. **13**(3), 229–239 (1984)
26. Verdegay, J.L.: A dual approach to solve the fuzzy linear programming problem. Fuzzy Sets Syst. **14**(2), 131–141 (1984)
27. Carlsson, C., Korhonen, P.: A parametric approach to fuzzy linear programming. Fuzzy Sets Syst. **20**(1), 17–30 (1986)
28. Gasimov, R.N., Yenilmez, K.: Solving fuzzy linear programming problems with linear membership functions. Turk. J. Math. **26**(4), 375–396 (2002)
29. Farhadinia, B.: Solving fuzzy linear programming problems with linear membership functions-revisited. Casp. J. Math. Sci. **3**(2), 317–396 (2014)
30. Villacorta, P.J., Rabelo, C.A., Pelta, D.A., Verdegay, J.L.: FuzzyLP: an R package for solving fuzzy linear programming problems. In: Granular, soft and fuzzy approaches for intelligent systems, Studies in Fuzziness and Soft Computing, pp. 209–230, vol. 344, Springer, Berlin (2017)
31. Sakawa, M., Yano, H., Nishizaki, N.: Linear and multiobjective programming with fuzzy stochastic extensions. In: International Series in Operations Research & Management Science, vol. 203, Springer, New York (2013)
32. Lai, Y.J., Hwang, C.L.: A new approach to some possibilistic linear programming problem. Fuzzy Sets Syst. **49**(2), 121–133 (1992)

33. Klir, G.J., Yuan, B.: Fuzzy Sets and Fuzzy Logic, Theory and Applications. Prentice-Hall, PTR, Englewood Cliffs (1995)
34. Zimmerman, H.J.: Fuzzy programming and linear programming with several objective functions. Fuzzy Sets Syst. **1**(1), 45–55 (1978)
35. Carlsson, C., Korhonen, P.: A parametric approach to fuzzy linear program-ming. Fuzzy Sets Syst. **20**(1), 17–30 (1986)
36. Sakawa, M., Yana, H.: Interactive decision making for multi-objective linear fractional programming problems with fuzzy parameters. Cybern. Syst. **16**, 377–397 (1985)
37. Azimov, A.Y., Gasimov, R.N.: On weak conjugacy, weak subdifferentials and duality with zero-gap in non-convex optimization. Int. J. Appl. Math. **1**, 171–192 (1999)
38. Lai, Y.J., Hwang, C.L.: IFLP-II: a decision support system. Fuzzy Sets Syst. **54**(1), 47–56 (1993)

Chapter 3
Linear Programming with Fuzzy Parameters: Simplex Based Approaches

3.1 Introduction

The aim of this chapter is to study the existing solution approaches for solving several kinds of LP problems with fuzzy parameters. As argued by Ebrahimnejad and Verdegay [1] the solution approaches for solving LP problems with fuzzy parameters can be classified into two general groups: Simplex based approach and Non-simplex based approach. In the simplex based approaches, the conventional simplex algorithms are generalized to obtain the optimal solution. In the non-simplex based approach, the FLP problems under consideration are first converted into equivalent crisp problems, which are then solved by the standard LP algorithms.

In this chapter, the simplex based approaches for solving several kinds of LP problems with fuzzy parameters are explored. In such approaches, the comparison of fuzzy numbers is done by use of linear ranking functions. Linear programming problems (LPPs) with fuzzy parameters that can be solved based on simplex approaches could be classified into the following five general groups:

- LPPs with fuzzy parameters of type-1: The FLP problems in this group involve fuzzy numbers for the coefficients of the decision variables in the objective function, termed also as LP with fuzzy cost coefficients.
- LPPs with fuzzy parameters of type-2: The FLP problems in this group involve fuzzy numbers for the decision variables and the right-hand side of the constraints, termed also as fuzzy variables linear programming (FVLP) problems.
- LPPs with fuzzy parameters of type-3: The FLP problems in this group involve symmetric fuzzy numbers for the decision variables, the coefficients of the decision variables in the objective function and the right-hand side of the constraints.

© Springer International Publishing AG, part of Springer Nature 2018
A. Ebrahimnejad and J. L. Verdegay, *Fuzzy Sets-Based Methods and Techniques for Modern Analytics*, Studies in Fuzziness and Soft Computing 364,
https://doi.org/10.1007/978-3-319-73903-8_3

- LPPs with fuzzy parameters of type-4: The FLP problems in this group, the so-called Symmetric Fully FLP (SFFLP) problems, involve symmetric fuzzy numbers for the decision variables, the coefficients of the decision variables in the objective function, the coefficients of the decision variables in the constraints and the right-hand side of the constraints.
- LPPs with fuzzy parameters of type-5: The FLP problems in this group involve non-negative triangular fuzzy numbers for the decision variables, the coefficients of the decision variables in the objective function, the coefficients of the decision variables in the constraints and the right-hand side of the constraints.

A comprehensive literature review on solving LPPs with fuzzy parameters based on simplex based approaches is presented in the next section.

3.2   A Brief Literature Review on Simplex Based FLP Models

In the LPPs with fuzzy parameters that can be solved by the use of simplex approaches, some or all parameters of the problems under consideration may be fuzzy numbers and the inequalities may be interpreted in terms of fuzzy rankings. An overview of the papers which studied LPPs with fuzzy parameters based on simplex based approaches and which discussed solution methodologies are summarized as follows.

Maleki et al. [2] proved the fuzzy analogues of some important theorems of the crisp LPPs to the LPPs with fuzzy parameters of type-1 and then extended the primal simplex algorithm in the fuzzy sense to obtain the optimal solution of this kind of fuzzy problems. After that, Nasseri and Ebrahimnejad [3] generalized the dual simplex algorithm in the fuzzy sense for solving the same problem based on duality results developed by Mahdavi-Amiri and Nasseri [4]. Ebrahimnejad [5] proved that in the absence of degeneracy, the fuzzy primal and fuzzy dual algorithms [2, 3] stop in a finite number of iterations. He also proved that if an LPP with fuzzy parameters of type-1 has an optimal solution, then it also has a basic optimal solution. Another approach, namely the primal-dual simplex algorithm in the fuzzy sense has been proposed by Ebrahimnejad [6] which, unlike the dual simplex algorithm, does not require a dual feasible solution to be basic. Using these algorithms, the sensitivity analysis for the same problem was discussed in [7]. Ebrahimnejad et al. [8] and Ebrahimnejad and Nasseri [9] extended the fuzzy primal and dual simplex algorithms for solving LPPs with fuzzy parameters of type-1 where some or all the variables are restricted to lie within lower and upper bounds. Furthermore, Ebrahimnejad and Nasseri [10] and Ebrahimnejad et al. [11] used bounded fuzzy primal and dual simplex algorithms for solving the minimum cost flow problem with fuzzy costs. Moreover, Ebrahimnejad [12] investigated the parametric analysis for the same problem to determine the optimal solution and the fuzzy optimal objective values as a function of parameters when the fuzzy cost coefficients are perturbed along a new fuzzy cost vector.

Maleki et al. [2] used the crisp solution of the LPPs with fuzzy parameters of type-1 as an auxiliary problem for finding the fuzzy solution of the LPPs with fuzzy parameters of type-2. Mahdavi-Amiri and Nasseri [13] showed that the auxiliary problem is indeed the dual of the LPPs with fuzzy parameters of type-2, and proved duality results by a natural extension of the results of crisp LP. Using these results, Mahdavi-Amiri and Nasseri [13] and Ebrahimnejad et al. [14] developed two new methods in the fuzzy sense, namely the fuzzy dual simplex algorithm and the fuzzy primal-dual simplex algorithm, respectively, for solving the same fuzzy problems directly and without any need for an auxiliary problem. Using these fuzzy algorithms, the sensitivity analysis for the same problem was discussed in [15]. Ebrahimnejad and Nasseri [16] applied the complementary slackness to solve the LPPs problem with fuzzy parameters of type-1 and type-2 without the need for fuzzy simplex tableaus. Nasseri and Ebrahimnejad [17] and Nasseri et al. [18] used the fuzzy primal algorithm for solving flexible LP and FLP problems, respectively. Ebrahimnejad [19] proved that if the LPPs problem with fuzzy parameters of type-2 has a fuzzy optimum feasible solution, then it has a fuzzy basic feasible solution that is optimal. Ebrahimnejad and Verdegay [20] and Ebrahimnejad [21] generalized the bounded simplex algorithms in the fuzzy sense for solving that kind of LPPs with fuzzy parameters of type-2 in which some or all variables are restricted to lie within fuzzy lower and fuzzy upper bounds. Ebrahimnejad and Tavana [22] proposed another approach to find the fuzzy solution of the LPPs with fuzzy parameters of type-2 with the help of an equivalent crisp LP problem and without solving any fuzzy problem. Yang et al. [23] focused on two kinds of parametric LPPs with fuzzy parameters of type-2 to describe the behaviors of the optimal value under parametric perturbations of the objective function coefficients and the right-hand side of constraint equations, respectively. Firstly, for these two kinds of parametric fuzzy problems, they investigated how to obtain an optimal basis of them from an optimal basis of their corresponding fuzzy problems based on the parameters in them. They then proposed two algorithms to solve these two parametric fuzzy problems.

Ganesan and Veeramani [24] introduced a type of fuzzy arithmetic for symmetric trapezoidal fuzzy numbers and then proposed a primal simplex method for solving the LPPs with fuzzy parameters of type-3. Ebrahimnejad [25] proved that in the absence of degeneracy, their fuzzy algorithm stops in a finite number of iterations, either with a fuzzy optimal fuzzy basic feasible solution or with the conclusion that the optimal value is unbounded. Then, he proposed a revised version of their method that is more efficient and robust in practice. Nasseri et al. [26]

discussed a concept of duality for the same fuzzy problems and derived the weak and strong duality theorems. Based on these duality results, Ebrahimnejad and Nasseri [27] proposed a duality approach for the same fuzzy problem for situations in which a fuzzy dual feasible solution is at hand. However, their algorithm for solving the LPPs with fuzzy parameters of type-3 also needs an initial dual basic feasible solution. To overcome this limitation, Ebrahimnejad [28] developed the fuzzified version of the conventional primal-dual method of LP problems in which any dual feasible solution, whether basic or not, is adequate to initiate this method. Then, Ebrahimnejad [29] proposed the tableau format of the fuzzy primal-dual method which is easy to apply in real life problems. Ebrahimnejad and Verdegay [30] and Kheirfam and Verdegay [31] have used these algorithms to deal with the sensitivity and parametric analysis of such fuzzy problems without converting them to the crisp LP ones. Moreover, Ebrahimnejad et al. [32] proposed a bounded fuzzy primal simplex algorithm for solving the bounded kind of the LPPs with fuzzy parameters of type-3. Ebrahimnejad and Tavana [33] proposed a novel method for solving the same fuzzy problems which is simpler and computationally more efficient than the above-mentioned algorithms.

Kheirfam and Verdegay [34] introduced a new type of fuzzy inverse and division for symmetric trapezoidal fuzzy to develop methods for solving the LPPs with fuzzy parameters of type-4. Ezaati et al. [35] proposed a fuzzy primal simplex algorithm to find the symmetric fuzzy optimal solution of the LPPs with fuzzy parameters of type-4. Kheirfam and Verdegay [34] generalized the duality results for the LPPs with fuzzy parameters of type-4 and proposed a symmetric fuzzy dual simplex method for solving such a problem without the need to convert it to a conventional LP problem. Khan et al. [36] proposed a novel technique for solving the LPPs with fuzzy parameters of type-5. They used a modified version of the well-known simplex method for solving that kind of fuzzy problems. To do this, they applied a ranking function together with the Gaussian elimination process. Bhardwaj and Kumar [37] claimed that there was error in the technique proposed by Khan et al. [36]. Then, Khan et al. [38] answered the issues raised in Bhardwaj and Kumar and proved that the results obtained in Khan et al. [36] are sound and correct and fulfilled all the necessary requirements of its scope and objectives.

3.3   LPPs with Fuzzy Parameters of Type-1

An LP problem with fuzzy parameters having fuzzy numbers for the coefficients of the decision variables in the objective function is termed an LPP with fuzzy parameters of type-1.

The general from of this kind of FLP problem is formulated as follows:

$$\max \tilde{z} \approx \sum_{j=1}^{n} \tilde{c}_j x_j$$

$$\text{s.t.} \quad \sum_{j=1}^{n} a_{ij} x_j \leq b_i, \quad i = 1, 2, \ldots, m, \tag{3.1}$$

$$x_j \geq 0, \quad j = 1, 2, \ldots, n.$$

The matrix form of this problem can be rewritten as follows:

$$\max \tilde{z} \approx \tilde{c} x$$

$$\text{s.t.} \quad Ax \leq b, \tag{3.2}$$

$$x \geq 0.$$

where $b = (b_1, b_2, \ldots, b_m) \in \mathbb{R}^m$, $\tilde{c} = (\tilde{c}_1, \tilde{c}_2, \ldots \tilde{c}_n) \in F(\mathbb{R})^n$, $A = (a_{ij})_{m \times n} \in \mathbb{R}^{m \times n}$ and $x = (x_1, x_1, \ldots, x_n) \in \mathbb{R}^n$.

Consider the system of equality constraints of (3.2) where A is a matrix of order $(m \times n)$ and $rank\ (A) = m$. Therefore, A can be partitioned as $[B, N]$, where $B_{m \times n}$ is a nonsingular matrix with $rank\ (B) = m$. Here B is called the basic matrix and N is called the non-basic matrix.

Definition 3.1 The basic solution $x = (x_B, x_N) = (B^{-1}b, 0)$ is called a basic solution of $Ax = b$. This basic solution is primal feasible, whenever $x_B \geq 0$. Furthermore, the corresponding fuzzy objective function value is obtained as $\tilde{z} \approx \tilde{c}_B x_B, \tilde{c}_B = (\tilde{c}_{B_1}, \tilde{c}_{B_2}, \ldots, \tilde{c}_{B_m})$.

Suppose J_N is the set of indices associated with the current non-basic variables. For each non-basic variable $x_j, j \in J_N$ the fuzzy variable $\tilde{z}_j$ is defined as $\tilde{z}_j \approx \tilde{c}_B B^{-1} a_j = \tilde{c}_B y_j$ where y_j is the solution of $By_j = a_j$.

In what follows, two solution approaches proposed by Maleki et al. [2] and Nasseri and Ebrahimnejad [3] are explored for solving the LPP with fuzzy parameters of type-1. According to these approaches, linear ranking functions are used for comparison of fuzzy numbers.

3.3.1 Fuzzy Primal Simplex Method

Maleki et al. [2] have proposed using the concept of fuzzy number comparisons with ranking functions to extend the primal simplex method for solving LPPs with fuzzy parameters of type-1. They stated the following important results in order to improve a feasible solution and to provide unbounded criteria and the optimality conditions for the FLP problem (3.2) with equality constraints.

Theorem 3.1 *If for a basic feasible solution with basis B and objective value $\tilde{z}$, $\tilde{z}_k \prec \tilde{c}_k$ for some non-basic variable x_k while $y_k \nleq 0$, then a new feasible solution can be obtained as follows with objective value $\tilde{z}_{new} = \tilde{z} - (\tilde{z}_k - \tilde{c}_k)x_k$, such that $\tilde{z} \preceq \tilde{z}_{new}$.*

$$
\begin{aligned}
x_{B_r} &= x_k = \theta = \frac{\bar{b}_r}{y_{rk}} = \min_{1 \le i \le m} \left\{ \frac{\bar{b}_i}{y_{ik}} \big| y_{ik} > 0 \right\}, \\
x_{B_i} &= \bar{b}_i - y_{ik} \frac{\bar{b}_r}{y_{rk}}, \quad i = 1, 2, \ldots, m, i \ne r, \\
x_j &= 0, \qquad\qquad\quad j \in J_N, j \ne k.
\end{aligned}
\tag{3.3}
$$

Proof Suppose that the solution $x = (x_B, x_N) = (B^{-1}b, 0)$ is the basic solution associated with the current basic matrix B. Both the basic variable and the objective function can be represented in terms of non-basic variables x_N as follows:

$$
x_B = B^{-1}b - B^{-1}Nx_N
\tag{3.4}
$$

$$
\begin{aligned}
\tilde{z} &= \tilde{c}_B x_B + \tilde{c}_N x_N = \tilde{c}_B(B^{-1}b - B^{-1}Nx_N) + \tilde{c}_N x_N \\
&= \tilde{c}_B B^{-1}b - (\tilde{c}_B B^{-1}N - \tilde{c}_N)x_N
\end{aligned}
\tag{3.5}
$$

The fuzzy objective function value associated with the new solution (3.3) is obtained as follows:

$$
\begin{aligned}
\tilde{z}_{new} &\approx \tilde{c}_B x_B + \tilde{c}_N x_N = \sum_{i=1, i \ne r}^{m} \left(\tilde{c}_{B_i}(\bar{b}_i - y_{ik} \frac{\bar{b}_r}{y_{rk}}) \right) + \tilde{c}_k \frac{\bar{b}_r}{y_{rk}} \\
&= \sum_{i=1, i \ne r}^{m} \left(\tilde{c}_{B_i}(\bar{b}_i - y_{ik}\theta) \right) + \tilde{c}_k \theta \\
&= \sum_{i=1}^{m} (c_{1B_i}, c_{2B_i}, c_{3B_i}, c_{4B_i})(\bar{b}_i - y_{ik}\theta) + (c_{1k}, c_{2k}, c_{3k}, c_{4k})\theta
\end{aligned}
\tag{3.6}
$$

Or

$$
\tilde{z}_{new} = (z_1 - \theta z'_{1k} + \theta c_{1k}, z_2 - \theta z'_{2k} + \theta c_{2k}, z_3 - \theta z'_{3k} + \theta c_{3k}, z_4 - \theta z'_{4k} + \theta c_{4k})
\tag{3.7}
$$

where

$$z_1 = \sum_{i=1}^{m} c_{1B_i}\bar{b}_i, \quad z_2 = \sum_{i=1}^{m} c_{2B_i}\bar{b}_i, \quad z_3 = \sum_{i=1}^{m} c_{3B_i}\bar{b}_i, \quad z_4 = \sum_{i=1}^{m} c_{4B_i}\bar{b}_i, \quad (3.8)$$

$$z'_{1k} = \sum_{i=1}^{m} c_{1B_i}y_{ik}, \quad z'_{2k} = \sum_{i=1}^{m} c_{2B_i}y_{ik}, \quad z'_{3k} = \sum_{i=1}^{m} c_{3B_i}y_{ik}, \quad z'_{4k} = \sum_{i=1}^{m} c_{4B_i}y_{ik}$$

$$(3.9)$$

Examining Eq. (3.7) it is clear that if $\theta = 0$, then $\tilde{z}_{new} = (z_1, z_2, z_3, z_4) = \tilde{z}$. On the other hand, in the case of $\theta > 0$ we show that $\Re(\tilde{z}_{new}) \geq \Re(\tilde{z})$, where is $\Re$ the linear ranking function given in Remark 1.15.

For each non-basic variable $x_j, j \in J_N$ we rewrite the fuzzy variable $\tilde{z}_j$ as

$$\tilde{z}_j \approx \tilde{c}_B B^{-1} a_j = \tilde{c}_B y_j = \sum_{i=1}^{m} (c_{1B_i}, c_{2B_i}, c_{3B_i}, c_{4B_i}) y_{ij}$$

$$= \sum_{\{i:y_{ij} \geq 0\}} (c_{1B_i}y_{ij}, c_{2B_i}y_{ij}, c_{3B_i}y_{ij}, c_{4B_i}y_{ij})$$

$$+ \sum_{\{i:y_{ij} < 0\}} (c_{4B_i}y_{ij}, c_{3B_i}y_{ij}, c_{2B_i}y_{ij}, c_{1B_i}y_{ij}) = (z_{1j}, z_{2j}, z_{3j}, z_{4j}).$$

$$(3.10)$$

where

$$z_{1j} = \sum_{\{i:y_{ij} \geq 0\}} c_{1B_i}y_{ij} + \sum_{\{i:y_{ij} < 0\}} c_{4B_i}y_{ij},$$

$$z_{2j} = \sum_{\{i:y_{ij} \geq 0\}} c_{2B_i}y_{ij} + \sum_{\{i:y_{ij} < 0\}} c_{3B_i}y_{ij}$$

$$z_{3j} = \sum_{\{i:y_{ij} \geq 0\}} c_{3B_i}y_{ij} + \sum_{\{i:y_{ij} < 0\}} c_{2B_i}y_{ij},$$

$$z_{4j} = \sum_{\{i:y_{ij} \geq 0\}} c_{4B_i}y_{ij} + \sum_{\{i:y_{ij} < 0\}} c_{1B_i}y_{ij}.$$

$$(3.11)$$

Now, regarding to Eqs. (3.9) and (3.11) we have:

$$z_{1k} + z_{4k} = \sum_{\{i:y_{ik} \geq 0\}} c_{1B_i} y_{ik} + \sum_{\{i:y_{ik} < 0\}} c_{4B_i} y_{ik} \sum_{\{i:y_{ik} \geq 0\}} c_{4B_i} y_{ik} + \sum_{\{i:y_{ik} < 0\}} c_{1B_i} y_{ik}$$

$$= \left(\sum_{\{i:y_{ik} \geq 0\}} c_{1B_i} y_{ik} + \sum_{\{i:y_{ik} < 0\}} c_{1B_i} y_{ik} \right) + \left(\sum_{\{i:y_{ik} \geq 0\}} c_{4B_i} y_{ik} + \sum_{\{i:y_{ik} < 0\}} c_{4B_i} y_{ik} \right)$$

$$= z'_{1k} + z'_{4k}. \tag{3.12}$$

In a similar way, we can conclude that

$$z_{2k} + z_{3k} = z'_{2k} + z'_{3k} \tag{3.13}$$

Hence, examining Eqs. (3.12) and (3.13), we get

$$z_{1k} + z_{2k} + z_{3k} + z_{4k} = z'_{1k} + z'_{2k} + z'_{3k} + z'_{4k} \tag{3.14}$$

On the other hand, since $\tilde{z}_k \prec \tilde{c}_k$ and thus $\Re(\tilde{z}_k) \leq \Re(\tilde{c}_k)$, with regard to Eq. (3.14) we conclude that

$$\frac{1}{4}[(c_{1k} - z'_{1k}) + (c_{2k} - z'_{2k}) + (c_{3k} - z'_{3k}) + (c_{4k} - z'_{4k})] \geq 0 \tag{3.15}$$

Now based on Eq. (3.7), we have

$$\Re(\tilde{z}_{new}) = \frac{1}{4}\left[z_1 - \theta z'_{1k} + \theta c_{1k} + z_2 - \theta z'_{2k} + \theta c_{2k} + z_3 - \theta z'_{3k} + \theta c_{3k} + z_4 - \theta z'_{4k} + \theta c_{4k}\right]$$

$$= \frac{1}{4}\left[(z_1 + z_2 + z_3 + z_4) + \theta\left((c_{1k} - z'_{1k}) + (c_{2k} - z'_{2k}) + (c_{3k} - z'_{3k}) + (c_{4k} - z'_{4k})\right)\right]$$

$$= \Re(\tilde{z}) + \frac{\theta}{4}\left[(c_{1k} - z'_{1k}) + (c_{2k} - z'_{2k}) + (c_{3k} - z'_{3k}) + (c_{4k} - z'_{4k})\right]$$

$$\tag{3.16}$$

Hence, with regard to Eq. (3.15), we conclude that $\Re(\tilde{z}_{new}) \geq \Re(\tilde{z})$, i.e. $\tilde{z}_{new} \succ \tilde{z}$. $\square$

Theorem 3.2 *If for a basic feasible solution with basis B and objective value $\tilde{z}$, $\tilde{z}_k \prec \tilde{c}_k$ for some non-basic variable x_k while $y_k \leq 0$, then the optimal solution of the FLP problem (3.2) is unbounded.*

Proof Examining Eq. (3.16), we have $\Re(\tilde{z}_{new}) = \Re(\tilde{z}) + \theta \Re(\tilde{c}_k - \tilde{z}_k)$. Also, since $\tilde{z}_k \prec \tilde{c}_k$, we have $\Re(\tilde{c}_k - \tilde{z}_k) > 0$. This indicates that in the case of $\theta > 0$, $\Re(\tilde{z}_{new})$ can be made arbitrarily large and hence the optimal solution of the FLP problem (3.2) is unbounded. $\square$

Theorem 3.3 *The basic solution* $x = (x_B, x_N) = (B^{-1}b, 0)$ *is an optimal solution for the FLP problem (3.2) if* $\tilde{z}_j \succeq \tilde{c}_j$ *for all* $j \in J_N$.

Proof We denote the fuzzy objective function value associated with the current feasible basic solution $x = (x_B, x_N) = (B^{-1}b, 0)$, by $\tilde{z}_0 = \tilde{c}_B B^{-1} b = \tilde{c}_B \bar{b}$. Examining Eq. (3.5), we get $\tilde{z} = \tilde{z}_0 - \sum_{j \in J_N} (\tilde{z}_j - \tilde{c}_j) x_j$. Since $\tilde{z}_j \succeq \tilde{c}_j$ i.e. $\Re(\tilde{z}_j) \geq \Re(\tilde{c}_j)$ for all $j \in J_N$, we conclude that $\Re(\tilde{z}) \leq \Re(\tilde{z}_0)$ or equivalently $\tilde{z} \preceq \tilde{z}_0$. $\qquad\qquad\square$

Now we are in a position to summarize the fuzzy primal simplex algorithm Maleki et al. [2] for solving the FLP problem (3.2).

Algorithm 3.1: Fuzzy primal simplex algorithm *(Maximization Problem)*

Initialization step
Choose a starting feasible basic solution with basis.

Main steps

(1) Solve the system $Bx_B = b$. Let $x_B = B^{-1}b$, $x_N = 0$, and $\tilde{z} = \tilde{c}_B x_B$.
(2) Solve the system $\tilde{w}B = \tilde{c}_B$. Let $\tilde{w} = \tilde{c}_B B^{-1}$.
(3) Calculate $\tilde{z}_j = \tilde{c}_B B^{-1} a_j = \tilde{w} a_j$ for all $j \in J_N$. Find the rank of $\tilde{z}_j - \tilde{c}_j$ for all $j \in J_N$ based on ranking function $\Re$ given in Remark 1.15 and let $\Re(\tilde{z}_k - \tilde{c}_k) = \min_{j \in J_N} \{ \Re(\tilde{z}_j - \tilde{c}_j) \}$. If $\Re(\tilde{z}_k - \tilde{c}_k) \geq 0$, then stop with the current basic solution as an optimal solution.
(4) Solve the system $By_k = a_k$ and let $y_k = B^{-1} a_k$. If $y_k \leq 0$ then stop with the conclusion that the problem is unbounded.
If $y_k \nleq 0$, then x_k enters the basis and x_{B_r} leaves the basis providing that

$$\frac{\bar{b}_r}{y_{rk}} = \min_{1 \leq i \leq m} \left\{ \frac{\bar{b}_i}{y_{ik}} | y_{rk} > 0 \right\}.$$

(5) Update the basis B where a_k replaces a_{B_r}, update the index set J_N and go to (1).

Example 3.1 [2] Consider the following FLP problem:

$$\begin{aligned}
Max\, \tilde{z} &= (3, 5, 8, 13)x_1 + (4, 6, 10, 16)x_2 \\
s.t. \quad 2x_1 + 3x_2 &\leq 6, \\
5x_1 + 4x_2 &\leq 10, \\
x_1, x_2 &\geq 0.
\end{aligned} \tag{3.17}$$

After introducing the slack variables x_3 and x_4 we obtain the following FLP problem:

$$Max\,\tilde{z} = (3,5,8,13)x_1 + (4,6,10,16)x_2$$
$$s.t.\quad 2x_1 + 3x_2 + x_3 = 6,$$
$$5x_1 + 4x_2 + x_4 = 10,\qquad\qquad (3.18)$$
$$x_1, x_2, x_3, x_4 \geq 0.$$

The steps of Algorithm 3.1 for solving the FLP problem (3.18) are presented as follows:

Iteration 1:

The starting feasible basic is $B = [a_3, a_4] = \begin{bmatrix} 1 & 0 \\ 0 & 1 \end{bmatrix}$. Thus, the non-basic matrix

N is $N = [a_1, a_2] = \begin{bmatrix} 2 & 3 \\ 5 & 4 \end{bmatrix}$.

Step 1: We find the basic feasible solution by solving the system $Bx_B = b$:

$$\begin{bmatrix} 1 & 0 \\ 0 & 1 \end{bmatrix}\begin{bmatrix} x_3 \\ x_4 \end{bmatrix} = \begin{bmatrix} 6 \\ 10 \end{bmatrix}$$

Solving this system leads to $x_{B_1} = x_3 = 6$ and $x_{B_2} = x_4 = 10$. The non-basic variables are $x_1 = 0$, $x_2 = 0$ and the fuzzy objective value $\tilde{z} = \tilde{c}_B x_B = (0,0,0,0)$.

Step 2: We find $\tilde{w}$ by solving the fuzzy system $\tilde{w}B = \tilde{c}_B$:

$$(\tilde{w}_1, \tilde{w}_2)\begin{bmatrix} 1 & 0 \\ 0 & 1 \end{bmatrix} = \begin{bmatrix} (0,0,0,0,) \\ (0,0,0,0,) \end{bmatrix} \Rightarrow \tilde{w}_1 = \tilde{w}_2 = (0,0,0,0,)$$

Step 3: We calculate $\tilde{z}_j - \tilde{c}_j = \tilde{w}a_j - \tilde{c}_j$ for all $j \in J_N = \{1, 2\}$:

$$\tilde{z}_1 - \tilde{c}_1 = \tilde{w}a_1 - \tilde{c}_1 = ((0,0,0,0),(0,0,0,0))\begin{bmatrix} 2 \\ 5 \end{bmatrix} - (3,5,8,13) = (-13,-5,-5,-3)$$

$$\tilde{z}_2 - \tilde{c}_2 = \tilde{w}a_2 - \tilde{c}_2 = ((0,0,0,0),(0,0,0,0))\begin{bmatrix} 3 \\ 4 \end{bmatrix} - (4,6,10,16) = (-16,-10,-6,-4)$$

Computing the rank of $\tilde{z}_1 - \tilde{c}_1$ and $\tilde{z}_2 - \tilde{c}_2$ based on the ranking function given in Remark 1.15 leads to $\Re(\tilde{z}_1 - \tilde{c}_1) = -\frac{29}{4}$ and $\Re(\tilde{z}_2 - \tilde{c}_2) = -9$. Therefore,

$$min\left\{\Re(\tilde{z}_1 - \tilde{c}_1) = -\frac{29}{2}, \Re(\tilde{z}_2 - \tilde{c}_2) = -9\right\} = \Re(\tilde{z}_2 - \tilde{c}_2) = -9$$

Step 4: We find y_2 by solving the system $By_2 = a_2$:

$$\begin{bmatrix} 1 & 0 \\ 0 & 1 \end{bmatrix} \begin{bmatrix} y_{12} \\ y_{22} \end{bmatrix} = \begin{bmatrix} 3 \\ 4 \end{bmatrix} \Rightarrow y_2 = \begin{bmatrix} y_{12} \\ y_{22} \end{bmatrix} = \begin{bmatrix} 3 \\ 4 \end{bmatrix}$$

Step 5: The variable x_{B_r} leaving the basis is determined by the following test:

$$min \begin{bmatrix} \dfrac{\bar{b}_1}{y_{12}}, \dfrac{\bar{b}_2}{y_{22}} \end{bmatrix} = min \left\{ \dfrac{6}{3}, \dfrac{10}{4} \right\} = \dfrac{6}{3} = 2$$

Hence, $x_{B_1} = x_3$ leaves the basis.

Step 6: We update the basis B where a_2 replaces $a_{B_1} = a_3$:

$$B = [a_2, a_4] = \begin{bmatrix} 3 & 0 \\ 4 & 1 \end{bmatrix}$$

Also, we update $J_N = \{1, 2\}$ as $J_N = \{1, 3\}$.

This process is continued and after two more iterations the optimal basis $B =$

$[a_2, a_1] = \begin{bmatrix} 3 & 2 \\ 4 & 5 \end{bmatrix}$ is found. The optimal solution and the optimal objective func-

tion value of the FLP problem (3.17) are therefore given by:

$$x^* = (x_1^*, x_2^*, x_3^*, x_4^*) = \left(\frac{6}{7}, \frac{6}{7}, \frac{10}{7}, \frac{10}{7} \right), \quad \tilde{z}^* = \left(\frac{58}{7}, \frac{90}{7}, \frac{148}{7}, \frac{244}{7}, \right) \quad (3.19)$$

3.3.2 Fuzzy Dual Simplex Method

The fuzzy primal simplex approach proposed by Maleki et al. [2] is efficient if an initial fuzzy feasible solution can be computed readily. In certain instances it is difficult to find a starting basic solution that is feasible. In these same instances it might be possible to find a starting basis, which is not necessarily feasible, but that is dual feasible. In such cases we develop a new approach, namely the fuzzy dual simplex method [3], for solving an LPP with fuzzy parameters of type-1 (3.2). To do this, the following important definitions and results for the FLP problem (3.2) with equality constraints are provided.

Definition 3.2 The basic matrix B is called a dual feasible basis of $Ax = b$ for a maximization problem whenever $\tilde{z}_j \succeq \tilde{c}_j$ for all $j = 1, 2, \ldots, n$, where $\tilde{z}_j = \tilde{w}a_j$ and $\tilde{w} = \tilde{c}_B B^{-1}$ is a solution to $w_B = \tilde{c}_B$. Furthermore, the corresponding fuzzy objective function value is obtained as $\tilde{z} \approx \tilde{c}_B x_B = \tilde{c}_B B^{-1}b = \tilde{w}b$.

Definition 3.3 The basis B is called a dual nondegenerate basis for a maximization problem whenever $\tilde{z}_j \succ \tilde{c}_j$ for all $j \in J_N$. The problem is called a dual nondegenerate problem if all bases of the problem are dual nondegenerate.

Suppose that we have a basic solution of the system $Ax = b$, $x \geq 0$ with basis B. Suppose that x_k is introduced into the basis and x_{B_r} is removed from the basis. Denote the new basis by $\hat{B}$. Then, $\hat{B} = BE$ and $\hat{B}^{-1} = E^{-1}B^{-1}$, where

$$
E = \begin{bmatrix}
1 & 0 & \cdots & y_{1k} & \cdots & 0 \\
0 & 0 & \cdots & y_{2k} & \cdots & 0 \\
\vdots & \vdots & \cdots & \vdots & \cdots & \vdots \\
0 & 0 & \cdots & y_{rk} & \cdots & 0 \\
\vdots & \vdots & \cdots & \vdots & \cdots & \vdots \\
0 & 0 & \cdots & y_{mk} & \cdots & 1
\end{bmatrix} \leftarrow r\text{th row}
$$
$$\underset{r\text{th column}}{\uparrow}$$

$$
E^{-1} = \begin{bmatrix}
1 & 0 & \cdots & \frac{-y_{1k}}{y_{rk}} & \cdots & 0 \\
0 & 0 & \cdots & \frac{-y_{2k}}{y_{rk}} & \cdots & 0 \\
\vdots & \vdots & \cdots & \vdots & \cdots & \vdots \\
0 & 0 & \cdots & \frac{1}{y_{rk}} & \cdots & 0 \\
\vdots & \vdots & \cdots & \vdots & \cdots & \vdots \\
0 & 0 & \cdots & \frac{-y_{mk}}{y_{rk}} & \cdots & 1
\end{bmatrix} \leftarrow r\text{th row}
$$
$$\underset{r\text{th column}}{\uparrow}$$

Theorem 3.4 *If for a basic solution with basis $B = [a_{B_1}, \ldots, a_{B_r}, \ldots, a_{B_m}]$, x_k, is introduced into the basis and x_{B_r} is removed from the basis. The new objective row for the new basis by $\hat{B} = [a_{B_1}, \ldots, a_k, \ldots, a_{B_m}]$ is given as*

$$
(\tilde{z}_j - \tilde{c}_j)_{new} = (\tilde{z}_j - \tilde{c}_j) - \frac{y_{rj}}{y_{rk}}(\tilde{z}_k - \tilde{c}_k) \tag{3.20}
$$

Proof According to the definition of $\tilde{z}_j$ we obtain:

$$(\tilde{z}_j - \tilde{c}_j)_{new} = \tilde{c}_B \hat{B}^{-1} a_j - \tilde{c}_j = (\tilde{c}_{B_1}, \ldots, \tilde{c}_k, \ldots, \tilde{c}_{B_m}) E^{-1} B^{-1} a_j - \tilde{c}_j$$

$$= (\tilde{c}_{B_1}, \ldots, \tilde{c}_{B_1} \frac{-y_{1k}}{y_{rk}} + \cdots + \tilde{c}_k \frac{1}{y_{rk}} + \cdots + \tilde{c}_{B_m} \frac{-y_{mk}}{y_{rk}}, \ldots, \tilde{c}_{B_m}) B^{-1} a_j - \tilde{c}_j$$

$$= (\tilde{c}_{B_1}, \ldots, \tilde{c}_{B_1} \frac{-y_{1k}}{y_{rk}} + \cdots + \tilde{c}_k \frac{1}{y_{rk}} + \cdots + \tilde{c}_{B_m} \frac{-y_{mk}}{y_{rk}}, \ldots, \tilde{c}_{B_m}) y_j - \tilde{c}_j$$

$$= (\tilde{c}_{B_1}, \ldots, \tilde{c}_{B_1} \frac{-y_{1k}}{y_{rk}} + \cdots + \tilde{c}_k \frac{1}{y_{rk}} + \cdots + \tilde{c}_{B_m} \frac{-y_{mk}}{y_{rk}}, \ldots, \tilde{c}_{B_m}) \begin{bmatrix} y_{1j} \\ \vdots \\ y_{rj} \\ \vdots \\ y_{mj} \end{bmatrix} - \tilde{c}_j$$

$$= \left(\tilde{c}_{B_1} y_{1j} + \cdots + \frac{-y_{rj}}{y_{rk}} (\tilde{c}_{B_1} y_{1k} + \cdots + \tilde{c}_k + \cdots + \tilde{c}_{B_m} y_{mk}) + \cdots \tilde{c}_{B_m} y_{mj} \right) - \tilde{c}_j$$

$$= \left[(\tilde{c}_{B_1} y_{1j} + \cdots + \tilde{c}_{B_r} y_{rj} + \cdots + \tilde{c}_{B_m} y_{mj}) - \tilde{c}_j \right]$$

$$\quad - \frac{y_{rj}}{y_{rk}} (\tilde{c}_{B_1} y_{1k} + \cdots + \tilde{c}_k + \cdots + \tilde{c}_{B_m} y_{mk})$$

$$= \left[\tilde{c}_{B_1} y_j - \tilde{c}_j \right] - \frac{y_{rj}}{y_{rk}} \left[(\tilde{c}_{B_1} y_{1k} + \cdots + \tilde{c}_{B_r} y_{rk} + \cdots + \tilde{c}_{B_m} y_{mk}) - \tilde{c}_k \right]$$

$$= (\tilde{z}_j - \tilde{c}_j) - \frac{y_{rj}}{y_{rk}} (\tilde{z}_k - \tilde{c}_k). \qquad \square$$

Theorem 3.5 *If for a dual feasible basic solution with basis B, the index of the entering variable, x_k, is determined by the following minimum ratio test and x_{B_r} is removed from the basis. The new basic solution is dual feasible.*

$$\frac{\Re(\tilde{z}_k - \tilde{c}_k)}{y_{rk}} = \max \left\{ \frac{\Re(\tilde{z}_j - \tilde{c}_j)}{y_{rj}}, y_{rj} < 0 \right\} \qquad (3.21)$$

Proof Regarding to the Definition 3.1, it is sufficient to prove that $(\tilde{z}_j - \tilde{c}_j)_{new} \succeq \tilde{0}$, i.e., $\Re(\tilde{z}_j - \tilde{c}_j)_{new} \geq 0$. Thus, with regard to Eq. (3.20), we show that

$$\Re(\tilde{z}_j - \tilde{c}_j)_{new} = \Re(\tilde{z}_j - \tilde{c}_j) - \frac{y_{rj}}{y_{rk}} \Re(\tilde{z}_k - \tilde{c}_k) \geq 0$$

If $y_{rj} > 0$, and since $\Re(\tilde{z}_k - \tilde{c}_k) \geq 0$ and $y_{rk} < 0$, then $\frac{y_{rj}}{y_{rk}} \Re(\tilde{z}_k - \tilde{c}_k) \leq 0$, and hence $\Re(\tilde{z}_j - \tilde{c}_j)_{new} \geq \Re(\tilde{z}_j - \tilde{c}_j)$. Since the previous solution was dual feasible, then $\Re(\tilde{z}_j - \tilde{c}_j) \geq 0$, and hence $\Re(\tilde{z}_j - \tilde{c}_j)_{new} \geq 0$. Next, consider the case where $y_{rj} < 0$. By Eq. (3.21), we have

$$\frac{\Re(\tilde{z}_k - \tilde{c}_k)}{y_{rk}} \geq \frac{\Re(\tilde{z}_j - \tilde{c}_j)}{y_{rj}}$$

Multiplying both sides by $y_{rj} < 0$, we have $\Re(\tilde{z}_j - \tilde{c}_j) - \frac{y_{rj}}{y_{rk}}\Re(\tilde{z}_k - \tilde{c}_k) \geq 0$, which is $\Re(\tilde{z}_j - \tilde{c}_j)_{new} \geq 0$. □

Theorem 3.6 *Suppose for a dual feasible basic solution with basis B, the index of the entering variable, x_k, is determined by the given minimum ratio test in Eq. (3.21) and $x_{B_r} = \bar{b}_r < 0$ is removed from the basis, where $y_{rj} \geq 0$ for all $j = 1, 2, \ldots, n$. Then the LPP with fuzzy parameters of type-1 (3.2) is infeasible.*

Proof Since $y_{rj} \geq 0$ for all $j = 1, 2, \ldots, n$, hence no variable is eligible to be the entering variable with regard to the minimum ratio test given in Eq. (3.21). In this case, from the row r, we have $\sum_{j=1}^{n} y_{rj}x_j = \bar{b}_r$. Since $y_{rj} \geq 0$ for all $j = 1, 2, \ldots, n$, and x_j is required to be nonnegative, then $\sum_{j=1}^{n} y_{rj}x_j \geq 0$ for any feasible solution. However, $\bar{b}_r < 0$. This contradiction shows that the LPP with fuzzy parameters of type-1 (3.2) is infeasible. □

Now we are in a position to summarize the fuzzy dual simplex algorithm for solving the FLP problem (3.2).

Algorithm 3.2: Fuzzy dual simplex algorithm *(Maximization Problem)*

Initialization step
Choose a starting dual feasible basic solution with basis B such that $\Re(\tilde{z}_j - \tilde{c}_j) \geq 0$ for all $j = 1, 2, \ldots, n$.

Main steps

(1) Solve the system $Bx_B = b$. Let $x_B = B^{-1}b = \bar{b}$, and $\tilde{z} = \tilde{c}_B x_B$.
(2) Let $\bar{b}_r = \min_{1 \leq i \leq m} \{\bar{b}_i\}$. If $\bar{b}_r \geq 0$, then stop with the current basic solution as an optimal solution.
(3) Solve the system $By_j = a_j$ for all $j \in J_N$ and let $y_j = B^{-1}a_j$. If $y_{rj} \geq 0$ for all $j \in J_N$ then stop with the conclusion that the FLP problem (3.2) is infeasible.
(4) Solve the system $\tilde{w}B = \tilde{c}_B$. Let $\tilde{w} = \tilde{c}_B B^{-1}$. Calculate $\tilde{z}_j = \tilde{w}a_j$ for all $j \in J_N$. If $y_{rj} \not\geq 0$ for all $J \in J_N$, then x_{B_r} leaves the basis and x_k enters the basis providing that

$$\frac{\Re(\tilde{z}_k - \tilde{c}_k)}{y_{rk}} = \max\left\{\frac{\Re(\tilde{z}_j - \tilde{c}_j)}{y_{rj}}, y_{rj} < 0\right\}$$

(5) Update the basis B where a_k replaces a_{B_r}, update the index set J_N and go to (1).

Theorem 3.7 *In the absence of dual degeneracy, the fuzzy dual simplex method (Algorithm 3.2) stops in a finite number of iterations, either with an optimal basic feasible solution or with the conclusion that the problem is infeasible.*

Proof Note that at each iteration of the fuzzy dual simplex method, one of the following three actions is done: (a) We may stop with an optimal solution if

$\bar{b}_r = \min_{1 \le i \le m} \{\bar{b}_i\} \ge 0$; (b) we may conclude that the problem is infeasible if $\bar{b}_r < 0$ and $y_{rj} \ge 0$ for all $j = 1, 2, \ldots, n$; or otherwise, (c) we generate a new dual feasible solution if $\bar{b}_r < 0$ and $y_{rj} \not\ge 0$.

First we show that the difference in the fuzzy dual objective value between two successive iterations is

$$\tilde{z}_{new} - \tilde{z} = -\frac{\bar{b}_r}{y_{rk}} (\tilde{z}_k - \tilde{c}_k) \tag{3.22}$$

According to the assumptions of Theorem 4.3, we have

$$\tilde{z}_{new} = \tilde{c}_{\hat{B}} \hat{B}^{-1} b = (\tilde{c}_{B_1}, \ldots, \tilde{c}_k, \ldots, \tilde{c}_{B_m}) E^{-1} B^{-1} b$$

$$= (\tilde{c}_{B_1}, \ldots, \tilde{c}_{B_1} \frac{-y_{1k}}{y_{rk}} + \cdots + \tilde{c}_k \frac{1}{y_{rk}} + \cdots + \tilde{c}_{B_m} \frac{-y_{mk}}{y_{rk}}, \ldots, \tilde{c}_{B_m}) \bar{b}$$

$$= (\tilde{c}_{B_1}, \ldots, \tilde{c}_{B_1} \frac{-y_{1k}}{y_{rk}} + \cdots + \tilde{c}_k \frac{1}{y_{rk}} + \cdots + \tilde{c}_{B_m} \frac{-y_{mk}}{y_{rk}}, \ldots, \tilde{c}_{B_m}) \begin{bmatrix} \bar{b}_1 \\ \vdots \\ \bar{b}_r \\ \vdots \\ \bar{b}_m \end{bmatrix}$$

$$= (\tilde{c}_{B_1} \bar{b}_1 + \cdots + \frac{-\bar{b}_r}{y_{rk}} (\tilde{c}_{B_1} y_{1k} + \cdots + \tilde{c}_k \frac{1}{y_{rk}} + \cdots + \tilde{c}_{B_m} y_{mk}) + \cdots + \tilde{c}_{B_m} \bar{b}_m)$$

$$= (\tilde{c}_{B_1} \bar{b}_1 + \cdots + \tilde{c}_{B_r} \bar{b}_r + \cdots + \tilde{c}_{B_m} \bar{b}_m) - \frac{\bar{b}_r}{y_{rk}} [(\tilde{c}_{B_1} y_{1k} + \cdots + \tilde{c}_{B_r} y_{rk} + \cdots + \tilde{c}_{B_m} y_{mk}) - \tilde{c}_k]$$

$$= \tilde{z} - \frac{\bar{b}_r}{y_{rk}} (\tilde{z}_k - \tilde{c}_k)$$

This means that $\tilde{z}_{new} - \tilde{z} = -\frac{\bar{b}_r}{y_{rk}} (\tilde{z}_k - \tilde{c}_k)$. Note that $\bar{b}_r < 0$, $y_{rk} < 0$, and $\Re(\tilde{z}_k - \tilde{c}_k) \ge 0$, and hence, $-\frac{\bar{b}_r}{y_{rk}} \Re(\tilde{z}_k - \tilde{c}_k) < 0$. In particular, in the case of $\Re(\tilde{z}_k - \tilde{c}_k) > 0$, the dual fuzzy objective strictly decreases. Hence no basis can be repeated and the algorithm must converge in a finite number of steps. $\quad\square$

Example 3.2 [3] Consider the following LPP with fuzzy parameter of type-1:

$$\begin{aligned} Max\, \tilde{z} &\approx (-5, -4, -2, -1)x_1 + (-6, -5, -3, -2)x_2 \\ s.t. \quad & x_1 - x_2 \ge 1, \\ & x_1 + 2x_2 \ge 2, \\ & x_1, x_2 \ge 0. \end{aligned} \tag{3.23}$$

After introducing the surplus variables x_3 and x_4 we obtain the following FLP problem to start the dual simplex method:

$$\begin{aligned}
Max\,\tilde{z} &\approx (-5,-4,-2,-1)x_1 + (-6,-5,-3,-2)x_2 \\
s.t. \quad -x_1 &+ x_2 + x_3 = -1, \\
-x_1 &- 2x_2 + x_4 = -2, \\
x_1, x_2, &x_3, x_4 \geq 0.
\end{aligned} \tag{3.24}$$

The steps of Algorithm 3.2 for solving the FLP problem (3.24) are presented as follows:

Iteration 1:

The starting dual feasible basic is $B = [a_3, a_4] = \begin{bmatrix} 1 & 0 \\ 0 & 1 \end{bmatrix}$. Thus, the non-basic

matrix N is $N = [a_1, a_2] = \begin{bmatrix} -1 & 1 \\ -1 & -2 \end{bmatrix}$.

Step 1: We solve the system $Bx_B = b$:

$$\begin{bmatrix} 1 & 0 \\ 0 & 1 \end{bmatrix} \begin{bmatrix} x_3 \\ x_4 \end{bmatrix} = \begin{bmatrix} -1 \\ -2 \end{bmatrix}$$

Solving this system leads to $\bar{b}_1 = x_{B_1} = x_3 = -1$ and $\bar{b}_2 = x_{B_2} = x_4 = -2$. The non-basic variables are $x_1 = 0$ and $x_4 = 0$, and the fuzzy objective value is $\tilde{z} = \tilde{c}_B x_B = (0,0,0,0)$.

Step 2: We have $\bar{b}_r = \bar{b}_2 = min\{\bar{b}_1 = -1, \bar{b}_2 = -2\}$.

Step 3: We solve the system $By_j = a_j$ for all $j \in J_N = \{1,2\}$:

$$\begin{bmatrix} 1 & 0 \\ 0 & 1 \end{bmatrix} \begin{bmatrix} y_{11} \\ y_{21} \end{bmatrix} = \begin{bmatrix} -1 \\ -1 \end{bmatrix} \Rightarrow y_1 = \begin{bmatrix} y_{11} \\ y_{21} \end{bmatrix} = \begin{bmatrix} -1 \\ -1 \end{bmatrix}$$

$$\begin{bmatrix} 1 & 0 \\ 0 & 1 \end{bmatrix} \begin{bmatrix} y_{12} \\ y_{22} \end{bmatrix} = \begin{bmatrix} 1 \\ -2 \end{bmatrix} \Rightarrow y_1 = \begin{bmatrix} y_{12} \\ y_{22} \end{bmatrix} = \begin{bmatrix} 1 \\ -2 \end{bmatrix}$$

Step 4: We find $\tilde{w}$ by solving the fuzzy system $\tilde{w}B = \tilde{c}_B$:

$$(\tilde{w}_1, \tilde{w}_2) \begin{bmatrix} 1 & 0 \\ 0 & 1 \end{bmatrix} = \begin{bmatrix} (0,0,0,0) \\ (0,0,0,0) \end{bmatrix} \Rightarrow \tilde{w}_1 = \tilde{w}_2 = (0,0,0,0)$$

Step 5: We calculate $\tilde{z}_j - \tilde{c}_j = \tilde{w}a_j - \tilde{c}_j$ for all $j \in J_N = \{1, 2\}$:

$$\tilde{z}_1 - \tilde{c}_1 = \tilde{w}a_1 - \tilde{c}_1 = ((0,0,0,0),(0,0,0,0))\begin{bmatrix} -1 \\ -1 \end{bmatrix} - (-5,-4,-2,-1) = (1,2,4,5)$$

$$\tilde{z}_2 - \tilde{c}_2 = \tilde{w}a_2 - \tilde{c}_2 = ((0,0,0,0),(0,0,0,0))\begin{bmatrix} 1 \\ -2 \end{bmatrix} - (-6,-5,-3,-2) = (2,3,4,5)$$

Computing the rank of $\tilde{z}_1 - \tilde{c}_1$ and $\tilde{z}_2 - \tilde{c}_2$ based on the ranking function given in Remark 1.15 leads to $\Re(\tilde{z}_1 - \tilde{c}_1) = 3$ and $\Re(\tilde{z}_2 - \tilde{c}_2) = 4$.

The variable $x_{B_r} = x_4$ leaves the basis and the entering variable is determined by the following test:

$$max\left\{ \frac{\Re(\tilde{z}_1 - \tilde{c}_1)}{y_{21}}, \frac{\Re(\tilde{z}_2 - \tilde{c}_2)}{y_{22}} \right\} = max\left\{ \frac{3}{-1}, \frac{4}{-2} \right\} = -2 = \frac{\Re(\tilde{z}_2 - \tilde{c}_2)}{y_{22}}$$

Hence, the variable $x_k = x_2$ enters the basis.

Step 6: We update the basis B where a_2 replaces $a_{B_2} = a_4$:

$$B = [a_3, a_2] = \begin{bmatrix} 1 & 1 \\ 0 & -2 \end{bmatrix}$$

Also, we update $J_N = \{1, 2\}$ as $J_N = \{1, 4\}$.

Iteration 2:

Step 1: We solve the system $Bx_B = b$:

$$\begin{bmatrix} 1 & 1 \\ 0 & -2 \end{bmatrix}\begin{bmatrix} x_3 \\ x_2 \end{bmatrix} = \begin{bmatrix} -1 \\ -2 \end{bmatrix}$$

Solving this system leads to $\bar{b}_1 = x_{B_1} = x_3 = -2$ and $\bar{b}_2 = x_{B_2} = x_2 = 1$. The non-basic variables are $x_1 = 0$ and $x_2 = 0$, and the fuzzy objective value is $\tilde{z} = \tilde{c}_B x_B = (-6, -5, -3, -2)$.

Step 2: We have $\bar{b}_r = \bar{b}_1 = min\{\bar{b}_1 = -2, \bar{b}_2 = 1\} = -2$.

Step 3: We solve the system $By_j = a_j$ for all $j \in J_N = \{1, 4\}$:

$$\begin{bmatrix} 1 & 0 \\ 0 & -2 \end{bmatrix} \begin{bmatrix} y_{11} \\ y_{21} \end{bmatrix} = \begin{bmatrix} -1 \\ -1 \end{bmatrix} \Rightarrow y_1 = \begin{bmatrix} y_{11} \\ y_{21} \end{bmatrix} = \begin{bmatrix} -\frac{3}{2} \\ \frac{1}{2} \end{bmatrix}$$

$$\begin{bmatrix} 1 & 1 \\ 0 & -2 \end{bmatrix} \begin{bmatrix} y_{14} \\ y_{24} \end{bmatrix} = \begin{bmatrix} 0 \\ 1 \end{bmatrix} \Rightarrow y_1 = \begin{bmatrix} y_{14} \\ y_{24} \end{bmatrix} = \begin{bmatrix} \frac{1}{2} \\ -\frac{1}{2} \end{bmatrix}$$

Step 4: We find $\tilde{w}$ by solving the fuzzy system $\tilde{w}B = \tilde{c}_B$:

$$(\tilde{w}_1, \tilde{w}_2) \begin{bmatrix} 1 & 1 \\ 0 & -2 \end{bmatrix} = \begin{bmatrix} (0,0,0,0) \\ (-6,-5,-3,-2) \end{bmatrix} \Rightarrow \tilde{w}_1 = (0,0,0,0), \tilde{w}_2$$

$$= \left(1, \frac{3}{2}, \frac{5}{2}, 3\right)$$

Step 5: We calculate $\tilde{z}_j - \tilde{c}_j = \tilde{w}a_j - \tilde{c}_j$ for all $j \in J_N = \{1, 4\}$:

$$\tilde{z}_1 - \tilde{c}_1 = \tilde{w}a_1 - \tilde{c}_1 = \left((0,0,0,0), \left(1, \frac{3}{2}, \frac{5}{2}, 3\right)\right) \begin{bmatrix} -1 \\ -1 \end{bmatrix}$$

$$- (-5, -4, -2, -1) = \left(-2, -\frac{1}{2}, \frac{5}{2}, 4\right)$$

$$\tilde{z}_4 - \tilde{c}_4 = \tilde{w}a_4 - \tilde{c}_4 = \left((0,0,0,0), \left(1, \frac{3}{2}, \frac{5}{2}, 3\right)\right) \begin{bmatrix} 0 \\ 1 \end{bmatrix} - (0,0,0,0) = \left(1, \frac{3}{2}, \frac{5}{2}, 3\right)$$

Computing the rank of $\tilde{z}_1 - \tilde{c}_1$ and $\tilde{z}_4 - \tilde{c}_4$ based on the ranking function given in Remark 1.15 leads to $\Re(\tilde{z}_1 - \tilde{c}_1) = 1$ and $\Re(\tilde{z}_4 - \tilde{c}_4) = 2$.

The variable $x_{Br} = x_3$ leaves the basis and the entering variable is determined by the following test:

$$max\left\{\frac{\Re(\tilde{z}_1 - \tilde{c}_1)}{y_{21}}\right\} = max\left\{\frac{1}{-\frac{3}{2}}\right\} = \frac{\Re(\tilde{z}_1 - \tilde{c}_1)}{y_{21}}$$

Hence, the variable $x_k = x_1$ enters the basis.

Step 6: We update the basis B where a_1 replaces $a_{B_1} = a_3$:

$$B = [a_1, a_2] = \begin{bmatrix} -1 & 1 \\ -1 & -2 \end{bmatrix}$$

Also, we update $J_N = \{1, 4\}$ as $J_N = \{3, 4\}$.

Iteration 3:

Step 1: We solve the system $B_{x_B} = b$:

$$\begin{bmatrix} -1 & 1 \\ -1 & -2 \end{bmatrix} \begin{bmatrix} x_1 \\ x_2 \end{bmatrix} = \begin{bmatrix} -1 \\ -2 \end{bmatrix}$$

Solving this system leads to $\bar{b}_1 = x_{B_1} = x_1 = \frac{4}{3}$ and $\bar{b}_2 = x_{B_2} = x_2 = \frac{1}{3}$. The non-basic variables are $x_3 = 0$, $x_4 = 0$ and the fuzzy objective value $\tilde{z} = \tilde{c}_B x_B = \left(-\frac{26}{3}, -\frac{21}{3}, -\frac{11}{3}, -\frac{6}{3}\right)$.

Step 2: We have $\bar{b}_r = min\{\bar{b}_1 = \frac{4}{3}, \bar{b}_2 = \frac{1}{3}\} = \bar{b}_2 = \frac{1}{3} \geq 0$. Hence, the current solution is the optimal solution.

3.3.3   Parametric Analysis of LP with Fuzzy Parameters of Type-1

Parametric analysis is a basic tool for studying perturbations in optimization problems, and is considered to be one of the most interesting research areas in the field of fuzzy optimization problems. It can be applied to study the effect of the continuous variations of several parameters in the objective function coefficients on constraints on the optimum solution of fuzzy optimization problems.

It should be noted that flexibility in the values involved in the problem is allowed in FLP problems. However, even after formulating the problem under consideration in a fuzzy environment, it is quite possible that the entered values are wrong or change due to time variations, and hence sensitivity analysis must be used for FLP problem. Although several approaches have been proposed in the literature to investigate the sensitivity analysis, there are nevertheless only few works investigating parametric analysis in FLP problems. The parametric analysis is a general case of sensitivity analysis in which one often finds a direction along which the objective function or the constraints are perturbed, and then seeks to move along this direction. Based on this motivation, we seek the optimal solutions to the LPP with fuzzy parameters of type-1, by perturbing the fuzzy cost coefficients vector along with a new fuzzy cost vector and the right-hand side vector along with another vector.

3.3.3.1   Parametric Fuzzy Cost Problem

In this subsection, we first state the steps of fuzzy primal simplex algorithm (Algorithm 3.1) in tableau format and then explore it with the help of an application example. The obtained results will be used for parametric fuzzy cost analysis [12].

Suppose that we have a starting basic feasible solution x with basis B. The LP with fuzzy parameters of type-1 (3.2) in the case of equality constraints, by decomposing A into $[B, N]$, $\tilde{c}$ into $[\tilde{c}_B, \tilde{c}_N]$ can be represented as follows

$$\max \quad \tilde{z}$$
$$\text{s.t.} \quad \tilde{z} - \tilde{c}_B x_B - \tilde{c}_N x_N = \tilde{0} \tag{3.25}$$

$$Bx_B + Nx_N = b, \tag{3.26}$$
$$x_B, x_N \geq 0.$$

From Eq. (3.26) we have

$$x_B + B^{-1}Nx_N = B^{-1}b \tag{3.27}$$

Multiplying (3.27) by $\tilde{c}_B$ and adding to Eq. (3.25), we obtain

$$\tilde{z} + \tilde{0}x_B + (\tilde{c}_B B^{-1}N - \tilde{c}_N)x_N = \tilde{c}_B B^{-1}b \tag{3.28}$$

Hence, from (3.27) and (3.28) we can conveniently represent the current basic feasible solution with basis B in Table 3.1.

In this case the tableau format of Algorithm 3.1 for solving FLP (3.1) is given as follows:

Initialization Step
Suppose an initial fuzzy basic feasible solution with basis B is at hand. Form the initial tableau similar to Table 3.1.

Main Step

(1) Calculate $\tilde{z}_j - \tilde{c}_j$ for all non-basic variables. Let $\Re(\tilde{z}_k - \tilde{c}_k) = \underset{j \in J_N}{Max}\{\Re(\tilde{z}_j - \tilde{c}_j)\}$

(2) Let $y_k = B^{-1}a_k$. If $y_k \leq 0$, then stop; the problem is unbounded. Otherwise, determine the index r of variable x_{B_r} leaving the basic as follows:

$$\frac{\bar{b}_r}{y_{rk}} = \underset{1 \leq i \leq m}{min}\left\{\frac{\bar{b}_i}{y_{ik}}\Big| y_{ik} > 0\right\}.$$

Table 3.1 The initial FLP simplex tableau

Basis	x_B	x_N		R.H.S.
$\Re(\tilde{z})$	0	$\Re(\tilde{z}_N - \tilde{c}_N) = \Re(\tilde{c}_B Y_N - \tilde{c}_N)$		$\Re(\tilde{z}) = \Re(\tilde{c}_B B^{-1}b)$
$\tilde{z}$	$\tilde{0}$	$\tilde{z}_N - \tilde{c}_N = \tilde{c}_B Y_N - \tilde{c}_N$		$\tilde{c}_B B^{-1}b$
x_B	I	Y_N		$\bar{b} = B^{-1}b$

(3) Update the tableau by pivoting at y_{rk}. Update the basic and non-basic variables where x_k enters the basis and x_{B_r} leaves the basis, and go to (1).

Here we solve an application example by use of the tableau format of Algorithm 3.1 and we apply this example to deal with the parametric analysis on the FLP problem throughout this subsection.

Example 3.3 A company produces two products P_1 and P_2 on two different machines M_1 and M_2. The time required for manufacturing one unit of each product and the daily capacity of the machines are given in Table 3.2.

Due to price variations, the profit for each product can be changed. At the same time the company wishes to maintain the profit somewhat close to 40 dollars for P_1 and close to 30 dollars for P_2. The company wants to determine the range of each product to be produced per day in order to maximize its profit. It is assumed that all the amounts produced are consumed in the market.

Since the profit from each product is uncertain, the problem will be an FLP problem. We use trapezoidal fuzzy numbers for each uncertain value. The profit of product P_1 which is close to 40 is modeled as $(35, 36, 42, 49)$ and the profit of product P_2 which is close to 30 is modeled as $(23, 28, 34, 35)$. Assume x_1 and x_2 are the daily amount of products P_1 and P_2 to determine the optimal combination, respectively. In this case, the problem is formulated as follows:

$$\max \quad \tilde{z} = (35, 36, 42, 49)x_1 + (23, 28, 34, 35)x_2$$
$$\text{s.t.} \quad x_1 + 2x_2 \leq 44$$
$$3x_1 + x_2 \leq 42 \tag{3.29}$$
$$x_1, x_2 \geq 0$$

We now construct the standard form of the FLP problem (3.29) as follows where x_3 and x_4 are the slack variables:

$$\max \quad \tilde{z} = (35, 36, 42, 49)x_1 + (23, 28, 34, 35)x_2$$
$$\text{s.t.} \quad x_1 + 2x_2 + x_3 = 44$$
$$3x_1 + x_2 + x_4 = 42 \tag{3.30}$$
$$x_1, x_2, x_3, x_4 \geq 0$$

Table 3.2 The data of Example 3.3

Time per unit (minutes)			
Machines	P_1	P_2	Machine capacity (min/day)
M_1	1	2	44
M_2	3	1	42

The problem (3.30) can be solved using the tableau format of Algorithm 3.1. Table 3.3 presents the initial primal simplex tableau.

In this table, x_1 and x_4 are the entering variable and leaving variable, respectively. The second simplex tableau presented in Table 3.4 is derived by pivoting on $y_{21} = 3$.

In this table, x_2 is an entering variable and x_3 is a leaving variable. Pivoting on $y_{12} = \frac{5}{3}$ gives the third simplex tableau presented in Table 3.5.

Table 3.5 is the optimal tableau because for all non-basic variables:

$$\Re(\tilde{z}_3 - \tilde{c}_3) = \Re\left(\frac{22}{5}, \frac{42}{5}, \frac{66}{5}, \frac{70}{5}\right) \geq 0$$

$$\Re(\tilde{z}_4 - \tilde{c}_4) = \Re\left(\frac{35}{5}, \frac{38}{5}, \frac{56}{5}, \frac{71}{5}\right) \geq 0$$

This shows that the company should produce eight units of product P_1 and eighteen units of product P_2 to obtain a profit close to 860 dollars. The membership function of the total profit is as follows:

Table 3.3 The initial fuzzy simplex tableau

Basis	x_1	x_2	x_3	x_4	R.H.S.
$\Re(\tilde{z})$	-40	-30	0	0	0
$\tilde{z}$	$(-47, -42, -36, 37)$	$(-35, -34, -28, 29)$	$(0, 0, 0, 0)$	$(0, 0, 0, 0)$	$(0, 0, 0, 0)$
x_3	1	2	1	0	44
x_4	3	1	0	1	42

Table 3.4 The first iteration simplex tableau

Basis	x_1	x_2	x_3	x_4	R.H.S.
$\Re(\tilde{z})$	0	$\frac{-50}{3}$	0	$\frac{40}{3}$	560
$\tilde{z}$	$(0,0,0,0)$	$\left(\frac{-70}{3}, -22, -14, \frac{-22}{3}\right)$	$(0,0,0,0)$	$\left(\frac{31}{3}, 12, 14, \frac{43}{3}\right)$	$(490, 504, 588, 658)$
x_3	0	$\frac{5}{3}$	1	$\frac{-1}{3}$	30
x_3	1	$\frac{1}{3}$	0	$\frac{1}{3}$	14

Table 3.5 The optimal tableau

Basis	x_1	x_2	x_3	x_4	R.H.S.
$\Re(\tilde{z})$	0	0	10	10	860
$\tilde{z}$	$(0,0,0,0)$	$(0,0,0,0)$	$\left(\frac{22}{5}, \frac{42}{5}, \frac{66}{5}, \frac{70}{5}\right)$	$\left(\frac{35}{5}, \frac{38}{5}, \frac{56}{5}, \frac{71}{5}\right)$	$(622, 756, 984, 1078)$
x_2	0	1	$\frac{3}{5}$	$\frac{-1}{5}$	18
x_1	1	0	$\frac{-1}{5}$	$\frac{2}{5}$	8

$$\mu_{\tilde{z}}(x) = \begin{cases} \frac{x-622}{134} & \text{for} \quad 622 \leq x \leq 756 \\ 1 & \text{for} \quad 756 \leq x \leq 984 \\ \frac{1078-x}{94} & \text{for} \quad 984 \leq x \leq 1078 \end{cases}$$

Now, assume that B is an optimal basis of LP with fuzzy parameters of type-1 (3.2). Suppose that the fuzzy cost vector $\tilde{c}$ is perturbed along the fuzzy cost direction $\tilde{c}'$, that is, $\tilde{c}$ is replaced by $\tilde{c} + \lambda \tilde{c}'$, where λ is a parameter. The problem to be solved is of the following form:

$$\begin{aligned} \max \quad & \tilde{z}_\lambda(x) = (\tilde{c} + \lambda \tilde{c}')x \\ s.t. \quad & Ax = b \\ & x \geq 0 \end{aligned} \tag{3.31}$$

The FLP (3.31) is called a parametric fuzzy cost problem (PFCP). We are interested in finding the optimal points and corresponding fuzzy objective values as a function of parameter λ.

Clearly B is an optimal basis of PFCP (3.31) when $\lambda = 0$. Decomposing A into $[B, N]$, $\tilde{c}$ into $[\tilde{c}_B, \tilde{c}_N]$, and $\tilde{c}'$ into $[\tilde{c}'_B, \tilde{c}'_N]$, we obtain:

$$\begin{aligned} \tilde{z} - (\tilde{c}_B + \lambda \tilde{c}'_B)x_B - (\tilde{c}_N + \lambda \tilde{c}'_N)x_N &\simeq \tilde{0} \\ Bx_B + Nx_N &= b \end{aligned} \tag{3.32}$$

Updating the tableau and denoting $\tilde{c}'_B y_j$ by $\tilde{z}'_j$, we obtain

$$\begin{aligned} \tilde{z} - \sum_{j \in J_N} \left[(\tilde{z}_j - \tilde{c}_j) + \lambda(\tilde{z}'_j - \tilde{c}'_j) \right] x_j &\simeq \tilde{c}_B \bar{b} + \lambda \tilde{c}'_B \bar{b} \\ x_B + \sum_{j \in J_N} y_j x_j &= \bar{b} \end{aligned} \tag{3.33}$$

Let the canonical tableau with respect to basis B be Table 3.6.

Now, we determine the range of parameter λ so that when the fuzzy cost coefficients are distributed, the optimal solution remains invariant.

It is evident that the basis B is an optimal basis for all values of parameter λ satisfying:

$$\tilde{z}_j(\lambda) - \tilde{c}_j(\lambda) = (\tilde{z}_j - \tilde{c}_j) + \lambda(\tilde{z}'_j - \tilde{c}'_j) \succeq \tilde{0}, \quad j \in J_N, \tag{3.34}$$

Table 3.6 The canonical tableau of the fuzzy parametric cost problem

Basis	$\tilde{x}_B$	$\tilde{x}_N$	R.H.S.
$\Re(\tilde{z}(\lambda))$	0	$\Re(\tilde{z}_N - \tilde{c}_N) + \Re(\tilde{z}'_N - \tilde{c}'_N)$	$\Re(\tilde{z}(\lambda)) = \Re(\tilde{c}_B \bar{b} + \lambda \tilde{c}'_B \bar{b})$
$\tilde{z}(\lambda)$	$\tilde{0}$	$(\tilde{z}_N - \tilde{c}_N) + \lambda(\tilde{z}'_N - \tilde{c}'_N)$	$\tilde{z}(\lambda) = \tilde{c}_B \bar{b} + \lambda \tilde{c}'_B \bar{b}$
$\tilde{x}_B$	I	Y_N	$\bar{b} = B^{-1}b$

This implies that if $\Re(\tilde{z}'_j - \tilde{c}'_j) > 0$ then

$$\lambda \geq \frac{-\Re(\tilde{z}_j - \tilde{c}_j)}{\Re(\tilde{z}'_j - \tilde{c}'_j)} \tag{3.35}$$

and if $\Re(\tilde{z}'_j - \tilde{c}'_j) < 0$, then

$$\lambda \leq \frac{-\Re(\tilde{z}_j - \tilde{c}_j)}{\Re(\tilde{z}'_j - \tilde{c}'_j)}. \tag{3.36}$$

Hence, $\underline{\lambda}_B \leq \lambda \leq \bar{\lambda}_B$ where

$$\underline{\lambda}_B = \begin{cases} \max\limits_{j \in J_N}\left\{\frac{-\Re(\tilde{z}_j - \tilde{c}_j)}{\Re(\tilde{z}'_j - \tilde{c}'_j)}\middle| \Re(\tilde{z}'_j - \tilde{c}'_j) > 0\right\} \\ -\infty, \quad \text{if for all } j \quad \Re(\tilde{z}'_j - \tilde{c}'_j) \leq 0 \end{cases} \tag{3.37}$$

and

$$\underline{\lambda}_B = \begin{cases} \min\limits_{j \in J_N}\left\{\frac{-\Re(\tilde{z}_j - \tilde{c}_j)}{\Re(\tilde{z}'_j - \tilde{c}'_j)}\middle| \Re(\tilde{z}'_j - \tilde{c}'_j) < 0\right\}, \\ -\infty, \quad \text{if for all } j \quad \Re(\tilde{z}'_j - \tilde{c}'_j) \geq 0. \end{cases} \tag{3.38}$$

The values $\underline{\lambda}_B$ and $\bar{\lambda}_B$ are called the lower and upper break point of basis B, respectively. In addition, the closed interval $\left[\underline{\lambda}_B, \bar{\lambda}_B\right]$ is called the optimality interval of basis B. Therefore, in the optimality interval of basis the optimal solution is the same. The only change will occur in the fuzzy optimal objective value that will be equal to:

$$\tilde{z}_\lambda(x) \simeq \tilde{c}_B \bar{b} + \lambda \tilde{c}'_B \bar{b} \tag{3.39}$$

For solving the PFCP (3.31) in the case of $\lambda > \bar{\lambda}_B$, suppose the upper optimality interval of the Basis B, i.e. $\bar{\lambda}_B$ is finite. Assume that $\bar{\lambda}_B = \frac{-\Re(\tilde{z}_k - \tilde{c}_k)}{\Re(\tilde{z}'_k - \tilde{c}'_k)}$ in (3.38). When λ exceeds $\bar{\lambda}_B$, the fuzzy relative coefficient of x_k becomes negative, i.e., $\tilde{z}_k(\lambda) - \tilde{c}_k(\lambda) \simeq (\tilde{z}_k - \tilde{c}_k) + \lambda(\tilde{z}'_k - \tilde{c}'_k) \prec \tilde{0}$. In this case, x_k enters into the basic variable. When this is done, two cases might occur.

Case 1: The PFCP (3.31) becomes unbounded, i.e., $y_{ik} \leq 0$ for all i. In this case, this problem is unbounded for all $\lambda > \bar{\lambda}_B$.

Case 2: Assuming x_k enters into the basic variables; the fuzzy primal simplex algorithm is continued for finding the new optimal solution.

In this case, assume x_r drops out of the basic variables and update the tableau by pivoting on y_{rk}. Let the new basis be $\hat{B}$. The same procedure can be repeated with this new basis and the optimal solution of (3.31) determined as λ increases further. Some important results are given using the following theorems.

Theorem 3.8 *Bases B and $\hat{B}$ are two alternative optimal bases for* (3.31) *when* $\lambda = \bar{\lambda}_B$.

Proof Note that $\bar{\lambda}_B = \frac{-\Re(\tilde{z}_k - \tilde{c}_k)}{\Re(\tilde{z}'_k - \tilde{c}'_k)}$. This implies that $\Re(\tilde{z}_k - \tilde{c}_k) + \lambda\Re(\tilde{z}'_k - \tilde{c}'_k) = 0$. Therefore, the relative cost coefficients of any variable x_j with respect to bases B and $\hat{B}$ are both equal when $\lambda = \bar{\lambda}_B$ as the entry in this row in the pivot column is zero. This completes the proof. $\square$

Theorem 3.9 *The new basis $\hat{B}$ is not an optimal basis for any* $\lambda < \bar{\lambda}_B$.

Proof The relative cost coefficient of the dropping variable x_r with respect to the new basis $\hat{B}$ is

$$\tilde{\hat{z}}_k(\lambda) - \tilde{\hat{c}}_k(\lambda) \simeq (\tilde{z}_k - \tilde{c}_k) + \lambda(\tilde{z}'_k - \tilde{c}'_k) \simeq \frac{(\tilde{z}_k - \tilde{c}_k) + \lambda(\tilde{z}'_k - \tilde{c}'_k)}{y_{rk}}$$

We note that the optimality interval of basis B is $\left[\underline{\lambda}_B, \bar{\lambda}_B\right]$. This implies that we have $(\tilde{z}_k - \tilde{c}_k) + \lambda(\tilde{z}'_k - \tilde{c}'_k) \succeq \tilde{0}$ for any $\lambda < \bar{\lambda}_B$. Now, noting that $y_{rk} > 0$, we conclude that $(\tilde{\hat{z}}_k - \tilde{\hat{c}}_k) + \lambda(\tilde{\hat{z}}'_k - \tilde{\hat{c}}'_k) \prec \tilde{0}$. This implies that the new basis $\tilde{B}$ is not an optimal basis for any $\lambda < \bar{\lambda}_B$. $\square$

Theorem 3.10 *The lower breaking point of new basis $\hat{B}$ is* $\lambda = \bar{\lambda}_B$.

Proof Based on Theorem 3.8, bases B and $\hat{B}$ are two alternative optimal bases for (3.31) when $\lambda = \bar{\lambda}_B$ and based on Theorem 3.9, the new basis $\hat{B}$ is not an optimal basis for any $\lambda < \bar{\lambda}_B$. These imply that the lower break point of new basis $\hat{B}$ is equal to the upper break point of basis B, i.e, $\bar{\lambda}_B = \underline{\lambda}_{\hat{B}}$. $\square$

Returning to the original basis B, suppose $\underline{\lambda}_B$ is finite. Assume that $\underline{\lambda}_B = \frac{-\Re(\tilde{z}_t - \tilde{c}_t)}{\Re(\tilde{z}'_t - \tilde{c}'_t)}$ in (3.36). In the case of $\lambda < \underline{\lambda}_B$, the fuzzy relative coefficient of x_t becomes negative, i.e., $\tilde{z}_t(\lambda) - \tilde{c}_t(\lambda) \simeq (\tilde{z}_t - \tilde{c}_t) + \lambda(\tilde{z}'_t - \tilde{c}'_t) \prec \tilde{0}$. In this case, x_t enters into the basic variable. When this is done, two cases might occur. The PFCP (3.31) becomes unbounded for all $\lambda < \underline{\lambda}_B$, if for all i, $y_{it} \leq 0$. Otherwise, the fuzzy primal simplex algorithm is continued in order to find the new optimal solution. In fact, x_r is dropped out of the basic variables and the tableau is updated by pivoting on y_{rt}. Let the new basis be $\hat{B}$. From results similar to the case of $\lambda > \bar{\lambda}_B$, it is evident that B and $\hat{B}$ are two alternative optimal bases for (3.31) when $\lambda = \bar{\lambda}_B$. The optimality interval of this new basis is another closed interval with its right end point equal to the left end point of the optimality interval of basis B. The same procedure can be repeated with this new basis and the optimal solution of (3.31) obtained as λ decreases further.

Example 3.4 Consider Example 3.3. Suppose that the management of the company now have the option of making trade-offs in the profitability of the two activities, whereby the objective function coefficient of x_1 can be increased (decreased) by simultaneously decreasing (increasing) the objective function coefficient of x_2.

In this case, the aim is to find the optimal solution and the fuzzy optimal objective values of the class problem whose fuzzy objective function $((\tilde{c} + \lambda \tilde{c}') \simeq ((35, 36, 42, 47) + \lambda(16, 17, 21, 26). + (23, 28, 34, 36) + \lambda(-18, -11, -7, -6))$ is as a function of λ; that is, the profits of products P_1 and P_2 are perturbed along the fuzzy profit vector $\tilde{c}' \simeq ((16, 17, 21, 26), (-16, -11, -7, -6))$. In fact, we should solve the following fuzzy cost parametric problem:

$$\max \tilde{z} \simeq ((35, 36, 42, 47) + \lambda(16, 17, 21, 26))x_1 + ((23, 28, 34, 35) + \lambda(-6, -11, -7, -6))x_2$$
$$\text{s.t.} \quad x_1 + 2x_2 \leq 44$$
$$3x_1 + x_2 \leq 42$$
$$x_1, x_2 \geq 0$$

$$(3.40)$$

The optimal solution for $\lambda = 0$ is given in Table 3.5. In order to find the range over which this tableau is optimal, we first find $(\tilde{z}_3' - \tilde{c}_3')$ and $(\tilde{z}_4' - \tilde{c}_4')$ as follows:

$$\left(\tilde{z}_3' - \tilde{c}_3'\right) \simeq \tilde{c}_B y_3 - \tilde{c}_3' \simeq ((16, 17, 21, 26), (-18, -11, -7, -6)) \begin{pmatrix} \frac{3}{5} \\ \frac{-1}{5} \end{pmatrix}$$

$$- (0, 0, 0, 0) \simeq \left(\frac{-74}{5}, \frac{-54}{5}, \frac{-38}{5}, \frac{-34}{5}\right)$$

$$\left(\tilde{z}_4' - \tilde{c}_4'\right) \simeq \tilde{c}_B y_4 - \tilde{c}_4' \simeq ((16, 17, 21, 26), (-18, -11, -7, -6)) \begin{pmatrix} \frac{-1}{5} \\ \frac{1}{5} \end{pmatrix}$$

$$- (0, 0, 0, 0) \simeq \left(\frac{38}{5}, \frac{41}{5}, \frac{51}{5}, \frac{66}{5}\right)$$

We note that $\Re\left(\tilde{z}_3' - \tilde{c}_3'\right) < 0$ and $\Re\left(\tilde{z}_4' - \tilde{c}_4'\right) > 0$. Hence, based on Eqs. (3.37) and (3.38) we have $\underline{\lambda}_B = -1$ and $\bar{\lambda}_B = 1$, respectively. This implies that the optimality interval of basis $B = [a_2, a_1]$ is $[\underline{\lambda}_B, \bar{\lambda}_B] = [-1, 1]$. The fuzzy optimal objective function value in this interval based on (3.39) is given as follows:

$$\tilde{z}_\lambda(x) = \tilde{c}_B \bar{b} + \lambda \tilde{c}_B' \bar{b} = (622, 756, 984, 1078) + \lambda((-16, -11, -7, -6), (16, 17, 21, 26)) \begin{pmatrix} 18 \\ 8 \end{pmatrix}$$

$$= (622, 756, 984, 1078) + \lambda(-160, -62, 42, 100)$$

In this case, the optimal solution in this optimality interval is given as Table 3.7.

Now we should solve the problem for $\lambda > \bar{\lambda}_B = 1$ and $\lambda < \underline{\lambda}_B = -1$. For $\lambda > \bar{\lambda}_B = 1$ the variable x_3 does not satisfy the optimality condition. In fact, at $\bar{\lambda}_B = 1$, the value of $\Re(\tilde{z}_3(\lambda) - \tilde{c}_3(\lambda)) = 10 - 10\lambda$ is zero, and x_3 is introduced into the basis leading to Table 3.8.

Table 3.7 The optimal tableau in the range of $[-1, 1]$

Basis	x_1	x_2	x_3	x_4	R.H.S.
$\Re(\tilde{z}(\lambda))$	0	0	$10 - 10\lambda$	$10 + 10\lambda$	$860 - 20\lambda$
$\tilde{z}(\lambda)$	$\tilde{0}$	$\tilde{0}$	$\left(\frac{22}{5}, \frac{42}{5}, \frac{66}{5}, \frac{70}{5}\right) +$ $\left(\frac{-74}{5}, \frac{-54}{5}, \frac{-38}{5}, \frac{-34}{5}\right)\lambda$	$\left(\frac{35}{5}, \frac{38}{5}, \frac{56}{5}, \frac{56}{5}\right) +$ $\left(\frac{38}{5}, \frac{41}{5}, \frac{53}{5}, \frac{68}{5}\right)\lambda$	$(622, 756, 984, 1078) +$ $(-160, -62, 42, 100)\lambda$
x_2	1	0	$\frac{3}{5}$	$\frac{-1}{5}$	18
x_1	0	1	$\frac{-1}{5}$	$\frac{2}{5}$	8

Table 3.8 The updated tableau for $\lambda > 1$

Basis	x_1	x_2	x_3	x_4	R.H.S.
$\Re(\tilde{z}(\lambda))$	0	$\frac{-50}{3} + \frac{50}{3}\lambda$	0	$\frac{40}{3} + \frac{20}{3}\lambda$	$560 + 280\lambda$
$\tilde{z}(\lambda)$	$\tilde{0}$	$\left(\frac{68}{3}, -22, -14, \frac{-62}{3}\right)$ $+ \left(\frac{34}{3}, \frac{38}{3}, \frac{54}{3}, \frac{74}{3}\right)\lambda$	$\tilde{0}$	$\left(\frac{127}{15}, \frac{52}{5}, \frac{78}{5}, \frac{283}{15}\right)$ $+ \left(\frac{40}{15}, \frac{69}{15}, \frac{121}{15}, \frac{170}{15}\right)\lambda$	$(202, 360, 732, 946) +$ $(44, 166, 366, 544)\lambda$
x_3	0	$\frac{5}{3}$	1	$\frac{-1}{3}$	30
x_1	1	$\frac{1}{3}$	0	$\frac{1}{3}$	14

In order to obtain the optimality interval of basis $\hat{B} = [a_3, a_1]$, we should have $\Re(\tilde{z}_2(\lambda) - \tilde{c}_2(\lambda)) \geq 0$ and $\Re(\tilde{z}_4(\lambda) - \tilde{c}_4(\lambda)) \geq 0$:

$$\Re(\tilde{z}_2(\lambda) - \tilde{c}_2(\lambda)) = \Re\left(\left(\frac{68}{3}, -22, -14, \frac{-62}{3}\right) + \left(\frac{34}{3}, \frac{38}{3}, \frac{54}{3}, \frac{74}{3}\right)\lambda\right)$$
$$= -\frac{50}{3} + \frac{50}{3}\lambda \geq 0 \Rightarrow \lambda \geq 1$$
$$\Re(\tilde{z}_4(\lambda) - \tilde{c}_4(\lambda)) = \Re\left(\left(\frac{127}{15}, \frac{52}{5}, \frac{78}{5}, \frac{283}{15}\right) + \left(\frac{40}{15}, \frac{69}{15}, \frac{121}{15}, \frac{170}{15}\right)\lambda\right)$$
$$= \frac{40}{3} + \frac{20}{3}\lambda \geq 0 \Rightarrow \lambda \geq -2$$

These imply that the basis $\hat{B} = [a_3, a_1]$ is optimal for any $\lambda \geq 1$. In fact, since $\Re(\tilde{z}_2' - \tilde{c}_2') = \Re\left(\frac{34}{3}, \frac{38}{3}, \frac{54}{3}, \frac{74}{3}\right) = \frac{50}{3} > 0$ and
$\Re(\tilde{z}_4' - \tilde{c}_4') = \Re\left(\frac{40}{15}, \frac{69}{15}, \frac{121}{15}, \frac{170}{15}\right) = \frac{20}{3} > 0$, based on Eqs. (3.37) and (3.38) we conclude that

$$\underline{\lambda}_{\hat{B}} = max\left\{\frac{-\frac{-50}{3}}{\frac{50}{3}}, \frac{-\frac{40}{3}}{\frac{20}{3}}\right\} = 1$$

and $\overline{\lambda}_{\hat{B}} = \infty$, respectively.

Table 3.9 The updated tableau for $\lambda < -1$

Basis	x_1	x_2	x_3	x_4	R.H.S.
$\Re(\tilde{z}(\lambda))$	$-25 - 25\lambda$	0	$15 - 5\lambda$	0	$660 - 220\lambda$
$\tilde{z}(\lambda)$	$\left(\frac{-71}{2}, -28, -19, \frac{-35}{2}\right) +$ $\left(\frac{-68}{2}, \frac{-53}{2}, \frac{-41}{2}, \frac{-38}{2}\right)\lambda$	$\tilde{0}$	$\left(\frac{79}{10}, \frac{61}{5}, \frac{94}{5}, \frac{211}{10}\right) +$ $\left(-11, \frac{-67}{10}, \frac{-23}{10}, 0\right)\lambda$	$\tilde{0}$	$(338, 532, 832, 938)$ $+(-432, -274, -122, -52)\lambda$
x_2	$\frac{1}{2}$	1	$\frac{1}{2}$	0	22
x_4	$\frac{5}{2}$	0	$\frac{-1}{2}$	1	20

As we see the left end point of the new basis $\hat{B} = [a_3, a_1]$ is equal to the right end point of the basis $B = [a_2, a_1]$. The fuzzy optimal objective function value in this new optimality interval is equal to $\tilde{z}_\lambda(x) \simeq (202, 360, 732, 946) + \lambda(44, 166, 366, 544)$.

Now we solve the problem for $\lambda < \underline{\lambda}_B = -1$. For $\lambda < \underline{\lambda}_B = -1$ the variable x_4 does not satisfy the optimality condition with regard to Table 3.7. In fact, at $\lambda = \underline{\lambda}_B = -1$, the value of $\Re(\tilde{z}_4(\lambda) - \tilde{c}_4(\lambda)) = 10 + 10\lambda$ is zero, and x_4 is introduced into the basis leading to Table 3.9.

In order to obtain the optimality interval of basis $B = [a_2, a_4]$, we should have $\Re(\tilde{z}_1(\lambda) - \tilde{c}_1(\lambda)) \geq 0$ and $\Re(\tilde{z}_3(\lambda) - \tilde{c}_3(\lambda)) \geq 0$:

$$\Re(\tilde{z}_1(\lambda) - \tilde{c}_1(\lambda)) = \Re\left(\left(\frac{-71}{2}, -28, -19, \frac{-35}{2}\right) + \lambda\left(\frac{-68}{2}, \frac{-53}{2}, \frac{-41}{2}, \frac{-38}{2}\right)\right)$$
$$= -25 - 25\lambda \geq 0 \Rightarrow \lambda \leq -1.$$

$$\Re(\tilde{z}_3(\lambda) - \tilde{c}_3(\lambda)) = \Re\left(\left(\frac{79}{10}, \frac{61}{5}, \frac{94}{5}, \frac{211}{10}\right) + \lambda\left(-11, \frac{-67}{10}, \frac{-23}{10}, 0\right)\right)$$
$$= 15 - 5\lambda \geq 0 \Rightarrow \lambda \leq 3.$$

These imply that the basis $\hat{B} = [a_2, a_4]$ is optimal for any $\lambda \leq -1$. In fact, since $\Re(\tilde{z}'_1 - \tilde{c}'_1) = \Re((\frac{-68}{2}, \frac{-53}{2}, \frac{-41}{2}, \frac{-38}{2})) = -25 \leq 0$ and $\Re(\tilde{z}'_3 - \tilde{c}'_3) = \Re((-11, \frac{-67}{10}, \frac{-23}{10}, 0)) = -5 \leq 0$, based on Eqs. (3.37) and (3.38) we conclude that $\underline{\lambda}_{\hat{B}} = -\infty$ and

$$\bar{\lambda}_{\hat{B}} = min\left\{\frac{25}{-25}, \frac{-15}{-5}\right\} = -1.$$

As we see the right end point of the new basis $\hat{B} = [a_2, a_4]$ is equal to the left end point of the basis $B = [a_2, a_1]$. The fuzzy optimal objective function value in the new optimality interval $(-\infty, -1]$ is equal to $\tilde{z}_\lambda(x) = (338, 532, 832, 938) + \lambda(-432, -274, -122, -52)$.

3.3.3.2 Parametric Right-Hand Side Problem

In this subsection, we first state the steps of the fuzzy dual simplex algorithm (Algorithm 3.2) in tableau format and then we apply the obtained results for parametric right-hand side (RHS) analysis.

Suppose that we have a starting basic dual feasible solution x with basis B. The tableau format of Algorithm 3.2 for solving the LPP with fuzzy parameters of type-1 (3.2) in the case of equality constraints, can be summarized as follows:

Initialization Step

Suppose the basis B is dual feasible basis for the FLP problem (3.2), i.e. assuming $\tilde{z}_j - \tilde{c}_j = (z_{1j}, z_{2j}, z_{3j}, z_{4j})$, $\Re(\tilde{z}_j - \tilde{c}_j) = \frac{z_{1j} + z_{2j} + z_{3j} + z_{4j}}{4} \geq 0$ for all $j = 1, 2, \ldots, n$. Form the initial tableau as Table 3.1.

Main Step

(1) Suppose $\bar{b} = B^{-1}b$. If $\bar{b} \geq 0$, then stop; the current solution is optimal. Otherwise, suppose $\bar{b}_r = \min\{\bar{b}_i, 1 \leq i \leq m\}$.

(2) If $y_{rj} \geq 0$ for all $j = 1, 2, \ldots, n$, then stop; the FLP problem (3.2) is infeasible. Otherwise, determine the index of pivot column k by the following test:

$$\frac{\Re(\tilde{z}_k - \tilde{c}_k)}{y_{rk}} = \max\left\{\frac{\Re(\tilde{z}_j - \tilde{c}_j)}{y_{rj}}, y_{rj} < 0\right\}$$

(3) Update the tableau by pivoting at y_{rk}. Update the basic and nonbasic variable where x_k enters the basic and x_{Br} leaves the basic and go to (1).

One needs to point out that the fuzzy dual simplex algorithm is used for a rapid re-optimization without the need to find new primal basic feasible solutions, i.e., it is particularly useful for re-optimizing the solution of a certain problem after some parameters have been changed, so that the previously optimal solution may no longer be feasible. Besides this, since FLP problems are almost omnipresent in real life applications, it is patent that the fuzzy dual simplex algorithm is of utmost importance in a wide variety of applications and real practical problems where the RHS vector of constraint is perturbed along a vector, as the algorithm facilitate to re-optimize without the need for iterating from the beginning with the consequent saving of computation time. Thus, in what follows the fuzzy dual simplex algorithm will be used for parametric analysis on the RHS vector of the FLP problem (3.2).

Assume that B is an optimal basis of the FLP problem (3.2). Suppose that the RHS vector b is perturbed along the direction b', that is, b is replaced by $b + \lambda b'$ where λ is a parameter. The problem to be solved is of the following form:

$$\max \tilde{z} = \tilde{c}x$$
$$s.t A x = b + \lambda b' \tag{3.41}$$
$$x \geq 0$$

Table 3.10 The canonical tableau of the fuzzy parametric RHS problem

Basic	x_B	x_N	R.H.S
$\Re(\tilde{z})$	0	$\Re((\tilde{z}_N - \tilde{c}_N))$	$\Re(\tilde{z}(\lambda)) = \Re(\tilde{c}_B\bar{b} + \lambda\tilde{c}_B\bar{b}')$
$\tilde{z}$	$\tilde{0}$	$(\tilde{z}_N - \tilde{c}_N)$	$\tilde{z}(\lambda) = \tilde{c}_B\bar{b} + \lambda\tilde{c}_B\bar{b}'$
x_B	I	Y_N	$\bar{b} + \lambda\bar{b}'$

The FLP (3.41) is called a parametric RHS problem. We are interested in finding the fuzzy objective values as a function of parameter λ.

We note that B is an optimal basis of the parametric RHS problem (3.41) for $\lambda = 0$. Decomposing A into $[B, N]$ and $\tilde{c}$ into $[\tilde{c}_B, \tilde{c}_N]$ we obtain:

$$\tilde{z} - \tilde{c}_B x_B - \tilde{c}_N x_N = \tilde{0}$$
$$Bx_B + Nx_N = b$$
(3.42)

Updating the tableau and denoting $B^{-1}b'$ by $\bar{b}'$, we obtain

$$\tilde{z} - \sum_{j \in J_N} (\tilde{z}_j - \tilde{c}_j)x_j \simeq \tilde{c}_B\bar{b} + \lambda\tilde{c}_B\bar{b}'$$
$$x_B + \sum_{j \in J_N} y_j x_j = \bar{b} + \lambda\bar{b}'$$
(3.43)

Let the canonical tableau with respect to basis B be Table 3.10.

Here, we determine the range of parameter λ so that when the right-hand side vector is distributed, the optimal solution remains invariant.

Since B is an optimal basis for the fuzzy parametric RHS problem (3.41), so it must be dual feasible. The parameter λ appears only in the RHS vector of problem (3.2) and, hence, dual feasibility of a basis for problem (3.2) is a property that is independent of λ. Hence, the basis B is an optimal basis for all values of parameter λ for which it is primal feasible, i.e., for all values of parameter λ satisfying:

$$\bar{b}_i + \lambda\bar{b}'_i \geq 0, \quad i = 1, 2, \ldots, m$$
(3.44)

This implies that

$$\lambda \begin{cases} \geq \frac{-\bar{b}_i}{\bar{b}'_i}, & \text{if } \bar{b}'_i > 0 \\ \leq \frac{-\bar{b}_i}{\bar{b}'_i}, & \text{if } \bar{b}'_i < 0 \end{cases}$$
(3.45)

Hence $\underline{\lambda}_B \leq \lambda \leq \bar{\lambda}_B$, where

$$\underline{\lambda}_B = \begin{cases} \max_{j \in T}\left\{\frac{-\bar{b}_i}{\bar{b}'_i} : \bar{b}'_i > 0\right\} \\ -\infty \quad \text{if } \bar{b}'_i \leq 0 \text{ for all } i \end{cases}$$
(3.46)

$$\bar{\lambda}_B = \begin{cases} \min\limits_{j \in T} \left\{ \frac{-\bar{b}_i}{\bar{b}'_i} : \bar{b}'_i < 0 \right\} \\ \infty \quad \text{if } \bar{b}' \geq 0 \text{ for all } i \end{cases} \tag{3.47}$$

The values $\underline{\lambda}_B$ and $\bar{\lambda}_B$ are called the lower and upper breaking point of basis B, respectively. In addition, the closed interval $[\underline{\lambda}_B, \bar{\lambda}_B]$ is called the optimality interval of basis B. Therefore, for all values of parameter λ in this optimality interval basis B is the optimal basis and the optimal solution as a parameter of λ is as follows:

$$x^* = \begin{pmatrix} x_B^* \\ x_N^* \end{pmatrix} = \begin{pmatrix} \bar{b} + \lambda \bar{b}' \\ 0 \end{pmatrix} \tag{3.48}$$

In addition, the fuzzy optimal objective value as a parameter of λ will be equal to:

$$\tilde{z}_\lambda(x) = \tilde{c}_B \bar{b} + \lambda \tilde{c}_B \bar{b}' \tag{3.49}$$

Thus, in the optimality interval, the fuzzy optimal objective value and the values of basic variables in the optimal solution vary linearly with λ.

For solving the parametric RHS problem (3.41) in the case of $\lambda > \bar{\lambda}_B$ suppose the upper point of optimal basis B i.e., $\bar{\lambda}_B$ is finite. Assume that $\bar{\lambda}_B = \frac{-\bar{b}_r}{\bar{b}'_r}$ in (3.47). When λ exceeds $\bar{\lambda}_B$, the value of basic variable x_{B_r} becomes negative, i.e., $\bar{b}_r + \lambda \bar{b}'_r < 0$. In this case, x_{B_r} leaves the basic variables. When this is done, two cases might occur.

Case 1: If $y_{rj} \geq 0$ for all j, the fuzzy parametric RHS problem (3.41) is infeasible for all $\lambda > \bar{\lambda}_B$. This is the primal infeasibility criterion.

Case 2: If the primal infeasibility criterion is not satisfied, assuming x_{B_r} leaves the basic variables; the fuzzy dual simplex algorithm is continued to find the new optimal solution.

In this case, assume x_k enters into the basic variables and update the tableau by pivoting on y_{rk}. Let the new basis be $\hat{B}$. The same procedure is repeated with this new basis until the optimal solutions of (3.41) are determined for all required $\lambda > \bar{\lambda}_B$.

Theorem 3.11 *Bases B and $\hat{B}$ are alternative optimal bases for problem* (3.41) *when $\lambda = \bar{\lambda}_B$.*

Proof Note that $\bar{\lambda}_B = \frac{-\bar{b}_r}{\bar{b}'_r}$ and $\bar{b}'_r < 0$. This implies that $\bar{b}_r + \bar{\lambda}_B \bar{b}'_r = 0$. Therefore, the value of any basic variable x_i with respect to bases B and $\hat{B}$ are both equal when $\lambda = \bar{\lambda}_B$ as the value of the leaving variable is zero. This completes the proof. $\square$

Theorem 3.12 *The new basis $\hat{B}$ is not a primal feasible (and then optimal) basis for any $\lambda < \bar{\lambda}_B$.*

Proof The value of the entering variable x_k with respect to the new basis $\hat{B}$ is $x_{\hat{B}_k} = \hat{b}_k + \lambda \hat{b}'_k = \frac{b_r + \lambda b'_r}{y_{rs}}$, where $\hat{b} = \hat{B}^{-1}b$. We note that the optimality interval of the basis B is $\left[\underline{\lambda}_B, \bar{\lambda}_B\right]$. This implies that we have $\bar{b}_r + \lambda \bar{b}'_r > 0$ for any $\lambda < \bar{\lambda}_B$. Now, noting that $y_{rk} < 0$, we conclude that $x_{\hat{B}_k} = \hat{b}_k + \lambda \hat{b}'_k < 0$. This implies that the new basis $\hat{B}$ is not a primal feasible basis for any $\lambda < \bar{\lambda}_B$.   □

Theorem 3.13 *The lower breaking point of basis $\hat{B}$ is $\lambda = \bar{\lambda}_B$.*

Proof Based on Theorem 3.11, bases B and $\hat{B}$ are two alternative optimal bases for (3.41) when $\lambda = \bar{\lambda}_B$ and based on Theorem 3.12, the new basis $\hat{B}$ is not an optimal basis for any $\lambda < \bar{\lambda}_B$. Thus, we conclude that the lower breaking point of the basis $\hat{B}$ is equal to the upper breaking point of basis B.   □

Returning to the original basis B, suppose $\underline{\lambda}_B$ is finite. Assume that $\underline{\lambda}_B = \frac{-\bar{b}_t}{\bar{b}'_t}$ in (3.46). When $\lambda < \underline{\lambda}_B$, the value of basic variable x_{B_t} becomes negative, i.e. $\bar{b}_t + \lambda \bar{b}'_t < 0$. In this case, x_{B_t} leaves the basic variables. When this is done, two cases might occur. If $y_{tj} \geq 0$ for all j, the fuzzy parametric RHS problem (3.41) is infeasible for all $\lambda < \underline{\lambda}_B$. Otherwise, assuming x_k enters into the basic variables the fuzzy dual simplex algorithm is continued to find the new optimal solution. In fact, x_{B_t} is dropped out of the basic variables and the tableau is updated by pivoting on y_{tk}. Let the new basis be $\tilde{B}$. From results similar to the case of $\lambda > \bar{\lambda}_B$, it is evident that B and $\tilde{B}$ are alternative optimal bases for (3.41) when $\lambda = \underline{\lambda}_B$. The optimality interval is another closed interval with its right end point equal to the left end point of the optimality interval of basis B. The same procedure can be repeated with this new basis and the optimal solution of (3.41) obtained as λ decreases further. We note that in each interval, the optimal feasible solution and the fuzzy optimal objective value vary linearly in the parameter of λ.

Example 3.5 It is desired to find the optimal solution and optimal basis as the right-hand-side is perturbed along the direction $\begin{pmatrix} 3 \\ -1 \end{pmatrix}$, that is, if $b = \begin{pmatrix} 44 \\ 42 \end{pmatrix}$ is replaced by $b + \lambda b' = \begin{pmatrix} 44 \\ 42 \end{pmatrix} + \lambda \begin{pmatrix} 3 \\ -1 \end{pmatrix}$. If fact, we should solve the following fuzzy parametric RHS problem:

$$
\begin{aligned}
Max\, \tilde{z} = (36, 42, 1, 5)x_1 &+ (28, 34, 5, 1)x_2 \\
s.t \quad x_1 + 2x_2 &\leq 44 + 3\lambda \\
3x_1 + x_2 &\leq 42 - \lambda \\
x_1, x_2 &\geq 0
\end{aligned}
\tag{3.50}
$$

The optimal solution for $\lambda = 0$ is given in Table 3.5. The basis $B = [a_2, a_1]$ is the optimal basis with $\lambda = 0$. In order to find the range over which this basis is optimal, we first find $\bar{b}'$ as follows:

$$\bar{b}' = B^{-1}b' = \begin{bmatrix} \frac{3}{5} & \frac{-1}{5} \\ \frac{-1}{5} & \frac{2}{5} \end{bmatrix} \begin{pmatrix} 3 \\ -1 \end{pmatrix} = \begin{pmatrix} 2 \\ -1 \end{pmatrix} = \begin{pmatrix} \bar{b}'_1 \\ \bar{b}'_2 \end{pmatrix}$$

Hence, based on (3.46) we have $\underline{\lambda}_B = -9$ and $\bar{\lambda}_B = 8$. This implies that the optimality interval of the basis $B = [a_2, a_1]$ is $[\underline{\lambda}_B, \bar{\lambda}_B] = [-9, 8]$. The optimal solution and the fuzzy optimal objective function value in this interval with regard to (3.48) and (3.49) are given as follows, respectively:

$$x^* = \begin{pmatrix} x_B^*(\lambda) \\ x_N^* \end{pmatrix} = \begin{pmatrix} \bar{b} + \lambda\bar{b}' \\ 0 \end{pmatrix} = \begin{pmatrix} x_2(\lambda) \\ x_1(\lambda) \\ x_3 \\ x_4 \end{pmatrix} = \begin{pmatrix} 18 + 2\lambda \\ 8 - \lambda \\ 0 \\ 0 \end{pmatrix}$$

$$\tilde{z}(\lambda) = \tilde{c}_B(\bar{b} + \lambda\bar{b}') = (622, 756, 984, 1078) + \lambda(-1, 114, 32, 35)$$

The optimal tableau over the interval $[\underline{\lambda}_B, \bar{\lambda}_B] = [-9, 8]$ is depicted in Table 3.11.

Now we should solve the problem for $\lambda > \bar{\lambda}_B = 8$ and $\lambda > \underline{\lambda}_B = -9$. For $\lambda > \bar{\lambda}_B = 8$ the variable x_1 does not satisfy the primal feasibility criteria. In fact, at $\lambda = \bar{\lambda}_B = 8$, the value of the basic variable x_1 is zero, and according to the fuzzy dual simplex algorithm, x_3 is introduced into the basis leading to Table 3.12.

Table 3.11 The optimal tableau for $\lambda = [-9, 8]$

Basis	x_1	x_2	x_3	x_4	RHS
z	0	0	10	10	$860 + 20\lambda$
$\tilde{z}$	$(0,0,0,0)$	$(0,0,0,0)$	$\left(\frac{22}{5}, \frac{42}{5}, \frac{66}{5}, \frac{70}{5}\right)$	$\left(\frac{35}{5}, \frac{38}{5}, \frac{56}{5}, \frac{56}{5}\right)$	$(622, 756, 984, 1078) + \lambda(-1, 14, 32, 35)$
x_2	0	1	$\frac{3}{5}$	$\frac{-1}{5}$	$18 + 2\lambda$
x_1	1	0	$\frac{-1}{5}$	$\frac{2}{5}$	$8 - \lambda$

Table 3.12 The updated tableau over the interval $[8, 42]$

Basis	x_1	x_2	x_3	x_4	RHS
z	50	0	0	30	$1260 - 30\lambda$
$\tilde{z}$	$(22, 42, 66, 70)$	$(0,0,0,0)$	$(0,0,0,0)$	$\left(\frac{79}{5}, \frac{122}{5}, \frac{188}{5}, \frac{211}{5}\right)$	$(798, 1092, 1512, 1638) + \lambda(-53, -34, -28, -5)$
x_2	-3	1	0	1	$42 - \lambda$
x_3	-5	0	1	-2	$-40 + 5\lambda$

In order to obtain the optimality interval of the basis $\hat{B} = [a_2, a_3]$, we should have $\bar{b}_2 + \lambda \bar{b}_2' \geq 0$ and $\bar{b}_3 + \lambda \bar{b}_3' \geq 0$:

$$\bar{b}_2 + \lambda \bar{b}_2' = 42 - \lambda \geq 0 \Rightarrow \lambda \leq 42,$$
$$\bar{b}_3 + \lambda \bar{b}_3' = -40 + 5\lambda \geq 0 \Rightarrow \lambda \geq 8,$$

These imply that the basis $\hat{B} = [a_2, a_3]$ is optimal over the optimality interval $[\underline{\lambda}_{\hat{B}}, \bar{\lambda}_{\hat{B}}] = [8, 40]$. As we see, the left end point of new basis $\hat{B} = [a_2, a_1]$ is equal to the right end point of the basis $B = [a_2, a_3]$. The optimal solution and the fuzzy optimal objective function value in this interval with regard to (3.48) and (3.49) are given as follows, respectively:

$$x^* = \begin{pmatrix} x_2(\lambda) \\ x_3(\lambda) \\ x_1 \\ x_4 \end{pmatrix} = \begin{pmatrix} 42 - \lambda \\ -40 + 5\lambda \\ 0 \\ 0 \end{pmatrix}$$

$$\tilde{z}(\lambda) = (798, 1092, 1512, 1638) + \lambda(-53, -34, -28, -5)$$

At $\lambda = 42$, x_2 drops to zero. A fuzzy dual simplex pivot is performed so that x_2 leaves the basis and x_1 enters the basis leading to Table 3.13.

In order to obtain the optimality interval of basis $\bar{B} = [a_1, a_3]$, we should have $\bar{b}_1 + \lambda \bar{b}_1' \geq 0$ and $\bar{b}_3 + \lambda \bar{b}_3' \geq 0$:

$$\bar{b}_1 + \lambda \bar{b}_1' = -14 + \frac{1}{3}\lambda \geq 0 \Rightarrow \lambda \geq 42,$$
$$\bar{b}_3 + \lambda \bar{b}_3' = -110 + \frac{20}{3}\lambda \geq 0 \Rightarrow \lambda \geq \frac{33}{2},$$

These imply that the basis $\bar{B} = [a_1, a_3]$ is optimal for all $\lambda \geq 42$. In fact, since $\bar{b}_1' = \frac{1}{3} \geq 0$ and $\bar{b}_3' = \frac{20}{3} \geq 0$, based on (3.46) we conclude that $\underline{\lambda}_{\bar{B}} = \max\left\{ \frac{-(-14)}{\frac{1}{3}} \quad \frac{-(-110)}{\frac{20}{3}} \right\} = 42$ and $\bar{\lambda}_B = \infty$. The fuzzy optimal objective function

Table 3.13 The updated tableau over the interval $[42, \infty)$

Basis	x_1	x_2	x_3	x_4	RHS
z	0	$\frac{50}{3}$	0	$\frac{140}{3}$	$1960 - \frac{140}{3}\lambda$
$\tilde{z}$	$(0,0,0,0)$	$\left(\frac{22}{3}, 14, 22, \frac{70}{3}\right)$	$(0,0,0,0)$	$\left(\frac{347}{15}, \frac{192}{5}, \frac{298}{5}, \frac{983}{15}\right)$	$(1106, 1680, 2436, 2618) + \lambda\left(\frac{-229}{3}, -56, -42, \frac{-37}{3}\right)$
x_1	1	$\frac{-1}{3}$	0	$\frac{-1}{3}$	$-14 + \frac{1}{3}\lambda$
x_3	0	$\frac{-5}{3}$	1	$\frac{-11}{3}$	$-110 + \frac{20}{3}\lambda$

Table 3.14 The updated tableau for $\lambda < -9$

Basis	x_1	x_2	x_3	x_4	RHS
z	0		40	0	$1760 + 120\lambda$
$\tilde{z}$	$(0,0,0,0)$	$(35,38,56,71)$	$\left(\frac{127}{5},\frac{156}{5},\frac{234}{5},\frac{283}{5}\right)$	$(0,0,0,0)$	$(1252,1440,1992,2356) +$ $\lambda(69,90,144,177)$
x_4	0	-5	-3	1	$-90 - 10\lambda$
x_1	1	2	1	0	$28 + 3\lambda$

value in the optimality interval $[42,\infty)$ is equal to $\tilde{z}(\lambda) = (1106, 1680,$ $2436, 2618) + \lambda(\frac{-229}{3}, -56, -42, \frac{-37}{3})$.

Now we solve the problem for $\lambda < \underline{\lambda}_B = -9$. For $\lambda < \underline{\lambda}_B = -9$ the variable x_2 does not satisfy the feasibility criteria with regard to Table 3.11. In fact, at $\lambda = \underline{\lambda}_B = -9$, the value of the basic variable x_2 is zero, and x_4 is introduced into the basis leading to Table 3.14.

In order to obtain the optimality interval of the basis $\tilde{B} = [a_4, a_1]$, we should have $\bar{b}_4 + \lambda\bar{b}'_4 \geq 0$ and $\bar{b}_1 + \lambda\bar{b}'_1 \geq 0$:

$$\bar{b}_4 + \lambda\bar{b}'_4 = -90 - 10\lambda \geq 0 \Rightarrow \lambda \leq -9,$$

$$\bar{b}_1 + \lambda\bar{b}'_1 = 28 + 3\lambda \geq 0 \Rightarrow \lambda \geq -\frac{28}{3},$$

These imply that the basis $\tilde{B} = [a_4, a_1]$ is optimal over the optimality interval $[\underline{\lambda}_{\tilde{B}}, \bar{\lambda}_{\tilde{B}}] = \left[\frac{-28}{3}, -9\right]$. As we see the left end point of the new basis $\tilde{B} = [a_4, a_1]$ is equal to the right end point of the basis $B = [a_2, a_1]$. The optimal solution and the fuzzy optimal objective function value in this interval with regard to (3.48) and (3.49) are given as follows, respectively:

$$x^* = \begin{pmatrix} x_4(\lambda) \\ x_1(\lambda) \\ x_2 \\ x_3 \end{pmatrix} = \begin{pmatrix} -90 - 10\lambda \\ 28 + 3\lambda \\ 0 \\ 0 \end{pmatrix}$$

$$\tilde{z}(\lambda) = (1252, 1440, 1992, 2356) + \lambda(69, 90, 144, 177)$$

At $\lambda = -\frac{28}{3}$, x_1 drops to zero. Since all entries in the x_1 row are nonnegative, we stop with the conclusion that for $\lambda < -\frac{28}{3}$ no feasible solution exists.

3.4   LPPs with Fuzzy Parameters of Type-2

The LP problem involving fuzzy numbers for the decision variables and the right-hand side of the constraints is called LPP with fuzzy parameters of type-2 or the fuzzy variables linear programming (FVLP) problem and is formulated as follows:

$$
\begin{aligned}
\min \tilde{z} \quad &= c\tilde{x} \\
s.t. \quad A\tilde{x} &\succeq \tilde{b}, \\
\tilde{x} &\succeq \tilde{0}.
\end{aligned}
\tag{3.51}
$$

Definition 3.4 Any fuzzy vector $\tilde{x}$ which satisfies the constraints and nonnegative restrictions of (3.51) is said to be a fuzzy feasible solution.

Definition 3.5 The fuzzy feasible solution $\tilde{x}_*$ is said to be a fuzzy optimum solution to (3.51) if $c\tilde{x}_* \preceq c\tilde{x}$ for any arbitrary fuzzy feasible solution $\tilde{x}$.

Definition 3.6 (*Fuzzy basic solution*) Suppose $\tilde{\tilde{x}} = (\tilde{\tilde{x}}_1, \tilde{\tilde{x}}_2, \ldots, \tilde{\tilde{x}}_n)$ solves $A\tilde{x} = \tilde{b}$. If all $\tilde{\tilde{x}}_j = (\tilde{x}_{1j}, \tilde{x}_{2j}, \tilde{x}_{3j}, \tilde{x}_{4j})$ satisfy that $\frac{\tilde{x}_{1j} + \tilde{x}_{2j} + \tilde{x}_{3j} + \tilde{x}_{4j}}{4} = 0$, then $\tilde{\tilde{x}}$ is said to be a fuzzy basic solution. Otherwise, $\tilde{\tilde{x}}$ has some non-zero components, say $\tilde{\tilde{x}}_1, \tilde{\tilde{x}}_2, \ldots, \tilde{\tilde{x}}_k, 1 \le k \le m$, i.e. $\Re(\tilde{\tilde{x}}_j) = \frac{\tilde{x}_{1j} + \tilde{x}_{2j} + \tilde{x}_{3j} + \tilde{x}_{4j}}{4} \ne 0$. Then $A\tilde{x} = \tilde{b}$ can be written as:

$$
a_1\tilde{\tilde{x}}_1 + a_2\tilde{\tilde{x}}_2, \cdots + a_k\tilde{\tilde{x}}_k + a_{k+1}(\tilde{x}_{1k+1}, \tilde{x}_{2k+1}, \tilde{x}_{3k+1}, \tilde{x}_{4k+1}) + \cdots + a_n(\tilde{x}_{1n}, \tilde{x}_{2n}, \tilde{x}_{3n}, \tilde{x}_{4n}) = \tilde{b}
$$

If the columns $a_1, a_2, \ldots, a_k$ corresponding to the non-zero components $\tilde{\tilde{x}}_1, \tilde{\tilde{x}}_2, \ldots, \tilde{\tilde{x}}_k$ are linearly independent, then $\tilde{\tilde{x}}$ is said to be fuzzy basic solution.

Remark 3.1 Consider the fuzzy system of constraints (3.51) where A is a matrix of order $(m \times n)$ and $rank(A) = m$. Any $(m \times m)$ matrix B formed by m linearly independent columns of A is known as a basis for this fuzzy system. The column vectors of A and the fuzzy variables in the problem; can be partitioned into the basic and the non-basic part with respect to this basis B. Each column vector of A, which is in the basis B, is known as a basic column vector. All the remaining column vectors of A are called the non-basic column vectors.

Remark 3.2 Let $\tilde{x}_B$ be the vector of the variables associated with the basic column vectors. The variables in $\tilde{x}_B$ are known as the fuzzy basic variables with respect to the basis B, and $\tilde{x}_B$ is the fuzzy basic vector. Also, let $\tilde{x}_N$ and N be the vector and the matrix of the remaining fuzzy variables and columns, which are called the fuzzy non-basic variables and non-basic matrix, respectively. In this case, the solution $\tilde{x} = (\tilde{x}_B, \tilde{x}_N) = (B^{-1}\tilde{b}, \tilde{0})$ is also a fuzzy basic solution.

Definition 3.7 Suppose $\tilde{\tilde{x}}$ is a fuzzy basic feasible solution of fuzzy system $A\tilde{x} = \tilde{b}, \tilde{x} \succeq \tilde{0}$. If the number of fuzzy positive variables $\tilde{\tilde{x}}$ is exactly m, then it is called a non-degenerate fuzzy basic feasible solution, i.e., $\tilde{\bar{x}} = (\tilde{\bar{x}}_1, \tilde{\bar{x}}_2, \ldots, \tilde{\bar{x}}_m) \succ (\tilde{0}, \tilde{0}, \ldots, \tilde{0})$. If the number of positive $\tilde{\bar{x}}$ is less than of m, then $\tilde{\bar{x}}$ is called a degenerate fuzzy basic feasible solution.

Theorem 3.14 [20] *If the LPP with fuzzy parameters of type-2 (3.51) in the standard form has a fuzzy optimum feasible solution, then it has a fuzzy basic feasible solution that is optimal.*

Proof Suppose $\tilde{\bar{x}} = (\tilde{\bar{x}}_1, \tilde{\bar{x}}_2, \ldots, \tilde{\bar{x}}_n)$ be a fuzzy optimum feasible solution in which $\tilde{\bar{x}}_j = (\tilde{\bar{x}}_{1j}, \tilde{\bar{x}}_{2j}, \tilde{\bar{x}}_{3j}, \tilde{\bar{x}}_{4j})$. Let $\{a_j | \Re(\tilde{\bar{x}}_j) = \frac{\bar{x}_{1j} + \bar{x}_{2j} + \bar{x}_{3j} + \bar{x}_{4j}}{4} > 0\} = \{a_1, a_2, \ldots, a_k\}$. So we have

$$
\begin{aligned}
a_1 \tilde{\bar{x}}_1 + a_2 \tilde{\bar{x}}_2, &\ldots + a_k \tilde{\bar{x}}_k \\
&+ a_{k+1}(\bar{x}_{1k+1}, \bar{x}_{2k+1}, \bar{x}_{3k+1}, \bar{x}_{4k+1}) + \cdots + a_n(\bar{x}_{1n}, \bar{x}_{2n}, \bar{x}_{3n}, \bar{x}_{4n}) = \tilde{\bar{b}}
\end{aligned}
\tag{3.52}
$$

If $\{a_1, a_2, \ldots, a_k\}$ is linearly independent, then $\tilde{\bar{x}}$ is a fuzzy basic feasible solution of (3.51) and we are done. Suppose this set is linearly dependent. So, $y = (y_1, y_2, \ldots, y_k) \neq (0, 0, \ldots, 0)$ exists such that

$$
a_1 y_1 + a_2 y_2 + \cdots + a_k y_k = 0
\tag{3.53}
$$

Let $\tilde{y}_j = (y_j, y_j, y_j, y_j)$ for $j = 1, 2, \ldots, k$. Thus, from (3.53) we have

$$
a_1 \tilde{y}_1 + a_2 \tilde{y}_2 + \cdots + a_k \tilde{y}_k = \tilde{0}
\tag{3.54}
$$

Using Eqs. (3.52) and (3.54), we get the following relation in which θ is a real number.

$$
\begin{aligned}
a_1 (\tilde{\bar{x}}_1 + \theta \tilde{y}_1) + a_2 (\tilde{\bar{x}}_2 + \theta \tilde{y}_2) &+ \cdots + a_k (\tilde{\bar{x}}_k + \theta \tilde{y}_k) \\
&+ a_{k+1}(\bar{x}_{1k+1}, \bar{x}_{2k+1}, \bar{x}_{3k+1}, \bar{x}_{4k+1}) \\
&+ \cdots + a_n(\bar{x}_{1n}, \bar{x}_{2n}, \bar{x}_{3n}, \bar{x}_{4n}) = \tilde{\bar{b}}
\end{aligned}
\tag{3.55}
$$

Define the fuzzy vector $\tilde{\bar{x}}(\theta) = (\tilde{\bar{x}}_1(\theta), \tilde{\bar{x}}_2(\theta), \ldots, \tilde{\bar{x}}_n(\theta)))$, where

$$
\tilde{\bar{x}}_j(\theta) = \begin{cases} \tilde{\bar{x}}_j + \theta \tilde{y}_j & j = 1, 2, \ldots, k \\ 0 & j = k+1, \ldots, n \end{cases}
\tag{3.56}
$$

Clearly, $\tilde{\bar{x}}(\theta)$ satisfies $A\tilde{x} = \tilde{b}$. Define

$$
\theta_1 = \max_{1 \le j \le n} \left\{ \Re(\tilde{\bar{x}}_j) = \frac{\bar{x}_{1j} + \bar{x}_{2j} + \bar{x}_{3j} + \bar{x}_{4j}}{4} \Big| y_j > 0 \right\}
\tag{3.57}
$$

$$
\theta_2 = \max_{1 \le j \le n} \left\{ \Re(\tilde{\bar{x}}_j) = \frac{\bar{x}_{1j} + \bar{x}_{2j} + \bar{x}_{3j} + \bar{x}_{4j}}{4} \Big| y_j < 0 \right\}
\tag{3.58}
$$

Since $y = (y_1, y_2, \ldots, y_k) \neq (0, 0, \ldots, 0)$, at least one of θ_1 or θ_2 must be finite. It is clear that $\theta_1 < 0$ and $\theta_2 > 0$. Let $0 < \varepsilon \leq \min\{|\theta_1|, \theta_2\}$. Therefore, $\tilde{\bar{x}}(\theta)$ is a fuzzy feasible solution for all satisfying $-\varepsilon \leq \theta \leq \varepsilon$. In addition, we have,

$$\tilde{z}(\tilde{\bar{x}}(\theta)) = \tilde{z}(\tilde{\bar{x}}) + \theta(c_1 \tilde{y}_1 + c_2 \tilde{y}_2 + \cdots + c_k \tilde{y}_k) \tag{3.59}$$

We now show that the assumption that $\tilde{\bar{x}}$ is fuzzy optimal implies that any $(y_1, y_2, \ldots, y_k)$ satisfying (3.54) must also satisfy

$$c_1 \tilde{y}_1 + c_2 \tilde{y}_2 + \cdots + c_k \tilde{y}_k = \tilde{0} \tag{3.60}$$

Suppose not. If $c_1 \tilde{y}_1 + c_2 \tilde{y}_2 + \cdots + c_k \tilde{y}_k \succ \tilde{0}$, let $\pi = \varepsilon$ and if $c_1 \tilde{y}_1 + c_2 \tilde{y}_2 + \cdots + c_k \tilde{y}_k \prec \tilde{0}$, then let $\pi = -\varepsilon$. Then, $\tilde{\bar{x}}(\pi)$ is a fuzzy solution of problem (3.51) in the standard form and $\tilde{z}(\tilde{\bar{x}}(\pi)) \succ \tilde{z}(\tilde{\bar{x}})$, which contradicts the assumption that $\tilde{\bar{x}}$ is an optimal fuzzy solution of problem (3.51) in the standard form. This means that the fuzzy feasible solution $\tilde{\bar{x}}(\pi)$ is a fuzzy optimal feasible solution of (3.51) in the standard form in which the number of its positive fuzzy variables is at least one less than the number of positive fuzzy variables of $\tilde{\bar{x}}$. In a similar way, it is possible to obtain another fuzzy feasible solution $\tilde{\bar{x}}(\pi)$ in which the number of positive variables is at least one less than the number of positive variables of $\tilde{\bar{x}}(\pi)$. By (3.60), any such fuzzy feasible $\tilde{\bar{x}}(\pi)$ that we obtain must also satisfy $\tilde{z}(\tilde{\bar{x}}(\pi)) = \tilde{z}(\tilde{\bar{x}})$. Hence, when this procedure is applied repeatedly, an optimal fuzzy basic feasible solution of (3.51) in the standard form will be obtained after at most $(k - 1)$ applications of the procedure. $\Box$

Example 3.6 Consider the following LPP with fuzzy parameters of type-2:

$$\min \quad \tilde{z} = -2\tilde{x}_1 - 2\tilde{x}_2$$
$$\text{s.t.} \quad \tilde{x}_1 + \tilde{x}_2 + \tilde{x}_3 = (1, 6, 14, 21)$$
$$2\tilde{x}_1 + \tilde{x}_2 + \tilde{x}_4 = (0, 4, 12, 16) \tag{3.61}$$
$$\tilde{x}_1, \tilde{x}_2, \tilde{x}_3, \tilde{x}_4 \succeq \tilde{0}$$

It should be noted that $\tilde{\bar{x}} = \begin{bmatrix} (1, 4, 8, 11) \\ (0, 2, 6, 8) \\ (0, 0, 0, 0) \\ (0, 2, 6, 8) \end{bmatrix}$ is a fuzzy optimal solution for the

LPP with fuzzy parameters (3.61) with optimal objective value $\tilde{z}^*(\tilde{\bar{x}}) = (-38, -28, -12, -2)$ and $\Re(\tilde{z}^*(\tilde{\bar{x}})) = -20$. We see that $\left\{ a_j \Big|^{\frac{\tilde{x}_{1j} + \tilde{x}_{2j} + \tilde{x}_{3j} + \tilde{x}_{4j}}{4}} > 0 \right\} = \left\{ a_1 = \begin{bmatrix} 1 \\ 0 \end{bmatrix}, a_2 = \begin{bmatrix} 1 \\ 1 \end{bmatrix}, a_4 = \begin{bmatrix} 0 \\ 1 \end{bmatrix} \right\}$ is linearly independent. So, $y = (y_1, y_2, y_4) \neq (0, 0, 0)$ exists such that $y_1 \begin{bmatrix} 1 \\ 0 \end{bmatrix} + y_2 \begin{bmatrix} 1 \\ 1 \end{bmatrix} + y_4 \begin{bmatrix} 0 \\ 1 \end{bmatrix} = \begin{bmatrix} 0 \\ 0 \end{bmatrix}$. Let $y = (y_1, y_2, y_4) = (1, -1, 1)$. Now, define $\tilde{y}_1 = (1, 1, 1, 1), \tilde{y}_2 = (-1, -1, -1, -1)$ and $\tilde{y}_4 = $

$(1, 1, 1, 1)$. By definition of θ_1 and θ_2 we have $\theta_1 = -4$ and $\theta_2 = 4$. Thus, we obtain $\varepsilon = 4$. In the case of $\pi = \varepsilon = 4$, we obtain a new fuzzy optimal solution as

$$\tilde{x}(\pi) = \tilde{\tilde{x}} + \pi \tilde{y} = \begin{bmatrix} (5, 8, 12, 15) \\ (-4, -2, 2, 4) \\ (0, 0, 0, 0,) \\ (4, 6, 10, 12) \end{bmatrix}. \quad \text{For this new fuzzy solution, we have}$$

$\tilde{z}^*(\tilde{\tilde{x}}(\pi)) = (-30, -24, -16, -10)$ and $\Re(\tilde{z}^*(\tilde{\tilde{x}}(\pi)) = -20$. This shows that the new fuzzy solution is optimal too.

In addition, for the new fuzzy solution we have $\left\{ a_j \Big| \frac{\tilde{x}_{1j} + \tilde{x}_{2j} + \tilde{x}_{3j} + \tilde{x}_{4j}}{4} > 0 \right\} =$

$\left\{ a_1 = \begin{bmatrix} 1 \\ 0 \end{bmatrix}, a_4 = \begin{bmatrix} 0 \\ 1 \end{bmatrix} \right\}$ which is linearly independent. Thus, the new fuzzy optimal solution is a fuzzy optimal basis feasible solution.

On the other hand, if we let $\pi = \varepsilon = -4$, we obtain another new fuzzy optimal solution as $\tilde{\tilde{x}}(\pi) = \tilde{\tilde{x}} - 4\tilde{y} = \begin{bmatrix} (-3, 0, 4, 7) \\ (4, 6, 10, 12) \\ (0, 0, 0, 0) \\ (-4, -2, 2, 4) \end{bmatrix}$. Also, for this new fuzzy solution,

we have $\tilde{z}^*(\tilde{\tilde{x}}(\pi)) = (-38, -28, -12, -2)$ and $\Re \left(\tilde{z}^* \left(\tilde{\tilde{x}}(\pi) \right) \right) = -20$. This shows that it is an alternative fuzzy optimal solution. In addition, for this fuzzy solution we have $\left\{ a_j \Big| \frac{\tilde{x}_{1j} + \tilde{x}_{2j} + \tilde{x}_{3j} + \tilde{x}_{4j}}{4} > 0 \right\} = \left\{ a_1 = \begin{bmatrix} 1 \\ 0 \end{bmatrix}, a_2 = \begin{bmatrix} 1 \\ 1 \end{bmatrix} \right\}$ which is linearly independent. Thus, this fuzzy optimal solution is another fuzzy optimal basic feasible solution.

In what follows, three solution approaches for solving the LPP with fuzzy parameters of type-2 or the FVLP problem (3.51) are explored. The other approaches can be found in [14, 39].

3.4.1   Maleki et al.'s Approach

Maleki et al. [2] introduced an auxiliary problem, having only fuzzy cost coefficients, for an LPP with fuzzy parameters of type-2 and then used its crisp solution for obtaining the fuzzy solution of the primary problem. In order to describe this approach, we reformulate the LP problem with fuzzy parameters of type-2 (3.51), with respect to the change of all the parameters and variables, as the following FVLP problem:

$$\min \tilde{u} = \tilde{w}b$$
$$\text{s.t.} \quad \tilde{w}A \succeq \tilde{c}, \tag{3.62}$$
$$\tilde{w} \succeq \tilde{0}.$$

Maleki et al. [2] considered the LP problem with fuzzy parameters of type-1 (3.2) as the fuzzy auxiliary problem for the LP problem with fuzzy parameters of type-2 (3.62) and studied the relation between the FLP problem (3.62) and the fuzzy auxiliary problem (3.2). In fact, they proved that if $x_B = B^{-1}b$ is an optimal basic feasible solution of the auxiliary problem (3.2) then $\tilde{w} = \tilde{c}_B B^{-1}$ is a fuzzy optimal solution of the LPP with fuzzy parameters of type-2 (3.62). In what follows, we review the relationships between the primary and auxiliary problems.

Theorem 3.15 *If $\tilde{w}_o$ is any fuzzy feasible solution for (3.62) and x_o is any feasible solution to (3.2), then $\tilde{c}x_o \succeq \tilde{w}_o b$.*

Proof Since $\tilde{w}_o$ and x_o are feasible solutions for (3.62) and (3.2), respectively, we have $Ax_o \leq b, x_o \geq 0, \tilde{w}_o A \preceq \tilde{c}$ and $\tilde{w} \succeq \tilde{0}$. Multiplying $Ax_o \leq b$ on the left by $\tilde{w}_o \succeq \tilde{0}$ and $\tilde{w}_o A \preceq \tilde{c}$ on the right by $x_o \geq 0$, we obtain $\tilde{w}_o b \preceq \tilde{w}_o Ax_o \preceq \tilde{c}x_o$.   □

Theorem 3.16 *If $\tilde{w}_o$ is a fuzzy feasible solution of (3.62) and x_o is a feasible solution of (3.2), such that $\tilde{c}x_o = \tilde{w}_o b$; then x_o is an optimal solution of (3.2) and $\tilde{w}_o$ is a fuzzy optimal solution of (3.62).*

Proof Let $\tilde{w}'$ be an arbitrary fuzzy feasible solution of (3.62). Regarding Theorem 3.15, we obtain $\tilde{c}x_o \succeq \tilde{w}'b$. On the other hand, by assumption we have $\tilde{c}x_o = \tilde{w}_o b$. Thus, $\tilde{w}_o b \succeq \tilde{w}'b$ and this means that $\tilde{w}_o$ is a fuzzy optimal solution of maximization problem (3.62). In a similar way, let x' be an arbitrary solution of (3.2). Regarding Theorem 3.16, we obtain $\tilde{c}x' \succeq \tilde{w}_o b$. On the other hand we have $\tilde{c}x_o = \tilde{w}_o b$. Thus, $\tilde{c}x' \succeq \tilde{c}x_o$ and this means that x' is an optimal solution of minimization problem (3.2).   □

Theorem 3.17 *If the auxiliary problem (3.2) has an optimal basic solution; then problem (3.62) has a fuzzy optimal solution.*

Proof The problem (3.2) can be transformed into the following form:

$$\max \tilde{z} \approx \tilde{c}x$$
$$\text{s.t.} \quad Ax + Is = b, \tag{3.63}$$
$$x, s \geq 0.$$

Let $x = (x_B, x_N) = (B^{-1}b, 0)$ with optimal basis B, and the objective value $\tilde{c}x = \tilde{c}_B B^{-1}b$ is an optimal basic solution of (3.63). Thus, according to Theorem 3.3, we obtain $\tilde{z}_j \succeq \tilde{c}_j$ or $\tilde{z}_j - \tilde{c}_j \succeq \tilde{0}$ for all $j = 1, 2, \ldots, n, n+1, \ldots, n+m$. Define $\tilde{w} = \tilde{c}_B B^{-1}$. For $j = 1, 2, \ldots, n$, we have

$$\tilde{z}_j - \tilde{c}_j = \tilde{c}_B B^{-1} a_j - c_j = \tilde{w} a_j - \tilde{c}_j$$

Hence, $\tilde{z}_j - \tilde{c}_j \succeq \tilde{0}$ for $j = 1, 2, \ldots, n$ implies that $\tilde{w} a_j - \tilde{c}_j \preceq \tilde{0}$ for $j = 1, 2, \ldots, n$, which in turn implies that $\tilde{w} A \preceq \tilde{c}$. Furthermore, note that $a_{n+i} = -e_i$ and $\tilde{c}_{n+i} = \tilde{0}$ for $i = 1, 2, \ldots, m$, and so we have

$$\tilde{z}_{n+i} - \tilde{c}_{n+i} = \tilde{w}a_{n+i} - \tilde{c}_{n+i} = \tilde{w}(-e_i) - \tilde{0} = -\tilde{w}_i$$

In addition, if $\tilde{z}_{n+i} - \tilde{c}_{n+i} \succeq \tilde{0}$ for $i = 1, 2, \ldots, m$, then $\tilde{w}_i \succeq \tilde{0}$ for $i = 1, 2, \ldots, m$, and so, $\tilde{w} \succeq \tilde{0}$. We have just shown that $\tilde{z}_j - \tilde{c}_j \succeq \tilde{0}$ for all $j = 1, 2, \ldots, n, n+1, \ldots, n+m$ implies that $\tilde{w}A \preceq \tilde{c}$ and $\tilde{w} \succeq \tilde{0}$. In other words, $\tilde{w} = \tilde{c}_B B^{-1}$ is a fuzzy feasible solution of problem (3.52). Also, the fuzzy objective value of (3.62) for $\tilde{w} = \tilde{c}_B B^{-1}$ is equal to $\tilde{w}b = \tilde{c}_B B^{-1}b$ and so we obtain $\tilde{w}b = \tilde{c}x$. Hence, regarding Theorem 3.16, we conclude that $\tilde{w} = \tilde{c}_B B^{-1}$ is the fuzzy optimal solution of problem (3.62). □

Theorem 3.18 *If the auxiliary problem* (3.2) *has an unbounded solution; then problem* (3.62) *has no fuzzy optimal basic solution.*

Proof According to Theorem 3.16, it is straightforward. □

Example 3.7 [2, 13] Consider the following FVLP problem:

$$
\begin{aligned}
Min \quad & \tilde{z} \approx 6\tilde{x}_1 + 10\tilde{x}_2 \\
s.t. \quad & 2\tilde{x}_1 + 5\tilde{x}_2 \succeq (3, 5, 8, 11), \\
& 3\tilde{x}_1 + 4\tilde{x}_2 \succeq (4, 6, 10, 16), \\
& \tilde{x}_1, \tilde{x}_2 \succeq \tilde{0}.
\end{aligned}
\tag{3.64}
$$

Now, we solve this problem using the solution approach proposed by Maleki et al. [2]. To do this, we rewrite the FVLP problem (3.64) with respect to the change of all the parameters and variables as follows:

$$
\begin{aligned}
Min \quad & \tilde{u} \approx 6\tilde{w}_1 + 10\tilde{w}_2 \\
s.t. \quad & 2\tilde{w}_1 + 5\tilde{w}_2 \succeq (3, 5, 8, 13), \\
& 3\tilde{w}_1 + 4\tilde{w}_2 \succeq (4, 6, 10, 16), \\
& \tilde{w}_1, \tilde{w}_2 \succeq \tilde{0}.
\end{aligned}
\tag{3.65}
$$

The fuzzy auxiliary problem corresponding to the FVLP problem (3.65) is defined as follows:

$$
\begin{aligned}
Max \quad & \tilde{z} \approx (5, 8, 2, 5)x_1 + (6, 10, 2, 6)x_2 \\
s.t. \quad & 2x_1 + 3x_2 \le 6, \\
& 5x_1 + 4x_2 \le 10, \\
& x_1, x_2 \ge 0.
\end{aligned}
\tag{3.66}
$$

The fuzzy auxiliary problem (3.66) is exactly the LP problem with fuzzy parameters of type-1 (3.17). Thus, the optimal solution given in (3.19) is its optimal solution. Now, since the basis $B = [a_1, a_2] = \begin{bmatrix} 3 & 2 \\ 4 & 5 \end{bmatrix}$ is the optimal basis of the fuzzy auxiliary problem (3.66) or (3.17), regarding Theorem 3.17, the fuzzy optimal solution of the FVLP problem (3.65) is obtained as follows based on the formulation $\tilde{w} = \tilde{c}_B B^{-1}$:

$$
\begin{aligned}
\tilde{w}^* = (\tilde{w}_1^*, \tilde{w}_2^*) &= ((4,6,10,16),(3,5,8,13)) \begin{bmatrix} \frac{5}{7} & \frac{-2}{7} \\ \frac{-4}{7} & \frac{3}{7} \end{bmatrix} \\
&= \left(\left(\frac{-32}{7}, \frac{-2}{7}, \frac{30}{7}, \frac{68}{7} \right), \left(\frac{-23}{7}, \frac{-5}{7}, \frac{12}{7}, \frac{31}{7} \right) \right)
\end{aligned}
\tag{3.67}
$$

This means that the optimal solution of the FVLP problem (3.64) is as follows:

$$
\tilde{x}^* = \begin{bmatrix} \tilde{x}_1^* \\ \tilde{x}_2^* \end{bmatrix} = \begin{bmatrix} \left(\frac{-32}{7}, \frac{-2}{7}, \frac{30}{7}, \frac{68}{7} \right) \\ \left(\frac{-23}{7}, \frac{-5}{7}, \frac{12}{7}, \frac{31}{7} \right) \end{bmatrix}
\tag{3.68}
$$

Unlike the non-simplex based approach, this approach does not increase the number of the constraint of the primary FLP problem. Another advantage of this approach is to produce the fuzzy optimal solution that provides possible outcomes with a certain degree of memberships to the decision maker. However, the fuzzy optimal solution by using this approach may be not non-negative. For instance the value of $\tilde{x}_1^*$ in the fuzzy optimal solution (3.68) is non-negative as there is a negative part in this fuzzy number.

3.4.2 Mahdavi-Amiri and Nasseri's Approach

Mahdavi-Amiri and Nasseri [13] proved that the fuzzy auxiliary problem (LP with fuzzy parameters of type-1) introduced by Maleki et al. [2] is the dual of the LP with fuzzy parameters of type-2. Hence, they introduced a fuzzy dual simplex algorithm for solving the LP with fuzzy parameters of type-2 directly on the primary problem and without solving any fuzzy auxiliary problem.

Mahdavi-Amiri and Nasseri [13] defined the dual problem of the LP with fuzzy parameters of type-2 (3.51) using the usual definitions of the dual in linear programming and applying a linear ranking function.

We can write (3.51) as follows:

$$
\begin{aligned}
&\min \tilde{z} = \sum_{j=1}^{n} c_j (\bar{x}_{1j}, \bar{x}_{2j}, \bar{x}_{3j}, \bar{x}_{4j}) \\
&s.t. \quad \sum_{j=1}^{n} a_{ij} (\bar{x}_{1j}, \bar{x}_{2j}, \bar{x}_{3j}, \bar{x}_{4j}) \succeq (b_{1i}, b_{2i}, b_{3i}, b_{4i}), \quad i = 1, 2, \ldots, m, \\
&\qquad\quad (\bar{x}_{1j}, \bar{x}_{2j}, \bar{x}_{3j}, \bar{x}_{4j}) \succeq \tilde{0}, \qquad\qquad\qquad j = 1, 2, \ldots, n.
\end{aligned}
\tag{3.69}
$$

Using the linear ranking function given in Remark 1.15, problem (3.69) is rewritten as follows:

$$\min z = \sum_{j=1}^{n} c_j \frac{\bar{x}_{1j} + \bar{x}_{2j} + \bar{x}_{3j} + \bar{x}_{4j}}{4}$$

$$s.t. \quad \sum_{j=1}^{n} a_{ij} \frac{\bar{x}_{1j} + \bar{x}_{2j} + \bar{x}_{3j} + \bar{x}_{4j}}{4} \geq \frac{b_{1i} + b_{2i} + b_{3i} + b_{4i}}{4}, \quad i = 1, 2, \ldots, m, \qquad (3.69)$$

$$\frac{\bar{x}_{1j} + \bar{x}_{2j} + \bar{x}_{3j} + \bar{x}_{4j}}{4} \geq 0, \qquad\qquad j = 1, 2, \ldots, n.$$

We can write (3.69) as follows:

$$\min z = \sum_{j=1}^{n} \left(\tfrac{1}{4} c_j \bar{x}_{1j} + \tfrac{1}{4} c_j \bar{x}_{2j} + \tfrac{1}{4} c_j \bar{x}_{3j} + \tfrac{1}{4} c_j \bar{x}_{4j} \right)$$

$$s.t. \quad \sum_{j=1}^{n} \left(\tfrac{1}{4} a_{ij} \bar{x}_{1j} + \tfrac{1}{4} a_{ij} \bar{x}_{2j} + \tfrac{1}{4} a_{ij} \bar{x}_{3j} + \tfrac{1}{4} a_{ij} \bar{x}_{4j} \right) \leq \frac{b_{1i} + b_{2i} + b_{3i} + b_{4i}}{4}, i = 1, 2, \ldots, m,$$

$$\frac{\bar{x}_{1j} + \bar{x}_{2j} + \bar{x}_{3j} + \bar{x}_{4j}}{4} \leq 0, \qquad\qquad j = 1, 2, \ldots, n.$$
$$\qquad\qquad (3.70)$$

The dual of (3.70), using the definition of the dual in linear programming is given as follows:

$$\max u = \sum_{i=1}^{m} w_i \frac{b_{1i} + b_{2i} + b_{3i} + b_{4i}}{4}$$

$$s.t. \quad \sum_{i=1}^{m} \tfrac{1}{4} a_{ij} w_i + \tfrac{1}{4} w_{m+1} = \tfrac{1}{4} c_j, \quad j = 1, 2, \ldots, n,$$

$$\sum_{i=1}^{m} \tfrac{1}{4} a_{ij} w_i + \tfrac{1}{4} w_{m+1} = \tfrac{1}{4} c_j, \qquad j = 1, 2, \ldots, n,$$

$$\sum_{i=1}^{m} \tfrac{1}{4} a_{ij} w_i + \tfrac{1}{4} w_{m+1} = \tfrac{1}{4} c_j, \qquad j = 1, 2, \ldots, n, \qquad (3.71)$$

$$\sum_{i=1}^{m} \tfrac{1}{4} a_{ij} w_i + \tfrac{1}{4} w_{m+1} = \tfrac{1}{4} c_j, \qquad j = 1, 2, \ldots, n,$$

$$w_i \geq 0, \qquad\qquad i = 1, 2, \ldots, m+1.$$

We may write problem (3.71) as follows:

$$\max u = \sum_{i=1}^{m} w_i \frac{b_{1i} + b_{2i} + b_{3i} + b_{4i}}{4}$$

$$s.t. \quad \sum_{i=1}^{m} a_{ij} w_i \leq c_j, \qquad\qquad j = 1, 2, \ldots, n, \qquad (3.72)$$

$$w_i \geq 0, \qquad\qquad i = 1, 2, \ldots, m+1.$$

The objective function of (3.72) is equivalent to $\sum_{i=1}^{m} w_i(b_{1i}, b_{2i}, b_{3i}, b_{4i}) = \sum_{i=1}^{m} w_i \tilde{b}_i$. Thus, we have

$$
\begin{aligned}
\max \tilde{u} &= \sum_{i=1}^{m} w_i \tilde{b}_i \\
s.t. \quad & \sum_{i=1}^{m} a_{ij} w_i \leq c_j, \quad j = 1, 2, \ldots, n, \\
& w_i \geq 0, \qquad\qquad i = 1, 2, \ldots, m.
\end{aligned}
\tag{3.73}
$$

Using matrix notations, we can write (3.73) as follows:

$$
\begin{aligned}
\min \tilde{u} &= w\tilde{b} \\
s.t. \quad wA &\geq c, \\
w &\geq 0.
\end{aligned}
\tag{3.74}
$$

We see that only the coefficients of the objective function of this problem are fuzzy numbers. Thus, it plays the role of the auxiliary problem for the LP with fuzzy parameters of type-2 (3.51).

The following important definitions and results are required to extend a fuzzy dual approach for solving the FLP problem (3.51).

Definition 3.8 The basic matrix B is called a dual feasible basis of $Ax = b$ for a minimization problem whenever $z_j - c_j \leq 0$.

Definition 3.9 A dual feasible basis B for the LP with fuzzy parameters of type-2 (3.51) is said to be dual degenerate if for at least one non-basic fuzzy variable, say $\tilde{x}_j$, we have $z_j - c_j = 0$. Otherwise, it is said to be dual non-degenerate. The LPP with fuzzy parameters of type-2 (3.51) is said to be totally dual non-degenerate, if for all fuzzy non-basic fuzzy variables with respect to any basis for (3.51), we have $z_j - c_j \leq 0$.

Theorem 3.19 *If for a basic solution with the basis $B = [a_{B_1}, \ldots, a_{B_r}, \ldots, a_{B_m}]$, $\tilde{x}_k$ is introduced into the basis and $\tilde{x}_{B_r}$ is removed from the basis. The new objective row for the new basis by $\hat{B} = [a_{B_1}, \ldots, a_k, \ldots, a_{B_m}]$ is given as*

$$
(z_j - c_j)_{new} = (z_j - c_j) - \frac{y_{rj}}{y_{rk}}(z_k - c_k)
\tag{3.75}
$$

Proof It is similar to the proof of Theorem 3.5.   □

Theorem 3.20 *If for a dual feasible basic solution with the basis B, the index of the entering variable, $\tilde{x}_k$, is determined by the following minimum ratio test and $\tilde{x}_{B_r}$ is removed from the basis. The new basic solution is dual feasible.*

$$\frac{z_k - c_k}{y_{rk}} = \min\left\{\frac{z_j - c_j}{y_{rj}}, y_{rj} < 0\right\} \tag{3.76}$$

Proof It is similar to the proof of Theorem 3.5.   □

Theorem 3.21 *Suppose for a dual feasible basic solution with the basis B, the index of the entering fuzzy variable, $\tilde{x}_k$, is determined by the given minimum ratio test in Eq. (3.21) and $\Re(\tilde{x}_{B_r}) = \Re(\tilde{b}) < 0$ is removed from the basis, where $y_{rj} \geq 0$ for all $j = 1, 2, \ldots, n$. Then the LPP with fuzzy parameters of type-2 (3.51) is infeasible.*

Proof It is similar to the proof of Theorem 3.6.   □

Using the above results, Mahdavi-Amiri and Nasseri [13] proposed a dual approach to solve the LP with fuzzy parameters of type-2 (3.51). A summary of this approach is given as follows:

Algorithm 3.3: Fuzzy dual simplex algorithm *(Minimization Problem)*
   Initialization step
   Choose a starting dual feasible basic solution with the basis B.
   Main steps

(1) Solve the system $B\tilde{x}_B = \tilde{b}$. Let $\tilde{x}_B = B^{-1}\tilde{b} = \tilde{b}$, and $\tilde{z} = c_B\tilde{x}_B$.

(2) Find the rank of $\tilde{b}_i$ for all $1 \leq i \leq m$ based on ranking function $\Re$ given in Remark 1.15 and let $\Re(\tilde{b}_r) = \min_{1 \leq i \leq m}\left\{\Re(\tilde{b}_i)\right\}$. If $\Re(\tilde{b}_r) \geq 0$, then stop with the current fuzzy basic solution as an optimal fuzzy solution.

(3) Solve the system $By_j = a_j$ for all $j \in J_N$ and let $y_j = B^{-1}a_j$. If $y_{rj} \geq 0$ for all $j \in J_N$ then stop with the conclusion that the LPP with fuzzy parameters of type-2 is infeasible.

(4) Solve the system $wB = c_B$. Let $w = c_B B^{-1}$. Calculate $z_j = wa_j$ for all $j \in J_N$. If $y_{rj} \not\geq 0$ for all $j \in J_N$, then $\tilde{x}_{B_r}$ leaves the basis and $\tilde{x}_k$ enters the basis providing

   that

$$\frac{z_k - c_k}{y_{rk}} = \min_{j \in J_N}\left\{\frac{z_j - c_j}{y_{rk}}|y_{rk} < 0\right\}.$$

(5) Update the basis B where a_k replaces a_{B_r}, update the index set J_N and go to (1).

Theorem 3.22 *In the absence of dual degeneracy, the fuzzy dual simplex method (Algorithm 3.3) stops in a finite number of iterations, either with an optimal fuzzy basic feasible solution or with the conclusion that the problem is infeasible.*

Proof It is similar to the proof of Theorem 3.7.   □

Example 3.8 Again, consider the LPP with fuzzy parameters of type-2 (3.64) given in Example 3.7. Now, we solve this problem using Algorithm 3.3. To do this, we rewrite (3.64) as follows in order to obtain a dual feasible solution with identity basis where $\tilde{x}_3$ and $\tilde{x}_4$ are fuzzy surplus variables:

$$
\begin{aligned}
Min \quad & \tilde{z} = 6\tilde{x}_1 + 10\tilde{x}_2 \\
s.t. \quad & -2\tilde{x}_1 - 5\tilde{x}_2 + \tilde{x}_3 = (-13, -8, -5, -3), \\
& -3\tilde{x}_1 - 4\tilde{x}_2 + \tilde{x}_4 = (-16, -10, -6, -4), \\
& \tilde{x}_1, \tilde{x}_2, \tilde{x}_3, \tilde{x}_4 \succeq \tilde{0}
\end{aligned}
\tag{3.77}
$$

The steps of Algorithm 3.3 for solving the LPP with fuzzy parameters of type-2 (3.77) are explored as follows:

Iteration 1:

The starting dual feasible basic is $B = [a_3, a_4] = \begin{bmatrix} 1 & 0 \\ 0 & 1 \end{bmatrix}$. Thus, the non-basic matrix N is $N = [a_1, a_2] = \begin{bmatrix} -2 & -5 \\ -3 & -4 \end{bmatrix}$.

Step 1: We find the fuzzy basic solution by solving the fuzzy system $B\tilde{x}_b = \tilde{b}$:

$$
\begin{aligned}
\begin{bmatrix} 1 & 0 \\ 0 & 1 \end{bmatrix} \begin{bmatrix} \tilde{x}_3 \\ \tilde{x}_4 \end{bmatrix} &= \begin{bmatrix} (-13, -8, -5, -3) \\ (-16, -10, -6, -4) \end{bmatrix} \Rightarrow \begin{bmatrix} \tilde{x}_3 \\ \tilde{x}_4 \end{bmatrix} = \begin{bmatrix} (-13, -8, -5, -3) \\ (-16, -10, -6, -4) \end{bmatrix} \\
&= \begin{bmatrix} \tilde{b}_1 \\ \tilde{b}_2 \end{bmatrix}
\end{aligned}
$$

The fuzzy objective function value is obtained based on the formulation $\tilde{z} = c_B\tilde{x}_B$:

$$
\tilde{z} = (0, 0) \begin{bmatrix} (-13, -8, -5, -3) \\ (-16, -10, -6, -4) \end{bmatrix} = (0, 0, 0, 0)
$$

Step 2: We find the rank of $\tilde{b}_1$ and $\tilde{b}_2$ by use of ranking function $\Re$ given in Remark 1.15. Hence we have $\Re(\tilde{b}_1) = -\frac{29}{4}$ and $\Re(\tilde{b}_2) = -9$. Therefore,

$$
\min\left\{ \Re(\tilde{b}_1) = -\frac{29}{4}, \Re(\tilde{b}_2) = -9 \right\} = \Re(\tilde{b}_2) = -9 \not\succeq 0
$$

Step 3: We find y_j for all $j \in J_N = \{1,2\}$ by solving the system $By_j = a_j$:

$$\begin{bmatrix} 1 & 0 \\ 0 & 1 \end{bmatrix} \begin{bmatrix} y_{11} \\ y_{21} \end{bmatrix} = \begin{bmatrix} -2 \\ -3 \end{bmatrix} \Rightarrow y_1 = \begin{bmatrix} y_{11} \\ y_{21} \end{bmatrix} = \begin{bmatrix} -2 \\ -3 \end{bmatrix}$$

$$\begin{bmatrix} 1 & 0 \\ 0 & 1 \end{bmatrix} \begin{bmatrix} y_{12} \\ y_{22} \end{bmatrix} = \begin{bmatrix} -5 \\ -4 \end{bmatrix} \Rightarrow y_2 = \begin{bmatrix} y_{12} \\ y_{22} \end{bmatrix} = \begin{bmatrix} -5 \\ -4 \end{bmatrix}$$

Step 4: We find w by solving the system $wB = c_B$:

$$(w_1, w_2) \begin{bmatrix} 1 & 0 \\ 0 & 1 \end{bmatrix} = \begin{bmatrix} 0 \\ 0 \end{bmatrix} \Rightarrow w_1 = w_2 = 0$$

Now, we calculate $z_j - c_j = wa_j - c_j$ for all $j \in J_N = \{1,2\}$:

$$z_1 - c_1 = wa_1 - c_1 = (0,0) \begin{bmatrix} -2 \\ -3 \end{bmatrix} - 6 = -6$$

$$z_2 - c_2 = wa_2 - c_2 = (0,0) \begin{bmatrix} -5 \\ -4 \end{bmatrix} - 10 = -10$$

The fuzzy variable $\tilde{x}_{B_2} = \tilde{x}_4$ leaves the basis and the fuzzy variable $\tilde{x}_k$ that enters the basis is determined by the following test:

$$\min\left\{ \frac{z_1 - c_1}{y_{21}}, \frac{z_2 - c_2}{y_{22}} \right\} = \min\left\{ \frac{-6}{-3}, \frac{-10}{-4} \right\} = \frac{-6}{-3} = 2$$

Hence, $\tilde{x}_k = \tilde{x}_1$ enters the basis.

Step 5: We update the basis B where a_1 replaces $a_{B_2} = a_4$:

$$B = [a_3, a_1] = \begin{bmatrix} 1 & -2 \\ 0 & -3 \end{bmatrix}$$

Also, we update $J_N = \{1,2\}$ as $J_N = \{2,4\}$.

Iteration 2:

Step 1: We find the fuzzy basic solution by solving the fuzzy system $B\tilde{x}_b = \tilde{b}$:

$$\begin{bmatrix} 1 & -2 \\ 0 & -3 \end{bmatrix}\begin{bmatrix} \tilde{x}_3 \\ \tilde{x}_1 \end{bmatrix} = \begin{bmatrix} (-13,-8,-5,-3) \\ (-16,-10,-6,-4) \end{bmatrix} \Rightarrow \begin{bmatrix} \tilde{x}_3 \\ \tilde{x}_1 \end{bmatrix} = \begin{bmatrix} (-\frac{31}{3},-4,\frac{5}{3},\frac{23}{3}) \\ (\frac{4}{3},2,\frac{10}{3},\frac{16}{3}) \end{bmatrix}$$
$$= \begin{bmatrix} \tilde{b}_1 \\ \tilde{b}_2 \end{bmatrix}$$

The fuzzy objective function value is obtained based on the formulation $\tilde{z} = c_B \tilde{x}_B$:

$$\tilde{z} = (0,6)\begin{bmatrix} (-\frac{31}{3},-4,\frac{5}{3},\frac{23}{3}) \\ (\frac{4}{3},2,\frac{10}{3},\frac{16}{3}) \end{bmatrix} = (8,12,20,32)$$

Step 2: We find the rank of $\tilde{b}_1$ and $\tilde{b}_2$ by use of the ranking function $\Re$ given in Remark 1.15. Hence we have $\Re(\tilde{b}_1) = -\frac{15}{12}$ and $\Re(\tilde{b}_2) = 3$. Therefore,

$$\min\left\{\Re(\tilde{b}_1) = -\frac{15}{12}, \Re(\tilde{b}_2) = 3\right\} = R(\tilde{b}_1) = \frac{15}{12} \not\geq 0$$

Step 3: We find y_j for all $j \in J_N = \{2,4\}$ by solving the system $By_j = a_j$:

$$\begin{bmatrix} 1 & -2 \\ 0 & -3 \end{bmatrix}\begin{bmatrix} y_{12} \\ y_{22} \end{bmatrix} = \begin{bmatrix} -5 \\ -4 \end{bmatrix} \Rightarrow y_2 = \begin{bmatrix} y_{12} \\ y_{22} \end{bmatrix} = \begin{bmatrix} \frac{-7}{3} \\ \frac{4}{3} \end{bmatrix}$$

$$\begin{bmatrix} 1 & -2 \\ 0 & -3 \end{bmatrix}\begin{bmatrix} y_{14} \\ y_{24} \end{bmatrix} = \begin{bmatrix} 0 \\ 1 \end{bmatrix} \Rightarrow y_4 = \begin{bmatrix} y_{14} \\ y_{24} \end{bmatrix} = \begin{bmatrix} \frac{-2}{3} \\ \frac{-1}{3} \end{bmatrix}$$

Step 4: We find w by solving the system $wB = c_B$:

$$(w_1, w_2)\begin{bmatrix} 1 & -2 \\ 0 & -3 \end{bmatrix} = \begin{bmatrix} 0 \\ 6 \end{bmatrix} \Rightarrow w_1 = 0, w_2 = -2$$

Now, we calculate $z_j - c_j = wa_j - c_j$ for all $j \in J_N = \{2,4\}$:

$$z_2 - c_2 = wa_2 - c_2 = (0,-2)\begin{bmatrix} -5 \\ -4 \end{bmatrix} - 10 = -2$$

$$z_4 - c_4 = wa_4 - c_4 = (0,-2)\begin{bmatrix} 0 \\ 1 \end{bmatrix} - 0 = -2$$

The variable $\tilde{x}_{B_1} = \tilde{x}_3$ leaves the basis and the variable $\tilde{x}_k$ that enters the basis is determined by the following test:

$$\min\left\{ \frac{z_2 - c_2}{y_{12}}, \frac{z_4 - c_4}{y_{14}} \right\} = \min\left\{ \frac{-2}{\frac{-7}{3}}, \frac{-2}{\frac{-2}{3}} \right\} = \frac{-2}{\frac{-7}{3}} = \frac{6}{7}$$

Hence, $\tilde{x}_k = \tilde{x}_2$ enters the basis.

Step 5: We update the basis B where a_2 replaces $a_{B_1} = a_3$:

$$B = [a_2, a_1] = \begin{bmatrix} -5 & -2 \\ -4 & -3 \end{bmatrix}$$

Also, we update $J_N = \{2,4\}$ as $J_N = \{3,4\}$.

Iteration 3:

Step 1: We find the basic solution by solving the fuzzy system $B\tilde{x}_b = \tilde{b}$:

$$\begin{bmatrix} -5 & -2 \\ -4 & -3 \end{bmatrix}\begin{bmatrix} \tilde{x}_2 \\ \tilde{x}_1 \end{bmatrix} = \begin{bmatrix} (-13,-8,-5,-3) \\ (-16,-10,-6,-4) \end{bmatrix} \Rightarrow \begin{bmatrix} \tilde{x}_2 \\ \tilde{x}_1 \end{bmatrix} = \begin{bmatrix} \left(\frac{-23}{7},\frac{-5}{7},\frac{12}{7},\frac{31}{7}\right) \\ \left(\frac{-32}{7},\frac{-2}{7},\frac{30}{7},\frac{68}{7}\right) \end{bmatrix}$$
$$= \begin{bmatrix} \tilde{b}_1 \\ \tilde{b}_2 \end{bmatrix}$$

The fuzzy objective function value is obtained based on the formulation $\tilde{z} = c_B\tilde{x}_B$:

$$\tilde{z} = (10,6)\begin{bmatrix} \left(\frac{-23}{7},\frac{-5}{7},\frac{12}{7},\frac{31}{7}\right) \\ \left(\frac{-32}{7},\frac{-2}{7},\frac{30}{7},\frac{68}{7}\right) \end{bmatrix} = \left(\frac{-422}{7},\frac{-62}{7},\frac{300}{7},\frac{718}{7}\right)$$

Step 2: We find the rank of $\tilde{\tilde{b}}_1$ and $\tilde{\tilde{b}}_2$ by use of ranking function $\Re$ given in Remark 1.15. Hence we have $\Re(\tilde{\tilde{b}}_1) = \frac{15}{28}$ and $\Re(\tilde{\tilde{b}}_2) = \frac{16}{7}$. Therefore,

$$\min\left\{\Re(\tilde{\tilde{b}}_1) = \frac{15}{28}, \Re(\tilde{\tilde{b}}_2) = \frac{16}{7}\right\} = \Re(\tilde{\tilde{b}}_2) = \frac{15}{28} \geq 0$$

Thus, we conclude that the current basis $B = [a_2, a_1]$ is optimal. The fuzzy optimal solution of the LPP problem with fuzzy parameters of type-2 (3.64) is therefore given as follows, which is matched with that given in (3.77) obtained by Maleki et al.'s approach [2]:

$$\tilde{x}^* = \begin{bmatrix} \tilde{x}_1^* \\ \tilde{x}_2^* \end{bmatrix} = \left[\begin{array}{c} \left(\frac{-32}{7}, \frac{-2}{7}, \frac{30}{7}, \frac{68}{7}\right) \\ \left(\frac{-23}{7}, \frac{-5}{7}, \frac{12}{7}, \frac{31}{7}\right) \end{array} \right] \tag{3.78}$$

It should be noted that the advantages and disadvantages of this approach are the same as with the Maleki et al.'s approach [2].

3.4.3　Ebrahimnejad and Tavana's Approach

In this subsection we show that it is possible to find the fuzzy solution of the LPP with fuzzy parameters of type-2 (3.51) with the help of an equivalent crisp LP problem and without solving any fuzzy problem.

As mentioned earlier, Maleki et al. [2] introduced an auxiliary problem, LPP with fuzzy parameters of type-1, for an LP with fuzzy parameters of type-2 and then used its crisp solution to obtain the fuzzy solution of the LPP with fuzzy parameters of type-2. It is worth noting that they generalized the primal simplex algorithm for solving the fuzzy auxiliary problem. In their algorithm, a ranking function based on Remark 1.15 is used for choosing the entering variable.

In a similar way, Mahdavi-Amiri and Nasseri [14] extended the dual simplex method for solving the LPP with fuzzy parameters of type-2 (3.51).In their algorithm, to choose the leaving variable a ranking function based on Remark 1.15 is also used to compare between fuzzy numbers. Ebeahimnejad and Tavana [22] proved that based on the ranking function given in Remark 1.15 it is possible to define a rank for each trapezoidal fuzzy number. In fact assuming that $\tilde{A} = (a_1, a_2, a_3, a_4)$ is a trapezoidal fuzzy number, then $A = \Re(\tilde{A}) = \frac{a_1 + a_2 + a_3 + a_4}{4}$. Therefore, based on this equation they converted the LPP with fuzzy parameters of type-2 (3.51) into a crisp LP problem. To do this, they substitute the rank of each fuzzy number instead of the corresponding fuzzy number in the problem under consideration. This leads to an equivalent crisp LP problem which can be solved by the standard methods. Since the decision variables of the LPP with fuzzy parameters of type-2 (3.51) are fuzzy numbers, a fuzzy solution is required, just as the equivalent crisp problem requires a crisp solution. However, if the basis B is the optimal basis of the equivalent crisp problem, then it will be the optimal basis of the corresponding LPP with fuzzy parameters of type-2. In this case, the solution $(B^{-1}\tilde{b}, \tilde{0})$ will be the optimal solution of the LPP with fuzzy parameters of type-2. In summary, once the ranking function is chosen, the fuzzy problem is converted into a crisp one, which is easily solved by the existing simplex methods. Therefore, it is possible to obtain the fuzzy optimal solution of the LPP with fuzzy parameters of type-2, without the need for any auxiliary problem and without any fuzzy dual approach. As a result, the computational effort is decreased significantly in the approach proposed by Ebrahimnejad and Tavana [22].

The main contribution of the approach proposed by Ebrahimnejad and Tavana [22] is the reduction in the computational complexity of the existing methods [2, 13, 39]. In particular, it is shown that their proposed method entails fewer elementary operations such as additions, subtractions, and comparisons as compared to the existing methods [2, 13, 39].

According to the method proposed by Maleki et al. [2], to carry out Step 2 of Algorithm 3.1 it is required to solve the fuzzy system $\tilde{w}B = \tilde{c}_B$ with m fuzzy equations and m fuzzy variables in all iterations. After solving this fuzzy system, the fuzzy value $\tilde{z}_j - \tilde{c}_j$ for each non-basic variable is calculated based on $\tilde{z}_j - \tilde{c}_j = \tilde{w}a_j - \tilde{c}_j$. Finally, the entering variable is determined according to the most positive rank of $\tilde{z}_j - \tilde{c}_j$. As we see, Step 2 and Step 3 of this algorithm require a large number of fuzzy additions and subtractions on trapezoidal fuzzy numbers for all iterations. Using the proposed method by Ebrahimnejad and Tavana [33], the entering variable is found without solving any fuzzy system and without any fuzzy arithmetic operations.

Moreover, according to the proposed fuzzy dual method by Mahdavi-Amiri and Nasseri [13], to carry out the Step 1 of the Algorithm 3.3 it is necessary to solve the fuzzy system $B\tilde{x}_B = \tilde{b}$ with m fuzzy equations and m fuzzy variables in all iterations. After solving this fuzzy system, the leaving variable is determined according to the most negative rank of $\tilde{\tilde{b}}_i$. As we see, the Step 1 and Step 2 of this algorithm require a large number of fuzzy additions and subtractions on trapezoidal fuzzy numbers for all iterations. Using the method proposed by Ebrahimnejad and Tavana [22], the leaving variable is found without solving any fuzzy system and without any fuzzy arithmetic operations. Also, the comparison of fuzzy numbers is done once and all arithmetic operations are done on real numbers in their proposed approach.

These results confirm that the method proposed by Ebrahimnejad and Tavana [22] is simpler and computationally more efficient than the existing methods [2, 13, 39].

Example 3.9 Again, consider the LPP with fuzzy parameters of type-2 (3.64) given in Example 3.7. Now, we solve this problem using the method proposed by Ebrahimnejad and Tavana [22].

Substituting the rank order of each fuzzy number for its corresponding fuzzy number in the LPP with fuzzy parameters of type-2 (3.64) gives the following crisp problem:

$$
\begin{aligned}
Min\, z = {}& 6x_1 + 10x_2 \\
s.t. \quad & 2x_1 + 5x_2 \geq \tfrac{29}{4} \\
& 3x_1 + 4x_2 \geq 9 \\
& x_1, x_2 \geq 0
\end{aligned}
\tag{3.79}
$$

We then construct the standard form of the LP problem (3.79) as follows where x_3 and x_4 are surplus variables:

$$
\begin{aligned}
Min\, z = {}& 6x_1 + 10x_2 \\
s.t. \quad & 2x_1 + 5x_2 - x_3 = \tfrac{29}{4} \\
& 3x_1 + 4x_2 - x_4 = 9 \\
& x_1, x_2, x_3, x_4 \geq 0
\end{aligned}
\tag{3.80}
$$

The problem (3.80) is a crisp LP problem and can be solved using the standard dual simplex method. By solving the LP problem (3.80) based on the classical dual simplex algorithm, we obtain the basis $B = [a_1, a_2] = \begin{bmatrix} 2 & 5 \\ 3 & 4 \end{bmatrix}$ as the optimal basis of both the LP problem (3.80) and the FVLP problem (3.64). Thus, as mentioned earlier, the fuzzy optimal solution of the LPP with fuzzy parameters of type-2 (3.64) is obtained as follows:

$$
\tilde{x}_B = \begin{pmatrix} \tilde{x}_1 \\ \tilde{x}_2 \end{pmatrix} = B^{-1}\tilde{b} = \begin{bmatrix} \tfrac{-4}{7} & \tfrac{5}{7} \\ \tfrac{3}{7} & \tfrac{-2}{7} \end{bmatrix} \begin{bmatrix} (3,5,8,13) \\ (4,6,10,16) \end{bmatrix} \begin{bmatrix} \tfrac{-32}{7} & \tfrac{-2}{7} & \tfrac{30}{7} & \tfrac{68}{7} \\ \tfrac{-23}{7} & \tfrac{-5}{7} & \tfrac{12}{7} & \tfrac{31}{7} \end{bmatrix}
\tag{3.81}
$$

As shown here the solution given in Eq. (3.81) obtained by the approach proposed by Ebrahimnejad and Tavana [22] is matched with the solution given in Eq. (3.68) derived from Maleki et al.'s method [2] and the solution given in Eq. (3.78) derived from Mahdavi-Amiri and Nasseri's method [13]. However, the method proposed in this subsection is far simpler and computationally more efficient than the methods proposed by Maleki et al. [2] and Mahdavi-Amiri and Nasseri [13].

Example 3.10 Find the fuzzy optimal solution of the following LPP with fuzzy parameters of type-2:

$$
\begin{aligned}
Max\, \tilde{u} = {}& 4\tilde{w}_1 + 6\tilde{w}_2 \\
s.t. \quad & 3\tilde{w}_1 + 4\tilde{w}_2 \preceq (2,4,6,12) \\
& 2\tilde{w}_1 + \tilde{w}_2 \preceq (1,2,3,6) \\
& \tilde{w}_1, \tilde{w}_2 \tilde{0}
\end{aligned}
\tag{3.82}
$$

In order to solve the LPP with fuzzy parameters of type-2 (3.82) using Maleki et al.'s method [2], we should solve the following fuzzy auxiliary problem:

$$
\begin{aligned}
Min\,\tilde{z} &= (2,4,6,12)x_1 + (1,2,3,4)x_2 \\
s.t. \quad & 3x_1 + 2x_2 \geq 4 \\
& 4x_1 + x_2 \geq 6 \\
& x_1, x_2 \geq 0
\end{aligned}
\tag{3.83}
$$

If we want to solve the fuzzy auxiliary problem (3.83) by using Maleki et al.'s method [2], we must first solve the following LP problem by introducing the surplus variables x_3 and x_4, and the artificial variables x_5 and x_6. This problem minimizes the sum of the artificial variables over the feasible space of the LP problem (3.82) in order to obtain an initial basic feasible solution:

$$
\begin{aligned}
Min \quad \tilde{z} &= (1,1,1,1)x_5 + (1,1,1,1)x_6 \\
s.t. \quad & 3x_1 + 2x_2 - x_3 + x_5 = 4 \\
& 4x_1 + x_2 - x_4 + x_6 = 6 \\
& x_1, x_2, x_3, x_4, x_5, x_6 \geq 0
\end{aligned}
\tag{3.84}
$$

After finding an initial fuzzy basic feasible solution by solving the problem (3.84) based on Algorithm 3.1, we must minimize the original objective function of the problem (3.83). Finally, with regard to the relation between the fuzzy auxiliary problem (3.83) and the LPP with fuzzy parameters of type-2 (3.84), we can find the optimal solution of the problem (3.82). This process is therefore time consuming and is not computationally efficient. However, this approach gives the basis $B =$

$[a_1, a_2] = \begin{bmatrix} 3 & 2 \\ 4 & 1 \end{bmatrix}$ as the optimal basis of problem (3.83). Thus, the fuzzy optimal solution of the LPP with fuzzy parameters of type-2 (3.82) is achieved as follows, based on the formulation $\tilde{w} = \tilde{c}_B B^{-1}$:

$$
\begin{aligned}
\tilde{w} = (\tilde{w}_1, \tilde{w}_2) &= ((2,4,6,12),(1,2,3,6)) \begin{bmatrix} \frac{-1}{5} & \frac{2}{5} \\ \frac{4}{5} & \frac{-3}{5} \end{bmatrix} \\
&= \left(\left(\frac{-8}{5}, \frac{2}{5}, \frac{8}{5}, \frac{24}{5} \right), \left(\frac{-14}{5}, \frac{-1}{5}, \frac{6}{5}, \frac{21}{5} \right) \right)
\end{aligned}
\tag{3.85}
$$

It is worth noting that the LPP with fuzzy parameters of type-2 (3.82) cannot be solved by the fuzzy dual approach proposed by Mahdavi-Amiri and Nasseri [13], because an initial feasible dual solution with identity basis is not at hand.

Now, to find the fuzzy optimal solution of the LPP with fuzzy parameters of type-2 (3.82) based on the approach proposed by Ebrahimnejad and Tavana [22], we first rewrite the LPP with fuzzy parameters of type-2 (3.82) with respect to the change of all the parameters and variables as follows:

$$Max\, \tilde{u} = 4\tilde{x}_1 + 6\tilde{x}_2$$
$$s.t. \quad 3\tilde{x}_1 + 4\tilde{x}_2 \preceq (2, 4, 6, 12)$$
$$2\tilde{x}_1 + \tilde{x}_2 \preceq (1, 2, 3, 6) \quad\quad (3.86)$$
$$\tilde{x}_1, \tilde{x}_2 \succeq 0$$

We then substitute the rank order of each fuzzy number for its corresponding fuzzy number in the LPP with fuzzy parameters of type-2 (3.86) to obtain the following crisp problem:

$$Max\, z = 4x_1 + 6x_2$$
$$s.t. \quad\quad 3x_1 + 4x_2 \le 6$$
$$2x_1 + x_2 \le 3 \quad\quad (3.87)$$
$$x_1, x_2 \ge 0$$

We then construct the standard form of the LP problem (3.86) as follows where x_3 and x_4 are slack variables:

$$Max\, z = 4x_1 + 6x_2$$
$$s.t. \quad\quad 3x_1 + 4x_2 + x_3 = 6$$
$$2x_1 + x_2 + x_4 = 6 \quad\quad (3.88)$$
$$x_1, x_2, x_3, x_4 \ge 0$$

The problem (3.88) is a crisp LP problem and can be solved using the standard primal simplex method. By solving the LP problem (3.88) based on the classical primal simplex algorithm, we obtain the basis $B = [a_1, a_2] = \begin{bmatrix} 3 & 4 \\ 2 & 1 \end{bmatrix}$ as the optimal basis of both the LP problem (3.88) and the LPP with fuzzy parameters of type-2 (3.86) or (3.82). Thus, the fuzzy optimal solution of the LPP with fuzzy parameters of type-2 (3.86) is obtained as follows:

$$\tilde{x}_B = \begin{pmatrix} \tilde{x}_1 \\ \tilde{x}_2 \end{pmatrix} = B^{-1}\tilde{b} \begin{bmatrix} \frac{-1}{5} & \frac{4}{5} \\ \frac{2}{5} & \frac{-3}{5} \end{bmatrix} \begin{bmatrix} (2, 4, 6, 12) \\ (1, 2, 3, 6) \end{bmatrix} = \begin{bmatrix} (\frac{-8}{5}, \frac{2}{5}, \frac{8}{5}, \frac{24}{5}) \\ (\frac{-14}{5}, \frac{-1}{5}, \frac{6}{5}, \frac{21}{5}) \end{bmatrix} \quad (3.89)$$

As shown here, the solution given in Eq. (3.89) is equivalent to the optimal solution (3.85) derived from Maleki et al.'s method [2]. However, it is much easier to apply the proposed method as compared to this existing method.

The overall results of the proposed approaches for solving an LP with fuzzy parameters of type-2 confirm that: (1) In order to solve an LPP with fuzzy parameters of type-2 according to the method proposed by Maleki et al. [2] is required to find the optimal solution for an auxiliary problem with fuzzy cost coefficients (LPP with fuzzy parameters of type-1). (2) For solving the corresponding auxiliary problem according to Algorithm 3.1 it is necessary to solve a fuzzy system with all the iterations. (3) Although Algorithm 3.3 proposed by Mahdavi-Amiri and Nasseri [13] gives the fuzzy optimal solution problem of an LP with fuzzy parameters of type-2 directly on the primary problem and without solving any auxiliary problem, it is also necessary to solve a fuzzy system for all the iterations. (4) Both Algorithm 3.1 and Algorithm 3.3 proposed by Maleki et al. [2] and Mahdavi-Amiri and Nasseri [13], respectively, require a large number of fuzzy additions, subtractions and comparisons on the fuzzy numbers for all the iterations. (5) If we want to solve an LPP with fuzzy parameters of type-2, based on a ranking function, then the approach proposed by Ebrahimnejad and Tavana [22] is more effective than the existing methods [2, 13, 39].

3.4.4 Bounded LP with Fuzzy Parameters of Type-2

In previous subsections, various approaches have been used to study the solution of the LPP with fuzzy parameters of type-2 by introducing and solving certain auxiliary problems, or directly by the use of a dual simplex method based on a certain linear ranking function. But these methods are not applicable for situations in which some or all the variables are restricted to lie within fuzzy lower and fuzzy upper bounds. In this subsection, as a natural extension of the results in crisp LP problems and based on a certain linear ranking function, the fuzzy bounded dual approach proposed by Ebrahimnjead [21] is explored for solving bounded LPPs with fuzzy parameters of type-2.

In a bounded LPP with fuzzy parameters of type-2, we have to optimize the objective function subject to fuzzy linear constraints and bounded restrictions on the fuzzy variables. Such a problem will have the following form:

$$\begin{aligned} \min \quad & \tilde{z} = c\tilde{x} \\ s.t. \quad & A\tilde{x} = \tilde{b} \\ & \tilde{l} \preceq \tilde{x} \preceq \tilde{u} \end{aligned} \tag{3.90}$$

where $\tilde{b} \in (F(\mathbb{R}))^m$, $c \in \mathbb{R}^n$, $\tilde{l}, \tilde{u} \in (F(\mathbb{R}))^n$ and $A \in \mathbb{R}^{m \times n}$ are given matrices and $\tilde{x} \in (F(\mathbb{R}))^n$ is to be determined.

Definition 3.10 Any fuzzy vector $\tilde{x} \in (F(\mathbb{R}))^n$ which satisfies the constraints and bounded restrictions of (3.90) is said to be a fuzzy feasible solution.

Definition 3.11 A fuzzy feasible solution $\tilde{x}^*$ is said to be a fuzzy optimal solution to (3.90) if $c\tilde{x}^* \preceq c\tilde{x}$ for any aribitrary fuzzy feasible solution.

Definition 3.12 A fuzzy feasible solution $\tilde{\tilde{x}}$ to the equations $A\tilde{x} \simeq \tilde{b}$ is a fuzzy basic feasible solution (BFS) of (3.90) if the set $\{a_j : j \in J\}$ is linearly independent, where $J = \{j : \tilde{l}_j \prec \tilde{\tilde{x}}_j \prec \tilde{u}_j\}$.

Remark 3.3 Based on Definition 3.12, corresponding to every fuzzy basic feasible solution of (3.90), matrix A can be partitioned into $[B, N_1, N_2]$ where the matrix B has rank m such that with $\tilde{x}$ partitioned accordingly as $(\tilde{x}_B, \tilde{x}_{N_1}, \tilde{x}_{N_2})$, we have $\tilde{x}_{N_1} \simeq \tilde{l}_{N_1}, \tilde{x}_{N_2} \simeq \tilde{u}_{N_2}$ and accordingly $\tilde{x}_B \simeq B^{-1}\tilde{b} - B^{-1}N_1\tilde{l}_{N_1} - B^{-1}N_2\tilde{u}_{N_2}$. The matrix B is called the (working) basis, $\tilde{x}_B$ are the basic variables, and $\tilde{x}_{N_1}$ and $\tilde{x}_{N_2}$ are the non-basic variables at their lower and upper bounds, respectively.

Without loss of generality suppose that a basis $B = [a_1, a_2, \ldots, a_m]$ exists and suppose the non-basic matrix is decomposed into N_1 and N_2, that is, $A = [B, N_1, N_2]$. Accordingly, the vectors $\tilde{x}$ and c are decomposed into $(\tilde{x}_B, \tilde{x}_{N_1}, \tilde{x}_{N_2})$ and (c_B, c_{N_1}, c_{N_2}), respectively. The fuzzy basic variables can be represented in terms of the fuzzy non-basic variables $\tilde{x}_{N_1}$ and $\tilde{x}_{N_2}$ as follows:

$$A\tilde{x} = \tilde{b} \Leftrightarrow B\tilde{x}_B + N_1\tilde{x}_{N_1} + N_2\tilde{x}_{N_2} = \tilde{b} \tag{3.91}$$

Thus, we have:

$$\tilde{x}_B = B^{-1}\tilde{b} - B^{-1}N_1\tilde{x}_{N_1} - B^{-1}N_2\tilde{x}_{N_2} \tag{3.92}$$

Also, the objective function can be represented in terms of the fuzzy non-basic variables $\tilde{x}_{N_1}$ and $\tilde{x}_{N_2}$ as follows:

$$\tilde{z} = c\tilde{x} = c_B\tilde{x}_B + c_{N_1}\tilde{x}_{N_1} + c_{N_2}\tilde{x}_{N_2}, \tag{3.93}$$

or

$$\tilde{z} = c_B(B^{-1}\tilde{b} - B^{-1}N_1\tilde{x}_{N_1} - B^{-1}N_2\tilde{x}_{N_2}) + c_{N_1}\tilde{x}_{N_1} + c_{N_2}\tilde{x}_{N_2}$$

Therefore, we have:

$$\tilde{z} = c_B B^{-1}\tilde{b} + (c_{N_1} - c_B B^{-1}N_1)\tilde{x}_{N_1} + (c_{N_2} - c_B - B^{-1}N_2)\tilde{x}_{N_2} \tag{3.94}$$

Since B is a basic matrix corresponding to A thus every non-basic column a_j can be written as a linear combination of columns of B, that is, $a_j = By_j$ or $y_j = B^{-1}a_j$. Define the fuzzy variable $\tilde{z} = \tilde{c}_B B^{-1}a_j = \tilde{c}_B y_j$. Now from Eq. (3.94), we obtain

$$\tilde{z} = c_B B^{-1}\tilde{b} - \sum_{j \in J_{N_1}} (z_j - c_j)\tilde{x}_j - \sum_{j \in J_{N_2}} (z_j - c_j)\tilde{x}_j \tag{3.95}$$

where J_{N_1} and J_{N_2} are the set of indices of non-basic variables at their lower and upper bounds, respectively.

Now, suppose that a current fuzzy basic solution to (3.90) is given by

$$\begin{pmatrix} \tilde{x}_B \\ \tilde{x}_{N_1} \\ \tilde{x}_{N_2} \end{pmatrix} = \begin{pmatrix} \tilde{\hat{b}} = B^{-1}\tilde{b} - B^{-1}N_1\tilde{l}_{N_1} - B^{-1}N_2\tilde{u}_{N_2} \\ \tilde{l}_{N_1} \\ \tilde{u}_{N_2} \end{pmatrix} \tag{3.96}$$

The fuzzy value of the objective function for this solution is

$$\tilde{z} = \sum_{i=1}^{m} c_{B_i}\tilde{\hat{b}}_i + \sum_{j \in R_1} c_j \tilde{l}_j + \sum_{j \in R_2} c_j \tilde{u}_j \tag{3.97}$$

The following theorem gives the optimality conditions for problem (3.90).

Theorem 3.23 (Optimality conditions) *If we have a fuzzy basic feasible solution as (3.96) such that $z_j - c_j \leq 0$ for all nonbasic variables $\tilde{x}_j$ at their lower bounds and $z_j - c_j \leq 0$ for all nonbasic variables $\tilde{x}_j$ at their upper bounds, then the current fuzzy basic feasible solution is optimal.*

Proof Suppose that the fuzzy value of only one fuzzy non-basic variable is modified while all other non-basic variables are fixed. Without loss of generality, suppose $\tilde{x}_k (k \in J_{N_1})$ is such a non-basic fuzzy variable. Let $\tilde{x}_k = \tilde{l}_k + \tilde{\Delta}_k$ where $\tilde{\Delta}_k$ is the fuzzy increase in $\tilde{x}_k$. On substituting $\tilde{x}_k = \tilde{l}_k + \tilde{\Delta}_k$ into (3.95), we obtain

$$\begin{aligned} \tilde{z}_{new} &= c_B B^{-1}\tilde{b} - \sum_{j \in J_{N_1}, j \neq k} (z_j - c_j)\tilde{x}_j - \sum_{j \in J_{N_2}} (z_j - c_j)\tilde{x}_j - (z_k - c_k)(\tilde{l}_k + \tilde{\Delta}_k) \\ &= c_B B^{-1}\tilde{b} - \sum_{j \in J_{N_1}} (z_j - c_j)\tilde{l}_j - \sum_{j \in J_{N_2}} (z_j - c_j)\tilde{u}_j - (z_k - c_k)\tilde{\Delta}_k \\ &= \tilde{z} - (z_k - c_k)\tilde{\Delta}_k \end{aligned}$$

$$(3.98)$$

Since $k \in J_{N_1}$ and $z_k - c_k \leq 0$, we conclude that $\tilde{z}_{new} \succeq \tilde{z}$ by (3.98). This means that in this case it is impossible to improve the fuzzy value of the objective function and then the current solution is optimal. In a similar way, if $k \in J_{N_2}$, then substituting $\tilde{x}_k = \tilde{u}_k - \tilde{\Delta}_k$, where $\tilde{\Delta}_k$ is the fuzzy decrease in $\tilde{x}_k$, into (3.95) we obtain $\tilde{z}_{new} = \tilde{z} + (z_k - c_k)\tilde{\Delta}_k$. Since $z_k - c_k \geq 0$, this concludes $\tilde{z}_{new} \succeq \tilde{z}$ and hence in this case the current solution is optimal too. □

Suppose that we have a working basis $B = [a_1, a_2, \ldots, a_m]$ for (3.90). Let $\tilde{x}_B = (\tilde{x}_1, \tilde{x}_2, \ldots, \tilde{x}_m)$ and $c_B = (c_1, c_2, \ldots, c_m)$ be the fuzzy basic vector and basic cost vector, respectively. Let J_N be the set of indices of current non-basic variables.

Define a solution $\tilde{\tilde{x}} = (\tilde{x}_j)$ for $A\tilde{x} \simeq \tilde{b}$ as follows:

Table 3.15 The initial fuzzy bounded dual tableau

Basic	$\tilde{x}_B$	$\tilde{x}_N$	R.H.S
z	0	$z_N - c_N$	$\tilde{\tilde{z}} = \bar{b} + \lambda \bar{b}'$
$\tilde{x}_B$	I	Y_N	$\bar{b} - \sum_{j \in J_N} \bar{y}_j \tilde{x}_j$

$$\begin{aligned}
\tilde{x}_j = \tilde{x}_j = \tilde{l}_j \quad &\text{if } j \in J_N, z_j - c_j < 0 \\
\tilde{x}_j = \tilde{x}_j = \tilde{u}_j \quad &\text{if } j \in J_N, z_j - c_j \geq 0 \\
\tilde{x}_i = \tilde{x}_i = \bar{b}_i - \sum_{j \in J_N} y_{ij} \tilde{x}_j \quad &i = 1, 2, \ldots, m
\end{aligned} \tag{3.99}$$

The fuzzy objective function value for the solution (3.99) is given as follows:

$$\tilde{z} = \tilde{\tilde{z}} \simeq \sum_{i=1}^{m} c_i \tilde{x}_i + \sum_{j \in J_N} c_j \tilde{x}_j \tag{3.100}$$

Let Table 3.15 be the initial simplex tableau corresponding to the basis B.

From the manner that the values of fuzzy non-basic variables are defined in (3.99), it is clear that this fuzzy solution satisfies the dual feasibility (or equivalent, the optimality conditions), but it may be infeasible to (3.90) since the fuzzy value of the basic variable in (3.100) may violate the bounded restrictions $\tilde{l}_i \preceq \tilde{x}_i \preceq \tilde{u}_i$. Similar to this solution, all the solution obtained during this algorithm satisfies the optimality conditions. Hence, the moment a solution obtained during this algorithm satisfies the primal feasibility criterion (specially the bounded restrictions), then that solution is an optimal solution for (3.90) and the method terminates. If the current solution does not satisfy the primal feasibility criterion, the algorithm is moved to the pivot row choose step.

Definition 3.13 Row i is said to be a pivot row if the value of the i th basic variable doesn't satisfy the bounded restrictions, i.e $\tilde{\tilde{x}}_i \prec \tilde{l}_i$ or $\tilde{\tilde{x}} \succ \tilde{u}_i$.

Now one pivot row is selected and then it is checked whether the problem (3.90) is infeasible or not using the following theorem.

Theorem 3.24 *Let row r be the pivot row and $y_r = (y_{r1}, y_{r2}, \ldots, y_{rm})$. The problem (3.90) is infeasible if at least one of the following conditions holds.*

(i) *If $\tilde{\tilde{x}}_r \prec \tilde{l}_r$ then $y_{rj} \geq 0$ for all $j \in J_N$ satisfying $\tilde{\tilde{x}}_j = \tilde{l}_j$ and $y_{rj} \leq 0$ for all $j \in J_N$ satisfying $\tilde{\tilde{x}}_j = \tilde{u}_j$.*

(ii) *If $\tilde{\tilde{x}}_r \succ \tilde{u}_r$ then $y_{rj} \leq 0$ for all $j \in J_N$ satisfying $\tilde{\tilde{x}}_j = \tilde{l}_j$ and $y_{rj} \geq 0$ for all $j \in R$ satisfying $\tilde{\tilde{x}}_j = \tilde{u}_j$.*

Proof Since $\tilde{x}_{B_r} = \tilde{\tilde{x}}_r$ is the rth basic variable, we have $y_{rj} = 0$ for $j \in \{1, 2, \ldots, r - 1, r+1, \ldots m\}$ and $y_{rr} = 1$. So, the constraint corresponding to the present pivot row is given by (3.101) in which J_{N_1} and J_{N_2} are the set of indices of non-basic variables at their lower and upper bounds respectively:

$$\tilde{x}_{B_r} = \tilde{\tilde{x}}_r = \tilde{\tilde{b}}_r - \sum_{j \in J_{N_1}} y_{rj}\tilde{\tilde{x}}_j - \sum_{j \in J_{N_2}} y_{rj}\tilde{\tilde{x}}_j \qquad (3.101)$$

Now if (i) holds, then the $\tilde{\tilde{x}}_r$ should be increased to improve the feasibility of the problem (3.90). To this end, it is required to modify the fuzzy value of only one fuzzy non-basic variable while all the other nonbasic variables are fixed. Suppose k is the index of such variable. If $k \in J_{N_1}$, then $\tilde{x}_k$ is at lower bound and we have to increase it. Let $\tilde{\tilde{x}}_k = \tilde{l}_k + \tilde{\Delta}_k$, where $\tilde{\Delta}_k \succ \tilde{0}$ is the increase in $\tilde{\tilde{x}}_k$. Substituting $\tilde{\tilde{x}}_k = \tilde{l}_k + \tilde{\Delta}_k$ in (3.101) gives the new fuzzy value of $\tilde{\tilde{x}}_r$ as follows:

$$\left(\tilde{x}_{B_r}\right)_{new} = \tilde{\tilde{b}}_r - \sum_{J \neq k, j \in J_{N_1}} y_{rj}\tilde{\tilde{x}}_j - \sum_{j \in J_{N_2}} y_{rj}\tilde{\tilde{x}}_j - y_{rk}(\tilde{l}_k + \tilde{\Delta}_k) = \tilde{x}_{B_r} - y_{rk}\tilde{\Delta}_k \quad (3.102)$$

Noting that $y_{rk} \geq 0$ and $\tilde{\Delta}_k \succ \tilde{0}$, from (3.102) we have $\left(\tilde{x}_{B_r}\right)_{new} \preceq \tilde{x}_{B_r}$ and thus the constraint $\tilde{\tilde{x}}_j \preceq \tilde{l}_r$ can never be satisfied in the problem (3.90).

In addition, if $k \in R_2$, then $\tilde{x}_k$ is at upper bound and we have to decrease it. Let $\tilde{\tilde{x}}_k = \tilde{u}_k + \tilde{\Delta}_k$, where $\tilde{\Delta}_k \succ \tilde{0}$ is the decrease in $\tilde{\tilde{x}}_k$. Substituting $\tilde{\tilde{x}}_k = \tilde{u}_r - \tilde{\Delta}_k$ in (3.101) gives the new fuzzy value of $\tilde{\tilde{x}}_r$ as follows:

$$\left(\tilde{x}_{B_r}\right)_{new} = \tilde{\tilde{b}}_r - \sum_{j \in J_{N_1}} y_{rj}\tilde{\tilde{x}}_j - \sum_{J \neq k, j \in J_{N_2}} y_{rj}\tilde{\tilde{x}}_j - y_{rk}(\tilde{u}_k - \tilde{\Delta}_k) = \tilde{x}_{B_r} + y_{rk}\tilde{\Delta}_k \quad (3.103)$$

Noting that $y_{rk} \leq 0$ and $\tilde{\Delta}_k \succ \tilde{0}$, from (3.103) we have $\left(\tilde{x}_{B_r}\right)_{new} \preceq \tilde{x}_{B_r}$ and thus the constraint $\tilde{\tilde{x}}_r \preceq \tilde{l}_r$ can never be satisfied in the problem (3.90). The case (ii) can be proved in a similar manner. □

If either condition (i) or (ii) holds, then the problem (3.90) is infeasible and the algorithm terminates. Otherwise, we select the r th basic variable in the current basic vector as an existing variable and go to the entering variable selection procedure.

Let r be the pivot row. The entering variable is selected according to a special procedure in order to satisfy the following properties:

a. The new solution obtained after the pivot step also satisfies the optimality conditions.
b. In the new solution, both the leaving variable $\tilde{x}_r$ and the entering variable satisfy the bounded restrictions.

Remark 3.4 It may be necessary to perform a series of pivots in immediate succession using the same row r as the pivot row until property (b) is satisfied. It needs to be noted that the effect of a series of pivots in the same row is the same as the effect of the last pivot in the series.

Remark 3.5 The special procedure allows us to perform the intermediate pivot steps updating the data in the pivot row and the fuzzy cost row only, and to update the inverse tableau only when the last pivot step in the series is reached.

In each step of the special procedure a tentative entering variable is selected such that it leads to a new solution satisfying properties (a) and (b), then the tentative entering variable in this step is declared as the final entering variable and the algorithm moves to the pivot step. Otherwise we move to the next step in the entering variable selection procedure itself. In all the steps of this special procedure the pivot row remains the same row (row r).

We note that each step of the special procedure replaces the current basic variable in the pivot row (row r) by a non-basic variable, and this continue until we reach a step in which either the updated pivot row establishes primal infallibility or the present solution satisfies properties (a) and (b). Therefore, the variables $\tilde{x}_i$ for $i \in \{1, 2, \ldots, r-1, r+1, \ldots, m\}$ remain as basic variables throughout this special procedure; their values in the solution are updated in each step of the procedure, but not used in executing the procedure (for more details see Murty [40]).

Now we are in a position to describe the general step in the special procedure. Let $\tilde{\hat{x}} = (\tilde{x}_j)$, $(\tilde{y}_{r1}, \ldots, \tilde{y}_m)$, $(\hat{z}_1 - \hat{c}_1, \ldots, \hat{z}_n - \hat{c}_n)$, $\hat{\tilde{b}}_r$, $\tilde{\hat{z}}$ and $\tilde{x}_\delta$ be the current fuzzy solution, updated pivot row (row r), updated cost row, updated r th right-hand side constant, fuzzy objective value and the fuzzy basic variable in the pivot row, respectively. So, the set of indices of current fuzzy non-basic variables is $\hat{J}_N = \{J_N \cup r\} \setminus \delta$. Since $\tilde{x}_\delta$ is the current basic variable in the pivot row, we have $\hat{y}_{r\delta} = 1$ and $\hat{y}_{rj} = 0$ for $j \in \{1, 2, \ldots, r-1, r+1, \ldots, m\}$. Thus, the constraint corresponding to the pivot row is as follows:

$$\tilde{x}_\delta + \sum_{j \in \hat{R}} \hat{y}_{ij}\tilde{x}_j = \hat{\tilde{b}}_r \tag{3.104}$$

or

$$\tilde{x}_\delta + \sum_{j \in \hat{J}_{N_1}} \hat{y}_{ij}\tilde{x}_j + \sum_{j \in \hat{J}_{N_2}} \hat{y}_{ij}\tilde{x}_j = \hat{\tilde{b}}_r \tag{3.105}$$

where $\hat{J}_N = \hat{J}_{N_1} \cup \hat{J}_{N_2}$ and $\hat{J}_{N_1}$ and $\hat{J}_{N_2}$ are the set of indices of current non-basic variables at their lower and upper bounds, respectively.

From (3.105) we see that if $\tilde{x}_\delta \prec \tilde{l}_\delta$, the value of $\tilde{x}_\delta$ can be increased by either increasing the value of a non-basic variable $\tilde{x}_j$ for $j \in \hat{J}_{N_1}$ and $\hat{y}_{rj} < 0$ or by decreasing the value of a non-basic $\tilde{x}_j$ for $j \in \hat{J}_{N_2}$ and $\hat{y}_{rj} > 0$. In this case the index t of the tentative entering variable is determined by the following test:

$$\frac{\hat{z}_t - \hat{c}_t}{\hat{y}_{rt}} = \min\left\{\min_{j \in \hat{J}_{N_1}}\left\{\frac{\hat{z}_j - \hat{c}_j}{\hat{y}_{rj}}; \hat{y}_{rj} < 0\right\}, \min_{j \in \hat{J}_{N_2}}\left\{\frac{\hat{z}_j - \hat{c}_j}{\hat{y}_{rj}}; \hat{y}_{rj} > 0\right\}\right\} \quad (3.106)$$

In a similar manner, if $\tilde{\tilde{x}}_\delta \succ \tilde{u}_\delta$, then from (3.105) the value of $\tilde{x}_\delta$ can be decreased by either increasing the value of a non-basic variable $\tilde{x}_j$ for $j \in \hat{J}_{N_1}$ and $\hat{y}_{rj} > 0$ or by decreasing the value of a non-basic $\tilde{x}_j$ for $j \in \hat{J}_{N_2}$ and $\hat{y}_{rj} > 0$. In this case the index t of the tentative entering variable is determined by the following test:

$$\frac{\hat{z}_t - \hat{c}_t}{\hat{y}_{rt}} = \max\left\{\max_{j \in \hat{J}_{N_1}}\left\{\frac{\hat{z}_j - \hat{c}_j}{\hat{y}_{rj}}; \hat{y}_{rj} > 0\right\}, \max_{j \in \hat{J}_{N_2}}\left\{\frac{\hat{z}_j - \hat{c}_j}{\hat{y}_{rj}}; \hat{y}_{rj} < 0\right\}\right\} \quad (3.107)$$

On this basis, the new fuzzy basis vector satisfies the optimality (dual feasibility) condition. We recall that in this special procedure we perform the intermediate pivot steps updating the data in the pivot row and the fuzzy cost row only, and update the inverse tableau only when the last pivot step in the series is reached. Thus in this case, $\bar{\bar{y}}_{rj} = \frac{\hat{y}_{rj}}{\bar{y}_{r\delta}} (j = 1, 2, \ldots, n)$, $\bar{\bar{b}}_r = \frac{\tilde{b}_r}{\bar{y}_{r\delta}}$ and $\bar{\bar{z}}_j - \bar{\bar{c}}_j = (\hat{z}_j - \hat{c}_j) - \bar{\bar{y}}_{rj}(\hat{z}_t - \hat{c}_t)$ are the updated entries in the pivot and cost rows after $\tilde{x}_t$ replaces $\tilde{x}_\delta$ as the basic variable in the pivot row r.

Let row r be the pivot row and $\tilde{x}_t$ be the tentative variable during the special procedure. The new fuzzy solution and new fuzzy objective value after pivoting will be as follows:

$$\tilde{x}'_j = \begin{cases} \tilde{\tilde{x}}_j & j \in \hat{R}\backslash\{t\} \\ \tilde{l}_\delta & j = \delta, \ \tilde{\tilde{x}}_\delta \prec \tilde{l}_\delta \\ \tilde{u}_\delta & j = \delta, \ \tilde{\tilde{x}}_\delta \succ \tilde{u}_\delta \\ \tilde{\tilde{x}}_t + \frac{\tilde{\tilde{x}}_\delta - \tilde{l}_\delta}{\hat{y}_{rt}} & j = t, \ \tilde{\tilde{x}}_\delta \prec \tilde{l}_\delta \\ \tilde{\tilde{x}}_t + \frac{\tilde{\tilde{x}}_\delta - \tilde{u}_\delta}{\hat{y}_{rt}} & j = t, \ \tilde{\tilde{x}}_\delta \succ \tilde{u}_\delta \end{cases} \quad (3.108)$$

$$\tilde{x}'_i = \tilde{\tilde{x}}_i + y_{it}(\tilde{\tilde{x}}_t - \tilde{x}'_t) \quad i = 1, 2, \ldots, r-1, r+1, \ldots, m \quad (3.109)$$

$$\tilde{z}' = \tilde{\tilde{z}} + (\hat{z}_t - \hat{c}_t)(\tilde{\tilde{x}}_t - \tilde{x}'_t) \quad (3.110)$$

Now if $\tilde{l} \leq \tilde{x}'_t \leq \tilde{u}_t$, the new solution obtained as a result of choosing $\tilde{x}_t$ as the entering variable in the pivot row satisfies both properties (a) and (b). Hence, $\tilde{x}_t$ is chosen as the final entering variable to replace $\tilde{x}_r$ from $\tilde{x}_B$ and we move to the pivot step. If either $\tilde{x}'_t \prec \tilde{l}_t$ or $\tilde{x}'_t \succ \tilde{u}_t$, then $\tilde{x}_t$ is only a tentative variable replacing $\tilde{x}_\delta$ as the basic variable in the pivot row. Then we check whether the primal infallibility conditions are satisfied by the present data in the pivot row and if so terminate. Otherwise, we go on to the next step in the special entering selection procedure.

Let $\tilde{x}_s$ be the final entering variable determined by the entering variable selection procedure. The pivot column is then $y_{rs} = B^{-1}a_{rs}$. The tableau except the right-hand side column is updated by pivoting at y_{rs}. The new solution is the one obtained in the final step of the entering variable selection procedure. We now check whether the new solution satisfies the primal feasibility conditions; otherwise we continue the algorithm in the same manner.

The steps of the bounded dual simplex method for solving the bounded LPP with fuzzy parameters of type-2 (3.90) are summarized as follow:

Step 1   Suppose B be a working basis. Define the solution corresponding to this basis as (3.99).

Step 2   If $\tilde{l}_i \preceq \tilde{x}_i \preceq \tilde{u}_i$ for each $i \in \{1, 2, \ldots, m\}$, then the current solution is optimal. Otherwsie, determine the index r of a pivot row by Definition 3.13.

Step 3   If at least one of the conditions (i) or (ii) in Theorem 3.24 hold, then stop; the problem (3.90) is infeasible. Otherwise, select the index t of a tentative entering variable by test (3.106) or (3.107) depending on the current basic variable being strictly less than its lower bound or strictly greater than its upper bound, respectively.

Step 4   Update entries in the pivot and cost rows by use of pivot element y_{rt} and obtain the new solution as (3.108) and (3.109) with fuzzy objective value as (3.110).

Step 5   If the current tentative entering variable doesn ot satisfy the bounded restriction, go to Step 3. Otherwise, let $\tilde{x}_s$ be the final entering variable.

Step 6   Update the tableau except the right-hand side column by pivoting at y_{rs}. Repeat Step 2.

Example 3.11 A farmer who raises chickens would like to determine the amounts of the available ingredients that would meet certain nutritional requirements. The available ingredients, their costs and the nutrients in the ingredients are summarized in Table 3.16.

The minimum daily requirement of nutrients protein and vitamins are approximately 200 and 120 units, respectively. In addition, this special mix should include at least close to 10 units of corn and at most close to 20 units of corn; at least close to 20 units of lime and at most close to 40 units of lime; at least close to 10 units of alfalfa and at most close to 20 units of alfalfa; and at least close to 10 units of soya

Table 3.16 Data of Example 3.11

Ingredient				
Nutrient	Core	Lime	Alfalfa	Soya
Protein	1	2	1	4
Vitamins	3	1	2	2
Cost	3	2	4	5

and at most close to 30 units of soya. The farmer wishes to find the daily minimum-cost mix.

This problem is evidently an uncertain optimization problem due to variations in the minimum daily requirements. So the amount of each unit of ingredients will be uncertain. Hence, we will model the problem as an FLP problem. We use trapezoidal fuzzy numbers for each uncertain value. The minimum daily requirement of nutrients protein and vitamins which are approximately 200 and 120 units, are modeled as $(195, 196, 202, 207)$ and $(116, 118, 120, 126)$ respectively. In a similar way, the other parameters are also modeled as trapezoidal fuzzy numbers taking into account the nature of the problem and the other requirements. So the problem is formulated as follows:

$$\min \quad \tilde{z} = 3\tilde{x}_1 + 2\tilde{x}_2 + 4\tilde{x}_3 + 5\tilde{x}_4$$
$$\text{s.t.} \quad \tilde{x}_1 + 2\tilde{x}_2 + \tilde{x}_3 + 4\tilde{x}_4 \succeq (195, 196, 202, 207)$$
$$3\tilde{x}_1 + \tilde{x}_2 + 2\tilde{x}_3 + 2\tilde{x}_4 \succeq (116, 118, 120, 126)$$
$$(7, 8, 12, 13) \preceq \tilde{x}_1 \preceq (16, 18, 22, 24)$$
$$(16, 18, 22, 24) \preceq \tilde{x}_2 \preceq (34, 36, 42, 48)$$
$$(7, 8, 12, 13) \preceq \tilde{x}_3 \preceq (16, 18, 22, 24)$$
$$(7, 8, 12, 13) \preceq \tilde{x}_4 \preceq (25, 26, 32, 37)$$

We first multiply both sides of the first and second constraints by -1 and then introduce the fuzzy slack variables $\tilde{x}_5$ and $\tilde{x}_6$. Thus, the problem reduces to the following standard form:

$$\min \tilde{z} = 3\tilde{x}_1 + 2\tilde{x}_2 + 4\tilde{x}_3 + 5\tilde{x}_4$$
$$\text{s.t.} \quad -\tilde{x}_1 - 2\tilde{x}_2 - \tilde{x}_3 - 4\tilde{x}_4 + \tilde{x}_5 = (-207, -202, -196, -195)$$
$$-3\tilde{x}_1 - \tilde{x}_2 - 2\tilde{x}_3 - 2\tilde{x}_4 + \tilde{x}_6 = (-126, -120, -118, -116)$$
$$(7, 8, 12, 13) \preceq \tilde{x}_1 \preceq (16, 18, 22, 24)$$
$$(16, 18, 22, 24) \preceq \tilde{x}_2 \preceq (34, 36, 42, 48)$$
$$(7, 8, 12, 13) \preceq \tilde{x}_3 \preceq (16, 18, 22, 24)$$
$$(7, 8, 12, 13) \preceq \tilde{x}_4 \preceq (25, 26, 32, 37)$$
$$\tilde{x}_5, \tilde{x}_6 \succeq \tilde{0}$$

Now we use the proposed algorithm to obtain the fuzzy optimal solution of the above bounded LPP with fuzzy parameters of type-2.

Iteration 1, Step 1:

Suppose $B = [a_5, a_6] = \begin{bmatrix} 1 & 0 \\ 0 & 1 \end{bmatrix}$ is the initial working basis. We first compute the fuzzy values of $z_j - c_j$ for all $j \in J_N = \{1, 2, 3, 4\}$ and then define the initial solution corresponding to this basis as (3.99):

Table 3.17 The first tableau

Basis	$\tilde{x}_1$ l	$\tilde{x}_2$ l	$\tilde{x}_3$ l	$\tilde{x}_4$ l	$\tilde{x}_5$	$\tilde{x}_6$	R.H.S.
z	-3	-2	-4	-5	0	0	$(116, 132, 188, 204)$
$\tilde{x}_5$	-1	-2	-1	-4	1	0	$(-133, -118, -80, -69)$
$\tilde{x}_6$	-3	-1	-2	-2	0	1	$(-61, -46, -12, -1)$

$$z_1 - c_1 = -3, z_2 - c_2 = -2, z_3 - c_3 = -4, z_4 - c_4 = -5$$

Since $z_j - c_j \leq 0$ for all $j \in J_N = \{1, 2, 3, 4\}$, the non-basic variables $\tilde{x}_1, \tilde{x}_2, \tilde{x}_3$ and $\tilde{x}_4$ are at their lower bounds as $\tilde{x}_1 = (7, 8, 12, 13), \tilde{x}_2 = (16, 18, 22, 24), \tilde{x}_3 = (7, 8, 12, 13)$ and $\tilde{x}_3 = (7, 8, 12, 13)$. Thus (3.99) implies that the fuzzy values of basic variables $\tilde{x}_5$ and $\tilde{x}_6$ and also the fuzzy value of the objective function are given by $\tilde{x}_5 = (-123, -118, -80, -69), \tilde{x}_6 = (-31, -46, -12, -1)$ and $\tilde{z} = (116, 132, 188, 204)$.

We may form the simplex tableau corresponding to this solution as Table 3.17.

It will be helpful to distinguish between non-basic variables at their lower and upper bounds during the proposed algorithm iterations. This is done by flagging the corresponding columns by l and u, respectively.

Step 2:

We note that $\tilde{x}_{B_1} = \tilde{x}_5 = (-133, -118, -80, -69) \prec \tilde{0}$ and $\tilde{x}_{B_2} = \tilde{x}_6 = (-61, -46, -12, -1) \prec \tilde{0}$. So the current solution does not satisfy the bounded restrictions and we choose the first row as the pivot row.

Step 3:

Now we use the test (3.106) to determine the tentative entering variable as follows:

$$\min\left\{\frac{-3}{-1}, \frac{-2}{-2}, \frac{-4}{-1}, \frac{-5}{-4}\right\} = \min\left\{3, 1, 4, \frac{5}{4}\right\} = 1 = \frac{z_2 - c_2}{y_{12}}$$

This means that $\tilde{x}_2$ is selected as tentative entering variable. Then $\tilde{x}_2$ enters and $\tilde{x}_5$ leaves.

Step 4:

In this step we first update entries in the pivot and cost rows by pivoting at $y_{12} = -2$. Then we update the values of the tentative entering variable, the leaving variable, the basic variable and the objective function.

The variable $\tilde{x}_5$ leaves the basis and increases to its lower bound, i.e. $\tilde{x}_5 = \tilde{0}$. Also, the proposed algorithm only modifies the fuzzy value of the current non-basic variable while all other fuzzy non-basic variables are fixed at their bounds. So, we have $\tilde{x}_1 = \tilde{x}_3 = \tilde{x}_4 = (7, 8, 12, 13)$.

Moreover, the variable $\tilde{x}_2$ enters and its fuzzy value is obtained as follows based on formulation (3.108):

$$\tilde{x}_2 = (16, 18, 22, 24) + \frac{(-133, -118, -80, -69) - (0, 0, 0, 0)}{-2}$$
$$= \left(\frac{101}{2}, 58, 81, \frac{181}{2}\right)$$

In addition, the basic variable $\tilde{x}_6$ is updated based on formulation (3.109) as follows:

$$\tilde{x}_6 = (-61, -46, -12, -1) + (-1)\left((16, 18, 22, 24) - \left(\frac{101}{2}, 58, 81, \frac{181}{2}\right)\right)$$
$$= \left(\frac{-69}{2}, -10, 51, \frac{147}{2}\right)$$

Finally, based on formulation (3.110) the new fuzzy objective function value is updated as follows:

$$\tilde{z} = (116, 132, 188, 204) + (-2)(16, 18, 22, 24) - \left(\frac{101}{2}, 58, 81, \frac{181}{2}\right))$$
$$= (169, 204, 314, 353)$$

These lead to Table 3.18.

Since the tentative entering variable $\tilde{x}_2$ does not satisfy the bounded restrictions, then we return to Step 3.

Iteration 2, Step 3:
Neither of the conditions (i) or (ii) in Theorem 3.24 hold. Then, we select the index t of a tentative entering variable by test (3.107) as follows, since the current basic variable is greater than its upper bound.

$$\max\left\{\frac{-2}{\frac{1}{2}}, \frac{-3}{\frac{1}{2}}, \frac{-1}{2}\right\} = \max\left\{-4, -6, -\frac{1}{2}\right\} = -\frac{1}{2} = \frac{z_4 - c_4}{y_{14}}$$

Table 3.18 The first iteration

Basis	l $\tilde{x}_1$	$\tilde{x}_2$	l $\tilde{x}_3$	l $\tilde{x}_4$	l $\tilde{x}_5$	$\tilde{x}_6$	R.H.S.
z	-3	-2	-4	-5	0	0	$(169, 204, 314, 353)$
$\tilde{x}_2$	$\frac{1}{2}$	1	$\frac{1}{2}$	2	$\frac{-1}{2}$	0	$\left(\frac{101}{2}, 58, 81, \frac{181}{2}\right)$
$\tilde{x}_6$	-3	-1	-2	-2	0	1	$\left(\frac{-69}{2}, -10, 51, \frac{147}{2}\right)$

This means that $\tilde{x}_4$ is selected as the entative entering variable. Then $\tilde{x}_4$ enters and $\tilde{x}_2$ leaves.

Step 4:
In this step we first update entries in the pivot and cost rows by pivoting at $y_{14} = 2$. Then we update the values of the tentative entering variable, the leaving variable, the basic variable and the objective function.

The variable $\tilde{x}_2$ leaves the basis and decreases to its upper bound, i.e. $\tilde{x}_2 = (34, 36, 42, 48)$. Also, as said before, the proposed algorithm only modifies the fuzzy value of the current non-basic variable while all other fuzzy non-basic variables are fixed at their bounds. So, we have $\tilde{x}_1 = \tilde{x}_3 = (7, 8, 12, 13), \tilde{x}_5 = (0, 0, 0, 0)$.

Moreover, the variable $\tilde{x}_4$ enters and its fuzzy value is obtained as follows based on formulation (3.108):

$$\tilde{x}_4 = (7, 8, 12, 13) + \frac{\left(\frac{101}{2}, 58, 81, \frac{181}{2}\right) - (34, 36, 42, 48)}{2} = \left(\frac{33}{4}, 16, \frac{69}{2}, \frac{165}{4}\right)$$

In addition, the basic variable $\tilde{x}_6$ is given based on formulation (3.109) as follows:

$$\tilde{x}_6 = \left(\frac{-69}{2}, -10, 51, \frac{147}{2}\right) + (-2)\left((7, 8, 12, 13) - \left(\frac{33}{4}, 16, \frac{69}{2}, \frac{165}{4}\right)\right)$$
$$= (-44, -2, 104, 142)$$

Finally, based on formulation (3.110) the new fuzzy objective function value is updated as follows:

$$\tilde{z} = (169, 204, 314, 353) + (-1)\left((16, 18, 22, 24) - \left(\frac{33}{4}, 16, \frac{69}{2}, \frac{165}{4}\right)\right)$$
$$= \left(\frac{657}{4}, 208, \frac{681}{2}, \frac{1549}{4}\right)$$

Thus, we obtain Table 3.19.

Step 5:
In ths case the tentative entering variable $\tilde{x}_4$ satisfies the bounded restrictions, then this variable will be the final entering variable and we go to Step 6.

Table 3.19 The second iteration

	l	u	l	l	l		
Basis	$\tilde{x}_1$	$\tilde{x}_2$	$\tilde{x}_3$	$\tilde{x}_4$	$\tilde{x}_5$	$\tilde{x}_6$	R.H.S.
z	$-\frac{7}{4}$	$\frac{-1}{2}$	$\frac{-11}{4}$	0	$\frac{-5}{2}$	0	$\left(\frac{657}{4}, 208, \frac{681}{2}, \frac{1549}{4}\right)$
$\tilde{x}_4$	$\frac{1}{4}$	$\frac{1}{2}$	$\frac{1}{4}$	1	$\frac{-1}{4}$	0	$\left(\frac{33}{4}, 16, \frac{69}{2}, \frac{165}{4}\right)$
$\tilde{x}_6$	-3	-1	-2	-2	0	1	$(-44, -2, 104, 142)$

Table 3.20 The final iteration

	l	u	l	l	l		
Basis	$\tilde{x}_1$	$\tilde{x}_2$	$\tilde{x}_3$	$\tilde{x}_4$	$\tilde{x}_5$	$\tilde{x}_6$	R.H.S.
z	$-\frac{7}{4}$	$\frac{-1}{2}$	$\frac{-11}{4}$	0	$\frac{-5}{2}$	0	$\left(\frac{657}{4}, 208, \frac{681}{2}, \frac{1549}{4}\right)$
$\tilde{x}_4$	$\frac{1}{4}$	$\frac{1}{2}$	$\frac{1}{4}$	1	$\frac{-1}{4}$	0	$\left(\frac{33}{4}, 16, \frac{69}{2}, \frac{165}{4}\right)$
$\tilde{x}_6$	$\frac{-5}{2}$	0	$-\frac{3}{2}$	0	$\frac{-1}{2}$	1	$(-44, -2, 104, 142)$

Step 6:

In this step, we update the tableaus except the right-hand side column by pivoting at $y_{14} = 2$ from Table 3.3. The new solution is that obtained in the final step of the entering variable selection procedure. Thus, we obtain Table 3.20.

We now check whether the new solution satisfies the primal feasibility conditions; otherwise we continue the algorithm in the same manner. Since both basic variables $\tilde{x}_4$ and $\tilde{x}_6$ satisfy the bounded restrictions, the current solution is optimal solution and the algorithm stops. The fuzzy optimal solution and objective value are as follows:

$$\tilde{x}_1 = \tilde{x}_3 = (7, 8, 12, 13), \tilde{x}_2 = (34, 36, 42, 48),$$
$$\tilde{x}_4 = \left(\frac{33}{4}, 16, \frac{69}{2}, \frac{165}{4}\right), \tilde{z} = \left(\frac{657}{4}, 208, \frac{681}{2}, \frac{1549}{4}\right)$$

In addition, the membership function of the constraints and the objective function can be formulated as:

$$\mu_{\tilde{x}_1}(x) = \mu_{\tilde{x}_3}(x) = \begin{cases} x - 7, & \text{for} \quad 7 \le x \le 8, \\ 1, & \text{for} \quad 8 \le x \le 12, \\ 13 - x, & \text{for} \quad 12 \le x \le 13, \\ 0, & \text{else.} \end{cases}$$

$$\mu_{\tilde{x}_2}(x) = \begin{cases} \frac{x-34}{2}, & \text{for} \quad 34 \le x \le 36, \\ 1, & \text{for} \quad 36 \le x \le 42, \\ \frac{48-x}{6}, & \text{for} \quad 42 \le x \le 48, \\ 0, & \text{else.} \end{cases}$$

$$\mu_{\tilde{x}_4}(x) = \begin{cases} \frac{4x-33}{31}, & \text{for} \quad \frac{33}{4} \le x \le 16, \\ 1, & \text{for} \quad 16 \le x \le \frac{69}{2}, \\ \frac{165-4x}{27}, & \text{for} \quad \frac{69}{2} \le x \le \frac{165}{4}, \\ 0, & \text{else.} \end{cases}$$

$$\mu_{\tilde{z}}(x) = \begin{cases} \frac{4x-657}{175}, & \text{for} \quad \frac{657}{4} \le x \le 208, \\ 1, & \text{for} \quad 208 \le x \le \frac{681}{2}, \\ \frac{1549-4x}{187}, & \text{for} \quad \frac{681}{2} \le x \le \frac{1549}{4}, \\ 0, & \text{else.} \end{cases}$$

It is pointed out that the dual simplex method [13] can be used to obtain the fuzzy solution of the LPP with fuzzy parameters of type-2 (3.51) for which there is no restriction on decision variables. Thus, solving the bounded LPP with fuzzy parameters of type-2 (3.90) by use of the dual simplex method increases both the number of equality constraints and the number of variables and thus the problem size and computational effort would increase significantly. The advantage of the bounded dual simplex method over the dual simplex method [13] that it avoids this increased effort by removing the bounded restriction from the equality constraints and treating them separately, essentially like non-negativity constraints. Moreover, it is pointed out that the bounded primal simplex method [20] can be used to obtain the fuzzy solution of (3.90) when a primal fuzzy feasible basic solution be easily at hand. The bounded dual simplex method proposed here is useful when a dual feasible basic solution is easily at hand. The fuzzy bounded primal simplex algorithm proposed by Ebrahimnejad and Verdegay [20] and the fuzzy bounded dual simplex algorithm presented by Ebrahimnejad [21] would be useful for sensitivity analysis on the bounded LP with fuzzy parameters of type-2 (3.90).

3.5 LPP with Fuzzy Parameters of Type-3

The LP problem having symmetric trapezoidal fuzzy numbers for the costs coefficients, decision variables and the values of the right-hand side and having real numbers for the elements of the coefficient matrix is termed as the LPP with fuzzy parameters of type-3.

Definition 3.14 For any symmetric trapezoidal fuzzy number $\tilde{a}$, we define $\tilde{a} \succeq \tilde{0}$ if $a \geq 0$ and $\alpha \geq 0$ exist such that $\tilde{a} \succeq (a - \alpha, a, a, a + \alpha)$. We also denote $(a - \alpha, a, a, a + \alpha)$ by $\tilde{0}$. Note that $\tilde{0}$ is equivalent to $\tilde{0} = (0, 0, 0, 0)$, i.e. $(a - \alpha, a, a, a + \alpha) \approx (0, 0, 0, 0) = \tilde{0}$. It is easy to see that if $\tilde{a} \succeq \tilde{b}$, then $\tilde{a} - \tilde{b} \succeq \tilde{0}$.

Definition 3.15 If $\tilde{a} \approx \tilde{0}$, then $\tilde{a}$ is said to be a zero symmetric trapezoidal fuzzy number. If $\tilde{a}$ is not equivalent to $\tilde{0}$, i.e., $\tilde{a} \not\approx \tilde{0}$, then $\tilde{a}$ is said to be a non-zero symmetric trapezoidal fuzzy number. If $\tilde{a} \succeq \tilde{0}$ and $\tilde{a} \not\approx \tilde{0}$, then $\tilde{a}$ is said to be a positive symmetric trapezoidal fuzzy number and is denoted by $\tilde{a} \succ \tilde{0}$.

Definition 3.16 Let $\tilde{a} = (a_2 - \alpha, a_2, a_3, a_3 + \alpha)$ and $\tilde{b} = (b_2 - \beta, b_2, b_3, b_3 + \beta)$ be two symmetric trapezoidal fuzzy numbers. The new type of multiplication on $\tilde{a}$ and $\tilde{b}$ has been defined in [24] as follows:

$$\tilde{a}\tilde{b} \approx (p - w - \gamma, p - w, p + w, p + w + \gamma)$$

where

$$p = \left(\frac{a_2 + a_3}{2}\right)\left(\frac{b_2 + b_3}{2}\right), w = \frac{t_2 - t_1}{2}, \ t_1 = min\{a_2b_2, a_2b_3, a_3b_2, a_3b_3\},$$

$$t_2 = max\{a_2b_2, a_2b_3, a_3b_2, a_3b_3\}, \ \gamma = |a_3\beta + b_3\alpha|.$$

Definition 3.17 For any two symmetric trapezoidal fuzzy numbers $\tilde{a} = (a_2 - \alpha, a_2, a_3, a_3 + \alpha)$ and $\tilde{b} = (b_2 - \beta, b_2, b_3, b_3 + \beta)$ the relations $\precsim$ and $\approx$ have been defined in [24] as follows:

$$\tilde{a} \precsim \tilde{b} \Leftrightarrow \frac{a_2 + a_3}{2} \leq \frac{b_2 + b_3}{2}$$

$$\tilde{a} \approx \tilde{b} \Leftrightarrow \frac{a_2 + a_3}{2} = \frac{b_2 + b_3}{2}$$

Remark 3.6 It should be noted that the ordering defined in Definition 3.17 is then same as with the ordering defined in terms of the linear ranking function in Remark 1.15. This means that the linear ranking function given in Remark 1.15 for any symmetric trapezoidal fuzzy numbers $\tilde{a} = (a_2 - \alpha, a_2, a_3, a_3 + \alpha)$ is reduced to

$$\Re(\tilde{a}) = \frac{a_2 - \alpha + a_2 + a_3 + a_3 + \alpha}{4} = \frac{a_2 + a_3}{2}$$

A typical LPP with fuzzy parameters of type-3 is formulated as follows:

$$\max \tilde{z} \approx \tilde{c}\tilde{x}$$
$$s.t. \quad A\tilde{x} \approx \tilde{b}, \tag{3.111}$$
$$\tilde{x} \succeq \tilde{0}.$$

Definition 3.18 Any $\tilde{x} = (\tilde{x}_1, \tilde{x}_2, \ldots, \tilde{x}_n)$, where each $\tilde{x}_1$ is a symmetric trapezoidal fuzzy number, which satisfies the constraints and non-negativity restrictions of (3.111) is said to be a fuzzy feasible solution to (3.111).

Definition 3.19 A fuzzy feasible solution $\tilde{\tilde{x}}$ is said to be a fuzzy optimum solution to (3.111) if $\tilde{c}\tilde{\tilde{x}} \succeq \tilde{c}\tilde{x}$ for each arbitrary fuzzy feasible solution of (3.111).

Definition 3.20 Suppose $\tilde{\tilde{x}} = (\tilde{\tilde{x}}_1, \tilde{\tilde{x}}_2, \ldots, \tilde{\tilde{x}}_n)$ solves $A\tilde{x} \approx \tilde{b}$. If all $\tilde{\tilde{x}}_j \approx (\bar{x}_j - \alpha, \bar{x}_j, \bar{x}_j, \bar{x}_j + \alpha)$ for some $\bar{x}_j \geq 0$ and $\alpha \geq 0$, then $\tilde{\tilde{x}}$ is said to be a fuzzy basic solution. If $\tilde{\tilde{x}}_j \not\approx (\bar{x}_j - \alpha, \bar{x}_j, \bar{x}_j, \bar{x}_j + \alpha_j)$ for $\bar{x}_j \geq 0$ and $\alpha \geq 0$, then $\tilde{\tilde{x}}$ has some non-zero components, say $\tilde{\tilde{x}}_1, \tilde{\tilde{x}}_2, \ldots, \tilde{\tilde{x}}_k, 1 \leq k \leq n$. Then $A\tilde{x} \approx \tilde{b}$ can be written as:

$$a_1\tilde{\tilde{x}}_1 + a_2\tilde{\tilde{x}}_2, \ldots + a_k\tilde{\tilde{x}}_k + a_{k+1}(\bar{x}_{k+1} - \alpha_{k+1}, \bar{x}_{k+1}, \bar{x}_{k+1}, \bar{x}_{k+1} + \alpha_{k+1}) + \cdots$$
$$+ a_n(\bar{x}_n - \alpha_n, \bar{x}_n, \bar{x}_n, \bar{x}_n + \alpha_n) = \tilde{b}$$

If the columns $a_1, a_2, \ldots, a_k$ corresponding to the non-zero components $\tilde{\tilde{x}}_1, \tilde{\tilde{x}}_2, \ldots, \tilde{\tilde{x}}_k$ are linearly independent, then $\tilde{\tilde{x}}$ is said to be fuzzy basic solution.

Remark 3.7 Consider a system of m simultaneous fuzzy linear equations involving symmetric trapezoidal fuzzy numbers in n unknowns $A\tilde{x} \approx \tilde{b}$. Let B be any matrix formed by m linearly independent of A. In this case A is partitioned as $A = [B, N]$. Here B and N are called basic and non-basic matrices and the solution $\tilde{x} = (\tilde{x}_B, \tilde{x}_N) = (B^{-1}\tilde{b}, \tilde{0})$ is a fuzzy basic solution.

Definition 3.21 Suppose $\tilde{\tilde{x}}$ is a fuzzy basic feasible solution of $A\tilde{x} \approx \tilde{b}$. If the number of fuzzy positive variables $\tilde{\tilde{x}}$ is exactly m, then $\tilde{\tilde{x}}$ is called a nondegenerate fuzzy basic feasible solution, i.e. $\tilde{\tilde{x}}_B = (\tilde{\tilde{x}}_1, \ldots, \tilde{\tilde{x}}_m) \succ (\tilde{0}, \tilde{0}, \ldots, \tilde{0})$. If the number of the fuzzy positive $\tilde{\tilde{x}}_j$ is less than of m, then $\tilde{\tilde{x}}$ is called a degenerate fuzzy basic feasible solution.

Example 3.12 Consider the following systems of equalities:

$$\tilde{x}_1 + \tilde{x}_2 + \tilde{x}_3 \approx (0, 3, 9, 12)$$
$$\tilde{x}_2 + \tilde{x}_4 \approx (0, 1, 3, 4)$$
$$\tilde{x}_1 + 2\tilde{x}_2 + \tilde{x}_5 \approx (0, 4, 12, 16)$$

Note that $A = [a_1, a_2, a_3, a_4, a_5] = \begin{bmatrix} 1 & 1 & 1 & 0 & 0 \\ 0 & 1 & 0 & 1 & 0 \\ 1 & 2 & 0 & 0 & 1 \end{bmatrix}$. In this case the fuzzy

basic feasible solution $\tilde{x}_B = \begin{bmatrix} \tilde{x}_1 \\ \tilde{x}_2 \\ \tilde{x}_3 \end{bmatrix} = \begin{bmatrix} (0,2,6,8) \\ (0,1,3,4) \\ (0,0,0,0) \end{bmatrix}$ corresponding to the basis $B =$

$[a_1, a_2, a_3]$ is degenerate since the fuzzy basic variable $\tilde{x}_3 \approx \tilde{0}$.

Theorem 3.25 *If the LPP with fuzzy parameters of type-3 (3.111) has a fuzzy optimum feasible solution, then it also has a fuzzy optimal basic feasible solution.*

Proof It can be proved similar to the proof of Theorem 3.14. □

In what follows, three solution approaches are reviewed for solving the LPP with fuzzy parameters of type-3 (3.111). The other approaches can be found in [22, 32].

3.5.1 Ganesan and Veeramani's Approach

Ganesan and Veeramani [24] proved fuzzy analogues of some important theorems of the LP leading to a simplex based method for solving (3.111) without converting it into a crisp LP problem.

Let $B = [a_1, a_2, \ldots, a_m]$ form a basis for the columns of A. Let $\tilde{x} = (\tilde{x}_B, \tilde{x}_N) = (B^{-1}\tilde{b}, \tilde{0})$ be a fuzzy basic feasible solution and the fuzzy value of the objective function $\tilde{z}$ is given by $\tilde{z} = \tilde{c}_B\tilde{x}_B$, where $\tilde{c}_B = (\tilde{c}_1, \tilde{c}_2, \ldots, \tilde{c}_m)$ is the cost vector corresponding to $\tilde{x}_B$. Assume that $a_j = By_j = \sum_{i=1}^{m} a_i y_{ij}$ and the symmetric trapezoidal fuzzy number $\tilde{z}_j = \tilde{c}_B y_j = \sum_{i=1}^{m} \tilde{c}_i y_{ij}$ are known for every column vector a_j in A. The next theorems shall examine the possibility of finding another fuzzy basic feasible solution with an improved fuzzy value of $\tilde{z}$ by replacing one of the columns of B by a_j.

Theorem 3.26 *Let $\tilde{x} = (\tilde{x}_B, \tilde{x}_N) = (B^{-1}\tilde{b}, \tilde{0})$ be a fuzzy basic feasible solution of (3.111). If for any nonbasic column a_j, $y_{rj} \geq 0$ for some $i, i \in \{1, 2, \ldots, m\}$, then it is possible to obtain a new fuzzy basic feasible solution by replacing one of the basic columns in B by a_j.*

Proof Suppose that $\tilde{x} = (\tilde{x}_B, \tilde{x}_N) = (\tilde{x}_1, \tilde{x}_2, \ldots, \tilde{x}_m, \tilde{0}, \ldots \tilde{0})$ is a fuzzy basic feasible solution with k positive components such that $B\tilde{x}_B = \tilde{b}$ where,

$$\tilde{x}_i = (x_{2i} - \alpha_i, x_{2i}, x_{3i}, x_{3i} + \alpha), x_{2i} \leq x_{3i}, \alpha_i \geq 0, \quad i = 1, 2, \ldots m$$

$$\frac{x_{2i} + x_{3i}}{2} > 0, \quad i = 1, 2, \ldots, k, \frac{x_{2i} + x_{3i}}{2} = 0, \quad i = k+1, k+2, \ldots, m.$$

That is

$$\tilde{x}_i \succ \tilde{0}, \quad i = 1, 2, \ldots, k, \tilde{x}_i = (x_i - \alpha_i, x_i, x_i, x_i + \alpha_i), \quad i = k+1, k+2, \ldots, m.$$

Now equation $B\tilde{x}_B = \tilde{b}$ becomes

$$\sum_{i=1}^{k} a_i \tilde{x}_i + \sum_{i=k+1}^{m} a_i(x_i - \alpha_i, x_i, x_i, x_i + \alpha_i) \approx \tilde{b} \tag{3.112}$$

Then for any non-basic column a_j of A, we write

$$a_j = \sum_{i=1}^{k} a_i y_{ij} = a_1 y_{1j} + a_2 y_{2j} + \cdots a_r y_{rj} + \cdots + a_m y_{mj} = By_j$$

We know that if the basis vector a_r for which $y_{rj} \neq 0$ is replaced by a_j of A, then the new set of vectors $(a_1, a_2, \ldots, a_{r-1}, a_j, a_{r+1}, \ldots, a_m)$ still forms a basis. Now for $y_{rj} \neq 0$ and $r \leq k$, we can write

$$a_r = \frac{a_j}{y_{rj}} - \sum_{i=1,i\neq r}^{m} a_i \frac{y_{ij}}{y_{rj}} = \frac{a_j}{y_{rj}} - \sum_{i=1,i\neq r}^{k} a_i \frac{y_{ij}}{y_{rj}} + \sum_{i=K+1}^{m} a_i \frac{y_{ij}}{y_{rj}} \tag{3.113}$$

Equation (3.112) becomes

$$\sum_{i=1,i\neq r}^{k} a_i \tilde{x}_i + a_r \tilde{x}_r + \sum_{i=k+1}^{m} a_i(x_i - \alpha_i, x_i, x_i, x_i + \alpha_i) \approx \tilde{b} \tag{3.114}$$

By substituting (3.113) on (3.114), we obtain

$$\sum_{i=1,i\neq r}^{k} a_i \tilde{x}_i + \frac{\tilde{x}_r}{y_{rj}} a_j - \frac{\tilde{x}_r}{y_{rj}} \sum_{i=1,i\neq r}^{k} a_i y_{ij} - \frac{\tilde{x}_r}{y_{rj}} \sum_{i=k+1}^{m} a_i y_{ij} + \sum_{i=k+1}^{m} a_i(x_i - \alpha_i, x_i, x_i, x_i + \alpha_i) \approx \tilde{b}$$

$$\Rightarrow \sum_{i=1,i\neq r}^{k} a_i \left(\tilde{x}_i - \frac{\tilde{x}_r}{y_{rj}} y_{ij} \right) + \frac{\tilde{x}_r}{y_{rj}} a_j + \sum_{i=k+1}^{m} a_i \left((x_i - \alpha_i, x_i, x_i, x_i + \alpha_i) - \frac{\tilde{x}_r}{y_{rj}} y_{ij} \right) \approx \tilde{b}$$

Since $\tilde{x}_i = (x_i - \alpha_i, x_i, x_i, x_i + \alpha_i)$, for $i = k+1, k+2, \ldots, m$, we have

$$\sum_{i=1,i\neq r}^{k} a_i\tilde{x}_i + \frac{\tilde{x}_r}{y_{rj}}a_j - \frac{\tilde{x}_r}{y_{rj}}\sum_{i=1,i\neq r}^{k} a_iy_{ij} - \frac{\tilde{x}_r}{y_{rj}}\sum_{i=k+1}^{m} a_iy_{ij} + \sum_{i=k+1}^{m} a_ix_i \approx \tilde{b}$$

$$\Rightarrow \sum_{i=1,i\neq r}^{m} a_i\left(\tilde{x}_i - \frac{\tilde{x}_r}{y_{rj}}y_{ij}\right) + \frac{\tilde{x}_r}{y_{rj}}a_j \approx \tilde{b} \Rightarrow \sum_{i=1,i\neq r}^{m} a_i\hat{\tilde{x}}_i + \hat{\tilde{x}}_r a_j \approx \tilde{b}$$

where

$$\hat{\tilde{x}}_i = \tilde{x}_i - \frac{\tilde{x}_r}{y_{rj}}y_{ij}, i \neq r, \hat{\tilde{x}}_r = \frac{\tilde{x}_r}{y_{rj}} \tag{3.115}$$

Thus, solution (3.115) gives a new fuzzy basic solution $A\tilde{x} \approx \tilde{b}$.

It is required to show that the new fuzzy basic solution (3.115) is also feasible. This requires that

$$\hat{\tilde{x}}_i = \tilde{x}_i - \frac{\tilde{x}_r}{y_{rj}}y_{ij} \succeq \tilde{0}, \quad i \neq r, \hat{\tilde{x}}_r = \frac{\tilde{x}_r}{y_{rj}} \succeq \tilde{0}$$

Select y_{rj} such that $\frac{\tilde{x}_r}{y_{rj}} = \min_i\left\{\frac{\tilde{x}_i}{y_{ij}} \Big| y_{ij} > 0\right\} = \min_i\left\{\frac{x_{2i}+x_{3i}}{2y_{ij}} \Big| y_{ij} > 0\right\}$. Then

$$\frac{\tilde{x}_r}{y_{rj}} \preceq \frac{\tilde{x}_i}{y_{ij}} \Rightarrow \frac{x_{2r}+x_{3r}}{2y_{rj}} \leq \frac{x_{2i}+x_{3i}}{2y_{ij}} \Rightarrow \frac{x_{2i}+x_{3i}}{2y_{ij}} - \frac{x_{2r}+x_{3r}}{2y_{rj}} \geq 0 \Rightarrow \frac{\tilde{x}_i}{y_{ij}} - \frac{\tilde{x}_r}{y_{rj}} \succeq \tilde{0}$$

$$\Rightarrow \tilde{x}_i - \frac{\tilde{x}_r}{y_{rj}}y_{ij} \succeq \tilde{0}$$

Hence the new fuzzy basic solution is a fuzzy basic feasible solution. $\qquad\square$

After the replacement of basis vectors, the new basis matrix is $\hat{B} = (a_1, a_2, \ldots, a_{r-1}, a_j, a_{r+1}, \ldots, a_m)$. The new fuzzy basic feasible solution is $\hat{\tilde{x}}_B$. In this case $\hat{\tilde{x}}_i = \tilde{x}_i - \frac{\tilde{x}_r}{y_{rj}}y_{ij}$ $i \neq r$, and $\hat{\tilde{x}}_r = \frac{\tilde{x}_r}{y_{rj}}$ are the new basic variables.

Theorem 3.27 Let $\tilde{x} = (\tilde{x}_B, \tilde{x}_N) = (B^{-1}\tilde{b}, \tilde{0})$ be a fuzzy basic feasible solution of (3.111) with $\tilde{z} = \tilde{c}_B\tilde{x}_B$ as the fuzzy value of the objective function and if $\hat{\tilde{x}} = (\hat{\tilde{x}}_B, \hat{\tilde{x}}_N) = (\hat{B}^{-1}\tilde{b}, \tilde{0})$ is another fuzzy basic feasible solution $\hat{\tilde{z}} = \tilde{c}_{\hat{B}}\hat{\tilde{x}}_{\hat{B}}$ obtained by admitting the non-basic column vector a_j in the basis for which $\tilde{z}_j - \tilde{c}_j \prec \tilde{0}$ and $y_{rj} > 0$ for some $i, i \in \{1, 2, \ldots, m\}$, then $\hat{\tilde{z}} \succeq \tilde{z}$.

Proof Since $\tilde{c}_{\hat{B}_i} = \tilde{c}_{B_i} = \tilde{c}_i, i \neq r$ and $\tilde{c}_{\hat{B}_r} = \tilde{c}_j$, the new fuzzy value of the objective function corresponding to the new fuzzy basic feasible solution (3.115) becomes

$$
\hat{\tilde{z}} = \tilde{c}_{\hat{B}} \hat{\tilde{x}}_{\hat{B}} = \sum_{i=1}^{m} \tilde{c}_{\hat{B}_i} \hat{\tilde{x}}_i = \sum_{i=1,i\neq r}^{m} \tilde{c}_i \hat{\tilde{x}}_i + \tilde{c}_j \hat{\tilde{x}}_r \approx \sum_{i=1,i\neq r}^{k} \tilde{c}_i \left(\tilde{x}_i - \frac{\tilde{x}_r}{y_{rj}} y_{ij} \right) + \tilde{c}_j \frac{\tilde{x}_r}{y_{rj}}
$$

$$
\approx \sum_{i=1,i\neq r}^{m} \tilde{c}_i \left(\tilde{x}_i - \frac{\tilde{x}_r}{y_{rj}} y_{ij} \right) + \tilde{c}_i \left(\tilde{x}_r - \frac{\tilde{x}_r}{y_{rj}} y_{rj} \right) + \tilde{c}_j \frac{\tilde{x}_r}{y_{rj}} \approx \sum_{i=1}^{m} \tilde{c}_i \left(\tilde{x}_i - \frac{\tilde{x}_r}{y_{rj}} y_{ij} \right) + \tilde{c}_j \frac{\tilde{x}_r}{y_{rj}}
$$

$$
\approx \sum_{i=1}^{m} \tilde{c}_i \tilde{x}_i - \frac{\tilde{x}_r}{y_{rj}} \sum_{i=1}^{m} \tilde{c}_i y_{ij} + \tilde{c}_j \frac{\tilde{x}_r}{y_{rj}} \approx \sum_{i=1}^{m} \tilde{c}_i \tilde{x}_i - \frac{\tilde{x}_r}{y_{rj}} \left(\sum_{i=1}^{m} \tilde{c}_i y_{ij} - \tilde{c}_j \right)
$$

$$
\approx \tilde{z} - \frac{\tilde{x}_r}{y_{rj}} (\tilde{z}_j - \tilde{c}_j)
$$

That is

$$
\hat{\tilde{z}} \approx \tilde{z} - \frac{\tilde{x}_r}{y_{rj}} (\tilde{z}_j - \tilde{c}_j) \tag{3.116}
$$

Since $y_{rj} > 0$, $\tilde{z}_j - \tilde{c}_j \prec \tilde{0}$ and $\frac{\tilde{x}_r}{y_{rj}} \succeq \tilde{0}$, let

$$
\frac{\tilde{x}_r}{y_{rj}} = \left(\frac{x_{2r} - \alpha_r}{y_{rj}}, \frac{x_{2r}}{y_{rj}}, \frac{x_{3r}}{y_{rj}}, \frac{x_{3r} + \alpha_r}{y_{rj}} \right) \succeq \tilde{0}, x_{2r} \leq x_{3r}, \alpha_r \geq 0, \frac{x_{2r} + x_{3r}}{y_{rj}} \geq 0
$$

and

$$
\tilde{z}_j - \tilde{c}_j = \left(t_{2j} - \alpha_j, t_{2j}, t_{3j}, t_{3j} + \alpha_j \right) \prec \tilde{0}, \quad t_j < 0, \alpha_j \geq 0, \frac{t_{2j} + t_{3j}}{2} \leq 0
$$

Now

$$
\frac{\tilde{x}_r}{y_{rj}} (\tilde{z}_j - \tilde{c}_j) \approx \left(\frac{x_{2r} - \alpha_r}{y_{rj}}, \frac{x_{2r}}{y_{rj}}, \frac{x_{3r}}{y_{rj}}, \frac{x_{3r} + \alpha_r}{y_{rj}} \right) \left(t_{2j} - \alpha_j, t_{2j}, t_{3j}, t_{3j} + \alpha_j \right)
$$

$$
\approx \left(\left(\frac{x_{2r} + x_{3r}}{2y_{rj}} \right) \left(\frac{t_{2j} + t_{3j}}{2} \right) - w - \gamma, \left(\frac{x_{2r} + x_{3r}}{2y_{rj}} \right) \left(\frac{t_{2j} + t_{3j}}{2} \right) \right.
$$

$$
\left. -w, \left(\frac{x_{2r} + x_{3r}}{2y_{rj}} \right) \left(\frac{t_{2j} + t_{3j}}{2} \right) + w, \left(\frac{x_{2r} + x_{3r}}{2y_{rj}} \right) \left(\frac{t_{2j} + t_{3j}}{2} \right) - w + \gamma \right)
$$

Since, $\frac{x_{2r} + x_{3r}}{y_{rj}} \geq 0$ and $\frac{t_{2j} + t_{3j}}{2} \leq 0$, we obtain

$$\frac{\left(\frac{x_{2r} + x_{3r}}{2y_{rj}}\right)\left(\frac{t_{2j} + t_{3j}}{2}\right) - w + \left(\frac{x_{2r} + x_{3r}}{2y_{rj}}\right)\left(\frac{t_{2j} + t_{3j}}{2}\right) + w}{2} = \left(\frac{x_{2r} + x_{3r}}{2y_{rj}}\right)\left(\frac{t_{2j} + t_{3j}}{2}\right) \leq 0$$

So Eq. (3.116) becomes $\hat{\tilde{z}} \succeq \tilde{z}$. Hence, the new fuzzy basic feasible solution gives the improved fuzzy value of the objective function. □

We have seen that for a non-basic column a_j, for which $\tilde{z}_j - \tilde{c}_j \prec \tilde{0}$ and $y_{rj} > 0$ for some $i, i \in 1, 2, \ldots, m$, is considered alone for inserting into the basis. Let us now discuss the situation when there exists an a_j such that $\tilde{z}_j - \tilde{c}_j \prec \tilde{0}$ and $y_{rj} < 0$, for all $i = \{1, 2, \ldots, m\}$. If $\tilde{x} = (x_2 - \alpha, x_2, x_3, x_3 + \alpha) \succeq \tilde{0}$ and $\theta > 0$, then $\theta \tilde{x} = \left(\theta(x_2 - \alpha), \theta x_2, \theta x_3, \theta(x_3 + \alpha) \succeq \tilde{0}\right)$. Now θ can be made sufficiently large so that $\theta \tilde{x} \succ \tilde{y}$ for any symmetric trapezoidal fuzzy number $\tilde{y}$. If $\tilde{z}_j - \tilde{c}_j \prec \tilde{0}$, then $\theta(\tilde{z}_j - \tilde{c}_j) \prec \tilde{0}$. Now the proof of the following theorem is straightforward.

Theorem 3.28 Let $\tilde{x} = (\tilde{x}_B, \tilde{x}_N) = \left(B^{-1}\tilde{b}, \tilde{0}\right)$ be a fuzzy basic feasible solution of (3.111). If an a_j exists for which $\tilde{z}_j - \tilde{c}_j \prec \tilde{0}$ and $y_{rj} \leq 0$ for all $i = \{1, 2, \ldots, m\}$, then the FLP (3.111) has an unbounded solution.

Similar to in the classical LP problems, we can easily prove the optimality conditions for the LPP with fuzzy parameters of type (3) given in (3.111). The following theorem gives the optimality conditions of (3.111).

Theorem 3.29 Let $\tilde{x} = (\tilde{x}_B, \tilde{x}_N) = \left(B^{-1}\tilde{b}, \tilde{0}\right)$ be a fuzzy basic feasible solution of (3.111). If $\tilde{z}_j - \tilde{c}_j \succeq \tilde{0}$ for all non-basic columns a_j, then $\tilde{x} = (\tilde{x}_B, \tilde{x}_N) = \left(B^{-1}\tilde{b}, \tilde{0}\right)$ is a fuzzy optimal solution to (3.111).

Now we are in a position to summarize the fuzzy primal simplex algorithm Ganesan and Veeramani [24] for solving the LPP with fuzzy parameters of type-3 (3.111).

Algorithm 3.4: Symmetric fuzzy primal simplex algorithm (*Maximization Problem*)

Initialization step
Choose a starting feasible basic solution with the basis B.

Main steps

(1) Solve the system $B\tilde{x}_B = \tilde{b}$. Let $\tilde{x}_B = B^{-1}\tilde{b}, \tilde{x}_N = \tilde{0}$, and $\tilde{z} = \tilde{c}_B\tilde{x}_B$.

(2) Solve the fuzzy system $\tilde{w}B = \tilde{c}_B$. Let $\tilde{w} = \tilde{c}_B B^{-1}$.

(3) Calculate $\tilde{z}_j - \tilde{c}_j = \tilde{c}_B B^{-1} a_j - \tilde{c}_j = \tilde{w}a_j - \tilde{c}_j$ for all $j \in J_N$. Suppose that $\tilde{z}_j - \tilde{c}_j = \left(t_{2j} - \alpha_j, t_{2j}, t_{3j}, t_{3j} + \alpha_j\right)$ and let $t_{2k} + t_{3k} = \min_{j \in J_N}\{t_{2j} + t_{3j}\}$. If $t_{2k} + t_{3k} \geq 0$, then stop with the current basic solution as a fuzzy optimal solution.

Table 3.21 Data of Example 3.13

Machines	Time per unit (minutes)			Machine capacity (min/day)
	P_1	P_2	P_3	
M_1	12	13	12	490
M_2	14	–	13	470
M_3	12	15	–	480

(4) Solve the system $By_k = a_k$ and let $y_k = B^{-1}a_k$. If $y_k \leq 0$ then stop with the conclusion that the problem is unbounded.

If $y_k \nleq 0$, then $\tilde{x}_k$ enters the basis and $\tilde{x}_{B_r}$ leaves the basis providing that

$$\frac{\bar{b}_{2r} + \bar{b}_{3r}}{y_{rk}} = \min_{1 \leq i \leq m} \left\{ \frac{\bar{b}_{2i} + \bar{b}_{3i}}{y_{ik}} \middle| y_{rk} > 0 \right\}.$$

(5) Update the basis B where a_k replaces a_{Br}, update the index set J_N and go to (1).

Example 3.13 [24] A company produces three products P_1, P_2 and P_3. These products are made on three different machines M_1, M_2 and M_3. The time required to manufacture one unit of each product and the daily capacity of the machines are given in Table 3.21.

Note that the time availability can vary from day to day due to break down of machines, overtime work etc. Finally the profit for each product can also vary due to price variations. At the same time the company wants to keep the profit somewhat close to Rs. 14 for P_1, Rs. 13 for P_2 and Rs. 16 for P_3. The company wants to determine the range of each product to be produced per day to maximize its profit. It is assumed that all the amounts produced are consumed in the market.

Since the profit from each product and the time availability on each machine are uncertain, the number of units to be produced of each product will also be uncertain. So we will model the problem as a fuzzy linear programming problem. We use symmetric tarpezoidal fuzzy numbers for each uncertain value.

Profit for P_1, which is close to 14, is modelled as $(11, 13, 15, 17)$. Similarly, the other parameters are also modeled as symmetric trapezoidal fuzzy numbers taking into account the nature of the problem and the other requirements. So the problem is formulated as follows:

$$\max \quad \tilde{z} \approx (11, 13, 15, 17)\tilde{x}_1 + (9, 12, 14, 17)\tilde{x}_2 + (13, 15, 17, 19)\tilde{x}_3$$

s.t.

$$12\tilde{x}_1 + 13\tilde{x}_2 + 12\tilde{x}_3 \preceq (469, 475, 505, 511)$$
$$14\tilde{x}_1 + 13\tilde{x}_3 \preceq (452, 460, 480, 488)$$
$$12\tilde{x}_1 + 15\tilde{x}_2 \preceq (460, 465, 495, 500) \tag{3.117}$$
$$\tilde{x}_1, \tilde{x}_2, \tilde{x}_3 \succeq \tilde{0}$$

Now the standard form of (3.117) becomes

$$\max \quad \tilde{z} \approx (11, 13, 15, 17)\tilde{x}_1 + (9, 12, 14, 17)\tilde{x}_2 + (13, 15, 17, 19)\tilde{x}_3$$

s.t.

$$
\begin{aligned}
12\tilde{x}_1 + 13\tilde{x}_2 + 12\tilde{x}_3 + \tilde{x}_4 &\approx (469, 475, 505, 511) \\
14\tilde{x}_1 + 13\tilde{x}_3 + \tilde{x}_5 &\approx (452, 460, 480, 488) \\
12\tilde{x}_1 + 15\tilde{x}_2 + \tilde{x}_6 &\approx (460, 465, 495, 500) \\
\tilde{x}_1, \ \tilde{x}_2, \ \tilde{x}_3, \tilde{x}_4, \tilde{x}_5, \tilde{x}_6 &\succeq \tilde{0}
\end{aligned}
\tag{3.118}
$$

where $\tilde{x}_4$, $\tilde{x}_5$ and $\tilde{x}_6$ are the slack fuzzy variables.

The steps of Algorithm 3.4 for solving the LPP with fuzzy parameters of type-3 (3.118) are presented as follows:

Iteration 1:

The starting feasible basic is $B = [a_4, a_5, a_6] = \begin{bmatrix} 1 & 0 & 0 \\ 0 & 1 & 0 \\ 0 & 0 & 1 \end{bmatrix}$. Thus, the non-basic

matrix N is $N = [a_1, a_2, a_3] = \begin{bmatrix} 12 & 13 & 12 \\ 14 & 0 & 13 \\ 12 & 15 & 0 \end{bmatrix}$.

Step 1: We find the basic feasible solution by solving the system $B\tilde{x}_B = \tilde{b}$:

$$
\begin{bmatrix} 1 & 0 & 0 \\ 0 & 1 & 0 \\ 0 & 0 & 1 \end{bmatrix}
\begin{bmatrix} \tilde{x}_4 \\ \tilde{x}_5 \\ \tilde{x}_6 \end{bmatrix}
=
\begin{bmatrix} (469, 475, 505, 511) \\ (452, 460, 480, 488) \\ (460, 465, 495, 500) \end{bmatrix}
\Rightarrow
\begin{bmatrix} \tilde{x}_4 \\ \tilde{x}_5 \\ \tilde{x}_6 \end{bmatrix}
$$

$$
=
\begin{bmatrix} (469, 475, 505, 511) \\ (452, 460, 480, 488) \\ (460, 465, 495, 500) \end{bmatrix}
=
\begin{bmatrix} \bar{b}_1 \\ \bar{b}_2 \\ \bar{b}_3 \end{bmatrix}
$$

The non-basic variables are $\tilde{x}_1 = \tilde{0}$, $\tilde{x}_2 = \tilde{0}$ and $\tilde{x}_3 = \tilde{0}$ the fuzzy objective value is $\tilde{z} = \tilde{c}_B \tilde{x}_B = (0, 0, 0, 0)$.

Step 2: We find $\tilde{w}$ by solving the fuzzy system $\tilde{w}B = \tilde{c}_B$:

$$
(\tilde{w}_1, \tilde{w}_2, \tilde{w}_3)
\begin{bmatrix} 1 & 0 & 0 \\ 0 & 1 & 0 \\ 0 & 0 & 1 \end{bmatrix}
=
\begin{bmatrix} (0,0,0,0) \\ (0,0,0,0) \\ (0,0,0,0) \end{bmatrix}
\Rightarrow \tilde{w}_1 = \tilde{w}_2 = \tilde{w}_3 = (0,0,0,0) = \tilde{0}
$$

Step 3: We calculate $\tilde{z}_j - \tilde{c}_j = \tilde{w}a_j - \tilde{c}_j$ for all $j \in J_N = \{1, 2, 3\}$:

$$\tilde{z}_1 - \tilde{c}_1 = \tilde{w}a_1 - \tilde{c}_1 = (\tilde{0}, \tilde{0}, \tilde{0}) \begin{bmatrix} 12 \\ 14 \\ 12 \end{bmatrix} - (11, 13, 15, 17) = (-17, -15, -13, -11)$$

$$\tilde{z}_2 - \tilde{c}_2 = \tilde{w}a_2 - \tilde{c}_2 = (\tilde{0}, \tilde{0}, \tilde{0}) \begin{bmatrix} 13 \\ 0 \\ 15 \end{bmatrix} - (9, 12, 14, 17) = (-17, -14, -12, -9)$$

$$\tilde{z}_3 - \tilde{c}_3 = \tilde{w}a_3 - \tilde{c}_3 = (\tilde{0}, \tilde{0}, \tilde{0}) \begin{bmatrix} 12 \\ 13 \\ 0 \end{bmatrix} - (13, 15, 17, 19) = (-19, -17, -15, -13)$$

Now we have:

$$t_{2k} + t_{3k} = \min_{j \in J_N}\{t_{2j} + t_{3j}\} = \min\{-15 - 13, -14 - 12, -17 - 15\} = -32$$
$$= t_{22} + t_{32}$$

Step 4: We find y_3 by solving the system $By_3 = a_3$:

$$\begin{bmatrix} 1 & 0 & 0 \\ 0 & 1 & 0 \\ 0 & 0 & 1 \end{bmatrix} \begin{bmatrix} y_{13} \\ y_{23} \\ y_{23} \end{bmatrix} = \begin{bmatrix} 12 \\ 13 \\ 0 \end{bmatrix} \Rightarrow y_3 = \begin{bmatrix} y_{13} \\ y_{23} \\ y_{23} \end{bmatrix} = \begin{bmatrix} 12 \\ 13 \\ 0 \end{bmatrix}$$

With regard to Step 3, $\tilde{x}_3$ enters the basis and the variable $\tilde{x}_{B_r}$ leaving the basis is determined by the following test:

$$\min\left\{\frac{\bar{b}_{21} + \bar{b}_{31}}{y_{13}}, \frac{\bar{b}_{22} + \bar{b}_{32}}{y_{23}}\right\} = \min\left\{\frac{475 + 505}{12}, \frac{460 + 490}{13}\right\} = \frac{460 + 490}{13}$$
$$= \frac{\bar{b}_{22} + \bar{b}_{32}}{y_{23}}$$

Hence, $\tilde{x}_{B_2} = \tilde{x}_5$ leaves the basis.

Step 5: We update the basis B where a_3 replaces $a_{B_2} = a_5$:

$$B = [a_4, a_3, a_6] = \begin{bmatrix} 1 & 12 & 0 \\ 0 & 13 & 0 \\ 0 & 0 & 1 \end{bmatrix}$$

Also, we update $J_N = \{1, 2, 3\}$ as $J_N = \{1, 2, 5\}$.

Iteration 2:

Step 1: We find the fuzzy basic feasible solution by solving the system $B\tilde{x}_B = \tilde{b}$:

$$\begin{bmatrix} 1 & 12 & 0 \\ 0 & 13 & 0 \\ 0 & 0 & 1 \end{bmatrix} \begin{bmatrix} \tilde{x}_4 \\ \tilde{x}_3 \\ \tilde{x}_6 \end{bmatrix} = \begin{bmatrix} (469, 475, 505, 511) \\ (452, 460, 480, 488) \\ (460, 465, 495, 500) \end{bmatrix} \Rightarrow \begin{bmatrix} \tilde{x}_4 \\ \tilde{x}_3 \\ \tilde{x}_6 \end{bmatrix}$$

$$= \begin{bmatrix} \left(\frac{241}{13}, \frac{415}{13}, \frac{1045}{13}, \frac{1219}{13}\right) \\ \left(\frac{452}{13}, \frac{460}{13}, \frac{480}{13}, \frac{488}{13}\right) \\ (460, 465, 495, 500) \end{bmatrix} = \begin{bmatrix} \bar{b}_1 \\ \bar{b}_2 \\ \bar{b}_3 \end{bmatrix}$$

The non-basic variables are $\tilde{x}_1 = \tilde{0}$, $\tilde{x}_2 = \tilde{0}$ and $\tilde{x}_5 = \tilde{0}$ the fuzzy objective value is $\tilde{z} = \tilde{c}_B \tilde{x}_B = \left(\frac{5794}{13}, \frac{6890}{13}, \frac{8150}{13}, \frac{9246}{13}\right)$.

Step 2: We find $\tilde{w}$ by solving the fuzzy system $\tilde{w}B = \tilde{c}_B$:

$$(\tilde{w}_1, \tilde{w}_2, \tilde{w}_3) \begin{bmatrix} 1 & 12 & 0 \\ 0 & 13 & 0 \\ 0 & 0 & 1 \end{bmatrix} = \begin{bmatrix} (0, 0, 0, 0) \\ (9, 12, 14, 17) \\ (0, 0, 0, 0) \end{bmatrix} \Rightarrow \tilde{w}_1 = \tilde{w}_3 = (0, 0, 0, 0) = \tilde{0}, \tilde{w}_2$$

$$\approx \left(\frac{13}{13}, \frac{15}{13}, \frac{17}{13}, \frac{19}{13}\right)$$

Step 3: We calculate $\tilde{z}_j - \tilde{c}_j = \tilde{w}a_j - \tilde{c}_j$ for all $j \in J_N = \{1, 3, 5\}$:

$$\tilde{z}_1 - \tilde{c}_1 = \tilde{w}a_1 - \tilde{c}_1 = \left(\frac{-39}{13}, \frac{15}{13}, \frac{69}{13}, \frac{123}{13}\right)$$

$$\tilde{z}_2 - \tilde{c}_2 = \tilde{w}a_2 - \tilde{c}_2 = (-17, -14, -12, -9)$$

$$\tilde{z}_5 - \tilde{c}_5 = \tilde{w}a_5 - \tilde{c}_5 = \left(\frac{13}{13}, \frac{15}{13}, \frac{17}{13}, \frac{19}{13}\right)$$

Now we have:

$$t_{2k} + t_{3k} = \min_{j \in J_N}\{t_{2j} + t_{3j}\} = \min\left\{\frac{15}{13} + \frac{69}{13}, -14 - 12, \frac{15}{13} + \frac{17}{13}\right\} = -26$$

$$= t_{22} + t_{32}$$

Step 4: We find y_2 by solving the system $By_2 = a_2$:

$$\begin{bmatrix} 1 & 12 & 0 \\ 0 & 13 & 0 \\ 0 & 0 & 1 \end{bmatrix} \begin{bmatrix} y_{13} \\ y_{23} \\ y_{23} \end{bmatrix} = \begin{bmatrix} 13 \\ 0 \\ 15 \end{bmatrix} \Rightarrow y_2 = \begin{bmatrix} y_{12} \\ y_{22} \\ y_{32} \end{bmatrix} = \begin{bmatrix} 13 \\ 0 \\ 15 \end{bmatrix}$$

With regard to Step 3, $\tilde{x}_2$ enters the basis and the fuzzy variable $\tilde{x}_{B_r}$ leaving the basis is determined by the following test:

$$\min\left\{\frac{\bar{b}_{21} + \bar{b}_{31}}{y_{12}}, \frac{\bar{b}_{23} + \bar{b}_{33}}{y_{32}}\right\} = \min\left\{\frac{\frac{415}{13} + \frac{1045}{13}}{13}, \frac{465 + 495}{15}\right\} = \frac{\frac{415}{13} + \frac{1045}{13}}{13}$$

$$= \frac{\bar{b}_{21} + \bar{b}_{31}}{y_{12}}$$

Hence, $\tilde{x}_{B_1} = \tilde{x}_4$ leaves the basis.

Step 5: We update the basis B where a_2 replaces $a_{B_1} = a_4$:

$$B = [a_2, a_3, a_6] = \begin{bmatrix} 13 & 12 & 0 \\ 0 & 13 & 0 \\ 15 & 0 & 1 \end{bmatrix}$$

Also, we update $J_N = \{1, 3, 5\}$ as $J_N = \{1, 4, 5\}$.

Iteration 3:

Step 1: We find the fuzzy basic feasible solution by solving the system $B\tilde{x}_B = \tilde{b}$:

$$\begin{bmatrix} 13 & 12 & 0 \\ 0 & 13 & 0 \\ 15 & 0 & 1 \end{bmatrix} \begin{bmatrix} \tilde{x}_2 \\ \tilde{x}_3 \\ \tilde{x}_6 \end{bmatrix} = \begin{bmatrix} (469, 475, 505, 511) \\ (452, 460, 480, 488) \\ (460, 465, 495, 500) \end{bmatrix} \Rightarrow \begin{bmatrix} \tilde{x}_2 \\ \tilde{x}_3 \\ \tilde{x}_6 \end{bmatrix}$$

$$= \begin{bmatrix} \left(\frac{241}{169}, \frac{415}{169}, \frac{1045}{169}, \frac{1219}{169}\right) \\ \left(\frac{241}{169}, \frac{415}{169}, \frac{1045}{169}, \frac{1219}{169}\right) \\ \left(\frac{59455}{169}, \frac{62910}{169}, \frac{77430}{169}, \frac{80885}{169}\right) \end{bmatrix} = \begin{bmatrix} \tilde{b}_1 \\ \tilde{b}_2 \\ \tilde{b}_3 \end{bmatrix}$$

The non-basic variables are $\tilde{x}_1 = \tilde{0}$, $\tilde{x}_4 = \tilde{0}$ and $\tilde{x}_5 = \tilde{0}$ and the fuzzy objective value is $\tilde{z} = \left(\frac{77,416}{169}, \frac{94,235}{169}, \frac{120,265}{169}, \frac{140,084}{169} \right)$.

Step 2: We find $\tilde{w}$ by solving the fuzzy system $\tilde{w}B = \tilde{c}_B$:

$$
(\tilde{w}_1, \tilde{w}_2, \tilde{w}_3) \begin{bmatrix} 13 & 12 & 0 \\ 0 & 13 & 0 \\ 15 & 0 & 1 \end{bmatrix} = \begin{bmatrix} (11,13,15,17) \\ (9,12,14,17) \\ (0,0,0,0) \end{bmatrix}
$$

$$
\Rightarrow \tilde{w}_1 = \left(\frac{9}{13}, \frac{12}{13}, \frac{14}{13}, \frac{17}{13} \right), \tilde{w}_2
$$

$$
\approx \left(\frac{-35}{169}, \frac{27}{169}, \frac{77}{169}, \frac{139}{169} \right), \tilde{w}_3 = (0,0,0,0) = \tilde{0}
$$

Step 3: We calculate $\tilde{z}_j - \tilde{c}_j = \tilde{w}a_j - \tilde{c}_j$ for all $j \in J_N = \{1,4,5\}$:

$$
\tilde{z}_1 - \tilde{c}_1 = \tilde{w}a_1 - \tilde{c}_1 = \left(\frac{-687}{13}, \frac{15}{169}, \frac{729}{169}, \frac{1467}{169} \right)
$$

$$
\tilde{z}_4 - \tilde{c}_4 = \tilde{w}a_4 - \tilde{c}_4 = \left(\frac{9}{13}, \frac{12}{13}, \frac{14}{13}, \frac{17}{13} \right)
$$

$$
\tilde{z}_5 - \tilde{c}_5 = \tilde{w}a_5 - \tilde{c}_5 = \left(\frac{-35}{169}, \frac{27}{169}, \frac{77}{169}, \frac{139}{169} \right)
$$

Now we have:

$$
t_{2k} + t_{3k} = \min_{j \in J_N} \{ t_{2j} + t_{3j} \} = \min \left\{ \frac{51}{169} + \frac{729}{169}, \frac{12}{13} + \frac{14}{13}, \frac{27}{169} + \frac{77}{169} \right\}
$$

$$
= \frac{27}{169} + \frac{77}{169} = t_{25} + t_{35}
$$

Since $t_{25} + t_{35} = \frac{27}{169} + \frac{77}{169} \geq 0$, then stop with the current basic solution as a fuzzy optimal solution.

Theorem 3.30 *In the absence of degeneracy,* Algorithm 3.4 *stops in a finite number of iterations, either with a fuzzy optimal fuzzy basic feasible solution or with the conclusion that the optimal value is unbounded.*

Proof In the absence of degeneracy, every fuzzy basic feasible solution has exactly m fuzzy positive components and has a unique associated basis. Also, at each iteration of the method, one of the following three actions is executed. It may be stop with an optimal fuzzy basic solution if $t_{2k} + t_{3k} \geq 0$, i.e. $\tilde{z}_k - \tilde{c}_k \succeq \tilde{0}$; it may stop with an unbounded solution if $t_{2k} + t_{3k} < 0$ and $y_k \leq 0$; or else it gives a new fuzzy basic feasible solution if $t_{2k} + t_{3k} < 0$ and $y_k \nleq 0$. In the absence of degeneracy, $\bar{b}_{2r} + \bar{b}_{3r} > 0$, i.e. $\tilde{\tilde{x}}_B = \tilde{\tilde{b}}_r \succ \tilde{0}$ and hence $\frac{\tilde{\tilde{x}}_{B_r}}{y_{rk}} \succ \tilde{0}$. By (3.116), the difference between the fuzzy objective values at the previous iteration and the current iteration is $\frac{\tilde{\tilde{x}}_{B_r}}{y_{rk}} (\tilde{z}_k - \tilde{c}_k) \prec \tilde{0}$. Thus, the fuzzy objective value increases strictly in each iteration. Hence, a basis that appears once in the course of the method can never reappear. Moreover, the total number of bases for (3.111) is less than or equal to $\binom{n}{m}$. Hence, the method would stop in a finite number of steps with a finite optimal fuzzy basic solution or with an unbounded optimal solution.   □

Ebrahimnejad [25] described the symmetric fuzzy primal method proposed by Ganesan and Veeramani [24] in tableau method. Suppose that we have a starting fuzzy basic solution $\tilde{x}$ with the basis B. The LPP with fuzzy parameters of type-3 (3.111) can be represented as follows.

$$\begin{aligned} \max \quad & \tilde{z} \\ s.t \quad & \tilde{z} - \tilde{c}_B \tilde{x}_B - \tilde{c}_N \tilde{x}_N \approx \tilde{0} \\ & B\tilde{x}_B + N\tilde{x}_N \approx \tilde{b} \\ & \tilde{x}_B, \tilde{x}_N \succeq \tilde{0} \end{aligned} \qquad (3.119)$$

From the second constraint of (3.119) we obtain

$$\tilde{x}_B + B^{-1} N \tilde{x}_N \approx B^{-1} \tilde{b} \qquad (3.120)$$

Multiplying (3.120) by $\tilde{c}_B$ and adding to $\tilde{z} - \tilde{c}_B \tilde{x}_B - \tilde{c}_N \tilde{x}_N \approx \tilde{0}$, we obtain

$$\tilde{z} + \tilde{0}\tilde{x}_B + (\tilde{c}_B B^{-1} N - \tilde{c}_N)\tilde{x}_N \approx \tilde{c}_B B^{-1} \tilde{b} \qquad (3.121)$$

Currently $\tilde{x}_N \approx 0$, and from (3.120) and (3.121) we obtain $\tilde{x}_B \approx B^{-1}\tilde{b}$ and $\tilde{z} \approx \tilde{c}_B B^{-1}\tilde{b}$. Also. from (3.120) and (3.121), we can represent the current fuzzy basic solution with the basis B in Table 3.22.

We note that in Table 3.22, there are two objective rows: the first gives the $\tilde{z}_j - \tilde{c}_j$ and the second is used to choose the pivoting column. Also, two right-hand side columns exist: the first column gives the current fuzzy basic variables values and the second column gives the real numbers corresponding to the fuzzy numbers given in the first column to choose the pivoting row. In fact, similar to the simplex method for solving LP problems in a crisp environment, we can design a fuzzy simplex method in tableau format for solving LPP with fuzzy parameters of type-3 (3.111). We now describe the symmetric fuzzy primal method, proposed by Ganesan and Veermani [24], in tableau format as follows:

Table 3.22 The initial symmetric simplex tableau

Basis	$\tilde{x}_1$	...	$\tilde{x}_r$	...	$\tilde{x}_m$	...	$\tilde{x}_j$	...	$\tilde{x}_k$	...	R.H.S	$\Re$
$\tilde{z}$	$\tilde{0}$	...	$\tilde{0}$	...	$\tilde{0}$	...	$\tilde{z}_j - \tilde{c}_j$	...	$\tilde{z}_k - \tilde{c}_k$	...	$\tilde{z} = \tilde{c}_B \tilde{b}$	–
$\Re(\tilde{z})$	0	...	0	...	0	...	$t_{2j} + t_{3j}$	...	$t_{2k} + t_{3k}$	...	–	$z_2 + z_3$
$\tilde{x}_1$	1	...	0	...	0	...	y_{1j}	...	y_{1k}	...	$\tilde{b}_1$	$\bar{b}_{21} + \bar{b}_{31}$
$\vdots$	$\vdots$	...	$\vdots$	...	$\vdots$	...	$\vdots$	...	$\vdots$	...	$\vdots$	
$\tilde{x}_r$	$\vdots$	...	1	...	0	...	y_{rj}	...	y_{rk}	...	$\tilde{b}_r$	$\bar{b}_{2r} + \bar{b}_{3r}$
$\vdots$	$\vdots$	...	$\vdots$	...	$\vdots$	...	$\vdots$	...	$\vdots$	...	$\vdots$	
$\tilde{x}_m$	0	...	0	...	1	...	y_{mj}	...	y_{mk}	...	$\tilde{b}_m$	$\bar{b}_{2m} + \bar{b}_{3m}$

Algorithm 3.5: Tableau Symmetric fuzzy primal simplex algorithm (Maximization Problem)

Initialization step

Choose a starting fuzzy feasible basic solution with the basis B. Form the initial tableau as in Table 3.22.

Main steps

(1) Let $t_{2k} + t_{3k} = \max_{1 \le j \le n} \{t_{2j} + t_{3j}\}$. If $t_{2k} + t_{3k} \ge 0$, then stop with the current basic solution as an optimal solution.

(2) If $y_k \le 0$ then stop with the conclusion that the problem is unbounded. Otherwise, $\tilde{x}_k$ enters the basis and $\tilde{x}_{B_r}$ leaves the basis, providing that

$$\frac{\bar{b}_{2r} + \bar{b}_{3r}}{y_{rk}} = \min_{1 \le i \le m} \left\{ \frac{\bar{b}_{2i} + \bar{b}_{3i}}{y_{ik}} \middle| y_{ik} > 0 \right\}$$

(3) Update the tableau by pivoting on y_{rk} and go to (1).

Table 3.23 The initial simplex tableau of Example 3.14

Basis	$\tilde{x}_1$	$\tilde{x}_2$	$\tilde{x}_3$	$\tilde{x}_4$	$\tilde{x}_5$	$\tilde{x}_6$	R.H.S	$\Re$
z	-28	-26	-32	0	0	0	–	0
$\tilde{z}$	$(-17, -15,$ $-13, -11)$	$(-17, -14,$ $-12, -9)$	$(-19, -17,$ $-15, -13)$	$\tilde{0}$	$\tilde{0}$	$\tilde{0}$	$\tilde{0}$	–
$\tilde{x}_4$	12	13	12	1	0	0	$(469, 475,$ $505, 511)$	980
$\tilde{x}_5$	14	0	13	0	1	0	$(452, 460,$ $480, 488)$	940
$\tilde{x}_6$	12	15	0	0	0	1	$(460, 465,$ $495, 500)$	960

Example 3.14 [25] Reconsider the LPP with fuzzy parameters of type-3 (3.117) given in Example 3.13.

After introducing the slack variables $\tilde{x}_4$, $\tilde{x}_5$ and $\tilde{x}_6$ we obtain the initial fuzzy primal simplex tableau as Table 3.23.

Now $\tilde{x}_3$ is an entering variable and $\tilde{x}_5$ is a leaving variable. Then by pivoting on $y_{23} = 13$, we obtain the next tableau as Table 3.24.

In this case $\tilde{x}_2$ is an entering variable and $\tilde{x}_4$ is a leaving variable. The new tableau is given as Table 3.25 by pivoting on $y_{12} = 13$.

This is the optimal tableau since $\tilde{z}_j - \tilde{c}_j \geq \tilde{0}$ for all non-basic fuzzy variables. The optimal solution is matched with the solution obtained by Ganesan and Veeramani [24], given as follows:

$$\tilde{x}_1 \approx \tilde{0}, \tilde{x}_2 = \left(\frac{241}{169}, \frac{415}{169}, \frac{1045}{169}, \frac{1219}{169}\right), \tilde{x}_3 = \left(\frac{452}{13}, \frac{460}{13}, \frac{480}{13}, \frac{488}{13}\right)$$

Note that the Algorithms 3.4 and 3.5 for solving the LPP with fuzzy parameters of type-3 (3.111) start with a fuzzy basic feasible solution and move to an improved fuzzy basic solution, until the optimal solution is reached or else unboundedness of the fuzzy objective function is verified. However, in order to initialize this method, a basis B with $B^{-1}b \succsim \tilde{0}$ must be available. Ebrahimnejad [25] described a procedure to obtain an initial fuzzy basic solution with a very simple basis, namely the identity.

Consider problem (3.111) where $\tilde{b} \succeq \tilde{0}$. Suppose that A has no identity submatrix. To illustrate the procedure of finding an identity initial basis, suppose that the restrictions by adding an fuzzy artificial vector $\tilde{x}_a$ are changed to the fuzzy system $A\tilde{x} + \tilde{x}_a \approx \tilde{b}, \tilde{x} \succeq \tilde{0}, \tilde{x}_a \succeq \tilde{0}$. Hence, an identity matrix corresponding to the fuzzy artificial vector is obtained. This gives an immediate fuzzy basic solution of the new fuzzy system, namely $\tilde{x}_a \approx \tilde{b}, \tilde{x} \approx \tilde{0}$. Thus, the fuzzy primal method can be applied. But, these artificial vectors must change to fuzzy zero, because $A\tilde{x} \approx \tilde{b}$ if and only if $A\tilde{x} + \tilde{x}_a \approx \tilde{b}$ with $\tilde{x}_a \approx \tilde{0}$. To eliminate the fuzzy artificial vectors, it is sufficient to mimimize the sum of the artificial vectors. In other words, the following LPP with fuzzy parameters of type-3 is solved starting with the fuzzy basic feasible solution $\tilde{x} \approx \tilde{0}$ and $\tilde{x}_a \approx \tilde{b}$.

$$\begin{aligned} \min \quad & \tilde{z} \approx \tilde{1}\tilde{x} \\ s.t. \quad & A\tilde{x} + x_a \approx \tilde{b} \\ & \tilde{x} \succeq \tilde{0}, \tilde{x}_a \succeq \tilde{0} \end{aligned} \qquad (3.122)$$

where $\tilde{1} = ((1, 1, 1, 1), \ldots, (1, 1, 1, 1))$.

If the original problem (3.111) has a fuzzy feasible solution, the optimal value of (3.122) is fuzzy zero, where all the fuzzy artificial variables drop to fuzzy zero. In this case, a fuzzy basic solution of the (3.111) is found and the fuzzy primal method can be started with the original objective $\tilde{c}\tilde{x}$. Otherwise, at optimality $\tilde{x}_a \not\approx \tilde{0}$ and the

Table 3.24 The first iteration

Basis	$\tilde{x}_1$	$\tilde{x}_2$	$\tilde{x}_3$	$\tilde{x}_4$	$\tilde{x}_5$	$\tilde{x}_6$	R.H.S	$\Re$
z	$\frac{84}{13}$	-26	$\frac{32}{13}$	0	0	0	–	$\frac{15,040}{13}$
$\tilde{z}$	$\left(\frac{-39}{13},\frac{15}{13},\frac{69}{13},\frac{123}{13}\right)$	$(-17,-14,-12,-9)$	$\tilde{0}$	$\tilde{0}$	$\left(\frac{13}{13},\frac{15}{13},\frac{17}{13},\frac{19}{13}\right)$	$\tilde{0}$	$\left(\frac{5794}{13},\frac{6890}{13},\frac{8150}{13},\frac{9246}{13}\right)$	–
$\tilde{x}_4$	$-\frac{12}{13}$	13	0	1	$-\frac{12}{13}$	0	$\left(\frac{241}{13},\frac{415}{13},\frac{1045}{13},\frac{1219}{13}\right)$	$\frac{1460}{13}$
$\tilde{x}_3$	$\frac{14}{13}$	0	1	0	$\frac{1}{13}$	0	$\left(\frac{452}{13},\frac{460}{13},\frac{480}{13},\frac{488}{13}\right)$	$\frac{940}{13}$
$\tilde{x}_6$	12	15	0	0	0	0	$(460, 465, 495, 500)$	960

Table 3.25 The optimal iteration

Basis	$\tilde{x}_1$	$\tilde{x}_2$	$\tilde{x}_3$	$\tilde{x}_4$	$\tilde{x}_5$	$\tilde{x}_6$	R.H.S	$\Re$
z	$\frac{789}{13}$	0	0	$\frac{26}{13}$	0	0	—	$\frac{214,500}{169}$
$\tilde{z}$	$\left(\frac{-687}{13}, \frac{51}{169}, \frac{729}{169}, \frac{1467}{169}\right)$	$\tilde{0}$	$\tilde{0}$	$\left(\frac{9}{13}, \frac{12}{13}, \frac{14}{13}, \frac{17}{13}\right)$	$\left(\frac{-35}{13}, \frac{27}{13}, \frac{62}{13}, \frac{139}{13}\right)$	$\tilde{0}$	$\left(\frac{77,416}{169}, \frac{94,235}{169}, \frac{120,265}{169}, \frac{140,084}{169}\right)$	—
$\tilde{x}_2$	$\frac{-12}{169}$	1	0	$\frac{1}{13}$	$-\frac{12}{169}$	0	$\left(\frac{241}{169}, \frac{415}{169}, \frac{1045}{169}, \frac{1219}{169}\right)$	$\frac{1460}{169}$
$\tilde{x}_3$	$\frac{14}{13}$	0	1	0	$\frac{1}{13}$	0	$\left(\frac{452}{13}, \frac{460}{13}, \frac{480}{13}, \frac{488}{13}\right)$	$\frac{940}{13}$
$\tilde{x}_6$	$\frac{1848}{169}$	0	0	$\frac{-15}{13}$	0	1	$\left(\frac{59,455}{169}, \frac{62,910}{169}, \frac{77,430}{169}, \frac{80,885}{169}\right)$	$\frac{140340}{169}$

(3.111) has no fuzzy solution, because if there is an $\tilde{x}_a \succeq \tilde{0}$ with $A\tilde{x} \approx \tilde{b}$, then $\begin{bmatrix} \tilde{x} \\ \tilde{0} \end{bmatrix}$ is a fuzzy feasible solution of the (19) and $\tilde{0}(\tilde{x}) + \tilde{1}(\tilde{0}) \approx \tilde{0} \prec \tilde{1}\tilde{x}_a$, violating optimality $\tilde{x}_a$.

Example 3.15 Consider the following problem:

$$\max \quad \tilde{z} \approx (-3, -2, 0, 1)\tilde{x}_1 + (-1, 1, 3, 5)\tilde{x}_2$$
$$\text{s.t}$$
$$\tilde{x}_1 + \tilde{x}_2 \succeq (-1, 1, 3, 5)$$
$$-\tilde{x}_1 + \tilde{x}_2 \succeq (-1, 0, 2, 3)$$
$$\tilde{x}_2 \preceq (1, 2, 4, 5)$$
$$\tilde{x}_1, \tilde{x}_2 \succeq \tilde{0}$$

After introducing the slack variables $\tilde{x}_3, \tilde{x}_4, \tilde{x}_5$ and the fuzzy artificial variables $\tilde{x}_6$ and $\tilde{x}_7$, the following FLP to eliminate fuzzy artificial variables is obtained.

$$\min \quad \tilde{z} \approx (1, 1, 1, 1)\tilde{x}_6 + (1, 1, 1, 1)\tilde{x}_7$$
$$\text{s.t}$$
$$\tilde{x}_1 + \tilde{x}_2 - \tilde{x}_3 + \tilde{x}_6 \approx (-1, 1, 3, 5)$$
$$-\tilde{x}_1 + \tilde{x}_2 - \tilde{x}_4 + \tilde{x}_7 \approx (-1, 0, 2, 3)$$
$$\tilde{x}_2 + \tilde{x}_5 \approx (1, 2, 4, 5)$$
$$\tilde{x}_1, \tilde{x}_2, \tilde{x}_3, \tilde{x}_4, \tilde{x}_5, \tilde{x}_6, \tilde{x}_7 \succeq \tilde{0}$$

Now we may rewrite the above FLP problem in the primal simplex tableau format as Table 3.26.

Multiply rows 1 and 2 by $(1, 1, 1, 1)$ and add to the row corresponding to $\tilde{z}$. Thus the new tableau is given as Table 3.27.

Table 3.26 The initial tabaleau

Basic	$\tilde{x}_1$	$\tilde{x}_2$	$\tilde{x}_3$	$\tilde{x}_4$	$\tilde{x}_5$	$\tilde{x}_6$	$\tilde{x}_7$	RHS	$\Re$
z	0	0	0	0	0	-2	-2	–	0
$\tilde{z}$	$\tilde{0}$	$\tilde{0}$	$\tilde{0}$	$\tilde{0}$	$\tilde{0}$	$(-1, -1, -1, -1,)$	$(-1, -1, -1, -1,)$	$\tilde{0}$	–
$\tilde{x}_6$	1	1	-1	0	0	1	0	$(-1, 1, 3, 5)$	4
$\tilde{x}_7$	-1	1	0	-1	0	0	1	$(-1, 0, 2, 3)$	2
$\tilde{x}_5$	0	1	0	0	1	0	0	$(1, 2, 4, 5)$	6

Table 3.27 The revised initial tableau

Basic	$\tilde{x}_1$	$\tilde{x}_2$	$\tilde{x}_3$	$\tilde{x}_4$	$\tilde{x}_5$	$\tilde{x}_6$	$\tilde{x}_7$	RHS	$\Re$
z	0	4	-2	-2	0	0	0	–	6
$\tilde{z}$	$\tilde{0}$	$(2, 2, 2, 2)$	$(-1, -1, -1, -1,)$	$(-1, -1, -1, -1,)$	$\tilde{0}$	$\tilde{0}$	$\tilde{0}$	$(1, 1, 5, 5)$	–
$\tilde{x}_6$	1	1	-1	0	0	1	0	$(-1, 1, 3, 5)$	4
$\tilde{x}_7$	-1	1	0	-1	0	0	1	$(-1, 0, 2, 3)$	2
$\tilde{x}_5$	0	1	0	0	1	0	0	$(1, 2, 4, 5)$	6

Table 3.28 First iteration

Basic	$\tilde{x}_1$	$\tilde{x}_2$	$\tilde{x}_3$	$\tilde{x}_4$	$\tilde{x}_5$	$\tilde{x}_6$	$\tilde{x}_7$	RHS	$\Re$
z	4	0	−2	2	0	0	−4	–	2
$\tilde{z}$	(2, 2, 2, 2)	$\tilde{0}$	(−1, −1, −1, −1)	(1, 1, 1, 1)	$\tilde{0}$	$\tilde{0}$	(−2, −2, −2, −2)	(−3, −3, 5, 5)	–
$\tilde{x}_6$	2	0	−1	1	0	1	−1	(−4, −1, 3, 6)	2
$\tilde{x}_2$	−1	1	0	−1	0	0	1	(−1, 0, 2, 3)	2
$\tilde{x}_5$	0	1	0	0	1	0	0	(−2, 0, 4, 6)	4

Table 3.29 First iteration

Basic	$\tilde{x}_1$	$\tilde{x}_2$	$\tilde{x}_3$	$\tilde{x}_4$	$\tilde{x}_5$	$\tilde{x}_6$	$\tilde{x}_7$	RHS	$\Re$
z	0	0	0	2	0	−2	−2	–	0
$\tilde{z}$	$\tilde{0}$	$\tilde{0}$	$\tilde{0}$	$\tilde{0}$	$\tilde{0}$	(−1, −1, −1, −1)	(−1, −1, −1, −1)	(−6, −6, 6, 6) ≈ $\tilde{0}$	–
$\tilde{x}_1$	1	0	$\frac{-1}{2}$	$\frac{1}{2}$	0	$\frac{1}{2}$	$\frac{-1}{2}$	$\left(\frac{-4}{2}, \frac{-1}{2}, \frac{3}{2}, \frac{6}{2}\right)$	1
$\tilde{x}_2$	0	1	$\frac{-1}{2}$	$\frac{-1}{2}$	0	$\frac{1}{2}$	$\frac{1}{2}$	$\left(\frac{-6}{2}, \frac{-1}{2}, \frac{7}{2}, \frac{12}{2}\right)$	3
$\tilde{x}_5$	0	0	$\frac{1}{2}$	$\frac{1}{2}$	1	$\frac{-1}{2}$	$\frac{-1}{2}$	$\left(\frac{-10}{2}, \frac{3}{2}, \frac{9}{2}, \frac{16}{2}\right)$	3

Now $\tilde{x}_2$ is an entering variable and $\tilde{x}_7$ is a leaving variable. Then by pivoting on $y_{22} = 1$, we obtain the next tableau as Table 3.28.

In this case $\tilde{x}_1$ is an entering variable and $\tilde{x}_6$ is a leaving variable. The new tableau is as Table 3.29.

Since all the fuzzy artificial variables are at fuzzy zero level, we have an initial fuzzy basic solution with $B = [a_1, a_2, a_5]$ for the original FLP problems. Now we can form the initial Table 3.22 with the basis B and then use the Algorithm 3.5 to solve the original FLP problem.

3.5.2 Ebrahimnejad's Approach

In the fuzzy simplex method (Algorithms 3.4 and 3.5), many computations are performed at every pivoting step. Every time a pivot is performed, it is carried out on every column of the tableau. This can be very time-consuming. Ebrahimnejad [25] proposed a fuzzy revised simplex method, where all necessary computations are carried out restricting the pivoting operations on inverse matrix. In fact, the fuzzy revised simplex method is a systematic procedure for implementing the steps of the fuzzy simplex method proposed by Ganesan and Veeramani [24] in a smaller array, thus saving storage space.

Suppose that a fuzzy basic feasible solution with a known B^{-1} is at hand. Table 3.30, called the fuzzy revised simplex tableau, is constructed where $\tilde{w} = \tilde{c}_B B^{-1}$ and $\bar{\tilde{b}} = B^{-1}\tilde{b}$.

Note that similar to the canonical tableau, there are two right-hand side columns: the first gives the current fuzzy basic variables values and the second gives the real numbers corresponding to fuzzy numbers to do a minimum ratio test. Also, the objective rows are used to check optimality conditions. In fact, the first piece of

Table 3.30 The fuzzy revised simplex tableau

Basic inverse	R.H.S	$\Re$
w	–	$\tilde{c}_B \bar{b}$
$\tilde{w}$	$\tilde{c}_B \tilde{\bar{b}}$	–
B^{-1}	$\tilde{\bar{b}}$	$\bar{b}$

information required is the values of $\tilde{z}_j - \tilde{c}_j$. Since $\tilde{w}$ is known, these values are calculated as in Step 3 of the Algorithm 3.4 (Step 1 of the Algorithm 3.5) to check for optimality. Suppose that $\tilde{z}_j - \tilde{c}_j \prec \tilde{0}$. Then, the column of $\tilde{x}_k$ needs to be updated. Using B^{-1}, the value of $y_k = B^{-1} a_k$ is computed. If $y_k \leq 0$, then stop with the indication that the optimal solution value is unbounded. Otherwise, the updated column of $\tilde{x}_k$ can be appended to the fuzzy revised simplex tableau as shown, while the rest of the tableau is still kept hidden (Table 3.31).

The index r of Step 4 in the Algorithm 3.4 (Step 2 in the Algorithm 3.5) can be calculated by the usual minimum test. Also, pivoting on y_{rk} gives as usual the new $\tilde{w}$, B^{-1}, $\tilde{\bar{b}}$ and $\tilde{c}_B \tilde{B}$ and the process is repeated. The following is a summary of the fuzzy revised simplex method.

Algorithm 3.6. The fuzzy revised simplex method

Initialization Step

Find an initial fuzzy basic feasible solution with the basis B. Calculate $\tilde{w} = \tilde{c}_B B^{-1}$, $\tilde{\bar{b}}$ and form the fuzzy revised simple tableau (Table 3.30).

Main step

(1) Calculate $\tilde{z}_j - \tilde{c}_j$ for all non-basic variables. Let $t_{2k} + t_{3k} = \min\limits_{1 \leq j \leq n} \{t_{2j} + t_{3j}\}$. If $t_{2k} + t_{3k} \geq 0$, then stop; the current solution is optimal. Otherwise go to Step 2.

(2) Let $y_k = B^{-1} a_k$. If $y_k \leq 0$, then stop; the problem is unbounded. Otherwise,

insert the column $\begin{bmatrix} t_{2k} + t_{3k} \\ \tilde{z}_k - \tilde{c}_k \\ y_k \end{bmatrix}$ to the right of the fuzzy revised simplex tableau

as in Table 3.31. Determine the index r as follows:

$$\frac{\bar{b}_{2r} + \bar{b}_{3r}}{y_{rk}} = \min\limits_{1 \leq i \leq m} \left\{ \frac{\bar{b}_{2i} + \bar{b}_{3i}}{y_{ik}} \middle| y_{ik} > 0 \right\}.$$

(3) Pivot at y_{rk}. This updates the fuzzy tableau. Now the column corresponding to $\tilde{x}_k$ is completely eliminated from the tableau and the main step is repeated.

Table 3.31 Adding the updated column of $\tilde{x}_k$

Basic inverse	R.H.S	$\Re$	$\tilde{x}_k$
w	–	$\tilde{c}_B \bar{b}$	$t_{2k} + t_{3k}$
$\tilde{w}$	$\tilde{c}_B \tilde{\bar{b}}$	–	$\tilde{z}_k - \tilde{c}_k$
B^{-1}	$\tilde{\bar{b}}$	$\bar{b}$	y_k

Table 3.32 The initial fuzzy revised simplex tableau

	Basic inverse			R.H.S.	$\Re$
$\Re(\tilde{z})$	0	0	0	–	0
$\tilde{z}$	$\tilde{0}$	$\tilde{0}$	$\tilde{0}$	$\tilde{0}$	–
$\tilde{x}_4$	1	0	0	$(469, 475, 505, 511$	980
$\tilde{x}_5$	0	1	0	$(452, 460, 480, 488)$	940
$\tilde{x}_6$	0	0	1	$(460, 465, 495, 500)$	960

Example 3.16 For an illustration of the fuzzy revised method, we solve the LPP with fuzzy parameters (3.117) given in Example 3.13 by the Algorithm 3.6.

Initial Step:
The initial basic is $\tilde{B} = [a_4, a_5, a_6]$. Also, $\tilde{w} = \tilde{c}_B B^{-1} = (\tilde{0}, \tilde{0}, \tilde{0})$ and $\bar{b} = [\tilde{b}_1, \tilde{b}_2, \tilde{b}_3] = [(469, 475, 505, 511), (450, 460, 480, 488), (460, 465, 495, 500)]$. The initial fuzzy revised simple tableau is given as Table 3.32.

Iteration 1:

Step 1: Noting that $\tilde{z}_j - \tilde{c}_j = \tilde{w}a_j - \tilde{c}_j$, we obtain

$$\tilde{z}_1 - \tilde{c}_1 = (-17, -15, -13, -11), \tilde{z}_2 - \tilde{c}_2 = (-17, -14, -12, -9),$$
$$\tilde{z}_3 - \tilde{c}_3 = (-19, -17, -15, -13)$$
$$t_{2k} + t_{3k} = \min_{1 \le j \le 3} \{t_{2j} + t_{3j}\} = \min\{-28, -26, -32\} = -32 = t_{23} + t_{33}$$

Thus $k = 3$ and $\tilde{x}_3$ enters the basis.

Step 2: We have

$$y_3 = B^{-1}a_3 = \begin{bmatrix} 1 & 0 & 0 \\ 0 & 1 & 0 \\ 0 & 0 & 1 \end{bmatrix} \begin{bmatrix} 12 \\ 13 \\ 0 \end{bmatrix} = \begin{bmatrix} 12 \\ 13 \\ 0 \end{bmatrix}$$

Insert the vector

$$\begin{bmatrix} t_{23} + t_{33} \\ \tilde{z}_3 - \tilde{c}_3 \\ y_3 \end{bmatrix} = \begin{bmatrix} -32 \\ (-19, -17, -15, -13) \\ 12 \\ 13 \\ 0 \end{bmatrix}$$

to the right of Table 3.32 leading to Table 3.33.

Table 3.33 Adding the updated column of $\tilde{x}_3$

	Basic inverse			R.H.S.	$\Re$	$\tilde{x}_3$
$\Re(\tilde{z})$	0	0	0	–	0	-32
$\tilde{z}$	$\tilde{0}$	$\tilde{0}$	$\tilde{0}$	$\tilde{0}$	–	$(-19, -17, -15, -13)$
$\tilde{x}_4$	1	0	0	$(469, 475, 505, 511)$	980	12
$\tilde{x}_5$	0	1	0	$(452, 460, 480, 488)$	940	13
$\tilde{x}_6$	0	0	1	$(460, 465, 495, 500)$	960	0

The index r of the leaving variable is determined as follows:

$$\frac{\bar{b}_{2r} + \bar{b}_{3r}}{y_{r3}} = \min_{1 \leqslant i \leqslant m} \left\{ \frac{\bar{b}_{2i} + \bar{b}_{3i}}{y_{i3}} \mid y_{i3} > 0 \right\} = \min \left\{ \frac{980}{12}, \frac{940}{13} \right\} = \frac{940}{13} = \frac{\bar{b}_{22} + \bar{b}_{32}}{y_{23}}$$

Thus $r = 2$ and $\tilde{x}_{B_2} = \tilde{x}_5$ leaves the basis.

Step 3:

Now pivoting at $y_{23} = 13$ gives the new solution as reported in Table 3.34.

Iteration 2:

Step 1: Here $\tilde{w} = \left(\tilde{0}, (\frac{13}{13}, \frac{15}{13}, \frac{17}{13}, \frac{19}{13}), \tilde{0} \right)$. Noting that $\tilde{z}_j - \tilde{c}_j = \tilde{w}a_j - \tilde{c}_j$, we obtain

$$\tilde{z}_1 - \tilde{c}_1 = \left(\frac{-39}{13}, \frac{15}{13}, \frac{69}{13}, \frac{123}{13} \right), \tilde{z}_2 - \tilde{c}_2 = (-17, -14, -12, -9), \tilde{z}_5 - \tilde{c}_5 = \left(\frac{13}{13}, \frac{15}{13}, \frac{17}{13}, \frac{19}{13} \right)$$

$$t_{2k} + t_{3k} = \min_{j=1,2,5} \{ t_{2j} + t_{3j} \} = \min \left\{ \frac{84}{13}, -26, \frac{32}{13} \right\} = -26 = t_{22} + t_{32}$$

Thus $k = 2$ and $\tilde{x}_2$ enters the basis.

Table 3.34 First iteration of the fuzzy revised algorithm

	Basic inverse			R.H.S.	$\Re$
$\Re(\tilde{z})$	0	$\frac{32}{13}$	0	–	$\frac{15{,}040}{13}$
$\tilde{z}$	$\tilde{0}$	$\left(\frac{13}{13}, \frac{15}{13}, \frac{17}{13}, \frac{19}{13} \right)$	$\tilde{0}$	$\left(\frac{5794}{13}, \frac{6890}{13}, \frac{8150}{13}, \frac{9246}{13} \right)$	–
$\tilde{x}_4$	1	$-\frac{12}{13}$	0	$\left(\frac{241}{13}, \frac{415}{13}, \frac{1045}{13}, \frac{1219}{13} \right)$	$\frac{1460}{13}$
$\tilde{x}_3$	0	$\frac{1}{13}$	0	$\left(\frac{452}{13}, \frac{460}{13}, \frac{480}{13}, \frac{488}{13} \right)$	$\frac{940}{13}$
$\tilde{x}_6$	0	0	1	$(460, 465, 495, 500)$	960

Step 2: We have

$$
y_2 = B^{-1}a_2 = \begin{bmatrix} 1 & -\frac{12}{13} & 0 \\ 0 & \frac{1}{13} & 0 \\ 0 & 0 & 1 \end{bmatrix} \begin{bmatrix} 13 \\ 0 \\ 15 \end{bmatrix} = \begin{bmatrix} 13 \\ 0 \\ 15 \end{bmatrix}
$$

Insert the vector

$$
\begin{bmatrix} t_{22} + t_{32} \\ \tilde{z}_2 - \tilde{c}_2 \\ y_2 \end{bmatrix} = \begin{bmatrix} -26 \\ (-17, -14, -12, -9) \\ 13 \\ 0 \\ 15 \end{bmatrix}
$$

to the right of Table 3.31 leading to Table 3.35.

The index r of the leaving variable is determined as follows:

$$
\frac{\bar{b}_{2r} + \bar{b}_{3r}}{y_{r2}} = \min_{1 \leqslant i \leqslant m} \left\{ \frac{\bar{b}_{2i} + \bar{b}_{3i}}{y_{i2}} | y_{i2} > 0 \right\} = \min\left\{ \frac{\frac{1460}{13}}{13}, \frac{960}{15} \right\} = \frac{\frac{1460}{13}}{13} = \frac{\bar{b}_{21} + \bar{b}_{31}}{y_{12}}
$$

Thus $r = 1$ and $\tilde{x}_{B_2} = \tilde{x}_4$ leaves the basis.

Step 3:

Now pivoting at $y_{12} = 13$ gives the new solution as reported in Table 3.36.

Iteration 3:

Step 1: Here $\tilde{w} = \left(\left(\frac{9}{13}, \frac{12}{13}, \frac{14}{13}, \frac{17}{13} \right), \left(\frac{-35}{13}, \frac{27}{13}, \frac{62}{13}, \frac{139}{13} \right), \tilde{0} \right)$. Noting that $\tilde{z}_j - \tilde{c}_j = \tilde{w}a_j - c_j$, we obtain

Table 3.35 Adding the updated column of $\tilde{x}_2$

	Basic inverse			R.H.S.	$\Re$	$\tilde{x}_2$
$\Re(\tilde{z})$	0	$\frac{32}{13}$	0	–	$\frac{15,040}{13}$	-26
$\tilde{z}$	$\tilde{0}$	$\left(\frac{13}{13}, \frac{15}{13}, \frac{17}{13}, \frac{19}{13} \right)$	$\tilde{0}$	$\left(\frac{5794}{13}, \frac{6890}{13}, \frac{8150}{13}, \frac{9246}{13} \right)$	–	$(-17, -14, -12, -9)$
$\tilde{x}_4$	1	$-\frac{12}{13}$	0	$\left(\frac{241}{13}, \frac{415}{13}, \frac{1045}{13}, \frac{1219}{13} \right)$	$\frac{1460}{13}$	13
$\tilde{x}_5$	0	$\frac{1}{13}$	0	$\left(\frac{452}{13}, \frac{460}{13}, \frac{480}{13}, \frac{488}{13} \right)$	$\frac{940}{13}$	0
$\tilde{x}_6$	0	0	1	$(460, 465, 495, 500)$	960	15

Table 3.36 Second iteration of the fuzzy revised algorithm

	Basic inverse			R.H.S.	$\Re$
$\Re(\tilde{z})$	$\frac{26}{13}$	$\frac{89}{13}$	0	$-$	$\frac{214,500}{169}$
$\tilde{z}$	$\left(\frac{9}{13}, \frac{12}{13}, \frac{14}{13}, \frac{17}{13}\right)$	$\left(\frac{-35}{13}, \frac{27}{13}, \frac{62}{13}, \frac{139}{13}\right)$	$\tilde{0}$	$\left(\frac{77,416}{169}, \frac{94,235}{169}, \frac{120,265}{169}, \frac{140,084}{169}\right)$	$-$
$\tilde{x}_4$	$\frac{1}{13}$	$-\frac{12}{169}$	0	$\left(\frac{241}{169}, \frac{415}{169}, \frac{1045}{169}, \frac{1219}{169}\right)$	$\frac{1460}{13}$
$\tilde{x}_3$	0	$-\frac{1}{13}$	0	$\left(\frac{452}{13}, \frac{460}{13}, \frac{480}{13}, \frac{488}{13}\right)$	$\frac{940}{13}$
$\tilde{x}_6$	$\frac{-15}{13}$	0	1	$\left(\frac{59,455}{169}, \frac{62,910}{169}, \frac{77,430}{169}, \frac{80,885}{169}\right)$	$\frac{140,340}{169}$

$$\tilde{z}_1 - \tilde{c}_1 = \left(\frac{-687}{13}, \frac{51}{169}, \frac{729}{169}, \frac{1467}{169}\right),$$

$$\tilde{z}_4 - \tilde{c}_4 = \left(\frac{9}{13}, \frac{12}{13}, \frac{14}{13}, \frac{17}{13}\right),$$

$$\tilde{z}_5 - \tilde{c}_5 = \left(\frac{-35}{13}, \frac{27}{13}, \frac{62}{13}, \frac{139}{13}\right)$$

$$t_{2k} + t_{3k} = \min_{j=1,4,5} \{t_{2j} + t_{3j}\} = 2$$

Since $t_{2k} + t_{3k} = 2 \geq 0$, the fuzzy basic solution presented in Table 3.36 is optimal.

3.5.3 Kheirfam and Verdegay's Approach

In the approach proposed by Ganesan and Veeramani [24] it is assumed that a fuzzy primal feasible basic solution of the LLP of type-3 (3.111) is at hand. For situation in which a primal fuzzy feasible basic solution is not at hand, but there is a fuzzy dual feasible basic solution, Kheirfam and Verdegay [31] (see also Ebrahimnejad and Nasseri [27]) generalized the fuzzy analogue of the dual simplex algorithm of the LP to the problem (3.111) without converting it to a crisp LP problem.

Consider the dual problem of (3.111) Nasseri et al. [26]:

$$\min \tilde{w}\tilde{b}$$
$$s.t. \tilde{w}A \succeq \tilde{c} \tag{3.123}$$

Let $\tilde{x}$ and $\tilde{w}$ be the primal and dual fuzzy feasible solutions of problems (3.111) and (3.123), respectively. It should be remembered that a necessary and sufficient condition for $\tilde{x}$ and $\tilde{w}$ to be optimal is (Theorem 5.4, Nasseri et al. [26]):

$$(\tilde{w}a_j - \tilde{c})\tilde{x}_j = \tilde{0}, \quad j = 1, 2, \ldots, n \tag{3.124}$$

Consider the LPP with fuzzy parameters of type-3 (3.111). Suppose that a basic solution for this problem is given by $\tilde{x}_B = B^{-1}\tilde{b}$ and $\tilde{x}_N = \tilde{0}$, with the basis matrix B. Consider Table 3.22. Suppose that for $j = 1, 2, \ldots, n$, we have

$$\tilde{z}_j - \tilde{c}_j = \tilde{c}_B B^{-1} a_j - \tilde{c}_j \succeq \tilde{0}, \quad \text{i.e., } t_{2j} + t_{3j} \geq 0, \quad j = 1, 2, \ldots, n \tag{3.125}$$

This means that, the optimality condition of the LPP with fuzzy parameters of type-3 (3.111) at $\tilde{x}$ holds true. We define $\tilde{w} = \tilde{c}_B B^{-1}$. In this way, from (3.125), we have $\tilde{w}A \succeq \tilde{c}$, that is, $\tilde{w}$ is a dual fuzzy feasible solution. If $\tilde{\tilde{b}}_r \succeq \tilde{0}$, for all $r = 1, 2, \ldots, m$, then we have a fuzzy feasible solution for the LPP with fuzzy parameters of type-3 (3.111). Moreover, we will have $\tilde{c}\tilde{x} = \tilde{c}_B\tilde{x}_B = \tilde{c}_B B^{-1}\tilde{b} = \tilde{w}\tilde{b}$, and thus, by Corollary 5.1 in Nasseri et al. [26], we establish the optimality of $\tilde{x}$ and $\tilde{w}$ for the problems (3.111) and (3.123), respectively. Therefore, we have the following theorem Kheirfam and Verdegay [31].

Theorem 3.31 *The optimality criteria* $\tilde{z}_j - \tilde{c}_j \succsim \tilde{0}$ *i.e.* $t_{2j} + t_{3j} \geq 0$, *for all* $j = 1, 2, \ldots, n$, *for the problem* (3.111) *is equivalent to the feasibility condition for the problem* (3.123). *If, in addition,* $\tilde{x}$ *corresponding to a basis B is primal fuzzy feasible then $\tilde{x}$ is optimal for the problem* (3.111) *and $\tilde{w} = \tilde{c}_B B^{-1}$ is optimal to the problem* (3.123).

Now, assume that the problem (3.123) is feasible and $\tilde{x}$, corresponding to a basis B, is dual feasible but primal infeasible. That is, we have $\tilde{z}_j - \tilde{c}_j \succeq \tilde{0}$ for all $j = 1, 2, \ldots, n$ and at least one r exists such that $\tilde{\tilde{b}}_r \prec \tilde{0}$, i.e. $\tilde{b}_{2r} + \tilde{b}_{3r} < 0$. Thus, according to duality theory, the problem (3.111) can be either infeasible (in which case, the problem (3.123) is unbounded), or it has a fuzzy optimal solution. Next, we will show how to work on row r of the Table 3.21 corresponding to the basis B, as the pivoting row, and either (1) detect the infeasibility of the problem (3.111) (or unboundedness of the problem (3.123)), or find a column k, as a pivoting column, to pivot on y_{rk} and obtain a new dual feasible table with a non-increasing primal objective value. We explain these cases below.

Theorem 3.32 *If in a dual feasible simplex table an r exists such that $\tilde{\tilde{b}}_r \prec \tilde{0}$ and $y_{rj} \geq 0$, for all $j = 1, 2, \ldots, n$, then the problem* (3.111) *is infeasible.*

Proof Suppose that Table 3.22 is a dual feasible table, and an r exists such that $\tilde{\tilde{b}}_r \prec \tilde{0}$ and $y_{rj} \geq 0$, for all $j = 1, 2, \ldots, n$. Corresponding to the row r of the table, we have $\tilde{x}_{B_r} + \sum_{j \in J_N} y_{rj}\tilde{x}_j \approx \tilde{\tilde{b}}_r$. Since, by assumption, $y_{rj} \geq 0$, for all $j \in J_N$ and $\tilde{x}_j \succeq \tilde{0}$, then $\tilde{x}_{B_r} + \sum_{j \in J_N} y_{rj}\tilde{x}_j \approx \tilde{\tilde{b}}_r$ for any fuzzy basic feasible solution. However, $\tilde{\tilde{b}}_r \prec \tilde{0}$ and this shows that the problem (3.111) problem is infeasible. $\qquad \square$

Theorem 3.33 *If in a dual feasible simplex table, an r exists such that $\tilde{\tilde{b}}_r \prec \tilde{0}$ and a non-basic index $k \in J_N$ exists such that $y_{rk} < 0$, then pivoting on y_{rk} will yield a dual feasible table with a corresponding non-increasing objective value.*

Proof Pivoting on the pivot y_{rk} will result in the new objective row as follows:

$$(\tilde{z}_j - \tilde{c}_j)_{new} = (\tilde{z}_j - \tilde{c}_j) - \frac{y_{rj}}{y_{rk}}(\tilde{z}_k - \tilde{c}_k), \quad j \in J_N$$

For the new table to be dual feasible we need to have $(\tilde{z}_j - \tilde{c}_j)_{new} \succeq \tilde{0}$, which results in

$$\frac{\tilde{z}_j - \tilde{c}_j}{y_{rj}} \preceq \frac{\tilde{z}_k - \tilde{c}_k}{y_{rk}}, \quad y_{rj} < 0 \tag{3.126}$$

To satisfy (3.126), it is sufficient to let

$$\frac{\tilde{z}_k - \tilde{c}_k}{y_{rk}} = \max\left\{\frac{\tilde{z}_j - \tilde{c}_j}{y_{rj}} \middle| y_{rj} < 0\right\} \tag{3.127}$$

To verify (3.127), we can use the following equivalent test:

$$\frac{t_{2k} + t_{3k}}{y_{rk}} = \max\left\{\frac{t_{2j} + t_{3j}}{y_{rj}} \middle| y_{rj} < 0\right\} \tag{3.128}$$

On the other hand, since $\tilde{\tilde{b}}_r \prec \tilde{0}, y_{rk} < 0$ and $\tilde{z}_k - \tilde{c}_k \succeq \tilde{0}$, we have

$$\tilde{\tilde{z}}_{new} = \tilde{c}_B B^{-1} \tilde{b} - \frac{\tilde{\tilde{b}}_r}{y_{rk}}(\tilde{z}_k - \tilde{c}_k) \preceq \tilde{c}_B B^{-1} \tilde{b} = \tilde{\tilde{z}} \tag{3.129}$$

From (3.129) we conclude that the new objective value is non-increasing.   □

A summary of the symmetric fuzzy dual method for a maximization problem in tableau format is given as follows.

Algorithm 3.7: Symmetric fuzzy dual simplex algorithm *(Maximization Problem)*

Initialization step
Choose a starting primal feasible basic solution with the basis B. Form the initial tableau as in Table 3.22. Suppose $\tilde{z}_j - \tilde{c}_j = (t_{2j} - \alpha_j, t_{2j}, t_{3j}, t_{3j} + \alpha_j)$, so we have $t_{2j} + t_{3j} \geq 0$ for all j.

Main steps

(1) Suppose $\tilde{\tilde{b}} = B^{-1}\tilde{b}$. If $\tilde{\tilde{b}} = B^{-1}\tilde{b} \geq 0$ then stop; the current fuzzy solution is optimal.

Otherwise, suppose $\tilde{\tilde{b}}_i = (\bar{b}_{2i} - \alpha_i, \bar{b}_{2i}, \bar{b}_{3i}, \bar{b}_{3i} + \alpha_j)$ and let

$$\bar{b}_{2r} + \bar{b}_{3r} = \min_{1 \le i \le m} \left\{ \bar{b}_{2i}, +\bar{b}_{3i} \right\}$$

(2) If $y_{rj} \ge 0$ for all j, then stop; the problem (3.111) is infeasible.

Otherwise, $\tilde{x}_{B_r}$ leaves the basis and $\tilde{x}_k$ enters the basis, providing that

$$\frac{t_{2k} + t_{3k}}{y_{rk}} = \max \left\{ \frac{t_{2j} + t_{3j}}{y_{rj}} \Big| y_{rj} < 0 \right\}$$

(3) Update the tableau by pivoting on y_{rk} and go to (1).

Example 3.17 [27] Consider the following LPP with fuzzy parameters of type-3:

$$Max\,\tilde{z} \approx -(11, 13, 15, 17)\tilde{x}_1 - (9, 12, 14, 17)\tilde{x}_2 - (13, 15, 17, 19)\tilde{x}_3$$
$$s.t. \quad 2\tilde{x}_1 + 3\tilde{x}_2 + 2\tilde{x}_3 \succeq (39, 45, 55, 61),$$
$$4\tilde{x}_1 + 3\tilde{x}_3 \succeq (52, 60, 80, 88), \tag{3.130}$$
$$2\tilde{x}_1 + 5\tilde{x}_2 \succeq (60, 65, 95, 100),$$
$$\tilde{x}_1, \tilde{x}_2, \tilde{x}_3 \succeq \tilde{0}.$$

Initial Step:

After introducing the surplus variables $\tilde{x}_4$, $\tilde{x}_5$ and $\tilde{x}_6$ the initial fuzzy dual simplex tableau is given as in Table 3.37.

Iteration 1:

Step 1: Here,

$$\bar{b}_{2r} + \bar{b}_{3r} = \min_{1 \le i \le 3} \left\{ \bar{b}_{2i} + \bar{b}_{3i} \right\} = min\{-100, -140, -160\} = -160 = \bar{b}_{23} + \bar{b}_{33}$$

Thus, $r = 3$ and $\tilde{x}_{B_3} = \tilde{x}_6$ leaves the basis.

Table 3.37 The initial tableau of Example 3.17

Basis	$\tilde{x}_1$	$\tilde{x}_2$	$\tilde{x}_3$	$\tilde{x}_4$	$\tilde{x}_5$	$\tilde{x}_6$	R.H.S	$\Re$
$\tilde{z}$	$(11, 13, 15, 17)$	$(9, 12, 14, 17)$	$(13, 15, 17, 19)$	$\tilde{0}$	$\tilde{0}$	$\tilde{0}$	$\tilde{0}$	$-$
$\Re(\tilde{z})$	28	26	32	0	0	0	$-$	0
$\tilde{x}_4$	-2	-3	-2	1	0	0	$-(39, 45, 55, 61)$	-100
$\tilde{x}_5$	-4	0	-3	0	1	0	$-(52, 60, 80, 88)$	-140
$\tilde{x}_6$	-2	-5	0	0	0	1	$-(60, 65, 95, 100)$	-160

Table 3.38 The first iteration tableau of Example 3.17

Basis	$\tilde{x}_1$	$\tilde{x}_2$	$\tilde{x}_3$	$\tilde{x}_4$	$\tilde{x}_5$	$\tilde{x}_6$	R.H.S	$\Re$
$\tilde{z}$	$\left(\frac{21}{5},\frac{37}{5},\frac{51}{5},\frac{67}{5}\right)$	$\tilde{0}$	$(13,15,17,19)$	$\tilde{0}$	$\tilde{0}$	$\left(\frac{9}{5},\frac{12}{5},\frac{14}{5},\frac{17}{5}\right)$	$-(8,153,263,334)$	–
$\Re(\tilde{z})$	$\frac{88}{5}$	26	32	0	0	$\frac{26}{5}$	–	416
$\tilde{x}_4$	$\frac{-4}{5}$	0	-2	1	0	$\frac{-3}{5}$	$(-25,-16,12,21)$	-4
$\tilde{x}_5$	-4	0	-3	0	1	0	$-(52,60,80,88)$	-140
$\tilde{x}_2$	$\frac{2}{5}$	1	0	0	0	$\frac{-1}{5}$	$(12,13,19,20)$	32

Step 2: We have

$$\frac{t_{2k}+t_{3k}}{y_{3k}} = \text{ma2}\left\{\frac{t_{2j}+t_{3j}}{y_{3j}}\,|y_{3j}<0\right\} = \max\left\{\frac{28}{-2},\frac{26}{-5}\right\} = -\frac{26}{5} = \frac{t_{22}+t_{32}}{y_{32}}$$

Thus, $k=2$ and $\tilde{x}_k = \tilde{x}_2$ enters the basis.

Step 3: The updated tableau by pivoting on $y_{32} = -5$ is given as in Table 3.38.

Iteration 2:

Step 1: Here,

$$\bar{b}_{2r}+\bar{b}_{3r} = \min_{1\leq i\leq 3}\left\{\bar{b}_{2i},+\bar{b}_{3i}\right\} = min\{-4,-140,32\} = -140 = \bar{b}_{22}+\bar{b}_{32}$$

Thus, $r=2$ and $\tilde{x}_{B_2} = \tilde{x}_5$ leaves the basis.

Step 2: We have

$$\frac{t_{2k}+t_{3k}}{y_{2k}} = \max\left\{\frac{t_{2j}+t_{3j}}{y_{2j}}\,|y_{2j}<0\right\} = \max\left\{\frac{88}{-4},\frac{32}{-3}\right\} = -\frac{88}{20} = \frac{t_{21}+t_{31}}{y_{21}}$$

Thus, $k=1$ and $\tilde{x}_k = \tilde{x}_1$ enters the basis.

Step 3: The updated tableau by pivoting on $y_{21} = -4$ is given as in Table 3.39.

Table 3.39 The optimal tableau of Example 3.17

Basis	$\tilde{x}_1$	$\tilde{x}_2$	$\tilde{x}_3$	$\tilde{x}_4$	$\tilde{x}_5$	$\tilde{x}_6$	R.H.S	$\Re$
$\tilde{z}$	$\tilde{0}$	$\tilde{0}$	$\left(\frac{37}{20},\frac{147}{20},\frac{229}{20},\frac{251}{20}\right)$	$\tilde{0}$	$\tilde{0}$	$\left(\frac{9}{5},\frac{12}{5},\frac{14}{5},\frac{17}{5}\right)$	$-\left(\frac{2181}{10},\frac{2495}{10},\frac{4755}{10},\frac{5069}{10}\right)$	–
$\Re(\tilde{z})$	0	0	32	0	0	$\frac{26}{5}$	–	725
$\tilde{x}_4$	1	0	$\frac{-7}{5}$	1	$\frac{-1}{5}$	$\frac{-3}{5}$	$\left(\frac{-68}{5},\frac{-20}{5},\frac{140}{5},\frac{148}{5}\right)$	24
$\tilde{x}_1$	1	0	$\frac{3}{4}$	0	$\frac{-1}{4}$	0	$(13,15,20,22)$	35
$\tilde{x}_2$	0	1	$\frac{-3}{10}$	0	$\frac{1}{10}$	$\frac{-1}{5}$	$\left(\frac{16}{5},\frac{25}{5},\frac{65}{5},\frac{74}{5}\right)$	19

Iteration 3:

Step 1: Here,

$$\bar{b}_{2r} + \bar{b}_{3r} = \min_{1 \le i \le 3}\{\bar{b}_{2i}, +\bar{b}_{3i}\} = \min\{24, 35, 19\} = 19 \ge 0$$

Thus, the current solution is optimal. Therefore, the optimal solution of the fuzzy problem obtained by the symmetric fuzzy dual method is $\tilde{x}_1 = (13, 15, 20, 22)$, $\tilde{x}_2 = \left(\frac{16}{5}, \frac{25}{5}, \frac{65}{5}, \frac{74}{5}\right)$ and $\tilde{x}_3 = \tilde{0}$ with the fuzzy objective value

$$\tilde{z} = -\left(\frac{2181}{10}, \frac{2495}{10}, \frac{4755}{10}, \frac{5069}{10}\right).$$

In contrast to the LPP with fuzzy parameters of type-2 considered in [2, 13], in the LLP with fuzzy paprameters of type-3, introduced by Ganesan and Veeramini [24], not only the decision variables and the right-hand side of the constraints are fuzzy, but also the coefficients of decision variables in the objective function are fuzzy. However, the proposed approaches for solving the FLP problem considered in [24, 25, 29, 31] are valid only for situations in which the parameters are represented by symmetric trapezoidal fuzzy numbers. But, the proposed methods [2, 13] for solving the LPP with fuzzy parameters of type-2 are applicable to both symmetric and non-symmetric trapezoidal fuzzy numbers. Also, unlike the non-simplex based approach and similar to the existing approaches [2, 13, 39] in solving the LPP with fuzzy parameters of type-2, this approach does not increase the number of the constraint of the primary FLP problem. In addition, although these approaches [24, 25, 29, 31] give the fuzzy optimal solution, yet, the fuzzy optimal solution by means of these approaches may be not non-negative.

3.5.4   Ebrahimnejad and Tavana's Approach

In this section we explore the approach proposed by Ebrahimnejad and Tavana [33] for solving the LPP with fuzzy parameters of type-3 (3.111) which is simpler and computationally more efficient than the three competing fuzzy methods mentioned in the previous subsections.

According to Definition 3.17, a rank for each symmetric trapezoidal fuzzy number for comparison purposes is defined. Assuming that $\tilde{A} = (a_2 - \alpha, a_2, a_3, a_3 + \alpha)$ is a symmetric trapezoidal fuzzy number, then $\Re(\tilde{A}) = \frac{a_2 + a_3}{2}$. This equation is used to convert the LPP with fuzzy parameters of type-3 (3.111) into a crisp LP problem. To do this, the rank order of each fuzzy number is substituted for the corresponding fuzzy number in the fuzzy problem under consideration. This leads to an equivalent crisp LP problem which can be solved with a standard method. However, the decision variables in the original problem are fuzzy numbers as well. Therefore, a fuzzy solution is required, while the equivalent crisp problem gives a crisp solution. To overcome this concern, the relation between the crisp and fuzzy problems is used to obtain a fuzzy solution for the FLP problem under consideration. Note that the main steps of the fuzzy simplex methods (Algorithms 3.4, 3.5, 3.6 and 3.7) and the crisp simplex method proposed here are identical. Thus, if B is the optimal basis of the equivalent crisp problem, then it will be the optimal basis of the corresponding FLP problem. In this case, the solution of the crisp problem $\tilde{x} = (\tilde{x}_B, \tilde{x}_N) = (B^{-1}\tilde{b}, \tilde{0})$ will be the optimal solution of the FLP problem. However, all arithmetic operations in the fuzzy methods (Algorithms 3.4, 3.5, 3.6 and 3.7) are performed on the fuzzy numbers, while in the method proposed by Ebrahimnejad and Tavana [33], all arithmetic operations are done on the crisp numbers. As a result, the computational effort is decreased significantly in the method proposed by Ebrahimnejad and Tavana [33]. In the following example, we use the FLP problem of Ganesan and Veeramani [24] to demonstrate the computational simplicity of the method proposed by Ebrahimnejad and Tavana [33].

Example 3.18 We find the fuzzy optimal solution of the LPP with fuzzy parameters of type-3 (3.117) given in Example 3.13.

We first substitute the rank order of each fuzzy number for its corresponding fuzzy number in the above FLP problem to obtain the following crisp problem:

$$\begin{aligned} \max z = {}& 14x_1 + 13x_2 + 16x_3 \\ s.t. \quad & 12x_1 + 13x_2 + 12x_3 \leq 490 \\ & 14x_1 + 13x_3 \leq 470 \\ & 12x_1 + 15x_3 \leq 480 \\ & x_1, x_2, x_3 \geq 0 \end{aligned} \tag{3.131}$$

We then construct the standard form of the LP problem (3.131) as follows where x_4, x_5 and x_6 are the slack variables:

$$\begin{aligned} \max z = {}& 14x_1 + 13x_2 + 16x_3 \\ s.t. \quad & 12x_1 + 13x_2 + 12x_3 + x_5 = 490 \\ & 14x_1 + 13x_3 + x_6 = 470 \\ & 12x_1 + 15x_3 + x_7 = 480 \\ & x_1, x_2, x_3, x_4, x_5, x_6 \geq 0 \end{aligned} \tag{3.132}$$

The problem (3.132) is a crisp LP problem and can be solved using the standard primal simplex method. Table 3.40 presents the first primal simplex tableau.

Table 3.40 Initial tableau

Basis	x_1	x_2	x_3	x_4	x_5	x_6	RHS
z	−14	−13	−16	0	0	0	0
x_4	12	13	12	1	0	0	490
x_5	14	0	13	0	1	0	470
x_6	12	15	0	0	0	1	480

Table 3.41 First iteration tableau

Basis	x_1	x_2	x_3	x_4	x_5	x_6	RHS
z	$\frac{42}{13}$	−13	0	0	$\frac{18}{13}$	0	$\frac{7520}{13}$
x_4	$\frac{-12}{13}$	13	0	1	$\frac{-12}{13}$	0	$\frac{730}{13}$
x_3	$\frac{14}{13}$	0	1	0	$\frac{1}{13}$	0	$\frac{470}{13}$
x_6	12	15	0	0	0	1	480

Table 3.42 Optimal solution tableau

Basis	x_1	x_2	x_3	x_4	x_5	x_6	RHS
z	$\frac{42}{13}$	0	0	1	$\frac{52}{169}$	0	$\frac{112,150}{169}$
x_2	$\frac{-12}{169}$	1	0	1	$\frac{-12}{13}$	0	$\frac{730}{169}$
x_3	$\frac{14}{13}$	0	1	0	$\frac{1}{13}$	0	$\frac{470}{169}$
x_6	$\frac{1848}{169}$	0	0	0	$\frac{-180}{169}$	1	$\frac{70,170}{169}$

In this table, x_3 is an entering variable and x_5 is a leaving variable. The second simplex tableau presented in Table 3.41 is derived by pivoting on $y_{23} = 13$.

In this table, x_2 is an entering variable and x_4 is a leaving variable. The third simplex tableau presented in Table 3.42 is derived by pivoting on $y_{12} = 13$.

Table 3.42 is the optimal tableau because $z_j - c_j \geq 0$ for all non-basic variables. We then use the obtained optimal basis for the crisp Problem (3.131) to find the fuzzy optimal solution for the FLP Problem (3.132). Here,

$$B^{-1} = \begin{bmatrix} \frac{1}{13} & \frac{-12}{169} & 0 \\ 0 & \frac{1}{13} & 0 \\ 0 & \frac{-180}{169} & 1 \end{bmatrix}$$

Thus, we have:

$$\tilde{x}_B = \begin{bmatrix} \tilde{x}_2 \\ \tilde{x}_3 \\ \tilde{x}_6 \end{bmatrix} = B^{-1}\tilde{b} = \begin{bmatrix} \frac{1}{13} & \frac{-12}{169} & 0 \\ 0 & \frac{1}{13} & 0 \\ 0 & \frac{-180}{169} & 1 \end{bmatrix} \begin{bmatrix} (469, 475, 505, 511) \\ (452, 460, 480, 488) \\ (460, 465, 495, 500) \end{bmatrix}$$

$$= \begin{bmatrix} \left(\frac{231}{169}, \frac{405}{169}, \frac{1045}{169}, \frac{1219}{169}\right) \\ \left(\frac{460}{13}, \frac{1480}{13}, \frac{8}{13}, \frac{8}{13}\right) \\ \left(\frac{59,455}{169}, \frac{62,910}{169}, \frac{77,430}{169}, \frac{80,885}{169}\right) \end{bmatrix}, \quad \tilde{x}_N = \begin{bmatrix} \tilde{x}_1 \\ \tilde{x}_4 \\ \tilde{x}_5 \end{bmatrix} = \begin{bmatrix} \tilde{0} \\ \tilde{0} \\ \tilde{0} \end{bmatrix}$$

As shown here, the fuzzy optimal solution of the FLP method proposed by Ebrahimnejad and Tavana [33] is equivalent to the optimal solution derived from Ganesan and Veeramani's method [24]. However, the method proposed by Ebrahimnejad and Tavana [33] is by far simpler and computationally more efficient than the FLP method proposed Ganesan and Veeramani [24].

3.6 LPP with Fuzzy Parameters of Type-4

The LP problem having symmetric trapezoidal fuzzy numbers for the costs coefficients, the decision variables, the elements of the coefficient matrix and the values of the right-hand side is termed as the LPP with fuzzy parameters of type-4.

The LPP with fuzzy parameters of type-4 is mathematically formulated as follows:

$$\begin{aligned} \max \tilde{z} &= \tilde{c}\tilde{x} \\ s.t. \quad \tilde{A}\tilde{x} &\approx \tilde{b}, \\ \tilde{x} &\succeq \tilde{0}. \end{aligned} \tag{3.133}$$

Kheirfam and Verdegay [34] introduced a new type of fuzzy inverse and division for symmetric trapezoidal fuzzy numbers.

Definition 3.22 Let $\tilde{a} = (a_2 - \alpha, a_2, a_3, a_3 + \alpha)$ and $\tilde{b} = (b_2 - \beta, b_2, b_3, b_3 + \beta)$ be the symmetric trapezoidal fuzzy number and symmetric non-zero trapezoidal fuzzy number, respectively. Two new types of arithmetic operations on $\tilde{a}$ and $\tilde{b}$ have been defined in [34] as follows:

- $\dfrac{1}{\tilde{b}} = \tilde{b}^{-1} \approx \left(\left[\dfrac{2}{b_2 + b_3} - w\right] - \beta, \left[\dfrac{2}{b_2 + b_3} - w\right], \left[\dfrac{2}{b_2 + b_3} + w\right], \left[\dfrac{2}{b_2 + b_3} + w\right] + \beta \right)$

  where $w = \frac{t_2 - t_1}{2}, t_2 = \max\limits_{j=2,3}\left\{\frac{1}{b_j}\right\}, t_1 = \min\limits_{j=2,3}\left\{\frac{1}{b_j}\right\}, \frac{1}{b_j} = \begin{cases} \frac{1}{b_j} & b_j \neq 0, \\ 0, & b_j = 0. \end{cases}$

- $\dfrac{\tilde{a}}{\tilde{b}} \approx \left(\left[\dfrac{a_2 + a_3}{b_2 + b_3} - w \right] - \gamma, \left[\dfrac{a_2 + a_3}{b_2 + b_3} - w \right], \left[\dfrac{a_2 + a_3}{b_2 + b_3} + w \right], \left[\dfrac{a_2 + a_3}{b_2 + b_3} + w \right] + \gamma \right)$

where $w = \frac{t_2 - t_1}{2}, t_2 = \underset{i,j=2,3}{max}\left\{ \frac{a_i}{b_j} \right\}, t_1 = \underset{i,j=2,3}{min}\left\{ \frac{a_i}{b_j} \right\}, \gamma = \left| a_3 \beta + \frac{\alpha}{b_3} \right|.$

A symmetric trapezoidal fuzzy matrix $\tilde{A} = \left[\tilde{a}_{ij} \right]_{m \times n}$ is any rectangular array of symmetric trapezoidal fuzzy numbers. For any values of indices i and j, the ijth minor of the square matrix $\tilde{A} = \left[\tilde{a}_{ij} \right]_{n \times n}$, denoted by $\tilde{A}_{ij}$, is the $(n-1) \times (n-1)$ sub-matrix of $\tilde{A}$ obtained by deleting the ith row and the jth column of $\tilde{A}$. In this case, determinant $\tilde{A}$, denoted by $|\tilde{A}|$, is computed as follows [34]:

- For $n = 1, |\tilde{A}| = \tilde{a}_{11}.$
- For $n = 2, |\tilde{A}| = \tilde{a}_{11}\tilde{a}_{22} - \tilde{a}_{12}\tilde{a}_{21}.$
- For $n > 2, |\tilde{A}| = \sum\limits_{j=1}^{n} (-1)^{i+j}\tilde{a}_{ij}|\tilde{A}_{ij}|$, for any value of index $i = 1, 2, \ldots, n.$

A square matrix $\tilde{A} = \left[\tilde{a}_{ij} \right]_{n \times n}$, is called singular if $|\tilde{A}| = \tilde{0}$. In the other case, it said to be a non-singular fuzzy matrix and its inverse is calculated by $\tilde{A}^{-1} = \dfrac{(1,1,0,0)}{|\tilde{A}|}\left[(-1)^{i+j}|\tilde{A}_{ij}| \right]_{n \times n}.$

The rank of matrix $\tilde{A} = \left[\tilde{a}_{ij} \right]_{m \times n}$ is the largest order of its minor $\tilde{A}_{ij}$ such that $|\tilde{A}_{ij}| \not\approx \tilde{0}.$

Now; we are in a position to review two existing methods solving the LPPs with fuzzy parameters of type-4.

3.6.1 Ezzati et al.'s Approach

In this subsection, the fuzzy primal approach proposed by Ezzati et al. [35] is reviewed for solving the problem (3.133).

Definition 3.23 Consider the LPP with fuzzy parameters of type-4 (3.133). Let $rank(\tilde{A}) = m$. Partition $\tilde{A}$ as $\left[\tilde{B}, \tilde{N} \right]$, where $\tilde{B}$ is an $(m \times n)$ nonsingular fuzzy matrix (it means that $|\tilde{B}| \not\approx \tilde{0}$). It is obvious that $rank(\tilde{B}) = m$. It is apparent that the fuzzy solution $\left(\tilde{x}_{\tilde{B}}, \tilde{x}_{\tilde{N}} \right) = \left(\tilde{B}^{-1}\tilde{b}, \tilde{0} \right)$ is a fuzzy solution of $\tilde{A}\tilde{x} \approx \tilde{b}$. We call $\tilde{x} = \left(\tilde{x}_{\tilde{B}}, \tilde{x}_{\tilde{N}} \right)$ a fuzzy basic solution corresponding to the basis $\tilde{B}$. If $\tilde{x}_{\tilde{B}} \succeq \tilde{0}$, then the fuzzy basic solution is feasible and the corresponding fuzzy objective value is $\tilde{z} = \tilde{c}_{\tilde{B}}\tilde{x}_{\tilde{B}} = \tilde{c}_{\tilde{B}}\tilde{B}^{-1}\tilde{b}.$

Suppose a fuzzy basic feasible solution of the problem (3.133) with the basis $\tilde{B}$ is at hand. Let $\tilde{y}_j$ and $\tilde{w}$ be the solutions to $\tilde{B}\tilde{y}_j = \tilde{a}_j$ and $\tilde{w}\tilde{B} = \tilde{c}_B$, respectively. Define $\tilde{z}_j = \tilde{c}_{\tilde{B}}\tilde{B}^{-1}\tilde{a}_j = \tilde{w}\tilde{a}_j.$

Theorem 3.34 *Let* $\left(\tilde{x}_{\tilde{B}}, \tilde{x}_{\tilde{N}}\right) = \left(\tilde{B}^{-1}\tilde{b}, \tilde{0}\right)$ *be a fuzzy basic feasible solution of the problem* (3.133). *If for any column* $\tilde{a}_j$ *in* $\tilde{A}$ *which is not in* $\tilde{B}$, *the condition* $\tilde{z}_j - \tilde{c}_j \prec \tilde{0}$ *holds and* $\tilde{y}_{ij} \succ \tilde{0}$ *for some* $i, j \in \{1, 2, \ldots, m\}$, *then it is possible to obtain a new fuzzy basic feasible solution by replacing one of the columns in* $\tilde{B}$ *by* $\tilde{a}_j$.

Theorem 3.35 *If* $\left(\tilde{\tilde{x}}_{\tilde{B}}, \tilde{x}_{\tilde{N}}\right) = \left(\tilde{B}^{-1}\tilde{b}, \tilde{0}\right)$ *be a fuzzy basic feasible solution of the problem* (3.133) *with* $\tilde{z} = \tilde{c}_{\tilde{B}}\tilde{x}_{\tilde{B}}$ *as the fuzzy value of the objective function and if* $\left(\tilde{\tilde{x}}_{\tilde{B}}, \tilde{x}_{\tilde{N}}\right) = \left(\tilde{B}^{-1}\tilde{b}, \tilde{0}\right)$ *is another fuzzy basic feasible solution with* $\tilde{z}_{new} = \tilde{c}_{\tilde{B}}\tilde{x}_{\tilde{B}}$ *obtained by admitting a non-basic column vector* $\tilde{a}_j$ *in the basis for which* $\tilde{z}_j - \tilde{c}_j \prec \tilde{0}$ *holds and* $\tilde{y}_{ij} \succ \tilde{0}$ *for some* $i, i, \in \{1, 2, \ldots m\}$, *then* $\tilde{z}_{new} \succeq \tilde{z}$.

Theorem 3.36 *If we have a fuzzy basic feasible solution with* $\tilde{z}_k - \tilde{c}_k \prec \tilde{0}$ *for some non-basic fuzzy variable* $\tilde{x}_k$, *and* $\tilde{y}_{ik} \prec \tilde{0}$ *for all* $i \in \{1, 2, \ldots, m\}$, *then the problem* (3.133) *has an unbounded fuzzy solution.*

Theorem 3.37 (Optimality conditions) *Assume the LPP with fuzzy parameters of type-4* (3.133) *is non-degenerate. A fuzzy basic feasible solution* $\left(\tilde{\tilde{x}}_{\tilde{B}}, \tilde{x}_{\tilde{N}}\right) = \left(\tilde{B}^{-1}\tilde{b}, \tilde{0}\right)$ *is optimal for the problem* (3.133) *if* $\tilde{z}_j - \tilde{c}_j \succeq \tilde{0}$ *for all* $j \in \{1, 2, \ldots, n\}$.

Using these theorems, Ezzati et al. [35] developed and presented for the first time a symmetric fully fuzzy primal simplex algorithm for solving the LPP with fuzzy parameters of type-4 (3.133). A summery of their method in tableau format is given as follows.

Algorithm 3.8: Symmetric fully fuzzy primal simplex algorithm *(Maximization Problem)*

Initialization step

Let an initial fuzzy basic feasible solution with the basis $\tilde{B}$ be at hand. Form the initial tableau as in Table 3.43.

Main steps

(1) Compute $\tilde{z}_j - \tilde{c}_j$ for all non-basic fuzzy variables. Let $\tilde{z}_j - \tilde{c}_j = \left(t_{2j} - \alpha_j, t_{2j}, t_{3j}, t_{3j} + \alpha_j\right)$. Suppose

$$t_{2k} + t_{3k} = \min_{j \in J_N}\left\{t_{2j} + t_{3j}\right\}.$$

(2) Let $\tilde{y}_k = \tilde{B}^{-1}\tilde{a}_k$. If $\tilde{y}_k \preceq \tilde{0}$, then stop; the problem is unbounded. Otherwise, suppose $\tilde{b}_i = \left(\bar{b}_{2i} - \alpha_i, \bar{b}_{2i}, \bar{b}_{3i}, \bar{b}_{3i} + \alpha_i\right)$ and $\tilde{y}_{ik} = \left(y_{2,ik} - \alpha_{ik}, y_{2,ik}, y_{3,ik}, y_{3,ik} + \alpha_{ik}\right)$ and obtain the index r as follows:

$$\frac{\bar{b}_{2r} + \bar{b}_{3r}}{y_{2,rk} + y_{3,rk}} = \min_{1 \leq i \leq m}\left\{\frac{\bar{b}_{2i} + \bar{b}_{3i}}{y_{2,ik} + y_{3,ik}} \middle| y_{2,ik} + y_{3,ik} > 0\right\}$$

Table 3.43 Initial tableau of the LPP with fuzzy parameters of type-4

Basis	$\tilde{x}_1$	$\cdots$	$\tilde{x}_r$	$\cdots$	$\tilde{x}_m$	$\cdots$	$\tilde{x}_j$	$\cdots$	$\tilde{x}_k$	$\cdots$	R.H.S	$\Re$
$\tilde{z}$	$\tilde{0}$	$\cdots$	$\tilde{0}$	$\cdots$	$\tilde{0}$	$\cdots$	$\tilde{z}_j - \tilde{c}_j$	$\cdots$	$\tilde{z}_k - \tilde{c}_k$	$\cdots$	$\tilde{z} = \tilde{c}_B \tilde{b}$	$-$
$\Re(\tilde{z})$	0	$\cdots$	0	$\cdots$	0	$\cdots$	$t_{2j} + t_{2j}$	$\cdots$	$t_{2k} + t_{2k}$	$\cdots$	$-$	$\bar{z}_2 + \bar{z}_3$
$\tilde{x}_1$	1	$\cdots$	0	$\cdots$	0	$\cdots$	$\tilde{y}_{1j}$	$\cdots$	$\tilde{y}_{1k}$	$\cdots$	$\tilde{\bar{b}}_1$	$\bar{b}_{12} + \bar{b}_{13}$
$\vdots$	$\vdots$	$\cdots$	$\vdots$	$\cdots$	$\vdots$	$\cdots$	$\vdots$	$\cdots$	$\vdots$	$\cdots$	$\vdots$	
$\tilde{x}_r$	$\vdots$	$\cdots$	1	$\cdots$	0	$\cdots$	$\tilde{y}_{rj}$	$\cdots$	$\tilde{y}_{rk}$	$\cdots$	$\tilde{\bar{b}}_r$	$\bar{b}_{r2} + \bar{b}_{r3}$
$\vdots$	$\vdots$	$\cdots$	$\vdots$	$\cdots$	$\vdots$	$\cdots$	$\vdots$	$\cdots$	$\vdots$	$\cdots$	$\vdots$	
$\tilde{x}_m$	0	$\cdots$	0	$\cdots$	1	$\cdots$	$\tilde{y}_{mj}$	$\cdots$	$\tilde{y}_{mk}$	$\cdots$	$\tilde{\bar{b}}_m$	$\bar{b}_{m2} + \bar{b}_{m3}$

(3) Update the tableau by pivoting at $\tilde{y}_{rk}$. Update the basic and non-basic fuzzy variables so that $\tilde{x}_k$ enters the basis and $\tilde{x}_{\tilde{B}_r}$ leaves the basis, and go to step (1).

Theorem 3.38 *In the absence of degeneracy, the symmetric fully fuzzy primal simplex algorithm* (Algorithm 3.8) *stops in a finite number of iterations, either with a fuzzy optimal basic feasible solution or with the conclusion that the optimal value is unbounded.*

Example 3.18 We find the fuzzy optimal solution of the following LPP with fuzzy parameters of type-4 using the Algorithm 3.8:

$$
\begin{aligned}
\max \tilde{z} =& (0, 2, 2, 4)\tilde{x}_1 + (-1, 0, 2, 3)\tilde{x}_2 \\
s.t. \quad & (0, 2, 4, 5)\tilde{x}_1 + (-1, 1, 3, 5)\tilde{x}_2 \preceq (2, 4, 4, 6) \\
& (-3, -2, 0, 1)\tilde{x}_1 + (-4, -2, 4, 6)\tilde{x}_2 \preceq (-2, 1, 5, 8) \\
& \tilde{x}_1, \tilde{x}_2 \succeq \tilde{0}
\end{aligned}
\tag{3.134}
$$

We then construct the standard form of the LP problem (3.134) as follows where $\tilde{x}_4$ and $\tilde{x}_6$ are the slack variables:

$$
\begin{aligned}
\max \tilde{z} =& (0, 2, 2, 4)\tilde{x}_1 + (-1, 0, 2, 3)\tilde{x}_2 \\
s.t. \quad & (0, 2, 4, 5)\tilde{x}_1 + (-1, 1, 3, 5)\tilde{x}_2 + (1, 1, 1, 1)\tilde{x}_3 = (2, 4, 4, 6) \\
& (-3, -2, 0, 1)\tilde{x}_1 + (-4, -2, 4, 6)\tilde{x}_2 + (1, 1, 1, 1)\tilde{x}_4 = (-2, 1, 5, 8) \\
& \tilde{x}_1, \tilde{x}_2, \tilde{x}_3, \tilde{x}_4 \succeq \tilde{0}
\end{aligned}
\tag{3.135}
$$

Initial Step:
Table 3.44 presents the initial symmetric fully fuzzy primal simplex tableau.

Table 3.44 The initial simplex tableau of Example 3.18

Basis	$\tilde{x}_1$	$\tilde{x}_2$	$\tilde{x}_3$	$\tilde{x}_4$	R.H.S	$\Re$
$\tilde{z}$	$-(0,2,2,4)$	$-(-1,0,2,3)$	$\tilde{0}$	$\tilde{0}$	$\tilde{0}$	–
$\Re(\tilde{z})$	-4	-2	0	0	–	0
$\tilde{x}_3$	$(0,2,4,5)$	$(-1,1,3,5)$	$(1,1,1,1)$	$\tilde{0}$	$(2,4,4,6)$	8
$\tilde{x}_4$	$(-3,-2,0,1)$	$(-4,-2,4,6)$	$\tilde{0}$	$(1,1,1,1)$	$(-2,1,5,8)$	6

Iteration 1:

Step 1: Here, $\tilde{z}_1 - \tilde{c}_1 - (0,2,2,4)$ and $\tilde{z}_2 - \tilde{c}_2 = -(-1,0,2,3)$. Thus,

$$t_{2k} + t_{3k} = min\{t_{21} + t_{31}, t_{22} + t_{32}\}$$
$$= min\{-2 - 2, -0 - 2\} = -4 = t_{21} + t_{31}$$

This means that $k = 1$ and $\tilde{x}_1$ is the variable entering the basis.

Step 2: The index r of the leaving variable is determined by the following test:

$$\frac{\bar{b}_{2r} + \bar{b}_{3r}}{y_{2,r1} + y_{3,r1}} = min\left\{\frac{\bar{b}_{21} + \bar{b}_{31}}{y_{2,11} + y_{3,11}}\right\} = min\left\{\frac{8}{6}\right\} = \frac{8}{6}$$

This means that $r = 1$ and $\tilde{x}_{\tilde{B}_1} = \tilde{x}_3$ is the leaving variable from the basis.

Step 3: Now pivoting at $\tilde{y}_1 = (0,2,4,5)$ gives the new solution as reported in Table 3.45.

Iteration 2:

Step 1: Here, $\tilde{z}_2 - \tilde{c}_2 = (-6.5, -4.5834, -0.0834, 1.8332)$ and $\tilde{z}_3 - \tilde{c}_3 = (-2.50, 0.4166, 0.9166, 3.8332)$. Thus,

$$t_{2k} + t_{3k} = min\{t_{22} + t_{32}, t_{23} + t_{33}\}$$
$$= min\{-4.5834, -0.0834, 0.4166 + 0.9166\} = t_{22} + t_{32}$$

This means that $k = 2$ and $\tilde{x}_2$ is the variable entering the basis.

Step 2: The index r of the leaving variable is determined by the following test:

Table 3.45 The first iteration tableau of Example 3.18

Basis	$\tilde{x}_1$	$\tilde{x}_2$	$\tilde{x}_3$	$\tilde{x}_4$	R.H.S	$\Re$
$\tilde{z}$	$\tilde{0}$	(−6.5, −4.5834, −0.0834, 1.8332)	(−2.50, 0.4166, 0.9166, 3.8332)	$\tilde{0}$	(−11.0, 1.6666, 3.6666, 16.3332)	–
$\Re(\tilde{z})$	0	−4.6668	1.3332	0	–	5.3332
$\tilde{x}_1$	(1, 1, 1, 1)	(−1.5917, −1.2917, −0.0417, 0.4583)	(−0.7917, 0.2083, 0.4583, 1.4583)	$\tilde{0}$	(−3.6667, 0.8333, 1.8333, 6.3333)	2.6666
$\tilde{x}_4$	$\tilde{0}$	(−6.9167, −3.9584, 4.6250, 7.5833)	(−2.5833, −0.1250, 0.7916, 3.2499)	(1, 1, 1, 1)	(−13.3333, 0.5, 8.1666, 21.9999)	8.6666

$$\frac{\bar{b}_{2r} + \bar{b}_{3r}}{y_{2,r2} + y_{3,r2}} = \frac{8.6666}{0.6666} = \frac{\bar{b}_{22} + \bar{b}_{32}}{y_{2,22} + y_{3,22}}$$

This means that $r = 2$ and $\tilde{x}_{\tilde{B}_2} = \tilde{x}_4$ is the variable leaving the basis.

Step 3: Now pivoting at $\tilde{y}_{22} = (−6.9167, −3.9584, 4.6250, 7.5833)$ gives the new solution as reported in Table 3.46.

Iteration 3:

Step 1: Here, $\tilde{z}_3 - \tilde{c}_3(−18.2915, 0.0669, 5.9331, 24.2915)$ and $\tilde{z}_4 - \tilde{c}_4 = (−20.0554, −0.2967, 14.2985, 34.0572)$. Thus,

$$t_{2k} + t_{3k} = min\{t_{23} + t_{33}, t_{24} + t_{34}\}$$
$$= min\{6, 14.0018\} = t_{23} + t_{33} = 6 > 0$$

This means the current fuzzy solution is optimal. Thus, the fuzzy optimal solution and fuzzy value of the objective function of the problem (3.134) are given as follows:

$$\tilde{x}_1 = (46.9286, 0.0991, 19.9033, 66.931),$$
$$\tilde{x}_2 = (16.0634, 11.0868, 14.9156, 42.0658)$$
$$\tilde{z} = (−167.4103, −1.7162, 67.7234, 233.4175)$$

Table 3.46 The second iteration tableau of Example 3.18

Basis	$\tilde{x}_1$	$\tilde{x}_2$	$\tilde{x}_3$	$\tilde{x}_4$	R.H.S	$\Re$
$\tilde{z}$	$\tilde{0}$	$\tilde{0}$	(−18.2915, 0.0669, 5.9331, 24.2915)	(−20.0554, −0.2967, 14.2985, 34.0572)	(−167.4103, −1.7162, 67.7234, 233.4175)	–
$\Re(\tilde{z})$	0	0	6	0	–	66.0072
$\tilde{x}_1$	$\tilde{1}$	$\tilde{0}$	(−5.1779, 0.1263, 1.8737, 7.1779)	(−5.4698, −0.0312, 4.0318, 9.4704)	(46.9286, 0.0991, 19.9033, 66.931)	20.0024
$\tilde{x}_2$	$\tilde{0}$	$\tilde{1}$	(−2.0589, 0.8144, 1.1856, 4.0589)	(0.1924, 2.7659, 3.2347, 6.193)	(16.0634, 11.0868, 14.9156, 42.0658)	26.0024

3.6.2   Kheirfam and Verdegay's Approach

In this subsection, the fuzzy dual approach proposed by Kheirfam and Verdegay [34] is reviewed for solving the problem (3.133).

Kheirfam and Verdegay [34] generalized the duality results for the LPP with fuzzy parameters of type-4 and proposed a symmetric fuzzy dual simplex

method for solving said problem without the need to convert it to a conventional LP problem.

Definition 3.24 Consider the following LPP with fuzzy parameters of type-4,

$$\max \tilde{z} = \tilde{c}\tilde{x}$$
$$s.t. \qquad \tilde{A}\tilde{x} \preceq \tilde{b}, \qquad\qquad (3.136)$$
$$\tilde{x} \succeq \tilde{0}.$$

The dual problem of the problem (3.136) is defined as follows:

$$\min \tilde{u} = \tilde{w}\tilde{b}$$
$$s.t. \qquad \tilde{w}\tilde{A} \succeq \tilde{c}, \qquad\qquad (3.137)$$
$$\tilde{w} \succeq \tilde{0}.$$

Theorem 3.39 *If $\tilde{x}$ and $\tilde{w}$ are fuzzy feasible solutions to the problems (3.136) and (3.137), respectively, then $\tilde{w}\tilde{b} \succeq \tilde{c}\tilde{x}$.*

Theorem 3.40 *If $\tilde{x}$ and $\tilde{w}$ are fuzzy feasible solutions to the problems (3.136) and (3.137), respectively, and $\tilde{w}\tilde{b} \approx \tilde{c}\tilde{x}$, then $\tilde{x}$ and $\tilde{w}$ are fuzzy optimal solutions to their respective problems.*

Definition 3.25 The problem (3.136) (or the dual problem (3.137)) is unbounded if fuzzy feasible solutions exist that the fuzzy objective value is increased (or decreased) indefinitely.

The following result relates unboundedness of one problem to infeasibility of the other.

Theorem 3.41 *If any one of the problems (3.136) or (3.137) is unbounded, then the other problem has no fuzzy feasible solution.*

Theorem 3.42 *If any one of the problems (3.136) or (3.137) has a fuzzy optimal solution, then both problems have fuzzy optimal solutions and the fuzzy optimal objective values are equal.*

Theorem 3.43 (Complementary slackness) *Let $\tilde{x}^*$ and $\tilde{w}^*$ be any fuzzy feasible solutions to the problems (3.136) and (3.137), respectively. Then $\tilde{x}^*$ and $\tilde{w}^*$ are fuzzy optimal solutions if and only if*

$$\left(\tilde{w}^*\tilde{A} - \tilde{c}\right)\tilde{x}^* \approx \tilde{0}, \ \tilde{w}^*\left(\tilde{b} - \tilde{A}\tilde{x}^*\right) \approx \tilde{0}$$

Proof Suppose that $\tilde{x}^*$ and $\tilde{w}^*$ are fuzzy feasible solutions to the problems (3.136) and (3.137), respectively. By Theorem 3.42, we have

$$\tilde{0} \approx \tilde{c}\tilde{x}^* - \tilde{w}^*\tilde{b} \approx \left(\tilde{c}\tilde{x}^* - \tilde{w}^*\tilde{A}\tilde{x}^*\right)$$
$$+ \left(\tilde{w}^*\tilde{A}\tilde{x}^* - \tilde{w}^*\tilde{b}\right) \approx \left(\tilde{c} - \tilde{w}^*\tilde{A}\right)\tilde{x}^* + \tilde{w}^*\left(\tilde{A}\tilde{x}^* - \tilde{b}\right)$$

On the other hand, we have $\left(\tilde{c} - \tilde{w}^*\tilde{A}\right)\tilde{x}^* \preceq \tilde{0}$ and $\tilde{w}^*\left(\tilde{A}\tilde{x}^* - \tilde{b}\right) \preceq \tilde{0}$. Therefore, we obtain $\left(\tilde{c} - \tilde{w}^*\tilde{A}\right)\tilde{x}^* \approx \tilde{0}$, and $\tilde{w}^*\left(\tilde{A}\tilde{x}^* - \tilde{b}\right) \approx \tilde{0}$. The converse of the theorem follows from the fact that $\left(\tilde{c} - \tilde{w}^*\tilde{A}\right)\tilde{x}^* \approx \tilde{0}$, and $\tilde{w}^*\left(\tilde{A}\tilde{x}^* - \tilde{b}\right) \approx \tilde{0}$. This implies that $\tilde{c}\tilde{x}^* \approx \tilde{w}^*\tilde{b}$. Therefore, optimality of $\tilde{x}^*$ and $\tilde{w}^*$ follows from Theorem 3.40.   □

Theorem 3.44 *The optimality criteria $\tilde{z}_j - \tilde{c}_j \succeq \tilde{0}$ for all $j \in \{1, 2, \ldots, n\}$, for the problem (3.136) is equivalent to the feasibility condition for the dual problem (3.137). If, in addition, $\tilde{x}$ corresponding to a basis $\tilde{B}$ is primal fuzzy feasible then $\tilde{x}$ is optimal for the problem (3.136) and $\tilde{w} = \tilde{c}_{\tilde{B}}\tilde{B}^{-1}$ is optimal for the problem (3.137).*

Now, assume that the problem (3.136) is feasible and $\tilde{x}$, corresponding to a basis $\tilde{B}$, is dual feasible but primal infeasible. That is, we have $\tilde{z}_j - \tilde{c}_j \succeq \tilde{0}$ for all $j \in \{1, 2, \ldots, n\}$ and at least one r exist such that $\tilde{b}_r \prec \tilde{0}$. Thus, the problem (3.136) can be either infeasible (in which case, the dual problem (3.137) is unbounded), or it has an optimal solution. In this case, it is possible to work on row r of the Tableau 3.43 corresponding to the basis $\tilde{B}$, as the pivoting row, and either (1) detect the infeasibility of the problem (3.136) (or unboundedness of the problem (3.137)), or (2) find a column k, as a pivoting column, to pivot on $\tilde{y}_{rk}$ and obtain a new dual feasible tableau with a non-increasing primal objective value. The following results explain these cases.

Theorem 3.45 *If in a dual feasible simplex tableau an r exists such that $\tilde{\tilde{b}}_r \prec \tilde{0}$ and $\tilde{y}_{rj} \succsim \tilde{0}$, for all $j \in \{1, 2, \ldots, n\}$, then the problem (3.136) is infeasible.*

Theorem 3.46 *If in a dual feasible simplex table, an r exists such that $\tilde{\tilde{b}}_r \prec \tilde{0}$ and a non-basic index $k \in J_N$ exist such that $\tilde{y}_{rk} \prec \tilde{0}$, then pivoting on $\tilde{y}_{rk}$ will yield a dual feasible table with a corresponding non-increasing objective value, where the index k can be found by the following test:*

$$\frac{t_{2k} + t_{3k}}{y_{2,rk} + y_{3,rk}} = \max_{1 \leq j \leq n}\left\{\frac{t_{2j} + t_{3j}}{y_{2,rj} + y_{3,rj}} \Big| y_{2,rj} + y_{3,rj} < 0\right\}$$

Kheirfam and Verdegay [34], using the above results, introduced a new dual algorithm to solve the problem (3.133) directly, making use of the dual feasible simplex tableau. A summary of their proposed method is given as follows.

Algorithm 3.9: Symmetric fully fuzzy dual simplex algorithm *(Maximization Problem)*

Initialization step

Let $\tilde{B}$ be a basis for the LPP with fuzzy parameters of type-4 (3.133) such that $\tilde{z}_j - \tilde{c}_j = (t_{2j} - \alpha_j, t_{2j}, t_{3j}, t_{3j} + \alpha_j)$ for all j, i.e. $t_{2j} + t_{3j} \geq 0$. Form the initial tableau as in Table 3.43.

Main steps

(1) Suppose $\tilde{\bar{b}} = \tilde{B}^{-1}\tilde{b}$. If $\tilde{\bar{b}} \succeq \tilde{0}$ then stop; the current fuzzy solution is optimal. Otherwsie, suppose $\tilde{\bar{b}}_i = (\bar{b}_{2i} - \alpha_i, \bar{b}_{2i}, \bar{b}_{3i}, \bar{b}_{3i} + \alpha_i)$ and let

$$\bar{b}_{2r} + \bar{b}_{3r} = \min_{1 \leq i \leq m} \{\bar{b}_{2i} + \bar{b}_{3i}\}$$

(2) If $\tilde{y}_{ik} = (y_{2,ik} - \alpha_{ik}, y_{2,ik}, y_{3,ik}, y_{3,ik} + \alpha_{ik}) \succeq \tilde{0}$ for all j, i.e. $y_{2,ik} + y_{3,ik} \geq 0$, then stop; the problem (3.133) is infeasible.
Else, $\tilde{x}_{B_r}$ leaves the basis and $\tilde{x}_k$ enters the basis providing that

$$\frac{t_{2k} + t_{3k}}{y_{2,rk} + y_{3,rk}} = \max_{1 \leq j \leq n} \left\{ \frac{t_{2j} + t_{3j}}{y_{2,rj} + y_{3,rj}} \middle| y_{2,rj} + y_{3,rj} < 0 \right\}$$

(3) Update the tableau by pivoting on $\tilde{y}_{rk}$ and go to (1).

Example 3.19 [34] Consider the following LPP with fuzzy parameters of type-4:

$$Max \quad \tilde{z} \approx (-4, -2, 2, 5)\tilde{x}_1 + (-3, -2, 0, 1)\tilde{x}_2$$
$$s.t. \quad (1, 2, 4, 5)\tilde{x}_1 + (-5, -3, -1, 1)\tilde{x}_2 + (1, 1, 1, 1)\tilde{x}_3 \approx (-6, -4, -4, 2),$$
$$(-3, -2, 0, 1)\tilde{x}_1 + (-4, -2, 4, 6)\tilde{x}_2 + (1, 1, 1, 1)\tilde{x}_4 \approx (-4, -1, 5, 8),$$
$$\tilde{x}_1, \tilde{x}_2, \tilde{x}_3, \tilde{x}_4 \succeq \tilde{0}$$

$$(3.138)$$

The first dual feasible simplex tableau is showen in Table 3.47.

Since $\bar{b}_{21} + \bar{b}_{31} < 0$, thus $\tilde{x}_{B_1} = \tilde{x}_3$ leaves the basis and according to the test given in step (2) of Algorithm 3.9, $\tilde{x}_2$ is the entering variable. By pivoting on $\tilde{y}_{12} = (-5, -3, -1, 1)$, the new tableau shown in Table 3.48 is obtained.

Since $\tilde{\bar{b}}_1 = (\frac{-28}{3}, \frac{2}{3}, \frac{10}{3}, \frac{40}{3}) \succ \tilde{0}$ and $\tilde{\bar{b}}_2 = (\frac{-308}{3}, -13, 13, \frac{308}{3}) \approx \tilde{0}$, the current solution is optimal. Thus, the optimal solution of the problem (3.138) is

$$\tilde{x}^* = \begin{bmatrix} \tilde{x}_1^* \\ \tilde{x}_2^* \end{bmatrix} = \begin{bmatrix} (0, 0, 0, 0) \\ (\frac{-28}{3}, \frac{2}{3}, \frac{10}{3}, \frac{40}{3}) \end{bmatrix}$$

$$(3.139)$$

In contrast to the LPP with fuzzy parameters of type-2 considered in [2, 13] and the LPP with fuzzy parameters of type-3 introduced by Ganesan and Veeramini

Table 3.47 The initial simplex tableau of Example 3.19

Basis	$\tilde{x}_1$	$\tilde{x}_2$	$\tilde{x}_3$	$\tilde{x}_4$	R.H.S	$\mathfrak{R}$
$\tilde{z}$	$(-4, -2, 2, 5)$	$(-3, -2, 0, 1)$	$\tilde{0}$	$\tilde{0}$	$\tilde{0}$	–
$\mathfrak{R}(\tilde{z})$	0	-2	0	0	–	0
$\tilde{x}_3$	$(1, 2, 4, 5)$	$(-5, -3, -1, 1)$	$(1, 1, 1, 1)$	$\tilde{0}$	$(-6, -4, -4, 2)$	-8
$\tilde{x}_4$	$(-3, -2, 0, 1)$	$(-4, -2, 4, 6)$	$\tilde{0}$	$(1, 1, 1, 1)$	$(-4, -1, 5, 5)$	4

Table 3.48 The optimal tableau of Example 3.19

Basis	$\tilde{x}_1$	$\tilde{x}_2$	$\tilde{x}_3$	$\tilde{x}_4$	R.H.S	$\mathfrak{R}$
$\tilde{z}$	$(-20, \frac{-23}{6}, \frac{41}{6}, \frac{138}{6})$	$\tilde{0}$	$(\frac{-25}{6}, \frac{-1}{3}, \frac{4}{3}, \frac{31}{6})$	$\tilde{0}$	$(\frac{-86}{6}, \frac{-16}{3}, \frac{4}{3}, \frac{74}{6})$	–
$\mathfrak{R}(\tilde{z})$	3	0	1	0	–	-4
$\tilde{x}_2$	$(\frac{-61}{6}, \frac{-19}{6}, \frac{1}{6}, \frac{43}{6})$	$(1, 1, 1, 1)$	$(\frac{-17}{6}, \frac{-5}{6}, \frac{-1}{6}, \frac{11}{6})$	$\tilde{0}$	$(\frac{-28}{3}, \frac{2}{3}, \frac{10}{3}, \frac{40}{3})$	4
$\tilde{x}_4$	$(\frac{-383}{12}, \frac{-29}{3}, \frac{32}{3}, \frac{484}{12})$	$\tilde{0}$	$\tilde{0}$	$(1, 1, 1, 1)$	$\tilde{0}$	0

[24], in the LPP with fuzzy parameters of type-4 considered by Kheirfam and Verdegay [34] all the parameters and decision variables are specified in terms of symmetric fuzzy data. In addition, the fuzzy primal approach proposed by Ezzati et al. [35] and the fuzzy dual approach proposed by Kheirfam and Verdegay [34], not only give the fuzzy optimal solution, but also does not increase the number of the constraints of the primary fuzzy problem. However, similar to the existing approaches in [2, 13, 24, 25], the fuzzy optimal solution given by use of these approaches may be not non-negative. For instance the value of $\tilde{x}_2^* = \left(\frac{-28}{3}, \frac{2}{3}, \frac{10}{3}, \frac{40}{3}\right)$ in the fuzzy optimal solution (3.139) is non-negative as there is a negative part in this fuzzy solution.

3.7 LPP with Fuzzy Parameters of Type-5

The LP problem having non-negative triangular fuzzy numbers for the costs coefficients, the decision variables, the elements of the coefficient matrix and the values of the right-hand side is termed as the LPP with fuzzy parameters of type-5.

Khan et al. [36] defined the LPP with fuzzy parameters of type-5 as follows:

$$\begin{aligned} \max \tilde{z} &= \tilde{c}\tilde{x} \\ s.t. \quad \tilde{A}\tilde{x} &\preceq \tilde{b}, \\ \tilde{x} &\succeq \tilde{0}. \end{aligned} \tag{3.140}$$

Here, $\tilde{A} = \left[\tilde{a}_{ij} = (a_{1,ij}, a_{2,ij}, a_{3,ij})\right]_{m \times n}$, $\tilde{b} = \left[\tilde{b}_i = (b_{1,i}, b_{2,i}, b_{3,i})\right]_{1 \times m}$, $\tilde{c} = \left[\tilde{c}_j = (c_{1,j}, c_{2,j}, c_{3,j})\right]_{1 \times n}$ and $\tilde{x} = \left[\tilde{x}_j = (x_{1,j}, x_{2,j}, x_{3,j})\right]_{n \times 1}$.

Khan et al. [36] used the following approximate arithmetic operations on the non-negative triangular numbers.

Definition 3.26 Let $\tilde{a} = (a_1, a_2, a_3)$ and $\tilde{b} = (b_1, b_2, b_3)$ be two non-negative triangular fuzzy numbers. The approximate multiplication and division on non-negative triangular fuzzy numbers are defined as follows:

- $\tilde{a}\tilde{b} = (a_1 b_1, a_2 b_2, a_3 b_3), a_1 b_1 \geq 0$
- $\frac{\tilde{a}}{\tilde{b}} = \left(\frac{a_1}{b_3}, \frac{a_2}{b_2}, \frac{a_3}{b_1} \right), a_1 \geq 0, b_1 > 0$

For ranking triangular fuzzy numbers, Khan et al. [36] employed the ranking function as given below:

$$\Re(\tilde{a}) = \Re(a_1, a_2, a_3) = a_1 + a_3 - \frac{a_3 - a_1}{6} \qquad (3.141)$$

Now, we are in a position to review the method proposed by Khan et al. [36] for solving the LPP with fuzzy parameters of type-5 (3.140). The use of the ranking function (3.141) together with the Gaussian elimination process has been used in solving the problem (3.140) in a fully uncertain environment.

The modified version of the well-known simplex method for solving the FLP problem (3.140) is summarized as follows:

Step 1: The constraints matrix of (3.140) can be augmented as

$$[\tilde{A}, \tilde{I}] \begin{bmatrix} \tilde{x} \\ \tilde{x}_s \end{bmatrix} = \tilde{A}\tilde{x} + \tilde{I}\tilde{x}_s = \tilde{b}$$

Here $\tilde{x}_s$ denotes the vector of slack variables. The identity matrix $\tilde{I}$ consists of the columns of the augmented matrix corresponding to the slack variables. The slack variables $\{\tilde{x}_{n+i}\}_{i=1}^m$ are basic and their set is denoted by $\tilde{B}$. The set of non-basic variables $\{\tilde{x}_i\}_{i=1}^m$ is denoted by $\tilde{N}$. Form the initial tableau as given in Table 3.49.

Step 2: Find the value of $\tilde{c}_j$ with the most negative ranking value. Let it be $\tilde{c}_k$.

Table 3.49 The initial tableau of the fuzzy simplex algorithm

	$[\tilde{x}_j]_{1\times n}$	$[\tilde{x}_s]_{1\times(n+m)}$	
$\tilde{z}$	$[-\tilde{c}_j]_{1\times n}$	$[\tilde{0}]_{1\times(n+m)}$	
$\tilde{x}_s$	$[\tilde{a}_{ij}]_{m\times n}$	$[\tilde{I}]_{(n+1)\times(n+m)}$	$[\tilde{b}_i]_{m\times 1}$

Step 3: Find the rank values $\Re(\tilde{b}_i), i = 1, 2, \ldots, m$, $\Re(\tilde{a}_{ik}), i = 1, 2, \ldots, m$ and the ranking ratio $\frac{\Re(\tilde{b}_i)}{\Re(\tilde{a}_{ik})}, i = 1, 2, \ldots, m$.

Step 4: Find the least ranking ratio from $\frac{\Re(\tilde{b}_i)}{\Re(\tilde{a}_{ik})}, i = 1, 2, \ldots, m$. Let it be related to the basic row variable $\tilde{x}_r$.

Step 5: Next create a vector $\tilde{1} = (1, 1, 1)$ at the position of pivotal element $\tilde{a}_{rk}$ and $\tilde{0} = (0, 0, 0)$ above and below it by using the Gaussian elimination method. Furthermore, the basic variable $\tilde{x}_r$ leaves the basis and $\tilde{x}_k$ enters the basis.

Step 6: Repeat Steps 2–5 until there is no negative ranking value of the objective function coefficient $\tilde{c}_j$.

Example 3.20 [36] Consider the following LPP with fuzzy parameters of type-5:

$$\max \tilde{z} = (2, 5, 8)\tilde{x}_1 + \left(3, \tfrac{37}{6}, 10\right)\tilde{x}_2 + \left(5, \tfrac{34}{3}, 15\right)\tilde{x}_3$$

$$\text{s.t.} \quad (2, 5, 8)\tilde{x}_1 + \left(3, \tfrac{41}{6}, 10\right)\tilde{x}_2 + \left(5, \tfrac{31}{3}, 18\right)\tilde{x}_3 \preceq \left(6, \tfrac{50}{3}, 30\right)$$

$$\left(4, \tfrac{32}{3}, 12\right)\tilde{x}_1 + \left(5, \tfrac{73}{6}, 20\right)\tilde{x}_2 + \left(7, \tfrac{105}{6}, 30\right)\tilde{x}_3 \preceq (10, 30, 50) \qquad (3.142)$$

$$(3, 5, 7)\tilde{x}_1 + (5, 15, 20)\tilde{x}_2 + (5, 10, 15)\tilde{x}_3 \preceq \left(2, \tfrac{145}{6}, 30\right)$$

$$\tilde{x}_1, \tilde{x}_2, \tilde{x}_3 \succeq \tilde{0}$$

Step 1: The initial tableau of the primal problem (3.142) is given in Table 3.50.

Step 2: Here, $\Re(\tilde{c}_1) = \Re(-(2, 5, 8)) = -11$, $\Re(\tilde{c}_2) = \left(-\left(3, \tfrac{37}{6}, 10\right)\right) = -\tfrac{85}{6}$ and $\Re(\tilde{c}_3) = \left(-\left(5, \tfrac{34}{3}, 15\right)\right) = -\tfrac{130}{6}$. The most negative ranking value corresponds to $\tilde{x}_3$, thus $k = 3$ and column $\tilde{x}_3$ is identified as the pivotal column.

Step 3: Here, $\Re(\tilde{b}_1) = \Re\left(\left(6, \tfrac{50}{3}, 30\right)\right) = \tfrac{192}{6}$, $\Re(\tilde{b}_2) = \Re((10, 30, 50)) = \tfrac{320}{6}$ and $\Re(\tilde{b}_2) = \Re\left(\left(2, \tfrac{145}{6}, 30\right)\right) = \tfrac{164}{6}$.

Also, $\Re(\tilde{a}_{13}) = \Re\left(\left(5, \tfrac{31}{3}, 18\right)\right) = \tfrac{125}{6}$, $\Re(\tilde{a}_{23}) = \Re\left(\left(7, \tfrac{105}{6}, 30\right)\right) = \tfrac{199}{6}$ and $\Re(\tilde{a}_{33}) = \Re((5, 10, 15)) = \tfrac{110}{6}$.

Table 3.50 Initial tableau of Example 3.20

	$\tilde{x}_1$	$\tilde{x}_2$	$\tilde{x}_3$	$\tilde{x}_4$	$\tilde{x}_5$	$\tilde{x}_6$	
$\tilde{z}$	$-(2,5,8)$	$-\left(3,\frac{37}{6},10\right)$	$-\left(5,\frac{34}{3},15\right)$	$\tilde{0}$	$\tilde{0}$	$\tilde{0}$	$\tilde{0}$
$\tilde{x}_4$	$(2,5,8)$	$\left(3,\frac{41}{6},10\right)$	$\left(5,\frac{31}{3},18\right)$	$\tilde{1}$	$\tilde{0}$	$\tilde{0}$	$\left(6,\frac{50}{3},30\right)$
$\tilde{x}_5$	$\left(4,\frac{32}{3},12\right)$	$\left(5,\frac{73}{6},20\right)$	$\left(7,\frac{105}{6},30\right)$	$\tilde{0}$	$\tilde{1}$	$\tilde{0}$	$(10,30,50)$
$\tilde{x}_6$	$(3,5,7)$	$(5,15,20)$	$(5,10,15)$	$\tilde{0}$	$\tilde{0}$	$\tilde{1}$	$\left(2,\frac{145}{6},30\right)$

Table 3.51 First iteration tableau of Example 3.20

	$\tilde{x}_1$	$\tilde{x}_2$	$\tilde{x}_3$	$\tilde{x}_4$	$\tilde{x}_5$	$\tilde{x}_6$	
$\tilde{z}$	$\left(-7,\frac{2}{3},19\right)$	$\left(\frac{-25}{3},\frac{4}{9},59\right)$	$\tilde{0}$	$\tilde{0}$	$\tilde{0}$	$\left(\frac{1}{3},\frac{34}{3},3\right)$	$\left(5,\frac{986}{36},90\right)$
$\tilde{x}_4$	$\left(\frac{-8}{5},\frac{-1}{6},1\right)$	$\left(-3,\frac{-281}{36},-10\right)$	$\tilde{0}$	$\tilde{1}$	$\tilde{0}$	$\left(\frac{-6}{5},\frac{-31}{30},-1\right)$	$\left(-12,\frac{-299}{36},0\right)$
$\tilde{x}_5$	$\left(-2,\frac{23}{12},\frac{11}{5}\right)$	$\left(-5,\frac{-101}{8},-5\right)$	$\tilde{0}$	$\tilde{0}$	$\tilde{1}$	$\left(-2,\frac{-7}{4},\frac{-7}{5}\right)$	$\left(-20,\frac{-885}{72},8\right)$
$\tilde{x}_6$	$\left(\frac{1}{5},\frac{1}{2},\frac{7}{5}\right)$	$\left(\frac{1}{3},\frac{17}{12},4\right)$	$\tilde{1}$	$\tilde{0}$	$\tilde{0}$	$\left(\frac{1}{15},\frac{1}{10},\frac{1}{5}\right)$	$\left(1,\frac{29}{12},6\right)$

Thus, the ranking ratio $\frac{\Re(\tilde{b}_1)}{\Re(\tilde{a}_{13})} = \frac{192}{125} = 1.53$, $\frac{\Re(\tilde{b}_2)}{\Re(\tilde{a}_{23})} = \frac{320}{199} = 1.60$ and $\frac{\Re(\tilde{b}_3)}{\Re(\tilde{a}_{33})} = \frac{325}{110} = 1.49$.

Step 4: Here,

$$\min\left\{\frac{\Re(\tilde{b}_1)}{\Re(\tilde{a}_{13})} = 1.53, \frac{\Re(\tilde{b}_2)}{\Re(\tilde{a}_{23})} = 1.60, \frac{\Re(\tilde{b}_3)}{\Re(\tilde{a}_{33})} = 1.49\right\} = \frac{\Re(\tilde{b}_3)}{\Re(\tilde{a}_{33})} = 1.49$$

Thus, $r = 3$ and the least ranking ratio is related to the basic row variable $\tilde{x}_6$.

Step 5: We create a vector $\tilde{1} = (1,1,1)$ at the position of pivotal element $\tilde{a}_{33} = (5,10,15)$ by the row operation $\left(\frac{1}{15},\frac{1}{10},\frac{1}{5}\right)\tilde{x}_6$ leading to the new row $\tilde{x}_3$.

Now, create a vector $\tilde{0} = (0,0,0)$ at the positions above and below the pivotal element by the row operations $\tilde{x}_4 - \left(5,\frac{31}{3},18\right)\tilde{x}_3, \tilde{x}_5 - \left(7,\frac{105}{6},30\right)\tilde{x}_3$ and $\tilde{z} + \left(5,\frac{34}{3},15\right)\tilde{x}_3$. In this case $\tilde{x}_6$ leaves the basis and $\tilde{x}_3$ enters the basis. The new tableau is given in Table 3.51.

The first iteration is complete. For the second iteration, there is no number in the objective function row with a negative ranking value, so the solution is optimal and is given below:

$$\tilde{x}_1 = (0,0,0), \tilde{x}_2 = (0,0,0), \tilde{x}_3 = \left(1,\frac{29}{12},6\right), \tilde{z} = \left(1,\frac{29}{12},6\right)$$

3.8   Conclusions

In this chapter we have classified the LPPs with fuzzy parameters into five different groups and discussed various approaches for solving problems belonging to each group based on simplex algorithms. In such approaches, the comparison of fuzzy numbers is carried out by the use of linear ranking functions. The main advantage of the fuzzy simplex algorithms is that the fuzzy problems under consideration are solved directly without converting them to their equivalent crisp problems.

References

1. Ebrahimnejad, A., Verdegay, J.L.: A survey on models and methods for solving fuzzy linear programming problems. In: Fuzzy Logic in Its 50th Year, published in Studies in Fuzziness and Soft Computing, vol. 341, pp. 327–368. Springer International Publishing, Switzerland (2016)
2. Maleki, H.R., Tata, M., Mashinchi, M.: Linear programming with fuzzy variables. Fuzzy Sets Syst. **109**(1), 21–33 (2000)
3. Nasseri, S.H., Ebrahimnejad, A.: A fuzzy dual simplex method for a fuzzy number linear programming problem. Adv. Fuzzy Sets Syst. **5**(2), 81–95 (2010)
4. Mahdavi-Amiri, N., Nasseri, S.H.: Duality in fuzzy number linear programming by use of a certain linear ranking function. Appl. Math. Comput. **180**(1), 206–216 (2006)
5. Ebrahimnejad, A.: Some new results in linear programming problems with fuzzy cost coefficients. Walailak J. Sci. Technol. **10**(2), 191–199 (2013)
6. Ebrahimnejad, A.: A primal–dual method for solving linear programming problems with fuzzy cost coefficients based on linear ranking functions and its applications. Int. J. Ind. Syst. Eng. **12**(2), 119–140 (2012)
7. Ebrahimnejad, A.: Sensitivity analysis in fuzzy number linear programming problems. Math. Comput. Model. **53**(9–10), 1878–1888 (2011)
8. Ebrahimnejad, A., Nasseri, S.H., Mansourzadeh, S.M.: Bounded primal simplex algorithm for bounded linear programming with fuzzy cost coefficients. Int. J. Oper. Res. Inf. Syst. **2**(1), 96–120 (2011)
9. Ebrahimnejad, A., Nasseri, S.H.: A dual simplex method for bounded linear programmes with fuzzy numbers. Int. J. Math. Oper. Res. **2**(5), 762–779 (2010)
10. Ebrahimnejad, A., Nasseri, S.H.: Extension of network primal simplex algorithm for solving minimum cost flow problem with fuzzy costs based on ranking functions. Ann. Fuzzy Math. Inform. **4**(1), 9–24 (2012)
11. Ebrahimnejad, A., Nasseri, S.H., Mansourzadeh, S.M.: Modified bounded dual network simplex algorithm for solving minimum cost flow problem with fuzzy costs based on ranking functions. J. Intell. Fuzzy Syst **24**(1), 191–198 (2013)
12. Ebrahimnejad, A.: Cost parametric analysis of linear programming problems with fuzzy cost coefficients based on ranking functions. Int. J. Math. Model. Numer. Optimisation **8**(1), 62–91 (2017)
13. Mahdavi-Amiri, N., Nasseri, S.H.: Duality results and a dual simplex method for linear programming problems with trapezoidal fuzzy variables. Fuzzy Sets Syst. **158**(17), 1961–1978 (2007)
14. Ebrahimnejad, A., Nasseri, S.H., HosseinzadehLotfi, F., Soltanifar, M.: A primal-dual method for linear programming problems with fuzzy variables. Eur. J. Ind. Eng. **4**(2), 189–209 (2010)

15. Nasseri, S.H., Ebrahimnejad, A.: Sensitivity analysis on linear programming problems with trapezoidal fuzzy variables. Int. J. Oper. Res. Inf. Syst. **2**(2), 22–39 (2011)
16. Ebrahimnejad, A., Nasseri, S.H.: Using complementary slackness property to solve linear programming with fuzzy parameters. Fuzzy Inf. Eng. **1**(3), 233–245 (2009)
17. Nasseri, S.H., Ebrahimnejad, A.: A fuzzy primal simplex algorithm and its application for solving flexible linear programming problems. Eur. J. Ind. Eng. **4**(3), 372–389 (2010)
18. Nasseri, S.H., Attari, H., Ebrahimnejad, A.: Revised simplex method and its application for solving fuzzy linear programming problems. Eur. J. Ind. Eng. **6**(3), 259–280 (2012)
19. Ebrahimnejad, A.: A constructive proof of fundamental theory for fuzzy variable linear programming problems. Appl. Appl. Math. **7**(2), 691–707 (2012)
20. Ebrahimnejad, A., Verdegay, J.L.: On solving bounded fuzzy variable linear program and its applications. J. Intell. Fuzzy Syst. **27**(5), 2265–2280 (2014)
21. Ebrahimnejad, A.: A duality approach for solving bounded linear programming problems with fuzzy variables based on ranking functions and its application in bounded transportation problems. Int. J. Syst. Sci. **46**(11), 2048–2060 (2015)
22. Ebrahimnejad, A., Tavana, M.: A novel method for solving linear programming problems with trapezoidal fuzzy variables. RAIRO-Operations Research, (Submitted) (2017)
23. Yang, Y., Jia, Y.-L., Zhong, Y.-H.: Parametric sensitivity analysis of linear programming with fuzzy variables. J. Intell. Fuzzy Syst. **33**(1), 145–158 (2017)
24. Ganesan, K., Veeramani, P.: Fuzzy linear programming with trapezoidal fuzzy numbers. Ann. Oper. Res. **143**(1), 305–315 (2006)
25. Ebrahimnejad, A.: Some new results in linear programs with trapezoidal fuzzy numbers: finite convergence of the Ganesan and Veeramani's method and a fuzzy revised simplex method. Appl. Math. Model. **35**(9), 4526–4540 (2011)
26. Nasseri, S.H., Ebrahimnejad, A., Mizuno, S.: Duality in fuzzy linear programming with symmetric trapezoidal numbers. Appl. Appl. Math. **5**(10), 1467–1482 (2010)
27. Ebrahimnejad, A., Nasseri, S.H.: Linear programmes with trapezoidal fuzzy numbers: a duality approach. Int. J. Oper. Res. **13**(1), 67–89 (2012)
28. Ebrahimnejad, A.: A primal-dual simplex algorithm for solving linear programming problems with symmetric trapezoidal fuzzy numbers. Appl. Math. (Irvine) **2**(6), 676–684 (2011)
29. Ebrahimnejad, A.: Tableau form of the fuzzy primal-dual simplex algorithm for solving linear programmes with trapezoidal fuzzy numbers. Int. J. Oper. Res. **18**(2), 123–139 (2013)
30. Ebrahimnejad, A., Verdegay, J.L.: A novel approach for sensitivity analysis in linear programs with trapezoidal fuzzy numbers. J. Intell. Fuzzy Syst. **27**(1), 173–185 (2014)
31. Kheirfam, B., Verdegay, J.L.: The dual simplex method and sensitivity analysis for fuzzy linear programming with symmetric trapezoidal numbers. Fuzzy Optim. Decis. Mak. **12**(2), 171–189 (2013)
32. Ebrahimnejad, A., Nasseri, S.H., Hosseinzadeh Lotfi, F.: Bounded linear programs with trapezoidal fuzzy numbers. Int. J. Uncertain. Fuzziness Knowl.-Based Syst. **18**(3), 269–286 (2010)
33. Ebrahimnejad, A., Tavana, M.: A novel method for solving linear programming problems with symmetric trapezoidal fuzzy numbers. Appl. Math. Model. **38**(17–18), 4388–4395 (2014)
34. Kheirfam, B., Verdegay, J.L.: Optimization and reoptimization in fuzzy linear programming problems. In: The 8th Conference of the European Society for Fuzzy Logic and Technology, pp. 527–533 (2013)
35. Ezzati, R., Khorram, E., Enayati, R.: A novel computational method to solve fully fuzzy linear programming problems using fuzzy primal simplex algorithm. Iran. J. Fuzzy Syst. (submitted)
36. Khan, I.U., Ahmad, T., Maan, N.: A simplified novel technique for solving fully fuzzy linear programming problems. J. Optim. Theory Appl. **159**(2), 536–546 (2013)
37. Bhardwaj, B., Kumar, A.: A note on the paper "A simplified novel technique for solving fully fuzzy linear programming problems". J. Optim. Theory Appl. **163**(2), 685–696 (2014)

38. Khan, I.U., Ahmad, T., Maan, N.: A reply to a note on the paper "A simplified novel technique for solving fully fuzzy linear programming problems". J. Optim. Theory Appl. **173** (1), 353–356 (2017)
39. Maleki, H.R.: Ranking functions and their applications to fuzzy linear programming. Far East J. Math. Sci. **4**(2), 283–301 (2002)
40. Murty, G.H.: Linear Programming. Wiley, New York (1983)

Chapter 4
Linear Programming with Fuzzy Parameters: Non-simplex Based Approaches

4.1 Introduction

In this chapter, the non-simplex based approaches for solving several kinds of LPPs with fuzzy parameters are explored. In such approaches, the fuzzy constraints are first converted to crisp ones based on arithmetic operations on fuzzy numbers and then the standard methods are used for solving the crisp problems. Such approaches increase the number of functional constraints and thus directly affect the computational time of the simplex method.

Linear programming problems with fuzzy parameters that can be solved based on non-simplex approaches can be classified into the following seven general groups:

- Type-1 LPPs with fuzzy parameters: The FLP problems in this group involve fuzzy numbers for the coefficients of the decision variables in the constraints and the right-hand side of the constraints.
- Type-2 LPPs with fuzzy parameters: The FLP problems in this group involve fuzzy numbers for the coefficients of the decision variables in the objective function.
- Type-3 LPPs with fuzzy parameters: The FLP problems in this group involve fuzzy numbers for the coefficients of the decision variables in the objective function and in the constraints.
- Type-4 LPPs with fuzzy parameters: The FLP problems in this group involve fuzzy numbers for the coefficients of the decision variables in the objective function, the coefficients of the decision variables in the constraints and the right-hand side of the constraints.
- Type-5 LPPs with fuzzy parameters: The FLP problems in this group involve fuzzy numbers for the decision variables and the right-hand side of the constraints.
- Type-6 LPPs with fuzzy parameters: The FLP problems in this group involve symmetric trapezoidal fuzzy numbers for the decision variables, the coefficients

© Springer International Publishing AG, part of Springer Nature 2018
A. Ebrahimnejad and J. L. Verdegay, *Fuzzy Sets-Based Methods and Techniques for Modern Analytics*, Studies in Fuzziness and Soft Computing 364,
https://doi.org/10.1007/978-3-319-73903-8_4

of the decision variables in the objective function and the right-hand side of the constraints.

- Type-7 LPPs with fuzzy parameters: The FLP problems in this group, so-called fully FLP (FFLP) problems, involve fuzzy numbers for the decision variables, the coefficients of the decision variables in the objective function, the coefficients of the decision variables in the constraints and the right-hand side of the constraints.

A comprehensive literature review on solving LPPs with fuzzy parameters based on non-simplex based approaches is presented in the next section.

4.2 A Brief Literature Review on Non-simplex Based FLP Models

The LPPs with fuzzy parameters that can be solved based on non-simplex approaches are categorized into two general groups: FLP problems with (1) crisp decision variables and (2) fuzzy decision variables. An overview of the papers which studied LPPs with fuzzy parameters based on non-simplex based approaches and discussed solution methodologies are summarized as follows.

Ramik and Rimanek [1] proposed an approach for solving Type-1 LPPs with fuzzy parameters. They used a ranking approach to define the fuzzy inequality constraints and converted each fuzzy constraint into four crisp constraints assuming the fuzzy parameters are in terms of flat fuzzy numbers. Tanaka et al. [2] proposed another approach based on the extension principle and alpha cuts for solving the same fuzzy problem assuming the fuzzy coefficients have symmetric triangular membership functions. They used the concept of comparison of intervals to define the fuzzy inequality and then converted each fuzzy constraint into two crisp constraints for each given alpha cut. Dubois [3] investigated Type-1 LPPs with fuzzy parameters in the case of fuzzy equality constraints and provided two cases of soft and hard equivalent constraints. Li and Gong [4] introduced a class of Type-1 LPPs with fuzzy parameters based on the possibility and necessity relations and measured the fulfillment of the constraints by using the degree of possibility and necessity. According to this ranking index, they obtained the bounds of optimal solution at different degrees of possibility and necessity. Lai and Hwang [5] used multiple objective linear programming (MOLP) techniques for solving Type-2 LPPs with fuzzy parameters in the case of triangular fuzzy numbers for the fuzzy objective function. They proposed three approaches based on MOLP techniques to obtain the optimal solution of the FLP under consideration. Wan and Dong [6] extended the approach proposed by Lai and Hwang [5] for situations in which the fuzzy numbers of Type-2 LPPs with fuzzy parameters are all trapezoidal fuzzy numbers. Rommenlfanger et al. [7] used a finite set of alpha cuts for each fuzzy coefficient and reduced Type-2 LPPs with fuzzy parameters to an equivalent MOLP problem. They then used Zimmerman's approach [8] for solving the resulting MOLP problem. Delgado et al. [9] also proposed an MOLP approach for solving Type-2 LPPs with fuzzy parameters assuming the imprecise objective coefficients have trapezoidal membership functions. Maeda [10] defined the concept of optimal solutions for Type-2 LPPs with fuzzy parameters and investigated their properties. Then he introduced three bi-criteria optimization problems in order to find all optimal solutions. Zhang et al. [11] developed Maeda's approach for solving problems involving fuzzy numbers with any form of membership functions. Li and Guo [12] proved that their approach is not applicable for solving problems involving LR flat fuzzy numbers with different shapes. Cai et al. [13] investigated that kind of Type-2 LPPs with fuzzy parameters involving LR fuzzy numbers with the same reference functions and translated it into MOLP problems with four objectives. Lai and Hwang [5] proposed an approach for solving Type-3 LPPs with fuzzy parameters with the help of the concept of the most likely value for the imprecise parameters with triangular membership functions. Wan and Dong [6] developed their approach and proposed a weighted average method to deal with the fuzzy constraints. Lai and Hwang [5] and Wan and Dong [6] combined the fuzzy ranking concepts with the strategy used for the fuzzy objective functions of Type-2 LPPs with fuzzy parameters to develop an approach for solving Type-4 LPPs with fuzzy parameters. Buckley [15, 16] presented a solution procedure for Type-4 LPPs with fuzzy parameters having no equality constraints. The solution approach involves first specifying the membership function for the objective function and then determining a compromise solution for the derision variables. Negi and Lee [17] used the exceedance possibility and the strict exceedance possibility indexes to compare the inequality constraints of Type-4 LPPs with fuzzy parameters. Fuller [18] used a ranking function approach to convert Type-4 LPPs with fuzzy parameters into a crisp LP problem. Hatami-Marbini et al. [19] developed a new model for solving Type-4 LPPs with fuzzy parameters that is inspired by the work of Li and Gong [4] on possibility and necessity relation and the work of Rommelfanger et al. [7] on solving Type-2 LPPs with fuzzy parameters.

Buckley and Feuring [20] obtained solutions of Type-7 LPPs with fuzzy parameters where all the parameters and variables are fuzzy numbers. They first changed the problem of maximizing a fuzzy number, the value of the objective function, into an MOLP problem. Then they designed an evolutionary algorithm to solve the fuzzy flexible problem. Saati et al. [21] defined some new definitions for non-negative trapezoidal fuzzy numbers and fuzzy variables and then developed a new solution method for solving Type-5 LPPs with fuzzy parameters without the use of fuzzy arithmetic. Kumar and Kaur [22] proposed a solution approach for solving Type-6 LPPs with fuzzy parameters by converting them into crisp LP ones based on fuzzy arithmetic for the fuzzy constraints and ranking function approach for the fuzzy objective function. Hashemi et al. [23] proposed a two-phase approach to find the optimal solutions of Type-7 LPPs with fuzzy parameters based on the comparison of the mean and standard deviation of fuzzy numbers. Hosseinzadeh Lotfi et al. [24] proposed a method for solving Type-7 LPPs with fuzzy parameters in the case of equality constraints by approximating all the coefficients into their nearest symmetric fuzzy numbers. Kumar et al. [26] proposed a solution technique for the same problem where, in contrast to the proposed method by Hosseinzaeh Lotfi et al. [24], the constraints of the fuzzy problem are satisfied exactly by the fuzzy optimal solution. Kumar and Kaur [25] proposed a similar approach for solving a general form of Type-7 LPP with fuzzy parameters in the cases of both fuzzy equality and inequality constraints. Ezzati et al. [27] proposed a novel algorithm for solving the same problem by converting it to its equivalent MOLP problem. Das et al. [28] developed their approach assuming that all fuzzy parameters and variables are non-negative trapezoidal fuzzy numbers. Cheng et al. [29] investigated the Type-7 LPP with fuzzy parameters in the case of fuzzy equality constraints. They converted the fuzzy equality constraints into crisp inequality ones using the measure of the similarity and then transformed the fuzzy objective into two crisp objectives by considering expected value and uncertainty of fuzzy objective. Since the feasibility degree of constraints is in conflict with the optimal value of the objective function, they finally constructed an auxiliary three-objective LPP, which is solved through a compromise programming approach. Hosseinzadeh and Edalatpanah [30] designed a method for solving the Type-7 LPP with fuzzy parameters by the lexicography method and LP methods.

4.3 Type-1 LPPs with Fuzzy Parameters

The general form of LPPs in which the right-hand side of the constraints and the coefficients of the constraint matrix are fuzzy numbers, can be formulated as follows:

$$
\begin{aligned}
\max z &= \sum_{j=1}^{n} c_j x_j \\
s.t. \quad &\sum_{j=1}^{n} \tilde{a}_{ij} x_j \preceq \tilde{b}_i, \quad i = 1, 2, \ldots, m, \\
&x_j \geq 0, \qquad j = 1, 2, \ldots, n.
\end{aligned}
\tag{4.1}
$$

In this section we will explore three approaches for solving Type-1 LPPs with fuzzy parameters (4.1).

4.3.1 Ramik and Rimanek's Approach

Let us assume that all fuzzy numbers are trapezoidal. Thus, $\tilde{a}_{ij}$ and $\tilde{b}_i$ are represented as $\tilde{a}_{ij} = (a_{1,ij}, a_{2,ij}, a_{3,ij}, a_{4,ij})$ and $\tilde{b}_i = (b_{1,i}, b_{2,i}, b_{3,i}, b_{4,i})$, respectively. Hence, the FLP problem (4.1) can be reformulated as follows:

$$\max z = \sum_{j=1}^{n} c_j x_j$$

$$s.t. \quad \sum_{j=1}^{n} (a_{1,ij}, a_{2,ij}, a_{3,ij}, a_{4,ij}) x_j \preceq (b_{1,i}, b_{2,i}, b_{3,i}, b_{4,i}), \quad i = 1, 2, \ldots, m, \quad (4.2)$$

$$x_j \geq 0, \quad j = 1, 2, \ldots, n.$$

Equivalently, with regard to the definitions of addition of trapezoidal fuzzy numbers and scalar multiplication, the FLP (4.2) can be rewritten as follows:

$$\max z = \sum_{j=1}^{n} c_j x_j$$

$$s.t. \quad \left(\sum_{j=1}^{n} a_{1,ij} x_j, \sum_{j=1}^{n} a_{2,ij} x_j, \sum_{j=1}^{n} a_{3,ij} x_j, \sum_{j=1}^{n} a_{4,ij} x_j \right) \preceq (b_{1,i}, b_{2,i}, b_{3,i}, b_{4,i}), \quad i = 1, 2, \ldots, m,$$

$$x_j \geq 0, \quad j = 1, 2, \ldots, n.$$

$$(4.3)$$

To solve this FLP problem, the fuzzy inequality $\preceq$ should be defined. To define this fuzzy inequality, Ramik and Rimanek [1] used the partial order given in Remark 1.21. Thus, with regard to Remark 1.21, the FLP (4.3) can be rewritten as follows:

$$\max z = \sum_{j=1}^{n} c_j x_j$$

$$s.t. \quad \sum_{j=1}^{n} a_{1,ij} x_j \leq b_{1,i}, \quad i = 1, 2, \ldots, m,$$

$$\sum_{j=1}^{n} a_{2,ij} x_j \leq b_{2,i}, \quad i = 1, 2, \ldots, m,$$

$$\sum_{j=1}^{n} a_{3,ij} x_j \leq b_{3,i}, \quad i = 1, 2, \ldots, m, \quad (4.4)$$

$$\sum_{j=1}^{n} a_{4,ij} x_j \leq b_{4,i}, \quad i = 1, 2, \ldots, m,$$

$$x_j \geq 0, \quad j = 1, 2, \ldots, n.$$

The optimal solution of the crisp LP problem (4.4) can be considered as the solution of the FLP problem (4.2).

Example 4.1 Consider the following FLP problem:

$$\max z = 5x_1 + 4x_2$$
$$s.t. \quad (2,4,5,6)x_1 + (2,5,6,7)x_2 \preceq (19,24,32,43),$$
$$(3,4,6,10)x_1 + (4,5,6,13)x_2 \preceq (6,12,19,24)$$
$$x_1, x_2 \geq 0.$$
(4.5)

This problem is converted into the following crisp LP problem with regard to the problem (4.4):

$$\max z = 5x_1 + 4x_2$$
$$s.t. \quad 2x_1 + 2x_2 \leq 19,$$
$$4x_1 + 5x_2 \leq 24,$$
$$5x_1 + 6x_2 \leq 32,$$
$$6x_1 + 7x_2 \leq 43,$$
$$3x_1 + 4x_2 \leq 6,$$
$$4x_1 + 5x_2 \leq 12,$$
$$6x_1 + 6x_2 \leq 19,$$
$$10x_1 + 13x_2 \leq 24,$$
$$x_1, x_2 \geq 0.$$
(4.6)

Solving this problem gives the optimal solution as $x^* = (x_1^*, x_1^*) = (1.5, 3)$ with the objective function $z^* = 19.5$.

The optimal solutions obtained by this approach are real numbers, which represent a compromise in terms of fuzzy numbers involved. In addition, this approach increases the number of functional constraints and thus this proposal needs a greater computation cost than the simplex based approaches.

4.3.2   *Tanaka et al.'s Approach*

Tanaka et al. [2] assumed that all fuzzy coefficients are symmetric triangular fuzzy numbers. Thus, $\tilde{a}_{ij}$ and $\tilde{b}_i$ are represented as $\tilde{a}_{ij} = (a_{1,ij}, a_{2,ij}, a_{3,ij})$ and $\tilde{b}_i = (b_{1,i}, b_{2,i}, b_{3,i})$, respectively, where $a_{2,ij} - a_{1,ij} = a_{3,ij} - a_{2,ij}$ and $b_{2,i} - b_{1,i} = b_{3,i} - b_{2,i}$. Based on the extension principle, they considered the FLP problem (4.1) as follows:

$$\max z = \sum_{j=1}^{n} c_j x_j$$

$$s.t. \quad \sum_{j=1}^{n} \tilde{a}_{ij} x_j \leq_\alpha \tilde{b}_i, \quad i = 1, 2, \ldots, m, \tag{4.7}$$

$$x_j \geq 0, \quad j = 1, 2, \ldots, n.$$

where α is determined by the decision maker.

To solve this FLP problem, the fuzzy inequality $\leq_\alpha$ should be defined. To define this fuzzy inequality, Tanaka et al. [2] asserted

$$\sum_{j=1}^{n} \tilde{a}_{ij} x_j \leq_\alpha \tilde{b}_i \Leftrightarrow \left[\sum_{j=1}^{n} \tilde{a}_{ij} x_j \right]_\alpha = \left[\left(\sum_{j=1}^{n} \tilde{a}_{ij} x_j \right)_\alpha^L, \left(\sum_{j=1}^{n} \tilde{a}_{ij} x_j \right)_\alpha^U \right] \leq \left[\tilde{b}_i \right]_\alpha$$
$$= \left[\left(\tilde{b}_i \right)_\alpha^L, \left(\tilde{b}_i \right)_\alpha^U \right]$$

This means that

$$\sum_{j=1}^{n} \tilde{a}_{ij} x_j \leq_\alpha \tilde{b}_i \Leftrightarrow \left(\sum_{j=1}^{n} \tilde{a}_{ij} x_j \right)_\alpha^L \leq \left(\tilde{b}_i \right)_\alpha^L, \left(\sum_{j=1}^{n} \tilde{a}_{ij} x_j \right)_\alpha^U \leq \left(\tilde{b}_i \right)_\alpha^U$$

Thus, the FLP (4.7) can be rewritten as follows:

$$\max z = \sum_{j=1}^{n} c_j x_j$$

$$s.t. \quad \left(\sum_{j=1}^{n} \tilde{a}_{ij} x_j \right)_\alpha^L \leq \left(\tilde{b}_i \right)_\alpha^L, \quad i = 1, 2, \ldots, m,$$

$$\left(\sum_{j=1}^{n} \tilde{a}_{ij} x_j \right)_\alpha^U \leq \left(\tilde{b}_i \right)_\alpha^U, \quad i = 1, 2, \ldots, m, \tag{4.8}$$

$$x_j \geq 0, \quad j = 1, 2, \ldots, n.$$

Note that the α-cuts of the symmetric triangular fuzzy numbers $\tilde{a}_{ij} = (a_{1,ij}, a_{2,ij}, a_{3,ij})$ and $\tilde{b}_i = (b_{1,i}, b_{2,i}, b_{3,i})$ are given as follows:

$$[\tilde{a}_{ij}]_\alpha = \left[(\tilde{a}_{ij})_\alpha^L, (\tilde{a}_{ij})_\alpha^U \right] = \left[\frac{\alpha}{2} a_{3,ij} + \left(1 - \frac{\alpha}{2} \right) a_{1,ij}, \left(1 - \frac{\alpha}{2} \right) a_{3,ij} + \frac{\alpha}{2} a_{1,ij} \right]$$

$$[\tilde{b}_i]_\alpha = \left[(\tilde{b}_i)_\alpha^L, (\tilde{b}_i)_\alpha^U \right] = \left[\frac{\alpha}{2} b_{3,i} + \left(1 - \frac{\alpha}{2} \right) b_{1,i}, \left(1 - \frac{\alpha}{2} \right) b_{3,i} + \frac{\alpha}{2} b_{1,i} \right]$$

Thus, with regard to the definition of addition on closed interval and scalar multiplication we have:

$$\left[\sum_{j=1}^{n} \tilde{a}_{ij}x_j\right]_\alpha = \left[\left(\sum_{j=1}^{n} \tilde{a}_{ij}x_j\right)_\alpha^L, \left(\sum_{j=1}^{n} \tilde{a}_{ij}x_j\right)_\alpha^U\right]$$

$$= \left[\frac{\alpha}{2}\sum_{j=1}^{n} a_{3,ij}x_j + \left(1 - \frac{\alpha}{2}\right)\sum_{j=1}^{n} a_{1,ij}x_j, \left(1 - \frac{\alpha}{2}\right)\sum_{j=1}^{n} a_{3,ij}x_j + \frac{\alpha}{2}\sum_{j=1}^{n} a_{1,ij}x_j\right]$$

By using the above results, the FLP (4.8) can be rewritten as follows:

$$\max z = \sum_{j=1}^{n} c_j x_j$$

$$s.t. \quad \frac{\alpha}{2}\sum_{j=1}^{n} a_{3,ij}x_j + \left(1 - \frac{\alpha}{2}\right)\sum_{j=1}^{n} a_{1,ij}x_j \leq \frac{\alpha}{2}b_{3,i} + \left(1 - \frac{\alpha}{2}\right)b_{1,i}, \quad i = 1,2,\ldots,m,$$

$$\left(1 - \frac{\alpha}{2}\right)\sum_{j=1}^{n} a_{3,ij}x_j + \frac{\alpha}{2}\sum_{j=1}^{n} a_{1,ij}x_j \leq \left(1 - \frac{\alpha}{2}\right)b_{3,i} + \frac{\alpha}{2}b_{1,i}, \quad i = 1,2,\ldots,m,$$

$$x_j \geq 0, \quad j = 1,2,\ldots,n.$$

$$(4.9)$$

Example 4.2 [14] Consider the following FLP problem:

$$\max z = 25x_1 + 18x_2$$
$$s.t. \quad (12,15,18)x_1 + (32,34,36)x_2 \preceq (750,800,850),$$
$$(19,20,21)x_1 + (7,10,13)x_2 \preceq (380,430,480)$$
$$x_1, x_2 \geq 0.$$

$$(4.10)$$

Assume that $\alpha = 0.4$ is given. Regarding to the problem (4.9), the equivalent crisp problem of (4.10) is given as follows:

$$\max z = 25x_1 + 18x_2$$
$$s.t. \quad 16.8x_1 + 35.2x_2 \leq 830,$$
$$13.2x_1 + 3.2x_2 \leq 770,$$
$$20.6x_1 + 11.8x_2 \leq 460,$$
$$19.4x_1 + 8.2x_2 \leq 400,$$
$$x_1, x_2 \geq 0.$$

$$(4.11)$$

The optimal solution of the crisp problem (4.11) is as $x^* = (x_1^*, x_1^*) = (12.14, 17.78)$ with the objective function $z^* = 623.54$.

4.3.3   Li and Gong's Approach

Based on the possibility and necessity relation, Li and Gong [4] proposed an approach for solving Type-1 LPP with fuzzy parameters (4.1). They considered the following FLP problems based on the relation of possibility and necessity:

$$\max z = \sum_{j=1}^{n} c_j x_j$$

$$s.t. \quad \sum_{j=1}^{n} \tilde{a}_{ij} x_j \preceq^{Pos} \tilde{b}_i, \ i = 1, 2, \ldots, m, \tag{4.12}$$

$$x_j \geq 0, \qquad j = 1, 2, \ldots, n.$$

$$\max z = \sum_{j=1}^{n} c_j x_j$$

$$s.t. \quad \sum_{j=1}^{n} \tilde{a}_{ij} x_j \preceq^{Nec} \tilde{b}_i, \ i = 1, 2, \ldots, m, \tag{4.13}$$

$$x_j \geq 0, \qquad j = 1, 2, \ldots, n.$$

With respect to the constraints of (4.12) and (4.13), the optimal solution is completely determined by degrees of possibility and necessity, so they proposed the following two crisp LP problems to obtain the optimal solution of (4.12) and (4.13) at α-cut levels and $(1 - \alpha)$-cut levels, respectively:

$$\max z = \sum_{j=1}^{n} c_j x_j$$

$$s.t. \quad Pos\left(\sum_{j=1}^{n} \tilde{a}_{ij} x_j \preceq \tilde{b}_i \right) \geq \alpha, \ i = 1, 2, \ldots, m, \tag{4.14}$$

$$x_j \geq 0, \quad j = 1, 2, \ldots, n.$$

$$\max z = \sum_{j=1}^{n} c_j x_j$$

$$s.t. \quad Nec\left(\sum_{j=1}^{n} \tilde{a}_{ij} x_j \prec \tilde{b}_i \right) \geq 1 - \alpha, \ i = 1, 2, \ldots, m, \tag{4.15}$$

$$x_j \geq 0, \quad j = 1, 2, \ldots, n.$$

By using the Theorem 1.17, the FLP (4.14) can be rewritten as follows:

$$\max z = \sum_{j=1}^{n} c_j x_j$$

$$s.t. \quad b_{3,i} - \left(\sum_{j=1}^{n} a_{2,ij} x_j \right) \ge (\alpha - 1) \left((b_{4,i} - b_{3,i}) + \sum_{j=1}^{n} (a_{2,ij} - a_{1,ij}) x_j \right), \ i = 1, 2, \ldots, m,$$

$$x_j \ge 0, \quad j = 1, 2, \ldots, n.$$

$$(4.16)$$

In a similar way, by using the Theorem 1.18, the FLP (4.15) can be rewritten as follows:

$$\max z = \sum_{j=1}^{n} c_j x_j$$

$$s.t. \quad b_{2,i} - \left(\sum_{j=1}^{n} a_{3,ij} x_j \right) \ge (1 - \alpha) \left((b_{2,i} - b_{1,i}) + \sum_{j=1}^{n} (a_{4,ij} - a_{3,ij}) x_j \right), \ i = 1, 2, \ldots, m,$$

$$x_j \ge 0, \quad j = 1, 2, \ldots, n.$$

$$(4.17)$$

We denote the feasible space and the optimal value of the problem (4.16) by X_α^{Pos} and Z^{Pos}, respectively, and the feasible space and the optimal value of the problem (4.17) by X_α^{Nec} and Z^{Nec}, respectively.

Theorem 4.1 *For the problems* (4.16) *and* (4.17), $X_\alpha^{Nec} \subseteq X_\alpha^{Pos}$ *and* $Z^{Nec} \le Z^{Pos}$.

Proof If $x \in X_\alpha^{Nec}$, then

$$b_{2,i} - \left(\sum_{j=1}^{n} a_{3,ij} x_j \right) \ge (1 - \alpha) \left((b_{2,i} - b_{1,i}) + \sum_{j=1}^{n} (a_{4,ij} - a_{3,ij}) x_j \right) \ (i = 1, 2, \ldots, m)$$

This means that

$$\sum_{j=1}^{n} a_{3,ij} x_j + (1 - \alpha) \sum_{j=1}^{n} (a_{4,ij} - a_{3,ij}) x_j \le b_{2,i} - (1 - \alpha)(b_{2,i} - b_{1,i}) \ (i = 1, 2, \ldots, m)$$

Since

$$\sum_{j=1}^{n} a_{3,ij} x_j + (1 - \alpha) \sum_{j=1}^{n} (a_{4,ij} - a_{3,ij}) x_j \ge \sum_{j=1}^{n} a_{2,ij} x_j - (1 - \alpha) \sum_{j=1}^{n} (a_{2,ij} - a_{1,ij}) x_j$$

and

$$b_{3,i} + (1 - \alpha)(b_{3,i} - b_{2,i}) \geq b_{2,i} - (1 - \alpha)(b_{2,i} - b_{1,i}),$$

then

$$\sum_{j=1}^{n} a_{2,ij}x_j - (1 - \alpha)\sum_{j=1}^{n}(a_{2,ij} - a_{1,ij})x_j \leq b_{3,i} + (1 - \alpha)(b_{3,i} - b_{2,i}),$$

Thus $x \in X_\alpha^{Pos}$. This means that $X_\alpha^{Nec} \subseteq X_\alpha^{Pos}$ and $Z^{Nec} \leq Z^{Pos}$.   □

Theorem 4.2 *For problems* (4.16) *and* (4.17), $0 \leq \alpha_1 \leq \alpha_2 \leq 1$ $X_{\alpha_2}^{Pos} \subseteq X_{\alpha_1}^{Pos}$, $X_{\alpha_2}^{Nec} \supseteq X_{\alpha_1}^{Nec}$, $z_{\alpha_2}^{Pos} \leq z_{\alpha_1}^{Pos}$ and $z_{\alpha_1}^{Nec} \leq z_{\alpha_2}^{Nec}$.

Proof If $x \in X_{\alpha_2}^{Pos}$, then

$$b_{3,i} - \sum_{j=1}^{n} a_{2,ij}x_j \geq (\alpha_2 - 1)\left[\sum_{j=1}^{n}(a_{2,ij} - a_{1,ij})x_j + (b_{3,i} - b_{2,i})\right]$$

Since $\sum_{j=1}^{n}(a_{2,ij} - a_{1,ij})x_j + (b_{3,i} - b_{2,i}) \geq 0$ and $\alpha_1 \leq \alpha_2$, we have

$$(\alpha_2 - 1)\left[\sum_{j=1}^{n}(a_{2,ij} - a_{1,ij})x_j + (b_{3,i} - b_{2,i})\right] \geq (\alpha_1 - 1)\left[\sum_{j=1}^{n}(a_{2,ij} - a_{1,ij})x_j + (b_{3,i} - b_{2,i})\right],$$

Thus, $b_{3,i} - \sum_{j=1}^{n} a_{2,ij}x_j \geq (\alpha_1 - 1)\left[\sum_{j=1}^{n}(a_{2,ij} - a_{1,ij})x_j + (b_{3,i} - b_{2,i})\right]$. This means that $x \in X_{\alpha_1}^{Pos}$ and so $X_{\alpha_2}^{Pos} \subseteq X_{\alpha_1}^{Pos}$ and $z_{\alpha_2}^{Pos} \leq z_{\alpha_1}^{Pos}$. In a similar way, we can obtain $X_{\alpha_2}^{Nec} \supseteq X_{\alpha_1}^{Nec}$ and $z_{\alpha_1}^{Nec} \leq z_{\alpha_2}^{Nec}$.   □

From Theorems 4.1 and 4.2, we conclude that that z_α^{Nec} and z_α^{Pos} constitute the lower and upper bounds of the fuzzy objective value at level α. For z_α^{Nec} is increasing and z_α^{Pos} is decreasing when α increases, so $z_\alpha = \left[z_\alpha^{Nec}, z_\alpha^{Pos}\right]$ is a nested set.

Example 4.3 [4] Consider the following FLP problem:

$$\begin{aligned}
\max z &= 3x_1 + 4x_2 \\
s.t. \quad &(2,3,4,5)x_1 + (1,2,3,4)x_2 \preceq (3,5,7,9), \\
&(1,2,3,4)x_1 + (3,4,5,6)x_2 \preceq (4,6,8,10) \\
&x_1, x_2 \geq 0.
\end{aligned} \tag{4.18}$$

Based on the possibility relation, with regard to the problem (4.16) we should solve the following crisp LP problem:

$$\max z = 3x_1 + 4x_2$$
$$s.t. \quad (\alpha + 2)x_1 + (\alpha + 1)x_2 \leq 9 - 2\alpha,$$
$$(\alpha + 1)x_1 + (\alpha + 3)x_2 \leq 10 - 2\alpha, \qquad (4.19)$$
$$x_1, x_2 \geq 0.$$

Based on the necessity relation, with regard to the problem (4.17) we should solve the following crisp LP problem:

$$\max z = 3x_1 + 4x_2$$
$$s.t. \quad (5 - \alpha)x_1 + (4 - \alpha)x_2 \leq 3 + 2\alpha,$$
$$(4 - \alpha)x_1 + (6 - \alpha)x_2 \leq 4 + 2\alpha, \qquad (4.20)$$
$$x_1, x_2 \geq 0.$$

For different α-cut levels, we can get different optimal values of the crisp LP problems (4.19) and (4.20) as reported in Table 4.1.

4.4   Type-2 LPP with Fuzzy Parameters

The general form of LPPs in which the coefficients of the decision variables in the objective function are fuzzy numbers, can be formulated as follows:

$$\max \tilde{z} = \sum_{j=1}^{n} \tilde{c}_j x_j$$
$$s.t. \quad \sum_{j=1}^{n} a_{ij} x_j \leq b_i, \, i = 1, 2, \ldots, m, \qquad (4.21)$$
$$x_j \geq 0, \qquad j = 1, 2, \ldots, n.$$

In this section, we shall explore several approaches for solving Type-2 LPP with fuzzy parameters (4.21).

4.4.1   Lai and Hwang's Approach

Lai and Hwang [5] assumed that the fuzzy numbers in the FLP (4.21) are all triangular fuzzy numbers. Thus, $\tilde{c}_j$ is represented by $\tilde{c}_j = (c_{1,j}, c_{2,j}, c_{3,j})$. Thus the FLP problem (4.21) can be rewritten as follows:

Table 4.1 The lower and upper bounds of the fuzzy objective values of Example 4.3

α	0.1	0.2	0.3	0.4	0.5	0.6	0.7	0.8	0.9	1.0
z_α^{Pos}	17.56604	16.28571	15.13559	14.09677	13.15385	12.29412	11.50704	10.78378	10.11688	9.5
z_α^{Nec}	2.912409	3.119403	3.335878	3.5625	2.8	4.049180	4.310924	4.586207	4.876106	5.181818

$$\max \tilde{z} = \sum_{j=1}^{n} (c_{1,j}, c_{2,j}, c_{3,j}) x_j = \left(\sum_{j=1}^{n} c_{1,j} x_j, \sum_{j=1}^{n} c_{2,j} x_j, \sum_{j=1}^{n} c_{3,j} x_j \right) = (c_1 x, c_2 x, c_3 x)$$

$$s.t. \quad \sum_{j=1}^{n} a_{ij} x_j \le b_i, \ i = 1, 2, \ldots, m,$$

$$x_j \ge 0, \quad j = 1, 2, \ldots, n.$$

$$\text{(4.22)}$$

Lai and Hwang [5] considered the fuzzy objective function of model (4.22) as a multiple objective linear function and used MOLP techniques for solving this FLP problem.

They first provided a most likely solution for the problem (4.22) by solving the following auxiliary LP problem:

$$\max z = \frac{1}{6} \sum_{j=1}^{n} (c_{1,j} + 4 c_{2,j} + c_{3,j}) x_j$$

$$s.t. \quad \sum_{j=1}^{n} a_{ij} x_j \le b_i, \quad i = 1, 2, \ldots, m,$$

$$x_j \ge 0, \quad\quad j = 1, 2, \ldots, n.$$

$$\text{(4.23)}$$

In another approach, assuming the minimal acceptable possibility, α, is given, Lai and Hwang [5] proposed the following problem:

$$\max z = \frac{1}{6} \sum_{j=1}^{n} (c_{1,j,\alpha} + 4 c_{2,j} + c_{3,j,\alpha}) x_j$$

$$s.t. \quad \sum_{j=1}^{n} a_{ij} x_j \le b_i, \quad i = 1, 2, \ldots, m,$$

$$x_j \ge 0, \quad\quad j = 1, 2, \ldots, n.$$

$$\text{(4.24)}$$

where $c_{1,j,\alpha} = c_{1,j} + \alpha(c_{1,j,\alpha} - c_{2,j})$ and $c_{3,j,\alpha} = c_{3,j} - \alpha(c_{3,j,\alpha} - c_{2,j})$ are the most pessimistic and most optimistic values of the acceptable events.

On the other hand, Lai and Hwang [5] explored the concept that the fuzzy objective can be defined by three corner points $(c_2 x, 1)$, $(c_2 x, 0)$ and $(c_3 x, 0)$, geometrically. Thus, maximizing the fuzzy objective function can be achieved by pushing these points in the direction of the right-hand side. In the Model (4.22), the fuzzy objective $\tilde{z} = (c_1 x, c_2 x, c_3 x)$ is a triangular fuzzy number. The most possible value of $\tilde{z}$ is $c_2 x$, the lower and upper values are $c_1 x$ and $c_3 x$, respectively. Intuitively, maximization of the fuzzy objective $\tilde{z}$ can be obtained by maximizing the lower value $c_1 x$, upper value $c_3 x$, and the most possible value $c_2 x$, simultaneously. However, the above three objective functions $c_1 x, c_2 x$ and $c_3 x$ should always preserve the form of the triangular fuzzy number $\tilde{z} = (c_1x, c_2x, c_3x)$ during the optimization process. Thus, in order to keep the TFN shape (normal and convex) of the possibility distribution, it is natural to maximize c_2x, minimize $c_2x - c_1x$ and maximize $c_3x - c_2x$. These three new objectives also guarantee that the triangular membership function of the fuzzy objective function is pushed in the direction of the right-hand side.

In this case, the Model (4.22) can be transformed into the following MOLP problem:

$$\min z_1 = \sum_{j=1}^{n} (c_{2,j} - c_{1,j})x_j$$

$$\max z_2 = \sum_{j=1}^{n} c_{2,j}x_j$$

$$\max z_3 = \sum_{j=1}^{n} (c_{3,j} - c_{2,j})x_j \tag{4.25}$$

$$s.t. \quad \sum_{j=1}^{n} a_{ij}x_j \leq b_i, \ i = 1, 2, \ldots, m,$$

$$x_j \geq 0, \quad j = 1, 2, \ldots, n.$$

In order to solve the MOLP problem (4.25), Lai and Hwang [5] suggested using Zimmerman's fuzzy programming method. According to this method, the positive ideal solution (PIS) and negative ideal solution (NIS) of the objective functions of (4.25) are provided by solving the following six LP problems

$$z_1^{PIS} = \min \sum_{j=1}^{n} (c_{2,j} - c_{1,j})x_j \qquad\qquad z_1^{NIS} = \max \sum_{j=1}^{n} (c_{2,j} - c_{1,j})x_j$$

$$s.t. \quad \sum_{j=1}^{n} a_{ij}x_j \leq b_i, \ i = 1, 2, \ldots, m, \qquad s.t. \quad \sum_{j=1}^{n} a_{ij}x_j \leq b_i, \ i = 1, 2, \ldots, m,$$

$$x_j \geq 0, \quad j = 1, 2, \ldots, n. \qquad\qquad x_j \geq 0, \quad j = 1, 2, \ldots, n.$$

$$z_2^{PIS} = \max \sum_{j=1}^{n} c_{2,j}x_j \qquad\qquad z_2^{NIS} = \min \sum_{j=1}^{n} c_{2,j}x_j$$

$$s.t. \quad \sum_{j=1}^{n} a_{ij}x_j \leq b_i, \ i = 1, 2, \ldots, m, \qquad s.t. \quad \sum_{j=1}^{n} a_{ij}x_j \leq b_i, \ i = 1, 2, \ldots, m,$$

$$x_j \geq 0, \quad j = 1, 2, \ldots, n. \qquad\qquad x_j \geq 0, \quad j = 1, 2, \ldots, n.$$

$$z_3^{PIS} = \min \sum_{j=1}^{n} (c_{3,j} - c_{2,j})x_j \qquad z_3^{NIS} = \max \sum_{j=1}^{n} (c_{3,j} - c_{2,j})x_j$$

$$s.t. \quad \sum_{j=1}^{n} a_{ij}x_j \leq b_i, \ i = 1, 2, \ldots, m, \qquad s.t. \quad \sum_{j=1}^{n} a_{ij}x_j \leq b_i, \ i = 1, 2, \ldots, m,$$

$$x_j \geq 0, \quad j = 1, 2, \ldots, n. \qquad\qquad x_j \geq 0, \quad j = 1, 2, \ldots, n.$$

$$(4.26)$$

The linear membership function of the objective functions $\tilde{z}_1$ and $\tilde{z}_i \, (i = 2, 3)$ can now be calculated as follows:

$$\mu_{\tilde{z}_1}(z_1) = \begin{cases} 1, & z_1 < z_1^{PIS} \\ \frac{z_1^{NIS} - z_1}{z_1^{NIS} - z_1^{PIS}} & z_1^{PIS} \leq z_1 \leq z_1^{NIS} \\ 0 & z_1 > z_1^{NIS} \end{cases} \qquad (4.27)$$

$$\mu_{\tilde{z}_i}(z_i) = \begin{cases} 1, & z_i > z_1^{PIS} \\ \frac{z_i - z_i^{NIS}}{z_i^{PIS} - z_i^{NIS}} & z_i^{PIS} \leq z_i \leq z_i^{NIS} \\ 0 & z_i < z_i^{NIS} \end{cases} \qquad (4.28)$$

Finally, the Zimmerman's equivalent single objective LP model that follows, is solved:

$$\max \alpha$$
$$s.t. \quad \mu_{\tilde{z}_i}(z_i) \geq \alpha, \quad i = 1, 2, 3,$$
$$\sum_{j=1}^{n} a_{ij}x_j \leq b_i, \ i = 1, 2, \ldots, m, \qquad (4.29)$$
$$x_j \geq 0, \quad j = 1, 2, \ldots, n.$$

Substituting the membership functions of (4.27) and (4.28) into the problem (4.29), we obtain the following problem:

$$\max \alpha$$
$$s.t. \quad z_1 \leq z_1^{NIS} - \alpha(z_1^{NIS} - z_1^{PIS}),$$
$$z_i \geq z_i^{NIS} + \alpha(z_i^{PIS} - z_i^{NIS}), \quad i = 2, 3,$$
$$\sum_{j=1}^{n} a_{ij}x_j \leq b_i, \ i = 1, 2, \ldots, m, \qquad (4.30)$$
$$x_j \geq 0, \quad j = 1, 2, \ldots, n.$$

The optimal solution of the problem (4.30) gives a satisfactory solution under the strategy of minimizing the risk of lower profit and maximizing the most likely value and the possibility of higher profit.

Example 4.4 Consider the following Type-2 LPP with fuzzy parameters:

$$\max \tilde{z} = (1, 2, 3)x_1 + (1, 3, 5)x_2$$
$$\begin{aligned} s.t. \quad & x_1 + x_2 \leq 4, \\ & -x_1 + x_2 \leq 2, \\ & x_1 + 2x_2 \geq 2, \\ & x_1, x_2 \geq 0. \end{aligned} \tag{4.31}$$

We should solve the following auxiliary LP problem in order to obtain the most likely solution for the problem (4.31) with regard to the problem (4.23):

$$\max z = 2x_1 + 3x_2$$
$$\begin{aligned} s.t. \quad & x_1 + x_2 \leq 4, \\ & -x_1 + x_2 \leq 2, \\ & x_1 + 2x_2 \geq 2, \\ & x_1, x_2 \geq 0. \end{aligned} \tag{4.32}$$

Solving the crisp problem (4.32), we obtain the optimal solution $x^* = (x_1^*, x_1^*) = (1, 3)$ with the objective function $z^* = 11$. Also, substituting the optimal solution $x^* = (x_1^*, x_1^*) = (1, 3)$ in the objective function of the problem (4.31), we obtain the optimum fuzzy objective function value $\tilde{z}^* = (4, 11, 18)$.

Now, let us assume that the minimal acceptable possibility $\alpha = 0.4$ is given. In this case, we should solve the following auxiliary LP with regard to the problem (4.24):

$$\max z = 1.9x_1 + 2.6x_2$$
$$\begin{aligned} s.t. \quad & x_1 + x_2 \leq 4, \\ & -x_1 + x_2 \leq 2, \\ & x_1 + 2x_2 \geq 2, \\ & x_1, x_2 \geq 0. \end{aligned} \tag{4.33}$$

Solving the crisp problem (4.33), we obtain the optimal solution $x^* = (x_1^*, x_2^*) = (1, 3)$ with the objective function $z^* = 9.7$.

Finally, according to the problem (4.25) we should solve the following MOLP problem:

$$\begin{aligned}
\min z_1 &= x_1 + 2x_1 \\
\max z_2 &= 2x_1 + 3x_2 \\
\max z_3 &= x_1 + 2x_1 \\
s.t. \quad x_1 + x_2 &\leq 4, \\
-x_1 + x_2 &\leq 2, \\
x_1 + 2x_2 &\geq 2, \\
x_1, x_2 &\geq 0.
\end{aligned} \qquad (4.34)$$

First, we obtain the PIS and NIS of the objective functions (4.34) by solving the following crisp problems with regard to (4.26):

$$\begin{array}{ll}
z_1^{PIS} = \min z_1 = x_1 + 2x_2 & z_1^{NIS} = \max z_1 = x_1 + 2x_2 \\
s.t. \quad x_1 + x_2 \leq 4, & s.t. \quad x_1 + x_2 \leq 4, \\
\quad -x_1 + x_2 \leq 2, & \quad -x_1 + x_2 \leq 2, \\
\quad x_1 + 2x_2 \geq 2, & \quad x_1 + 2x_2 \geq 2, \\
\quad x_1, x_2 \geq 0. & \quad x_1, x_2 \geq 0. \\
z_2^{PIS} = \max z_2 = 2x_1 + 3x_2 & z_2^{NIS} = \max z_2 = 2x_1 + 3x_2 \\
s.t. \quad x_1 + x_2 \leq 4, & s.t. \quad x_1 + x_2 \leq 4, \\
\quad -x_1 + x_2 \leq 2, & \quad -x_1 + x_2 \leq 2, \\
\quad x_1 + 2x_2 \geq 2, & \quad x_1 + 2x_2 \geq 2, \\
\quad x_1, x_2 \geq 0. & \quad x_1, x_2 \geq 0. \\
z_3^{PIS} = \max z_3 = x_1 + 2x_2 & z_3^{NIS} = \min z_3 = x_1 + 2x_2 \\
s.t. \quad x_1 + x_2 \leq 4, & s.t. \quad x_1 + x_2 \leq 4, \\
\quad -x_1 + x_2 \leq 2, & \quad -x_1 + x_2 \leq 2, \\
\quad x_1 + 2x_2 \geq 2, & \quad x_1 + 2x_2 \geq 2, \\
\quad x_1, x_2 \geq 0. & \quad x_1, x_2 \geq 0.
\end{array} \qquad (4.35)$$

By the graphic technique, we obtain $z_1^{PIS} = 2, z_1^{NIS} = 7, z_2^{PIS} = 11, z_2^{NIS} = 3$, $z_3^{PIS} = 7$ and $z_3^{NIS} = 2$. Thus, we solve the following problem with regard to the problem (4.30):

$$\max \alpha$$

$$
\begin{aligned}
s.t. \quad & x_1 + 2x_2 \leq 7 - 5\alpha \\
& 2x_1 + 3x_2 \geq 3 + 4\alpha \\
& x_1 + 2x_2 \geq 2 + 5\alpha \\
& x_1 + x_2 \leq 4, \\
& -x_1 + x_2 \leq 2, \\
& x_1 + 2x_2 \geq 2, \\
& x_1, x_2 \geq 0.
\end{aligned}
\tag{4.36}
$$

The optimal solution of the problem (4.36) is $(\alpha^*, x_1^*, x_2^*) = (0.3, 0, 1.75)$. Thus, we have $z_1^* = 3.5, z_2^* = 5.25$ and $z_3^* = 3.5$. Therefore, the optimal value of the objective function has a triangular membership function as $z^* = (z_2^* - z_1^*, z_2^*, z_2^* + z_3^*) = (1.75, 3.5, 8.75)$.

4.4.2   Wan and Dong's Approach

Wan and Dong [6] extended the proposed approach by Lai and Hwang [5] for the situation in which the fuzzy numbers in the FLP (4.21) are all trapezoidal fuzzy numbers. Thus, $\tilde{c}_j$ is represented by $\tilde{c}_j = (c_{1,j}, c_{2,j}, c_{3,j}, c_{4,j})$. Therefore, the FLP problem (4.21) can be rewritten as follows:

$$
\begin{aligned}
\max \tilde{z} &= \sum_{j=1}^{n} (c_{1,j}, c_{2,j}, c_{3,j}, c_{4,j}) x_j \\
&= \left(\sum_{j=1}^{n} c_{1,j} x_j, \sum_{j=1}^{n} c_{2,j} x_j, \sum_{j=1}^{n} c_{3,j} x_j, \sum_{j=1}^{n} c_{4,j} x_j \right) = (c_1 x, c_2 x, c_3 x, c_4 x)
\end{aligned}
\tag{4.37}
$$

$$
s.t. \quad \sum_{j=1}^{n} a_{ij} x_j \leq b_i, \ i = 1, 2, \ldots, m,
$$

$$
x_j \geq 0, \quad j = 1, 2, \ldots, n.
$$

In the Model (4.37), the fuzzy objective $\tilde{z} = (c_1 x, c_2 x, c_3 x, c_4 x)$ is a trapezoidal fuzzy number. The most likely value of $\tilde{z}$ is the interval $[c_2 x, c_3 x]$, the lower and upper values are $c_1 x$ and $c_4 x$, respectively. Intuitively, maximization of the fuzzy objective $\tilde{z}$ can be obtained by maximizing the lower value $c_1 x$, the upper value $c_4 x$, and the two endpoints $c_2 x$ and $c_3 x$ of the core $[c_2 x, c_3 x]$, simultaneously. However, the above four objective functions $c_1 x, c_2 x, c_3 x$ and $c_4 x$ should always preserve the form of the trapezoidal fuzzy number $\tilde{z} = (c_1 x, c_2 x, c_3 x, c_4 x)$ during the optimization process. Thus, in order to keep the TrFN shape (normal and convex) of the possibility distribution, it is natural to maximize $[c_2 x, c_3 x]$. According to the definition of maximization problems with the interval objective functions introduced by Ishibuchi and Tanaka [31], in order to maximize the interval $[c_2x, c_3x]$, it is required to maximize the left endpoint c_2x and maximize the middle point $\frac{1}{2}[c_2x + c_3x]$ of this interval simultaneously. For the lower and upper limits of $\tilde{z} = (c_1x, c_2x, c_3x, c_4x)$, it is required to minimize $[c_2x - c_1x]$ and maximize $[c_4x - c_3x]$ instead of maximizing the lower c_1x and the upper c_4x, respectively. Based on this discussion, the problem (4.37) is transformed into the following MOLP problem:

$$\min z_1 = \sum_{j=1}^{n} (c_{2,j} - c_{1,j})x_j$$

$$\max z_2 = \sum_{j=1}^{n} c_{2,j}x_j$$

$$\max z_3 = \sum_{j=1}^{n} \frac{1}{2}[c_{2,j}x_j + c_{3,j}x_j] \tag{4.38}$$

$$\max z_4 = \sum_{j=1}^{n} (c_{4,j} - c_{3,j})x_j$$

$$s.t. \quad \sum_{j=1}^{n} a_{ij}x_j \leq b_i, \ i = 1, 2, \ldots, m,$$

$$x_j \geq 0, \quad j = 1, 2, \ldots, n.$$

Wan and Dong [6] proposed three kinds of approaches to solving the MOLP problem (4.38).

Let $z_i^{\max}$ and x_i^* be the maximum objective value and the optimal solution, respectively, for the following single objective LP model:

$$\max z_i = z_i(x)$$

$$s.t. \quad \sum_{j=1}^{n} a_{ij}x_j \leq b_i, \ i = 1, 2, \ldots, m, \tag{4.39}$$

$$x_j \geq 0, \quad j = 1, 2, \ldots, n.$$

Then, set $z_i^{\min} = \min\{z_i(x_1^*), z_i(x_2^*), z_i(x_3^*), z_i(x_4^*)\}$ $(i = 1, 2, 3, 4)$. The linear membership function of the objective functions $\tilde{z}_1$ and $\tilde{z}_i$ $(i = 2, 3, 4)$ can now be calculated as follows:

$$\mu_{\tilde{z}_1}(z_1) = \begin{cases} 1, & z_1 < z_1^{\min} \\ \frac{z_1^{\max} - z_1}{z_1^{\max} - z_1^{\min}} & z_1^{\min} \leq z_1 \leq z_1^{\max} \\ 0 & z_1 > z_1^{\max} \end{cases} \tag{4.40}$$

$$\mu_{\bar{z}_i}(z_i) = \begin{cases} 1, & z_i > z_i^{\max} \\ \frac{z_i - z_i^{\min}}{z_i^{\max} - z_i^{\min}}, & z_i^{\min} \leq z_i \leq z_i^{\max} \\ 0, & z_i < z_i^{\min} \end{cases} \tag{4.41}$$

Thus, the problem (4.38) can be solved by the following LP model according to a kind of pessimistic approach which shows that the decision maker (DM) is very conservative:

$$\max \alpha$$

$$s.t. \quad 4\mu_{\bar{z}_i} + \sum_{i=1}^{4} \mu_{\bar{z}_i} \geq 8\alpha, \ i = 1, 2, 3, 4,$$

$$\sum_{j=1}^{n} a_{ij}x_j \leq b_i, \ i = 1, 2, \ldots, m, \tag{4.42}$$

$$x_j \geq 0, \quad j = 1, 2, \ldots, n.$$

Also, the problem (4.38) can be solved by the following LP model according to a kind of optimistic approach which shows that the DM is very aggressive:

$$\max \alpha$$

$$s.t. \quad 4\mu_{\bar{z}_i} + \sum_{i=1}^{4} \mu_{\bar{z}_i} \leq 8\alpha, i = 1, 2, 3, 4,$$

$$\sum_{j=1}^{n} a_{ij}x_j \leq b_i, \ i = 1, 2, \ldots, m, \tag{4.43}$$

$$x_j \geq 0, \quad j = 1, 2, \ldots, n.$$

Finally, the problem (4.38) can be solved by the following LP model which is the linear sum approach based on the membership function:

$$\max \sum_{i=1}^{4} w_i \mu_{\bar{z}_i}$$

$$s.t. \quad \sum_{j=1}^{n} a_{ij}x_j \leq b_i, \ i = 1, 2, \ldots, m, \tag{4.44}$$

$$x_j \geq 0, \quad j = 1, 2, \ldots, n.$$

where $w = (w_1, w_2, w_3, w_4)$ is the weight vector of objective z_i $(i = 1, 2, 3, 4)$ satisfies that $w_i \geq 0$, $(i = 1, 2, 3, 4)$ and $\sum_{i=1}^{4} w_i = 1$.

Example 4.5 Consider the following Type-2 LPP with fuzzy parameters:

$$\max \tilde{z} = (1, 2, 4, 6)x_1 + (1, 3, 5, 6)x_2$$
$$s.t. \quad x_1 + x_2 \le 3,$$
$$\quad -x_1 + x_2 \le 1, \tag{4.45}$$
$$\quad x_1, x_2 \ge 0.$$

The problem (4.45) is transformed into the following MOLP problem with regard to the model (4.38):

$$\min z_1 = x_1 + 2x_2$$
$$\max z_2 = 2x_1 + 3x_2$$
$$\max z_3 = 3x_1 + 4x_2$$
$$\max z_4 = 2x_1 + x_2 \tag{4.46}$$
$$s.t. \quad x_1 + x_2 \le 3,$$
$$\quad -x_1 + x_2 \le 1,$$
$$\quad x_1, x_2 \ge 0.$$

We first solve the following single objective LP models in order to obtain $z_i^{\max}$ $(i = 1, 2, 3, 4)$ and x_i^* $(i = 1, 2, 3, 4)$:

$$\max z_1 = x_1 + 2x_2$$
$$s.t. \quad x_1 + x_2 \le 3,$$
$$\quad -x_1 + x_2 \le 1, \tag{4.47}$$
$$\quad x_1, x_2 \ge 0.$$

$$\max z_2 = 2x_1 + 3x_2$$
$$s.t. \quad x_1 + x_2 \le 3,$$
$$\quad -x_1 + x_2 \le 1, \tag{4.48}$$
$$\quad x_1, x_2 \ge 0.$$

$$\max z_3 = 3x_1 + 4x_2$$
$$s.t. \quad x_1 + x_2 \le 3,$$
$$\quad -x_1 + x_2 \le 1, \tag{4.49}$$
$$\quad x_1, x_2 \ge 0.$$

$$\max z_4 = 2x_1 + x_2$$
$$s.t. \quad x_1 + x_2 \leq 3,$$
$$-x_1 + x_2 \leq 1, \tag{4.50}$$
$$x_1, x_2 \geq 0.$$

The optimal solution of the problem (4.47) is $x_1^* = (1, 2)$ with the maximum objective value $z_1^{\max} = 5$. Also, the optimal solution of the problem (4.48) is $x_2^* = (1, 2)$ with the maximum objective value $z_2^{\max} = 8$. Moreover, the optimal solution of the problem (4.49) is $x_3^* = (1, 2)$ with the maximum objective value $z_3^{\max} = 11$. Finally, the optimal solution of the problem (4.50) is $x_4^* = (3, 0)$ with the maximum objective value $z_4^{\max} = 6$. In this case, we have

$$z_1^{\min} = \min\{z_1(x_1^*), z_1(x_2^*), z_1(x_3^*), z_1(x_4^*)\} = 3$$
$$z_2^{\min} = \min\{z_2(x_1^*), z_2(x_2^*), z_2(x_3^*), z_2(x_4^*)\} = 6$$
$$z_3^{\min} = \min\{z_3(x_1^*), z_3(x_2^*), z_3(x_3^*), z_3(x_4^*)\} = 9$$
$$z_4^{\min} = \min\{z_4(x_1^*), z_4(x_2^*), z_4(x_3^*), z_4(x_4^*)\} = 4$$

Now, according to the relations (4.40) and (4.41), the linear membership function of the objective functions $\tilde{z}_1$ and $\tilde{z}_i$ ($i = 2, 3, 4$) are calculated as follows:

$$\mu_{\tilde{z}_1}(z_1) = \begin{cases} 1, & z_1 < 3 \\ \frac{5 - z_1}{2} & 3 \leq z_1 \leq 5 \\ 0 & z_1 > 5 \end{cases} \tag{4.51}$$

$$\mu_{\tilde{z}_2}(z_2) = \begin{cases} 1, & z_2 > 8 \\ \frac{z_2 - 6}{2}, & 6 \leq z_2 \leq 8 \\ 0, & z_2 < 6 \end{cases} \tag{4.52}$$

$$\mu_{\tilde{z}_3}(z_3) = \begin{cases} 1, & z_3 > 11 \\ \frac{z_3 - 9}{2}, & 9 \leq z_3 \leq 11 \\ 0, & z_3 < 9 \end{cases} \tag{4.53}$$

$$\mu_{\tilde{z}_4}(z_4) = \begin{cases} 1, & z_4 > 6 \\ \frac{z_4 - 4}{2}, & 4 \leq z_4 \leq 6 \\ 0, & z_4 < 4 \end{cases} \tag{4.54}$$

According to the problem (4.42), by adopting the pessimistic approach the following problem is solved:

$$\max \alpha$$
$$
\begin{aligned}
s.t. \quad & x_1 - x_2 - 8\alpha \geq -3 \\
& 7x_1 + 9x_2 - 8\alpha \geq 19 \\
& 9x_1 + 11x_2 - 8\alpha \geq 25 \\
& 7x_1 + 5x_2 - 8\alpha \geq 15 \\
& x_1 + x_2 \leq 3, \\
& -x_1 + x_2 \leq 1, \\
& x_1, x_2 \geq 0.
\end{aligned}
\tag{4.55}
$$

The optimal solution of the problem (4.55) is $(\alpha^*, x_1^*, x_2^*) = (0.5, 1, 2)$. Thus, we have $z_1^* = 5$, $z_2^* = 8$, $z_3^* = 11$, and $z_4^* = 4$. Therefore, the optimal value of the objective function has a trapezoidal membership function as $z^* = (z_2^* - z_1^*, z_2^*, 2z_3^* - z_2^*, z_3^* + z_4^*) = (3, 8, 14, 22)$.

According to the problem (4.43), by adopting the optimistic approach the following problem is solved:

$$\max \alpha$$
$$
\begin{aligned}
s.t. \quad & x_1 - x_2 - 8\alpha \leq -3 \\
& 7x_1 + 9x_2 - 8\alpha \leq 19 \\
& 9x_1 + 11x_2 - 8\alpha \leq 25 \\
& 7x_1 + 5x_2 - 8\alpha \leq 15 \\
& x_1 + x_2 \leq 3, \\
& -x_1 + x_2 \leq 1, \\
& x_1, x_2 \geq 0.
\end{aligned}
\tag{4.56}
$$

The optimal solution of the problem (4.55) is $(\alpha^*, x_1^*, x_2^*) = (0.5, 1, 2)$. Thus, we have $z_1^* = 5$, $z_2^* = 8$, $z_3^* = 11$, and $z_4^* = 4$. Therefore, the optimal value of the objective function has a trapezoidal membership function as $z^* = (z_2^* - z_1^*, z_2^*, 2z_3^* - z_2^*, z_3^* + z_4^*) = (3, 8, 14, 22)$. This means that the most possible total value is between [8, 14]; the upper and lower limits of the objective value are 3 and 22, respectively.

4.4.3   Rommelfanger et al.'s Approach

Assume that the imprecise coefficients of the objective function of the model (4.21) have convex membership functions. Rommenlfanger et al. [7] used a finite set of α_i-cut ($i = 1, 2, \ldots, r$) for each fuzzy coefficient and reduced the problem (4.21) to the following MOLP problem:

$$\max \tilde{z} = \left[\sum_{j=1}^{n} \left[\tilde{c}_j \right]_{\alpha_1} x_j, \ldots, \sum_{j=1}^{n} \left[\tilde{c}_j \right]_{\alpha_r} x_j \right]$$

$$s.t. \quad \sum_{j=1}^{n} a_{ij} x_j \le b_i, \ i = 1, 2, \ldots, m,$$

$$x_j \ge 0, \quad j = 1, 2, \ldots, n. \tag{4.57}$$

Assuming $\left[\tilde{c}_j \right]_{\alpha_i} = \left[\tilde{c}_{j,\alpha_i}^L, \tilde{c}_{j,\alpha_i}^U \right]$, the problem (4.57) can be rewritten as follows:

$$\max \tilde{z} = \left[\sum_{j=1}^{n} \left[\tilde{c}_{j,\alpha_1}^L, \tilde{c}_{j,\alpha_1}^U \right] x_j, \ldots, \sum_{j=1}^{n} \left[\tilde{c}_{j,\alpha_r}^L, \tilde{c}_{j,\alpha_r}^U \right] x_j \right]$$

$$s.t. \quad \sum_{j=1}^{n} a_{ij} x_j \le b_i, \ i = 1, 2, \ldots, m,$$

$$x_j \ge 0, \quad j = 1, 2, \ldots, n. \tag{4.58}$$

Rommenlfanger et al. [7] used Zimmerman's approach [8] for solving the model (4.58).

Let $z_{i,\max}^L$ and $x_{i,\max}^L$ be the maximum objective value and the optimal solution, respectively, for the following single objective LP model:

$$z_{i,\max}^L = \max z_i^L(x) = \sum_{j=1}^{n} \tilde{c}_{j,\alpha_i}^L x_j$$

$$s.t. \quad \sum_{j=1}^{n} a_{ij} x_j \le b_i, \ i = 1, 2, \ldots, m,$$

$$x_j \ge 0, \quad j = 1, 2, \ldots, n. \tag{4.59}$$

Also, let $z_{i,\max}^U$ and $x_{i,\max}^U$ be the maximum objective value and the optimal solution, respectively, for the following single objective LP model:

$$z_{i,\max}^U = \max z_i^U(x) = \sum_{j=1}^{n} \tilde{c}_{j,\alpha_i}^U x_j$$

$$s.t. \quad \sum_{j=1}^{n} a_{ij} x_j \le b_i, \ i = 1, 2, \ldots, m,$$

$$x_j \ge 0, \quad j = 1, 2, \ldots, n. \tag{4.60}$$

Now, set $z_{i,\min}^L = z_i^L(x_{i,\max}^U)$ and $z_{i,\min}^U = z_i^U(x_{i,\min}^L)$ $(i = 1, 2, \ldots, r)$. The decision maker then identifies his/her objective with the membership functions $\mu_{z_i^L}$ and $\mu_{z_i^U}$ for each α_i-cut which express his/her satisfaction with the obtained objective values.

The linear membership function of the objective functions z_i^L and z_i^U can now be calculated as follows:

$$
\mu_{z_i^L}(x) = \begin{cases} 1, & z_i^L(x) > z_{i,\max}^L \\ \frac{z_i^L(x) - z_{i,\min}^L}{z_{i,\max}^L - z_{i,\min}^L}, & z_{i,\min}^L \le z_i^L(x) \le z_{i,\max}^L \\ 0, & z_i^L(x) < z_{i,\min}^L \end{cases}
\tag{4.61}
$$

$$
\mu_{z_i^U}(x) = \begin{cases} 1, & z_i^U(x) > z_{i,\max}^U \\ \frac{z_i^U(x) - z_{i,\min}^U}{z_{i,\max}^U - z_{i,\min}^U}, & z_{i,\min}^U \le z_i^U(x) \le z_{i,\max}^U \\ 0, & z_i^U(x) < z_{i,\min}^U \end{cases}
\tag{4.62}
$$

In this case, the following fuzzy optimization system is formulated to take into consideration all the information concerning all the α_i:

$$
\begin{aligned}
\max & \left[\mu_{z_1^L}(x), \mu_{z_1^U}(x), \ldots, \mu_{z_r^L}(x), \mu_{z_r^U}(x) \right] \\
s.t. \quad & \sum_{j=1}^{n} a_{ij} x_j \le b_i, \ i = 1, 2, \ldots, m, \\
& x_j \ge 0, \quad j = 1, 2, \ldots, n.
\end{aligned}
\tag{4.63}
$$

Assuming $\alpha = \min \left[\mu_{z_1^L}(x), \mu_{z_1^U}(x), \ldots, \mu_{z_r^L}(x), \mu_{z_r^U}(x) \right]$, the model (4.63) can be solved by solving the following linear model:

$$
\begin{aligned}
\max \ & \alpha \\
s.t. \quad & \mu_{z_i^L}(x) \ge \alpha, \ i = 1, 2, \ldots, r \\
& \mu_{z_i^U}(x) \ge \alpha, \ i = 1, 2, \ldots, r \\
& \sum_{j=1}^{n} a_{ij} x_j \le b_i, \ i = 1, 2, \ldots, m, \\
& x_j \ge 0, \quad j = 1, 2, \ldots, n.
\end{aligned}
\tag{4.64}
$$

Substituting the membership functions of (4.61) and (4.62) into the problem (4.64), the following problem is obtained:

$$
\begin{aligned}
\max \ & \alpha \\
s.t. \quad & z_i^L(x) \ge z_{i,\min}^L - \alpha \left(z_{i,\max}^L - z_{i,\min}^L \right), \ i = 1, 2, \ldots, r \\
& z_i^U(x) \ge z_{i,\min}^U + \alpha \left(z_{i,\max}^U - z_{i,\min}^U \right), \ i = 1, 2, \ldots, r \\
& \sum_{j=1}^{n} a_{ij} x_j \le b_i, \ i = 1, 2, \ldots, m, \\
& x_j \ge 0, \quad j = 1, 2, \ldots, n.
\end{aligned}
\tag{4.65}
$$

The arithmetic average of the objective value can be calculated as follows where x^* is the optimal solution of the model (4.64):

$$z^* = \frac{1}{2r} \sum_{i=1}^{r} \left(z_i^L(x^*) + z_i^U(x^*) \right) \tag{4.66}$$

Example 4.6 [14] Consider the following Type-2 LPP with fuzzy parameters:

$$
\begin{aligned}
\max \tilde{\tilde{z}} = & \tilde{c}_1 x_1 + \tilde{c}_2 x_2 \\
s.t. \quad & x_1 + 4x_2 \leq 100, \ x_1 + 3x_2 \leq 76 \\
& x_1 + 2x_2 \leq 53, \ 3x_1 + 5x_2 \leq 138 \\
& 3x_1 + 4x_2 \leq 120, \ 7x_1 + 8x_2 \leq 260 \\
& x_1 + x_2 \leq 36, \ 3x_1 + 2x_2 \leq 103 \\
& 2x_1 + x_2 \leq 68, \ x_1, x_2 \geq 0.
\end{aligned}
\tag{4.67}
$$

where $\tilde{c}_1$ and $\tilde{c}_2$ are fuzzy numbers and their membership functions are depicted in Fig. 4.1.

We take into consideration four cuts involving $\{0, 0.25, 0.5, 0.75\}$ for analyzing this problem. The interval values for two fuzzy coefficients of the objective function with respect to the aforementioned cuts are presented in Table 4.2.

We first solve the following LP problems to obtain $z_{i,\max}^L$ and $x_{i,\max}^L$ ($i = 1, 2, 3, 4$) with regard to the problem (4.59):

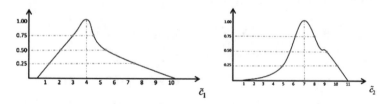

Fig. 4.1 The membership functions of $\tilde{c}_1$ and $\tilde{c}_2$

Table 4.2 α_i-cuts of $\tilde{c}_1$ and $\tilde{c}_2$

α_i	0	0.25	0.5	0.75
$[\tilde{c}_1]_{\alpha_i}$	[0.5, 10]	[1, 7]	[2, 5]	[3.2, 4.5]
$[\tilde{c}_2]_{\alpha_i}$	[1.4, 11]	[5, 10]	[6, 9]	[6.5, 7.5]

$$z_{1,\max}^L = \max z_1^L(x) = 0.5x_1 + 1.4x_2 \qquad z_{2,\max}^L = \max z_2^L(x) = x_1 + 5x_2$$

s.t.   s.t.

$$x_1 + 4x_2 \le 100, x_1 + 3x_2 \le 76 \qquad\qquad x_1 + 4x_2 \le 100, x_1 + 3x_2 \le 76$$

$$x_1 + 2x_2 \le 53, \ 3x_1 + 5x_2 \le 138 \qquad\qquad x_1 + 2x_2 \le 53, \ 3x_1 + 5x_2 \le 138$$

$$3x_1 + 4x_2 \le 120, 7x_1 + 8x_2 \le 260 \qquad 3x_1 + 4x_2 \le 120, 7x_1 + 8x_2 \le 260$$

$$x_1 + x_2 \le 36, \ 3x_1 + 2x_2 \le 103 \qquad\qquad x_1 + x_2 \le 36, \ 3x_1 + 2x_2 \le 103$$

$$2x_1 + x_2 \le 68, \ x_1, x_2 \ge 0. \qquad\qquad 2x_1 + x_2 \le 68, \ x_1, x_2 \ge 0.$$

$$z_{3,\max}^L = \max z_3^L(x) = 2x_1 + 6x_2 \qquad z_{4,\max}^L = \max z_4^L(x) = 3.2x_1 + 6.5x_2$$

s.t.   s.t.

$$x_1 + 4x_2 \le 100, x_1 + 3x_2 \le 76 \qquad\qquad x_1 + 4x_2 \le 100, x_1 + 3x_2 \le 76$$

$$x_1 + 2x_2 \le 53, \ 3x_1 + 5x_2 \le 138 \qquad\qquad x_1 + 2x_2 \le 53, \ 3x_1 + 5x_2 \le 138$$

$$3x_1 + 4x_2 \le 120, 7x_1 + 8x_2 \le 260 \qquad 3x_1 + 4x_2 \le 120, 7x_1 + 8x_2 \le 260$$

$$x_1 + x_2 \le 36, \ 3x_1 + 2x_2 \le 103 \qquad\qquad x_1 + x_2 \le 36, \ 3x_1 + 2x_2 \le 103$$

$$2x_1 + x_2 \le 68, \ x_1, x_2 \ge 0. \qquad\qquad 2x_1 + x_2 \le 68, \ x_1, x_2 \ge 0.$$

$$(4.68)$$

After solving the LP problems given in (4.68) we obtain the results reported in Table 4.3.

Now, we solve the following LP problems to obtain $z_{i,\max}^U$ and $x_{i,\max}^U$ ($i = 1, 2, 3, 4$) with regard to problem (4.60):

$$z_{1,\max}^U = \max z_1^U(x) = 10x_1 + 11x_2 \qquad z_{2,\max}^U = \max z_2^U(x) = 7x_1 + 10x_2$$

s.t.   s.t.

$$x_1 + 4x_2 \le 100, x_1 + 3x_2 \le 76 \qquad\qquad x_1 + 4x_2 \le 100, x_1 + 3x_2 \le 76$$

$$x_1 + 2x_2 \le 53, \ 3x_1 + 5x_2 \le 138 \qquad\qquad x_1 + 2x_2 \le 53, \ 3x_1 + 5x_2 \le 138$$

$$3x_1 + 4x_2 \le 120, 7x_1 + 8x_2 \le 260 \qquad 3x_1 + 4x_2 \le 120, 7x_1 + 8x_2 \le 260$$

$$x_1 + x_2 \le 36, \ 3x_1 + 2x_2 \le 103 \qquad\qquad x_1 + x_2 \le 36, \ 3x_1 + 2x_2 \le 103$$

$$2x_1 + x_2 \le 68, \ x_1, x_2 \ge 0. \qquad\qquad 2x_1 + x_2 \le 68, \ x_1, x_2 \ge 0.$$

Table 4.3 The results of the problems (4.68)	i	1	2	3	4
	$x_{i,\max}^L$	(7, 23)	(0, 25)	(7, 23)	(7, 23)
	$z_{i,\max}^L$	35.7	125	152	171.9

$$z_{3,\max}^U = \max z_3^U(x) = 5x_1 + 9x_2 \qquad z_{4,\max}^U = \max z_4^U(x) = 4.5x_1 + 7.5x_2$$

s.t. $\qquad\qquad\qquad\qquad\qquad\qquad$ s.t.

$$x_1 + 4x_2 \le 100, x_1 + 3x_2 \le 76 \qquad x_1 + 4x_2 \le 100, x_1 + 3x_2 \le 76$$
$$x_1 + 2x_2 \le 53, \ 3x_1 + 5x_2 \le 138 \qquad x_1 + 2x_2 \le 53, \ 3x_1 + 5x_2 \le 138$$
$$3x_1 + 4x_2 \le 120, 7x_1 + 8x_2 \le 260 \qquad 3x_1 + 4x_2 \le 120, 7x_1 + 8x_2 \le 260$$
$$x_1 + x_2 \le 36 \ 3x_1 + 2x_2 \le 103 \qquad x_1 + x_2 \le 36, \ 3x_1 + 2x_2 \le 103$$
$$2x_1 + x_2 \le 68, \ x_1, x_2 \ge 0. \qquad 2x_1 + x_2 \le 68, \ x_1, x_2 \ge 0.$$

$$(4.69)$$

After solving the LP problems given in (4.69) we get the results reported in Table 4.4.

Now, by setting $z_{i,\min}^L = z_i^L (x_{i,\max}^U)$ and $z_{i,\min}^U = z_i^U (x_{i,\min}^L)$ $(i = 1, 2, \ldots, r)$, we obtain the results given in Table 4.5:

In this case, the following fuzzy optimization system is formulated with regard to the problem (3.64) and Tables 3.4, 3.5 and 3.6:

$$\max \alpha$$

s.t. $\quad 0.5x_1 + 1.4x_2 \ge 25.2 + 10.5\alpha$

$\qquad x_1 + 5x_2 \ge 106 + 19\alpha$

$\qquad 2x_1 + 6x_2 \ge 148 + 4\alpha$

$\qquad 3.2x_1 + 6.5x_2 \ge 168.2 + 3.7\alpha$

$\qquad 10x_1 + 11x_2 \ge 223 + 145\alpha$

$\qquad 7x_1 + 10x_2 \ge 259 + 42\alpha$

$\qquad 5x_1 + 9x_2 \ge 236 + 8\alpha$ $\hspace{3cm}(4.70)$

$\qquad 4.5x_1 + 7.5x_2 \ge 204 + 3\alpha$

$\qquad 7x_1 + 4x_2 \le 100, x_1 + 3x_2 \le 76$

$\qquad x_1 + 2x_2 \le 53, \ 3x_1 + 5x_2 \le 138$

$\qquad 3x_1 + 4x_2 \le 120, 7x_1 + 8x_2 \le 260$

$\qquad x_1 + x_2 \le 36, \ 3x_1 + 2x_2 \le 103$

$\qquad 2x_1 + x_2 \le 68, \ x_1, x_2 \ge 0.$

Table 4.4 The results of the problems (4.69)

i	1	2	3	4
$x_{i,\max}^U$	(28, 8)	(16, 18)	(11, 21)	(11, 21)
$z_{i,\max}^U$	368	299	244	207

Table 4.5 The lower limits of objective functions

i	1	2	3	4
$z_{i,\min}^L$	25.2	106	148	168.2
$z_{i,\min}^U$	233	259	236	204

Table 4.6 Degrees of satisfaction

z_i^L	z_1^L	z_2^L	z_3^L	z_4^L
$\mu_{z_i^L}(x^*)$	0.96	0.68	0.5	0.97
z_i^U	z_1^U	z_2^U	z_3^U	z_4^U
$\mu_{z_i^U}(x^*)$	0.75	0.79	0.87	0.5

The solution of the problem (4.70) is $\alpha^* = 0.5$ and $x^* = (9, 22)$ leading to the following average objective value with regard to (4.66):

$$z^* = \frac{1}{8} \sum_{i=1}^{4} \left(z_i^L(x^*) + z_i^U(x^*) \right) = 192.45 \tag{4.71}$$

The corresponding realized degrees of satisfaction are summarized in Table 4.6.

4.4.4 Cai et al.'s Approach

Assuming the coefficients of objective functions of the problem (4.21) are triangular fuzzy numbers, Maeda [10] defined three types of two-objective LP problem to find all optimal solutions of the Type-2 LPP problem with fuzzy parameters (4.21). Subsequently, Zhang et al. [11] developed some theorems so as to convert the problem (4.21) to an MOLP problem with four-objective functions. Li and Guo [12] by means of an anti-example demonstrated that there is an error in that paper and added some constraint condition to the theorems so that the conclusions of the paper [10] still hold. Cai et al. [13], proved the problem (4.21) can be translated into MOLP problems with four objectives if fuzzy coefficients have the same reference functions, and provided an example to explain that this conclusion does not hold if fuzzy coefficients have a different shape. Then, they presented the methods to find the optimal solutions. In what follows, we explore their proposed method in detail.

Let $x = (x_1, x_2, \ldots, x_n)$ and $y = (y_1, y_2, \ldots, y_n) \in R^n$. We may write $x \underset{=}{>} y$ if and only if $x_i \underset{=}{>} y_i$, $\forall i = 1, 2, \ldots, n$; $x \geq y$ if and only if $x \underset{=}{>} y$ and $x \neq y$; and $x > y$ if and only if $x_i > y_i$, $\forall i = 1, 2, \ldots, n$.

Definition 4.1 For any LR fuzzy intervals $\tilde{a}, \tilde{b} \in F_{LR}$, a fuzzy max order $\underset{=}{\succ}$, a strict fuzzy max order $\succeq$ and a strong fuzzy max order $\succ$, respectively, are defined as follows:

$$\tilde{a} \underset{=}{\succ} \tilde{b} \text{ iff } \left((\tilde{a})_\alpha^L, (\tilde{a})_\alpha^R \right) \underset{=}{\geq} \left((\tilde{b})_\alpha^L, (\tilde{b})_\alpha^R \right), \ \forall \alpha \in [0, 1]$$

$$\tilde{a} \succeq \tilde{b} \text{ iff } \left((\tilde{a})_\alpha^L, (\tilde{a})_\alpha^R \right) \geq \left((\tilde{b})_\alpha^L, (\tilde{b})_\alpha^R \right), \ \forall \alpha \in [0, 1]$$

$$\tilde{a} \succ \tilde{b} \text{ iff } \left((\tilde{a})_\alpha^L, (\tilde{a})_\alpha^R \right) > \left((\tilde{b})_\alpha^L, (\tilde{b})_\alpha^R \right), \ \forall \alpha \in [0, 1]$$

Now, assume that coefficients of objective functions of the problem (4.21) belong to the set of all finite LR-fuzzy intervals or fuzzy numbers with the same reference functions on $\mathbb{R}^n$. We first review the concepts of optimal solution to the problem (4.21) proposed by Maeda [10]. For the sake of simplicity, we set

$$X = \left\{ x = (x_1, \ldots, x_n) \,\middle|\, \sum_{j=1}^{n} a_{ij} x_j \leq b_i, \ (i = 1, 2, \ldots, m), \ x_j \geq 0, \ (j = 1, 2, \ldots, n) \right\} \text{ and }$$

assume that X is compact.

Definition 4.2 A point $x^* \in X$ is said to be an optimal solution to the problem (4.21) if it holds that $\sum_{j=1}^{n} \tilde{c}_j x_j^* \succeq \sum_{j=1}^{n} \tilde{c}_j x_j$ for all $x \in X$.

Definition 4.3 A point $x^* \in X$ is said to be a non-dominated solution to the problem (4.21) if there is no $x \in X$ such that $\sum_{j=1}^{n} \tilde{c}_j x_j^* \succeq \sum_{j=1}^{n} \tilde{c}_j x_j$ holds.

Definition 4.4 A point $x^* \in X$ is said to be a weak non-dominated solution to the problem (4.21) if there is no $x \in X$ such that $\sum_{j=1}^{n} \tilde{c}_j x_j^* \succ \sum_{j=1}^{n} \tilde{c}_j x_j$ holds.

We denote the sets of all non-dominated solutions and all weak non-dominated solutions to the problem (4.21) by X^F and X^{wF}, respectively. Associated with the problem (4.21), we consider the following MOLP problem with four objective functions:

$$\max \left(\sum_{j=1}^{n} (\tilde{c}_j)_0^L x_j, \ \sum_{j=1}^{n} (\tilde{c}_j)_1^L x_j, \ \sum_{j=1}^{n} (\tilde{c}_j)_0^U x_j, \ \sum_{j=1}^{n} (\tilde{c}_j)_1^U x_j \right)$$

$$s.t. \quad \sum_{j=1}^{n} a_{ij} x_j \leq b_i, \ i = 1, 2, \ldots, m, \tag{4.72}$$

$$x_j \geq 0, \quad j = 1, 2, \ldots, n.$$

In the following, we introduce the concepts of optimal solutions of the MOLP problem (4.72).

Definition 4.5 [13] A point $x^* \in X$ is said to be a complete optimal solution to the MOLP problem (4.72) if for all $x \in X$ we have:

$$\left(\sum_{j=1}^{n} (\tilde{c}_j)_0^L x_j^*, \ \sum_{j=1}^{n} (\tilde{c}_j)_1^L x_j^*, \ \sum_{j=1}^{n} (\tilde{c}_j)_0^U x_j^*, \ \sum_{j=1}^{n} (\tilde{c}_j)_1^U x_j^* \right)$$

$$\geq \left(\sum_{j=1}^{n} (\tilde{c}_j)_0^L x_j, \ \sum_{j=1}^{n} (\tilde{c}_j)_1^L x_j, \ \sum_{j=1}^{n} (\tilde{c}_j)_0^U x_j, \ \sum_{j=1}^{n} (\tilde{c}_j)_1^U x_j \right)$$

Definition 4.6 [13] A point $x^* \in X$ is said to be a Pareto optimal solution to the MOLP problem (4.72) if there is no $x \in X$ such that:

$$
\left(\sum_{j=1}^{n} (\tilde{c}_j)_0^L x_j^*, \sum_{j=1}^{n} (\tilde{c}_j)_1^L x_j^*, \sum_{j=1}^{n} (\tilde{c}_j)_0^U x_j^*, \sum_{j=1}^{n} (\tilde{c}_j)_1^U x_j^* \right)
$$

$$
\leq \left(\sum_{j=1}^{n} (\tilde{c}_j)_0^L x_j, \sum_{j=1}^{n} (\tilde{c}_j)_1^L x_j, \sum_{j=1}^{n} (\tilde{c}_j)_0^U x_j, \sum_{j=1}^{n} (\tilde{c}_j)_1^U x_j \right)
$$

Definition 4.7 [13] A point $x^* \in X$ is said to be a Pareto optimal solution to the MOLP problem (4.72) if there is no $x \in X$ such that:

$$
\left(\sum_{j=1}^{n} (\tilde{c}_j)_0^L x_j^*, \sum_{j=1}^{n} (\tilde{c}_j)_1^L x_j^*, \sum_{j=1}^{n} (\tilde{c}_j)_0^U x_j^*, \sum_{j=1}^{n} (\tilde{c}_j)_1^U x_j^* \right)
$$

$$
< \left(\sum_{j=1}^{n} (\tilde{c}_j)_0^L x_j, \sum_{j=1}^{n} (\tilde{c}_j)_1^L x_j, \sum_{j=1}^{n} (\tilde{c}_j)_0^U x_j, \sum_{j=1}^{n} (\tilde{c}_j)_1^U x_j \right)
$$

Lemma 4.1 [13] Let intervals $\tilde{a} = (a_1, a_2, a_3, a_4)_{LR}$ and $\tilde{b} = (b_1, b_2, b_3, b_4)_{LR}$ be two LR-fuzzy intervals with the same reference functions. If $(\tilde{a})_0^L \geqq (\tilde{b})_0^L$, $(\tilde{a})_1^L \geqq (\tilde{b})_1^L$, $(\tilde{a})_0^U \geqq (\tilde{b})_0^U$ and $(\tilde{a})_1^U \geqq (\tilde{b})_1^U$ then $\tilde{a} \succ \tilde{b}$.

According Lemma 4.1, we obtain the solution relationship between the problem (4.21) and the MOLP problem (4.72) by the following theorems.

Theorem 4.3 [13] *Let a point $x^* \in X$ be a feasible solution to the problem (4.21). Then x^* is an optimal solution to the problem if and only if x^* is a complete optimal solution to the MOLP problem (4.72).*

Proof If x^* is an optimal solution to the problem (4.21), then for any $x \in X$, we have $\sum_{j=1}^{n} \tilde{c}_j x_j^* \succeq \sum_{j=1}^{n} \tilde{c}_j x_j$. Therefore, for any $\alpha \in [0, 1]$, according to Definition 4.1 we have:

$$
\left(\sum_{j=1}^{n} \tilde{c}_j x_j^* \right)_\alpha^L \geqq \left(\sum_{j=1}^{n} \tilde{c}_j x_j \right)_\alpha^L \text{ and } \left(\sum_{j=1}^{n} \tilde{c}_j x_j^* \right)_\alpha^U \geqq \left(\sum_{j=1}^{n} \tilde{c}_j x_j \right)_\alpha^U
$$

That is:

$$
\sum_{j=1}^{n} (\tilde{c}_j)_\alpha^L x_j^* \geq \sum_{j=1}^{n} (\tilde{c}_j)_\alpha^L x_j \text{ and } \sum_{j=1}^{n} (\tilde{c}_j)_\alpha^U x_j^* \geq \sum_{j=1}^{n} (\tilde{c}_j)_\alpha^U x_j
$$

This means that x^* is a complete optimal solution to the MOLP problem (4.72).

If x^* is a complete optimal solution to the MOLP problem (4.72), for all $x \in X$, we have

$$
\left(\sum_{j=1}^{n} (\tilde{c}_j)_0^L x_j^*, \sum_{j=1}^{n} (\tilde{c}_j)_1^L x_j^*, \sum_{j=1}^{n} (\tilde{c}_j)_0^U x_j^*, \sum_{j=1}^{n} (\tilde{c}_j)_1^U x_j^* \right)
$$

$$
\geq \left(\sum_{j=1}^{n} (\tilde{c}_j)_0^L x_j, \sum_{j=1}^{n} (\tilde{c}_j)_1^L x_j, \sum_{j=1}^{n} (\tilde{c}_j)_0^U x_j, \sum_{j=1}^{n} (\tilde{c}_j)_1^U x_j \right)
$$

Since $\tilde{c}_j$ $(j = 1, 2, \ldots, n)$ are LR fuzzy intervals or fuzzy numbers with the same reference functions, $\sum_{j=1}^{n} \tilde{c}_j x_j^*$ and $\sum_{j=1}^{n} \tilde{c}_j x_j$ are also LR fuzzy intervals or fuzzy numbers with the same reference functions. According to Lemma 4.1, we have $\sum_{j=1}^{n} \tilde{c}_j x_j^* \succeq \sum_{j=1}^{n} \tilde{c}_j x_j$. Therefore, x^* is an optimal solution to the problem (4.21). $\square$

Theorem 4.4 [13] *Let a point $x^* \in X$ be a feasible solution to the problem* (4.21). *Then x^* is a non-dominated solution to the problem if and only if x^* is a Pareto optimal solution to the MOLP problem* (4.72).

Proof Let x^* be a non-dominated solution to the problem (4.21). On the contrary, we suppose that there exists a $\bar{x} \in X$ such that

$$
\left(\sum_{j=1}^{n} (\tilde{c}_j)_0^L x_j^*, \sum_{j=1}^{n} (\tilde{c}_j)_1^L x_j^*, \sum_{j=1}^{n} (\tilde{c}_j)_0^U x_j^*, \sum_{j=1}^{n} (\tilde{c}_j)_1^U x_j^* \right)
$$

$$
\leq \left(\sum_{j=1}^{n} (\tilde{c}_j)_0^L \bar{x}_j, \sum_{j=1}^{n} (\tilde{c}_j)_1^L \bar{x}_j, \sum_{j=1}^{n} (\tilde{c}_j)_0^U \bar{x}_j, \sum_{j=1}^{n} (\tilde{c}_j)_1^U \bar{x}_j \right)
$$

This means that the following inequalities hold and at least one of these inequalities holds strictly:

$$
\sum_{j=1}^{n} (\tilde{c}_j)_0^L x_j^* \leq \sum_{j=1}^{n} (\tilde{c}_j)_0^L \bar{x}_j, \quad \sum_{j=1}^{n} (\tilde{c}_j)_1^L x_j^* \leq \sum_{j=1}^{n} (\tilde{c}_j)_1^L \bar{x}_j,
$$

$$
\sum_{j=1}^{n} (\tilde{c}_j)_0^U x_j^* \leq \sum_{j=1}^{n} (\tilde{c}_j)_0^U \bar{x}_j, \quad \sum_{j=1}^{n} (\tilde{c}_j)_1^U x_j^* \leq \sum_{j=1}^{n} (\tilde{c}_j)_1^U \bar{x}_j,
$$

According to Lemma 4.1, we have $\sum_{j=1}^{n} \tilde{c}_j \bar{x}_j \succeq \sum_{j=1}^{n} \tilde{c}_j x_j^*$. However, this contradicts the assumption that x^* is a non-dominated solution to the problem (4.21).

Conversely, let x^* be a Pareto optimal solution to the problem (4.72). On the contrary, if x^* is not a non-dominated solution to the problem (4.21), there exists a

$\bar{x} \in X$ such that $\sum_{j=1}^{n} \tilde{c}_j \bar{x}_j \succeq \sum_{j=1}^{n} \tilde{c}_j x_j^*$. Therefore for any $\alpha \in [0, 1]$, according to Definition 4.1 we have:

$$\left[\left(\sum_{j=1}^{n} \tilde{c}_j \bar{x}_j \right)_{\alpha}^{L}, \left(\sum_{j=1}^{n} \tilde{c}_j \bar{x}_j \right)_{\alpha}^{U} \right] \geq \left[\left(\sum_{j=1}^{n} \tilde{c}_j x_j^* \right)_{\alpha}^{L}, \left(\sum_{j=1}^{n} \tilde{c}_j x_j^* \right)_{\alpha}^{U} \right]$$

This means that

$$\left[\sum_{j=1}^{n} (\tilde{c}_j)_{\alpha}^{L} \bar{x}_j, \sum_{j=1}^{n} (\tilde{c}_j)_{\alpha}^{U} \bar{x}_j \right] \geq \left[\sum_{j=1}^{n} (\tilde{c}_j)_{\alpha}^{L} x_j^*, \sum_{j=1}^{n} (\tilde{c}_j)_{\alpha}^{U} x_j^* \right]$$

Hence, for $\alpha = 0$ and $\alpha = 1$, we have

$$\left[\sum_{j=1}^{n} (\tilde{c}_j)_{0}^{L} \bar{x}_j, \sum_{j=1}^{n} (\tilde{c}_j)_{0}^{U} \bar{x}_j \right] \geq \left[\sum_{j=1}^{n} (\tilde{c}_j)_{0}^{L} x_j^*, \sum_{j=1}^{n} (\tilde{c}_j)_{0}^{U} x_j^* \right]$$

$$\left[\sum_{j=1}^{n} (\tilde{c}_j)_{1}^{L} \bar{x}_j, \sum_{j=1}^{n} (\tilde{c}_j)_{1}^{U} \bar{x}_j \right] \geq \left[\sum_{j=1}^{n} (\tilde{c}_j)_{1}^{L} x_j^*, \sum_{j=1}^{n} (\tilde{c}_j)_{1}^{U} x_j^* \right]$$

This means that

$$\left[\sum_{j=1}^{n} (\tilde{c}_j)_{0}^{L} \bar{x}_j, \sum_{j=1}^{n} (\tilde{c}_j)_{1}^{L} \bar{x}_j, \sum_{j=1}^{n} (\tilde{c}_j)_{0}^{U} \bar{x}_j, \sum_{j=1}^{n} (\tilde{c}_j)_{1}^{U} \bar{x}_j \right]$$
$$\geq \left[\sum_{j=1}^{n} (\tilde{c}_j)_{0}^{L} x_j^*, \sum_{j=1}^{n} (\tilde{c}_j)_{1}^{L} x_j^*, \sum_{j=1}^{n} (\tilde{c}_j)_{0}^{U} x_j^*, \sum_{j=1}^{n} (\tilde{c}_j)_{1}^{U} x_j^* \right]$$

which contradicts the assumption that x^* is a Pareto optimal solution to the MOLP problem (4.72). □

Theorem 4.5 [13] *Let a point $x^* \in X$ be a feasible solution to the problem (4.21). Then x^* is a weak non-dominated solution to the problem if and only if x^* is a weak Pareto optimal solution to the MOLP problem (4.72).*

Proof See Theorem 4.4. □

From Theorems 4.3, 4.4 and 4.5, in order to find all optimal or non-dominated or all weak non-dominated solutions to the Type-2 LPP with fuzzy parameters (4.21), it suffices to find all complete or Pareto or weak Pareto optimal solutions to the MOLP problem (4.72). Now, associated with the MOLP problem (4.72), we consider the following weighting LP problem:

$$\max w_0^L \sum_{j=1}^n (\tilde{c}_j)_0^L x_j + w_1^L \sum_{j=1}^n (\tilde{c}_j)_1^L x_j + w_0^U \sum_{j=1}^n (\tilde{c}_j)_0^U x_j + w_1^U \sum_{j=1}^n (\tilde{c}_j)_1^U x_j$$

$$\text{s.t.} \quad \sum_{j=1}^n a_{ij} x_j \leq b_i, \; i = 1, 2, \ldots, m, \tag{4.73}$$

$$x_j \geq 0, \quad j = 1, 2, \ldots, n.$$

where $w = \left(w_0^L, w_1^L, w_0^U, w_1^U \right) \geqq (0, 0, 0, 0)$.

Theorem 4.6 [13] *Let a point $x^* \in X$ be a feasible solution to the problem (4.21). If it is an optimal solution of the problem (4.73) for some $w > 0$, then it is a non-dominated solution to the problem (4.21).*

Proof If an optimal solution $x^* \in X$ to the problem (4.73) is not a non-dominated solution to the problem (4.21), from Theorem 4.4, it is not a Pareto optimal solution to the MOLP problem (4.72), thus there exists a $\bar{x} \in X$ such that

$$\left(\sum_{j=1}^n (\tilde{c}_j)_0^L x_j^*, \sum_{j=1}^n (\tilde{c}_j)_1^L x_j^*, \sum_{j=1}^n (\tilde{c}_j)_0^U x_j^*, \sum_{j=1}^n (\tilde{c}_j)_1^U x_j^* \right)$$

$$\leq \left(\sum_{j=1}^n (\tilde{c}_j)_0^L \bar{x}_j, \sum_{j=1}^n (\tilde{c}_j)_1^L \bar{x}_j, \sum_{j=1}^n (\tilde{c}_j)_0^U \bar{x}_j, \sum_{j=1}^n (\tilde{c}_j)_1^U \bar{x}_j \right)$$

This means that the following inequalities hold and at least one of these inequalities holds strictly:

$$\sum_{j=1}^n (\tilde{c}_j)_0^L x_j^* \leqq \sum_{j=1}^n (\tilde{c}_j)_0^L \bar{x}_j, \quad \sum_{j=1}^n (\tilde{c}_j)_1^L x_j^* \leqq \sum_{j=1}^n (\tilde{c}_j)_1^L \bar{x}_j,$$

$$\sum_{j=1}^n (\tilde{c}_j)_0^U x_j^* \leqq \sum_{j=1}^n (\tilde{c}_j)_0^U \bar{x}_j, \quad \sum_{j=1}^n (\tilde{c}_j)_1^U x_j^* \leqq \sum_{j=1}^n (\tilde{c}_j)_1^U \bar{x}_j,$$

Noting that $w = \left(w_0^L, w_1^L, w_0^U, w_1^U \right) > (0, 0, 0, 0)$, we have

$$w_0^L \sum_{j=1}^n (\tilde{c}_j)_0^L x_j^* + w_1^L \sum_{j=1}^n (\tilde{c}_j)_1^L x_j^* + w_0^U \sum_{j=1}^n (\tilde{c}_j)_0^U x_j^* + w_1^U \sum_{j=1}^n (\tilde{c}_j)_1^U x_j^*$$

$$\leqq w_0^L \sum_{j=1}^n (\tilde{c}_j)_0^L \bar{x}_j + w_1^L \sum_{j=1}^n (\tilde{c}_j)_1^L \bar{x}_j + w_0^U \sum_{j=1}^n (\tilde{c}_j)_0^U \bar{x}_j + w_1^U \sum_{j=1}^n (\tilde{c}_j)_1^U \bar{x}_j$$

However, this contradicts the assumption that $x^* \in X$ is an optimal solution to the problem (4.73) for some $w > 0$. $\square$

Theorem 4.7 [13] *Let a point $x^* \in X$ be a feasible solution to the problem (4.21). If it is a non-dominated solution to the problem, then it is an optimal solution of the problem (4.73) for some $w > 0$.*

Proof If $x^* \in X$ is a non-dominated solution to the problem (4.21), then it is a Pareto optimal solution to the MOLP problem (4.72) from Theorem 4.4. By using Theorem 3.3 of [20], it is an optimal solution to the problem (4.73) for some $w > 0$.   □

Theorem 4.8 [13] *Let a point $x^* \in X$ be a feasible solution to the problem (4.21). Then it is a weak non-dominated solution to the problem (4.21), then it is an optimal solution of the problem (4.73) for some $w \geqq 0$.*

Example 4.7 [13] Consider the following Type-2 LPP with fuzzy parameters:

$$\max \tilde{z} = (5, 6, 7, 20)x_1 + (16, 17, 18, 40)x_2$$
$$\begin{aligned} s.t. \quad & x_1 + 4x_2 \leq 14, \\ & 4x_1 + 10x_2 \leq 38, \\ & 28x_1 - 5x_2 \leq 56 \\ & x_1, x_2 \geq 0. \end{aligned} \tag{4.74}$$

Note that $[\tilde{c}_1]_\alpha = \left[(\tilde{c}_1)^L_\alpha, (\tilde{c}_1)^U_\alpha\right] = [5 + \alpha, 20 - 13\alpha]$ and $[\tilde{c}_2]_\alpha = \left[(\tilde{c}_2)^L_\alpha, (\tilde{c}_2)^U_\alpha\right] = [16 + \alpha, 40 - 22\alpha]$. Therefore, $\left[(\tilde{c}_1)^L_0, (\tilde{c}_1)^U_0\right] = [5, 20]$, $\left[(\tilde{c}_1)^L_1, (\tilde{c}_1)^U_1\right] = [6, 7]$, $\left[(\tilde{c}_2)^L_0, (\tilde{c}_2)^U_0\right] = [16, 40]$ and $\left[(\tilde{c}_2)^L_1, (\tilde{c}_2)^U_1\right] = [17, 18]$.

Associated with the problem (4.74) and with regard to the MOLP problem (4.73), we consider the following MOLP problem:

$$\max[5x_1 + 16x_2, 6x_1 + 17x_2, 20x_1 + 40x_2, 7x_1 + 18x_2]$$
$$\begin{aligned} s.t. \quad & x_1 + 4x_2 \leq 14, \\ & 4x_1 + 10x_2 \leq 38, \\ & 28x_1 - 5x_2 \leq 56 \\ & x_1, x_2 \geq 0. \end{aligned} \tag{4.75}$$

Now, associated with the MOLP problem (4.75) and with regard to the problem (4.73), we consider the following weighting LP problem:

$$\max w^L_0(5x_1 + 16x_2) + w^L_1(6x_1 + 17x_2) + w^U_0(20x_1 + 40x_2) + w^U_1(7x_1 + 18x_2)$$
$$\begin{aligned} s.t. \quad & x_1 + 4x_2 \leq 14, \\ & 4x_1 + 10x_2 \leq 38, \\ & 28x_1 - 5x_2 \leq 56 \\ & x_1, x_2 \geq 0. \end{aligned} \tag{4.76}$$

Obviously, the solution of the problem depends on the choice of the weights for the objective functions. For example, for $w_0^L = w_1^L = \frac{2}{6}$ and $w_0^U = w_1^U = \frac{1}{6}$, we have the following crisp single objective LP problem:

$$\max z = \frac{49}{6}x_1 + \frac{124}{6}x_2$$
$$s.t. \quad x_1 + 4x_2 \le 14,$$
$$4x_1 + 10x_2 \le 38, \qquad (4.77)$$
$$28x_1 - 5x_2 \le 56$$
$$x_1, x_2 \ge 0.$$

The optimal solution of the problem (4.77) is $(x_1^*, x_2^*) = (2, 3)$. This means that $(x_1^*, x_2^*) = (2, 3)$ is a non-dominated solution of the problem (4.74) and then the membership function of the objective function is given as follows

$$\tilde{z} = (5, 6, 7, 20)x_1^* + (16, 17, 18, 40)x_2^* = (58, 63, 68, 160)$$

In a similar way for $w_0^L = w_1^L = w_0^U = w_1^U = \frac{1}{4}$, we have the following crisp single objective LP problem:

$$\max z = \frac{38}{4}x_1 + \frac{91}{4}x_2$$
$$s.t. \quad x_1 + 4x_2 \le 14,$$
$$4x_1 + 10x_2 \le 38, \qquad (4.78)$$
$$28x_1 - 5x_2 \le 56$$
$$x_1, x_2 \ge 0.$$

The optimal solution of the problem (4.78) is $(x_1^*, x_2^*) = (2.5, 2.8)$. This means that $(x_1^*, x_2^*) = (2.5, 2.8)$ is a non-dominated solution of the problem (4.74) and then the membership function of the objective function is given as follows

$$\tilde{z} = (5, 6, 7, 20)x_1^* + (16, 17, 18, 40)x_2^* = (57.3, 62.6, 67.9, 162).$$

4.5 Type-3 LPP with Fuzzy Parameters

The general form of LPPs in which the coefficients of the decision variables in the objective function and in the constraints are fuzzy numbers, can be formulated as follows [14]:

$$\max \tilde{z} = \sum_{j=1}^{n} \tilde{c}_j x_j$$

$$\text{s.t.} \quad \sum_{j=1}^{n} \tilde{a}_{ij} x_j \le b_i, \quad i = 1, 2, \ldots, m, \tag{4.79}$$

$$x_j \ge 0, \quad j = 1, 2, \ldots, n.$$

In this section, we shall explore two approaches for solving the Type-3 LPP with fuzzy parameters (4.79).

4.5.1   Lai and Hwang's Approach

Lai and Hwang [5] assumed that the fuzzy numbers in the FLP (4.79) are all triangular fuzzy numbers. Thus, $\tilde{c}_j$ and $\tilde{a}_{ij}$ are represented by $\tilde{c}_j = (c_{1,j}, c_{2,j}, c_{3,j})$ and $\tilde{a}_{ij} = (a_{1,ij}, a_{2,ij}, a_{3,ij})$, respectively. Thus the FLP problem (4.79) can be rewritten as follows:

$$\max \tilde{z} = \left(\sum_{j=1}^{n} c_{1,j} x_j, \sum_{j=1}^{n} c_{2,j} x_j, \sum_{j=1}^{n} c_{3,j} x_j \right)$$

$$\text{s.t.} \quad \sum_{j=1}^{n} (a_{1,ij}, a_{2,ij}, a_{3,ij}) x_j \le b_i, \quad i = 1, 2, \ldots, m, \tag{4.80}$$

$$x_j \ge 0, \quad j = 1, 2, \ldots, n.$$

To deal with the fuzzy coefficient $(a_{1,ij}, a_{2,ij}, a_{3,ij})$, Lai and Hwang [5] applied the concept of the most likely value which is $\frac{(a_{1,ij,\beta} + 4a_{2,ij} + a_{3,ij,\beta})}{6}$ where β is a minimal acceptable possibility, $a_{1,ij,\beta} = a_{1,ij} + \beta(a_{2,ij} - a_{1,ij})$ and $a_{3,ij,\beta} = a_{3,ij} - \beta(a_{3,ij} - a_{2,ij})$. Then, they obtained the auxiliary problem as follows with regard to the problem (4.25):

$$\min z_1 = \sum_{j=1}^{n} (c_{2,j} - c_{1,j}) x_j$$

$$\max z_2 = \sum_{j=1}^{n} c_{2,j} x_j$$

$$\max z_3 = \sum_{j=1}^{n} (c_{3,j} - c_{2,j}) x_j \tag{4.81}$$

$$\text{s.t.} \quad \sum_{j=1}^{n} \frac{(a_{1,ij,\beta} + 4a_{2,ij} + a_{3,ij,\beta})}{6} x_j \le b_i, \quad i = 1, 2, \ldots, m,$$

$$x_j \ge 0, \quad j = 1, 2, \ldots, n.$$

Note that the constraints of the problem (4.81) are nonlinear. However, if β is initially given by the DM, then the problem (4.81) is an MOLP model, which can be similarly solved by the method of solving the MOLP problem (4.25).

4.5.2 Wan and Dong's Approach

Assume that the fuzzy numbers in the FLP (4.79) are all trapezoidal fuzzy numbers. Thus, $\tilde{c}_j$ and $\tilde{a}_{ij}$ are represented by $\tilde{c}_j = (c_{1,j}, c_{2,j}, c_{3,j}, c_{4,j})$ and $\tilde{a}_{ij} = (a_{1,ij}, a_{2,ij}, a_{3,ij}, a_{4,ij})$. Therefore, the FLP problem (4.79) can be rewritten as follows:

$$
\max \tilde{z} = \left(\sum_{j=1}^{n} c_{1,j} x_j, \sum_{j=1}^{n} c_{2,j} x_j, \sum_{j=1}^{n} c_{3,j} x_j, \sum_{j=1}^{n} c_{4,j} x_j \right)
$$

$$
s.t. \quad \sum_{j=1}^{n} (a_{1,ij}, a_{2,ij}, a_{3,ij}, a_{4,ij}) x_j \le b_i, \quad i = 1, 2, \ldots, m, \tag{4.82}
$$

$$
x_j \ge 0, \quad j = 1, 2, \ldots, n.
$$

Wan and Dong [6] proposed a weighted average method to deal with the fuzzy constraints $\sum_{j=1}^{n} (a_{1,ij}, a_{2,ij}, a_{3,ij}, a_{4,ij}) x_j \le b_i$, $(i = 1, 2, \ldots, m)$. That is,

$$
\sum_{j=1}^{n} (w_1 a_{1,ij,\beta} + w_2 a_{2,ij} + w_3 a_{2,ij} + w_4 a_{4,ij,\beta}) x_j \le b_i, \quad \text{where} \quad w_i \ge 0 \ (i = 1, 2, \ldots, m),
$$

$\sum_{i=1}^{4} w_i = 1$, β is a minimal acceptable possibility, $a_{1,ij,\beta} = a_{1,ij} + \beta(a_{2,ij} - a_{1,ij})$ and $a_{3,ij,\beta} = a_{4,ij} - \beta(a_{4,ij} - a_{3,ij})$. Then, they obtained the auxiliary problem as follows with regard to the problem (4.38):

$$
\min z_1 = \sum_{j=1}^{n} (c_{2,j} - c_{1,j}) x_j
$$

$$
\max z_2 = \sum_{j=1}^{n} c_{2,j} x_j
$$

$$
\max z_3 = \sum_{j=1}^{n} \frac{1}{2} [c_{2,j} x_j + c_{3,j} x_j]
$$

$$
\max z_4 = \sum_{j=1}^{n} (c_{4,j} - c_{3,j}) x_j \tag{4.83}
$$

$$
s.t. \quad \sum_{j=1}^{n} \frac{1}{6} (a_{1,ij,\beta} + 2a_{2,ij} + 2a_{3,ij} + a_{4,ij,\beta}) x_j \le b_i, \quad i = 1, 2, \ldots, m,
$$

$$
x_j \ge 0, \quad j = 1, 2, \ldots, n.
$$

In this model $a_{1,ij,\beta}$ is relative to the lower limit and is too pessimistic, $a_{4,ij,\beta}$ is relative to the upper limit and is too optimistic, lower weights should be assigned to them. However, $a_{2,ij}$ and $a_{3,ij}$ are the most possible values, which can provide the most important and valuable information. Thus, it is very natural to assign greater weights to them. If β is initially given by the DM, then the problem (4.83) is an MOLP model, which can be similarly solved by the method of solving the MOLP problem (4.38).

4.6 Type-4 LPP with Fuzzy Parameters

The general form of LPPs in which the coefficients of the decision variables in the objective function, the coefficients of the decision variables in the constraints and the right-hand side of the constraints are all fuzzy numbers, can be formulated as follows [14]:

$$
\begin{aligned}
\max \tilde{z} &= \sum_{j=1}^{n} \tilde{c}_j x_j \\
s.t. \quad &\sum_{j=1}^{n} \tilde{a}_{ij} x_j \leq \tilde{b}_i, \ i = 1, 2, \ldots, m, \\
&x_j \geq 0, \quad j = 1, 2, \ldots, n.
\end{aligned}
\tag{4.84}
$$

In this section, we shall explore several approaches for solving the Type-4 LPP with fuzzy parameters (4.84).

4.6.1 Lai and Hwang's Approach

Lai and Hwang [14] combined the fuzzy ranking concepts with the strategy used for Type-2 LPPs with fuzzy parameters, when the fuzzy numbers in the FLP (4.84) are all triangular fuzzy numbers. Thus, $\tilde{b}_i$ is represented by $(b_{1,i}, b_{2,i}, b_{3,i})$. Thus, they obtained the following auxiliary problem:

$$
\begin{aligned}
\min z_1 &= \sum_{j=1}^{n} (c_{2,j} - c_{1,j}) x_j \\
\max z_2 &= \sum_{j=1}^{n} c_{2,j} x_j
\end{aligned}
$$

$$\max z_3 = \sum_{j=1}^{n} (c_{3,j} - c_{2,j})x_j$$

$$\text{s.t.} \quad \sum_{j=1}^{n} a_{1,ij,\beta}x_j \leq b_{1,i,\beta}, \ i = 1, 2, \ldots, m,$$

$$\sum_{j=1}^{n} a_{2,ij}x_j \leq b_{2,i}, \ i = 1, 2, \ldots, m, \qquad (4.85)$$

$$\sum_{j=1}^{n} a_{3,ij,\beta}x_j \leq b_{3,i,\beta}, \ i = 1, 2, \ldots, m,$$

$$x_j \geq 0, \quad j = 1, 2, \ldots, n.$$

where $b_{1,i,\beta} = b_{1,i} + \beta(b_{2,i} - b_{1,i})$ and $b_{3,i,\beta} = b_{3,i} - \beta(b_{3,i} - b_{2,i})$.

If the minimal acceptable possibility, β, is given, then the problem (4.85) is an MOLP model, which can be similarly solved by the method of solving the MOLP problem (4.25).

4.6.2   Wan and Dong's Approach

On the basis of the strategy for the fuzzy objective function, Wan and Dong [6] combined the fuzzy ranking concepts to handle the fuzzy constraints assuming that the fuzzy numbers in the FLP (4.79) are all trapezoidal fuzzy numbers. Thus, $\tilde{b}_i$ is represented by $(b_{1,i}, b_{2,i}, b_{3,i}, b_{4,i})$.

Therefore, they obtained the following auxiliary problem:

$$\min z_1 = \sum_{j=1}^{n} (c_{2,j} - c_{1,j})x_j$$

$$\max z_2 = \sum_{j=1}^{n} c_{2,j}x_j$$

$$\max z_3 = \sum_{j=1}^{n} \frac{1}{2}[c_{2,j}x_j + c_{3,j}x_j]$$

$$\max z_4 = \sum_{j=1}^{n} (c_{4,j} - c_{3,j})x_j$$

$$\text{s.t.} \quad \sum_{j=1}^{n} a_{1,ij,\beta}x_j \leq b_{1,i,\beta}, i = 1, 2, \ldots, m,$$

$$\sum_{j=1}^{n} a_{2,ij}x_j \leq b_{2,i}, \ i = 1, 2, \ldots, m,$$

$$\sum_{j=1}^{n} a_{3,ij} x_j \leq b_{3,i}, \quad i = 1, 2, \ldots, m,$$

$$\sum_{j=1}^{n} a_{4,ij,\beta} x_j \leq b_{4,i,\beta}, \quad i = 1, 2, \ldots, m, \tag{4.86}$$

$$x_j \geq 0, \quad j = 1, 2, \ldots, n.$$

If β is given, then the problem (4.86) is an MOLP model, which can be similarly solved by the method of solving the MOLP problem (4.38).

4.6.3   Negi and Lee's Approach

Negi and Lee [17] used the exceedance possibility and the strict exceedance possibility indexes, proposed by Dubois and Prade [33], to compare the inequality constraints of the problem (4.84).

Consider two trapezoidal fuzzy numbers $\tilde{a} = (a_1, a_2, a_3, a_4)$ and $\tilde{b} = (b_1, b_2, b_3, b_4)$. The exceedance possibility denoted by $Poss(\tilde{b}^U \geq \tilde{a}^L)$ is defined as follows (see point $E = (x_1, \delta^1)$ in Fig. 4.2):

$$Poss(\tilde{b}^U \geq \tilde{a}^L) = \begin{cases} 1, & b_3 > a_2 \\ \delta^1 & b_3 \leq a_2, a_1 \leq b_4 \\ 0 & a_1 > b_4 \end{cases} \tag{4.87}$$

where $\delta^1 = \frac{b_4 - a_1}{(b_4 - b_3) + (a_2 - a_1)}$.

Similarly, the strict exceedance possibility denoted by $Poss(\tilde{b}^U \geq \tilde{a}^U)$ is defined as follows (see point $S = (x_2, \delta^2)$ in Fig. 4.2):

$$Poss(\tilde{b}^U \geq \tilde{a}^U) = \begin{cases} 1, & b_3 > a_4 \\ \delta^2 & b_3 \leq a_4, a_3 \leq b_4 \\ 0 & a_3 > b_4 \end{cases} \tag{4.88}$$

Fig. 4.2 Comparison of trapezoidal fuzzy number

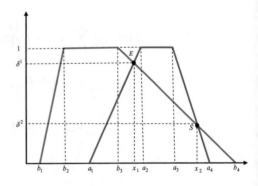

where $\delta^2 = \frac{b_4 - a_3}{(b_4 - b_3) + (a_4 - a_3)}$.

Since the addition or subtraction of trapezoidal fuzzy numbers results in trapezoidal fuzzy number [33], the inequality constraints in the problem (4.84) can be treated as a comparison of fuzzy numbers. Thus, according to the Eq. (4.87), the exceedance possibility that constraint i is satisfied can be expressed as:

$$Poss(x \in F_i) = \begin{cases} 1, & b_{3,i} > \sum_{j=1}^{n} a_{2,ij}x_j \\ \delta_i^1 & b_{3,i} \leq \sum_{j=1}^{n} a_{2,ij}x_j, \sum_{j=1}^{n} a_{1,ij}x_j \leq b_{4,i} \\ 0 & \sum_{j=1}^{n} a_{1,ij}x_j > b_{4,i} \end{cases} \quad (4.89)$$

where $\delta_i^1 = \frac{b_{4,i} - \sum_{j=1}^{n} a_{1,ij}x_j}{(b_{4,i} - b_{3,i}) + (\sum_{j=1}^{n} a_{2,ij}x_j - \sum_{j=1}^{n} a_{1,ij}x_j)}$.

Since we are generally not interested in a zero possibility, the decision space is two first conditions in the Eqs. (4.89). Thus, the only constraints that are needed to restrict the results in this decision space are:

$$\sum_{j=1}^{n} a_{1,ij}x_j \leq b_{4,i}, \quad i = 1, 2, \ldots, m \quad (4.90)$$

In order to satisfy all the constraints, we must have:

$$Poss(x \in F) = \min_{1 \leq i \leq m} \{Poss(x \in F_i)\} = \min\{\delta_1^1, \delta_2^1, \ldots, \delta_m^1\} \quad (4.91)$$

The fuzzy objective function of the problem (4.84) can be rewritten as follows:

$$\tilde{z} = \left(\sum_{j=1}^{n} c_{1,j}x_j, \sum_{j=1}^{n} c_{2,j}x_j, \sum_{j=1}^{n} c_{3,j}x_j, \sum_{j=1}^{n} c_{4,j}x_j \right) \quad (4.92)$$

Thus, the conditional possibility for the objective function z, given that the x satisfies the exceedance possibility for the constraints, is:

$$Poss(Z = z|x) = \begin{cases} \theta_1 = \frac{z - \sum_{j=1}^{n} c_{1,j}x_j}{\sum_{j=1}^{n} c_{2,j}x_j - \sum_{j=1}^{n} c_{1,j}x_j}, & \sum_{j=1}^{n} c_{1,j}x_j \leq z \leq \sum_{j=1}^{n} c_{2,j}x_j \\ 1, & \sum_{j=1}^{n} c_{2,j}x_j \leq z \leq \sum_{j=1}^{n} c_{3,j}x_j \\ \theta_2 = \frac{\sum_{j=1}^{n} c_{4,j}x_j - z}{\sum_{j=1}^{n} c_{4,j}x_j - \sum_{j=1}^{n} c_{3,j}x_j}, & \sum_{j=1}^{n} c_{3,j}x_j \leq z \leq \sum_{j=1}^{n} c_{4,j}x_j \end{cases} \quad (4.93)$$

Since we wish to maximize the objective function, the decision space we are interested in is θ_2. In order to keep the results in this decision space, the problem must satisfy the following constraint:

$$\sum_{j=1}^{n} c_{3,j} x_j \leq z \leq \sum_{j=1}^{n} c_{4,j} x_j \qquad (4.94)$$

Thus, the unconditional possibility distribution for the objective function for the space given by (4.90) and (4.94), can be represented by:

$$Poss(Z = z) = \min\{\min\{\delta_1^1, \delta_2^1, \ldots, \delta_m^1\}, \theta_2\} = \min\{\delta_1^1, \delta_2^1, \ldots, \delta_m^1, \theta_2\} \quad (4.95)$$

Since the equations δ_i^1 and θ_2 are nonlinear functions, the problem (4.95) is a nonlinear one. However, if the desired possibilities are known and given, this problem can be reduced to a linear one. Two different approaches can be formulated depending on whether we have the participation of the DM. If we have the interaction of the DM and the DM supplies a desired possibility level, say, a level of α, then the problem (4.95) becomes:

$$
\begin{aligned}
&\max z \\
&s.t. \quad \delta_1^1, \delta_2^1, \ldots, \delta_m^1, \theta_2 \geq \alpha \\
&\qquad \sum_{j=1}^{n} a_{1,ij} x_j \leq b_{4,i}, \; i = 1, 2, \ldots, m \\
&\qquad \sum_{j=1}^{n} c_{3,j} x_j \leq z \leq \sum_{j=1}^{n} c_{4,j} x_j \\
&\qquad x_j \geq 0, \quad j = 1, 2, \ldots, n.
\end{aligned}
\qquad (4.96)
$$

If we do not have the participation of the DM, the most reasonable value for the possibility level would be one. With all δ_i^1 and θ_2 equal to one, the fuzzy problem is reduced to the following crisp LP problem:

$$
\begin{aligned}
&\max z \\
&s.t. \quad \sum_{j=1}^{n} a_{2,ij} x_j \leq b_{3,i}, \; i = 1, 2, \ldots, m \\
&\qquad \sum_{j=1}^{n} c_{3,j} x_j \leq z \leq \sum_{j=1}^{n} c_{4,j} x_j \\
&\qquad x_j \geq 0, \quad j = 1, 2, \ldots, n.
\end{aligned}
\qquad (4.97)
$$

Similarly, if we adapt the strict exceedance possibility, according to the Eq. (4.88), the strict exceedance possibility that constraint i is satisfied can be expressed as:

$$Poss(x \in F_i) = \begin{cases} 1, & b_{3,i} > \sum_{j=1}^{n} a_{4,ij}x_j \\ \delta_i^2 & b_{3,i} \le \sum_{j=1}^{n} a_{4,ij}x_j, \sum_{j=1}^{n} a_{3,ij}x_j \le b_{4,i} \\ 0 & \sum_{j=1}^{n} a_{3,ij}x_j > b_{4,i} \end{cases} \qquad (4.98)$$

where $\delta_i^2 = \dfrac{b_{4,i} - \sum_{j=1}^{n} a_{3,ij}x_j}{(b_{4,i} - b_{3,i}) + (\sum_{j=1}^{n} a_{4,ij}x_j - \sum_{j=1}^{n} a_{3,ij}x_j)}$.

Similar to the previous discussion, the only constraints that are needed to restrict the results in the decision space are:

$$\sum_{j=1}^{n} a_{3,ij}x_j \le b_{4,i}, \ i = 1, 2, \ldots, m \qquad (4.99)$$

In order to satisfy all the constraints, we must have:

$$Poss(x \in F) = \min_{1 \le i \le m} \{Poss(x \in F_i)\} = \min\{\delta_1^2, \delta_2^2, \ldots, \delta_m^2\} \qquad (4.100)$$

Thus, the unconditional possibility distribution for the objective function for the space given by (4.94) and (4.99), can be represented by:

$$Poss(Z = z) = \min\{\min\{\delta_1^2, \delta_2^2, \ldots, \delta_m^2\}, \theta_2\} = \min\{\delta_1^2, \delta_2^2, \ldots, \delta_m^2, \theta_2\} \quad (4.101)$$

Now, if we have the interaction of the DM and the DM supplies a desired possibility level of α, then the problem (4.101) becomes:

$$\begin{aligned} \max \ &z \\ s.t. \ &\delta_1^2, \delta_2^2, \ldots, \delta_m^2, \theta_2 \ge \alpha \\ &\sum_{j=1}^{n} a_{3,ij}x_j \le b_{4,i}, \ i = 1, 2, \ldots, m \\ &\sum_{j=1}^{n} c_{3,j}x_j \le z \le \sum_{j=1}^{n} c_{4,j}x_j \\ &x_j \ge 0, \quad j = 1, 2, \ldots, n. \end{aligned} \qquad (4.102)$$

If we do not have the participation of the DM, the fuzzy problem is reduced to the following crisp LP problem:

$$
\max z
$$
$$
s.t. \quad \sum_{j=1}^{n} a_{4,ij} x_j \leq b_{3,i}, \quad i = 1, 2, \ldots, m
$$
$$
\sum_{j=1}^{n} c_{3,j} x_j \leq z \leq \sum_{j=1}^{n} c_{4,j} x_j \tag{4.103}
$$
$$
x_j \geq 0, \quad j = 1, 2, \ldots, n.
$$

Example 4.8 [17] Consider the following Type-4 LPP with fuzzy parameters:

$$
\max \tilde{z} = (1, 2, 3, 4)x_1 + (2, 3, 4, 6)x_2
$$
$$
s.t. \quad (0, 1, 2, 3)x_1 + (1, 2, 3, 4)x_2 \leq (3, 5, 6, 8),
$$
$$
(1, 2, 3, 6)x_1 + (0, 1, 2, 3)x_2 \leq (2, 4, 6, 7), \tag{4.104}
$$
$$
x_1, x_2 \geq 0.
$$

We first adopt the exceedance possibility to solve the problem (4.104). Thus, from (4.89), we obtain:

$$
Poss(x \in F_1) = \begin{cases} 1, & x_1 + 2x_2 \leq 8 \\ \delta_1^1 = \frac{8 - x_2}{2 + x_1 + x}, & x_1 + 2x_2 \geq 6, x_2 \leq 8 \\ 0, & x_2 \geq 8 \end{cases} \tag{4.105}
$$

$$
Poss(x \in F_2) = \begin{cases} 1, & 2x_1 + x_2 \leq 6 \\ \delta_1^1 = \frac{7 - x_1}{1 + x_1 + x_2}, & 2x_1 + x_2 \geq 6, x_1 \leq 7 \\ 0, & x_1 \geq 7 \end{cases} \tag{4.106}
$$

According to (4.93), we obtain:

$$
Poss(Z = z|x) = \begin{cases} \theta_1 = \frac{z - (x_1 + 2x_2)}{x_1 + x_2}, & x_1 + 2x_2 \leq z \leq 2x_1 + 3x_2 \\ 1, & 2x_1 + 3x_2 \leq z \leq 3x_1 + 4x_2 \\ \theta_2 = \frac{4x_1 + 6x_2 - z}{x_1 + 2x_2}, & 3x_1 + 4x_2 \leq z \leq 4x_1 + 6x_2 \end{cases} \tag{4.107}
$$

Based on the relations (4.105), (4.106), (4.107) and with regard to the problem (4.96), we should solve the following problem:

$$\max z$$
$$\text{s.t. } 8 - x_2 \geq \alpha(2 + x_1 + x_2)$$
$$7 - x_1 \geq \alpha(1 + x_1 + x_2)$$
$$4x_1 + 6x_2 - z \geq \alpha(x_1 + 2x_2)$$
$$x_2 \leq 8 \tag{4.108}$$
$$x_1 \leq 7$$
$$3x_1 + 4x_2 \leq z \leq 4x_1 + 6x_2$$
$$x_1, x_2 \geq 0.$$

The solution of the problem (4.108) for $\alpha = 0.5$ is $x_1 = 3.125, x_2 = 3.625$ and $z = 29.0625$.

Similarly, with $\alpha = 0.5$ and by the use of the strict exceedance possibility, the problem is reduced to the following crisp LP problem with regard to the problem (4.102):

$$\max z$$
$$\text{s.t. } 8 - 2x_1 - 3x_2 \geq 0.5(2 + x_1 + x_2)$$
$$7 - 3x_1 - 2x_2 \geq 0.5(1 + 3x_1 + x_2)$$
$$4x_1 + 6x_2 - z \geq \alpha(x_1 + 2x_2)$$
$$2x_1 + 3x_2 \leq 8 \tag{4.109}$$
$$3x_1 + 2x_2 \leq 7$$
$$3x_1 + 4x_2 \leq z \leq 4x_1 + 6x_2$$
$$x_1, x_2 \geq 0.$$

The solution of the problem (4.109) is $x_1 = 0.7234, x_2 = 1.2979$ and $z = 9.0213$.

4.6.4 Fuller's Approach

To solve the problem (4.84), Fuller [18] used a ranking function to convert the fuzzy problem under consideration into a crisp LP one.

Assume that the parametric forms of fuzzy numbers $\tilde{a}_{ij}, \tilde{c}_j$ and $\tilde{b}_i$ are presented by $\left[\tilde{a}_{ij}^L(r), \tilde{a}_{ij}^U(r)\right]$, $\left[\tilde{c}_j^L(r), \tilde{c}_j^U(r)\right]$ and $\left[\tilde{b}_i^L(r), \tilde{b}_i^U(r)\right]$, respectively. In this case, Fuller [18] used the following ranking function to convert the fuzzy numbers into real ones:

$$a_{ij} = \int_0^1 \left(\tilde{a}_{ij}^L(r) + \tilde{a}_{ij}^U(r) \right) dr \tag{4.110}$$

$$c_j = \int_0^1 \left(\tilde{c}_j^L(r) + \tilde{c}_j^U(r) \right) dr \tag{4.111}$$

$$b_i = \int_0^1 \left(\tilde{b}_i^L(r) + \tilde{b}_i^U(r) \right) dr \tag{4.112}$$

According to the relations (4.110), (4.111) and (4.112), Fuller [18] proposed the following auxiliary crisp LP problem to solve the problem (4.84):

$$
\begin{aligned}
\max z &= \sum_{j=1}^n c_j x_j \\
s.t. \quad &\sum_{j=1}^n a_{ij} x_j \le b_i, \ i = 1, 2, \ldots, m, \\
&x_j \ge 0, \qquad j = 1, 2, \ldots, n.
\end{aligned}
\tag{4.113}
$$

In particular, assume that the fuzzy numbers in the FLP (4.84) are all trapezoidal fuzzy numbers. In this case, we have

$$\left[\tilde{a}_{ij}^L(r), \tilde{a}_{ij}^U(r) \right] = \left[a_{1,ij} + r(a_{1,ij} - a_{2,ij}), a_{4,ij} + r(a_{3,ij} - a_{4,ij}) \right] \tag{4.114}$$

$$\left[\tilde{b}_i^L(r), \tilde{b}_i^U(r) \right] = \left[b_{1,i} + r(b_{1,i} - b_{2,i}), b_{4,i} + r(b_{3,i} - b_{4,i}) \right] \tag{4.115}$$

$$\left[\tilde{c}_j^L(r), \tilde{c}_j^U(r) \right] = \left[c_{1,j} + r(c_{1,j} - c_{2,j}), c_{4,j} + r(c_{3,j} - c_{4,j}) \right] \tag{4.116}$$

Substituting the relations (4.114), (4.115) and (4.116) into the relations (4.110), (4.111) and (4.112), respectively, we obtain:

$$a_{ij} = \frac{1}{3} \left[(a_{1,ij} + a_{3,ij}) + \frac{1}{2}(a_{2,ij} + a_{4,ij}) \right] \tag{4.117}$$

$$b_i = \frac{1}{3} \left[(b_{1,i} + b_{3,i}) + \frac{1}{2}(b_{2,i} + b_{4,i}) \right] \tag{4.118}$$

$$c_j = \frac{1}{3} \left[(c_{1,j} + c_{3,j}) + \frac{1}{2}(c_{2,j} + c_{4,j}) \right] \tag{4.119}$$

Thus, we can easily solve the problem (4.84) with trapezoidal fuzzy numbers by use of the relations (4.117)–(4.119) and the problem (4.113).

Example 4.9 We solve the Type-4 LPP with fuzzy parameters (4.104) given in the Example 4.8 according to Fuller's approach. By use of the relations (4.117)–(4.119), we obtain the following crisp LP problem:

$$
\begin{aligned}
\max z = {}& 2.334x_1 + 3.5x_2 \\
\text{s.t.} \quad & 1.334x_1 + 2.334x_2 \le 5.167, \\
& 2.667x_1 + 1.334x_2 \le 3.389, \\
& x_1, x_2 \ge 0.
\end{aligned}
\tag{4.120}
$$

The optimal solution of the crisp problem (4.120) is $x_1 = 0.228818$, $x_2 = 2.083015$.

4.6.5 Hatami-Marbini et al.'s Approach

To solve the problem (4.84), Hatami-Marbini et al. [19] developed a new stepwise FLP model with fuzzy parameters in both the objective function and the technical coefficients. In the first step, they used the possibility and necessity relations for fuzzy constraints without considering the fuzzy objective function according to the work of Li and Gong [4]. In the subsequent step, they extended the proposed method to the fuzzy objective function similar to Rommelfanger's approach [7].

Based on possibility and necessity concepts, the model (4.84) can be transformed into the models (4.121) and (4.122) as follows:

$$
\begin{aligned}
\max \tilde{Z}^p(x^p) = {}& \sum_{j=1}^{n} \tilde{c}_j x_j^p \\
\text{s.t.} \quad & \sum_{j=1}^{n} \tilde{a}_{ij} x_j^p \preceq^{Pos} \tilde{b}_i, \quad i = 1, 2, \ldots, m, \\
& x_j^p \ge 0, \quad j = 1, 2, \ldots, n.
\end{aligned}
\tag{4.121}
$$

$$
\begin{aligned}
\max \tilde{Z}^n(x^n) = {}& \sum_{j=1}^{n} \tilde{c}_j x_j^n \\
\text{s.t.} \quad & \sum_{j=1}^{n} \tilde{a}_{ij} x_j^n \preceq^{Nec} \tilde{b}_i, \quad i = 1, 2, \ldots, m, \\
& x_j^n \ge 0, \quad j = 1, 2, \ldots, n.
\end{aligned}
\tag{4.122}
$$

Similar to the problems (4.14) and (4.15), the following two crisp LP problems are proposed to obtain the optimal solution of (4.121) and (4.122), respectively:

$$\max \tilde{Z}^p(x^p) = \sum_{j=1}^{n} \tilde{c}_j x_j^p$$

$$s.t. \quad Pos\left(\sum_{j=1}^{n} \tilde{a}_{ij} x_j^p \preceq \tilde{b}_i\right) \geq \alpha, \ i = 1, 2, \ldots, m, \tag{4.123}$$

$$x_j^p \geq 0, \quad j = 1, 2, \ldots, n.$$

$$\max \tilde{Z}^n(x^n) = \sum_{j=1}^{n} c_j x_j^n$$

$$s.t. \quad Nec\left(\sum_{j=1}^{n} \tilde{a}_{ij} x_j^n \prec \tilde{b}_i\right) \geq 1 - \alpha, \ i = 1, 2, \ldots, m, \tag{4.124}$$

$$x_j^n \geq 0, \quad j = 1, 2, \ldots, n.$$

By using the Theorems 1.17 and 1.18, the problems (4.123) and (4.124), respectively, can be rewritten as follows in the case of using trapezoidal membership functions for the imprecise coefficients:

$$\max \tilde{Z}^p(x^p) = \sum_{j=1}^{n} \tilde{c}_j x_j^p$$

$$s.t. \quad b_{3,i} - \left(\sum_{j=1}^{n} a_{2,ij} x_j^p\right) \geq (\alpha - 1)\left((b_{4,i} - b_{3,i}) + \sum_{j=1}^{n} (a_{2,ij} - a_{1,ij}) x_j^p\right), \ i = 1, 2, \ldots, m,$$

$$x_j^p \geq 0, \quad j = 1, 2, \ldots, n.$$

$$\tag{4.125}$$

$$\max \tilde{Z}^n(x^n) = \sum_{j=1}^{n} \tilde{c}_j x_j^n$$

$$s.t. \quad b_{2,i} - \left(\sum_{j=1}^{n} a_{3,ij} x_j^n\right) \geq (1 - \alpha)\left((b_{2,i} - b_{1,i}) + \sum_{j=1}^{n} (a_{4,ij} - a_{3,ij}) x_j^n\right) \ i = 1, 2, \ldots, m,$$

$$x_j^n \geq 0, \quad j = 1, 2, \ldots, n.$$

$$\tag{4.126}$$

We denote the feasible spaces of the problems (4.126) and (4.126), respectively, by X_α^{Pos} and X_α^{Nec}, respectively.

Note that the problems (4.125) and (4.126) belong to the group of Type-2 LPPs with fuzzy parameters and can be solved by any of the approaches mentioned in Sect. 4.2. Thus, all the approaches discussed in Sect. 4.2 can be applied here.

Hatami-Marbini et al. [19] have used the approach proposed by Rommenlfanger et al. [7] in order to deal with the fuzzy objective function of the problems (4.125) and (4.126). When a α-cut is applied to the fuzzy objective function, the interval model (4.125) can be obtained as follows, where $[\tilde{c}_j]_\alpha = [\tilde{c}_{j,\alpha}^L, \tilde{c}_{j,\alpha}^U]$:

$$\max Z_\alpha^p(x^p) = \sum_{j=1}^{n} [\tilde{c}_{j,\alpha}^L, \tilde{c}_{j,\alpha}^U] x_j^p$$

$$s.t. \quad x^p \in X_\alpha^{Pos}$$

(4.127)

Let $z_{\alpha,\max}^{L,p}$ and $x_{\alpha,\max}^{L,p}$ be the maximum objective value and the optimal solution for the following single objective LP model, respectively:

$$z_{\alpha,\max}^{L,p} = \max Z_\alpha^{L,p}(x^p) = \sum_{j=1}^{n} \tilde{c}_{j,\alpha}^L x_j^p$$

$$s.t. \quad x^p \in X_\alpha^{Pos}$$

(4.128)

Also, let $z_{\alpha,\max}^{U,p}$ and $x_{\alpha,\max}^{U,p}$ be the maximum objective value and the optimal solution for the following single objective LP model, respectively:

$$z_{\alpha,\max}^{U,p} = \max Z_\alpha^{U,p}(x^p) = \sum_{j=1}^{n} \tilde{c}_{j,\alpha}^U x_j^p$$

$$s.t. \quad x^p \in X_\alpha^{Pos}$$

(4.129)

Now, set $z_{\alpha,\min}^{L,p} = z_\alpha^{L,p}(x_{\alpha,\max}^{U,p})$ and $z_{\alpha,\min}^{U,p} = z_\alpha^{U,p}(x_{\alpha,\min}^{L,p})$. The decision maker then identifies his/her objective with the membership functions $\mu_{z_\alpha^{L,p}}$ and $\mu_{z_\alpha^{U,p}}$ for each α-cut which express his/her satisfaction with the obtained objective values. The linear membership function of the objective functions $z_\alpha^{L,p}$ and $z_\alpha^{U,p}$ can now be calculated as follows:

$$\mu_{z_\alpha^{L,p}}(x) = \begin{cases} 1, & z_\alpha^{L,p}(x) > z_{\alpha,\max}^{L,p} \\ \frac{z_\alpha^{L,p}(x) - z_{\alpha,\min}^{L,p}}{z_{\alpha,\max}^{L,p} - z_{\alpha,\min}^{L,p}}, & z_{\alpha,\min}^{L,p} \leq z_\alpha^{L,p}(x) \leq z_{\alpha,\max}^{L,p} \\ 0, & z_\alpha^{L,p}(x) < z_{\alpha,\min}^{L,p} \end{cases}$$

(4.130)

$$\mu_{z_\alpha^{U,p}}(x) = \begin{cases} 1, & z_\alpha^{U,p}(x) > z_{\alpha,\max}^{U,p} \\ \frac{z_\alpha^{U,p}(x) - z_{\alpha,\min}^{U,p}}{z_{\alpha,\max}^{U,p} - z_{\alpha,\min}^{U,p}}, & z_{\alpha,\min}^{U,p} \leq z_\alpha^{U,p}(x) \leq z_{\alpha,\max}^{U,p} \\ 0, & z_\alpha^{U,p}(x) < z_{\alpha,\min}^{U,p} \end{cases}$$

(4.131)

It is important to note that if more information is available, it can be incorporated into the model by using other membership functions describing the decision maker's satisfaction. Next, the following fuzzy optimization system is formulated to take into consideration all information concerning all α:

$$\max \left[\mu_{z_\alpha^{L,p}}(x^p), \mu_{z_\alpha^{U,p}}(x^p) \right]$$
$$\text{s.t.} \quad x^p \in X_\alpha^{Pos} \tag{4.132}$$

Assuming $\omega^p = \min \left[\mu_{z_\alpha^{L,p}}(x^p), \mu_{z_\alpha^{U,p}}(x^p) \right]$, the model (4.132) can be solved by solving the following linear model:

$$\max \omega^p$$
$$\text{s.t.} \quad \mu_{z_\alpha^{L,p}}(x^p)) \geq \omega^p,$$
$$\mu_{z_\alpha^{U,p}}(x^p) \geq \omega^p, \tag{4.133}$$
$$x^p \in X_\alpha^{Pos}.$$

Substituting the membership functions of (4.130) and (4.131) into the problem (4.133) and with regard to the feasible space X_α^{Pos}, the following equivalent problem is obtained

$$\max \omega^p$$
$$\text{s.t.} \quad z_\alpha^{L,p}(x) \geq z_{\alpha,\min}^{L,p} + \omega^p \left(z_{\alpha,\max}^{L,p} - z_{\alpha,\min}^{L,p} \right),$$
$$z_\alpha^{U,p}(x) \geq z_{\alpha,\min}^{U,p} + \omega^p \left(z_{\alpha,\max}^{U,p} - z_{\alpha,\min}^{U,p} \right),$$
$$b_{3,i} - \left(\sum_{j=1}^n a_{2,ij} x_j^p \right) \geq (\alpha - 1) \left((b_{4,i} - b_{3,i}) + \sum_{j=1}^n (a_{2,ij} - a_{1,ij}) x_j^p \right), \quad i = 1, 2, \ldots, m,$$
$$x_j^p \geq 0, \quad j = 1, 2, \ldots, n. \tag{4.134}$$

In a similar way, by changing the index p to n in the models (4.127)–(4.133), the following crisp linear problem is obtained in order to deal with the fuzzy objective of the problem (4.126):

$$\max \omega^n$$
$$\text{s.t.} \quad z_\alpha^{L,n}(x) \geq z_{\alpha,\min}^{L,n} + \omega^p \left(z_{\alpha,\max}^{L,n} - z_{\alpha,\min}^{L,n} \right),$$
$$z_\alpha^{U,n}(x) \geq z_{\alpha,\min}^{U,n} + \omega^p \left(z_{\alpha,\max}^{U,n} - z_{\alpha,\min}^{U,n} \right),$$
$$b_{2,i} - \left(\sum_{j=1}^n a_{3,ij} x_j^n \right) \geq (1 - \alpha) \left((b_{2,i} - b_{1,i}) + \sum_{j=1}^n (a_{4,ij} - a_{3,ij}) x_j^n \right), \quad i = 1, 2, \ldots, m,$$
$$x_j^n \geq 0, \quad j = 1, 2, \ldots, n. \tag{4.135}$$

The arithmetic average of the objective value can be calculated as follows where x^{p*} and x^{n*} are respectively, the optimal solutions of the models (4.134) and (4.135) and r is the number of the α-cut:

$$z^* = \frac{1}{4r} \sum_{i=1}^{r} \left(z_{\alpha_i}^{L,n}(x^{n*}) + z_{\alpha_i}^{U,n}(x^{n*}) + z_{\alpha_i}^{L,p}(x^{p*}) + z_{\alpha_i}^{U,p}(x^{p*}) \right) \qquad (4.136)$$

In order to identify the stability of the result in the proposed method, they used the bounded objective values for each α-cut using a linear membership function and proposed the following scaling functions for the possibility and necessity relations:

$$\mu_{z_{\alpha}^{k,p}}(x^{p*}) = \frac{z_{\alpha}^{k,p}(x^{p*}) - z_{\alpha,\min}^{k,p}}{z_{\alpha,\max}^{k,p} - z_{\alpha,\min}^{k,p}}, \quad k = L, U \qquad (4.137)$$

$$\mu_{z_{\alpha}^{k,n}}(x^{n*}) = \frac{z_{\alpha}^{k,n}(x^{n*}) - z_{\alpha,\min}^{k,n}}{z_{\alpha,\max}^{k,n} - z_{\alpha,\min}^{k,n}}, \quad k = L, U \qquad (4.138)$$

The scaling metric corresponds to a linear approximation of the satisfaction level of the decision maker about the objective function values for each α-cut.

Example 4.10 Consider the following Type-4 LPP with fuzzy parameters:

$$\max \tilde{z} = \tilde{c}_1 x_1 + \tilde{c}_2 x_2$$
$$s.t. \quad (1,2,3,5)x_1 + (-1,1,2,3)x_2 \preceq (1,2,3,5) \qquad (4.139)$$
$$\text{Constraints of Problem (4.63)}$$

where the membership functions of fuzzy numbers $\tilde{c}_1$ and $\tilde{c}_2$ are depicted in Fig. 4.1.

We take into consideration five cuts involving $\{0, 0.25, 0.5, 0.75, 1\}$ when analyzing this problem. Table 4.7 presents the interval value for two fuzzy coefficients of the objective function with respect to the aforementioned cuts.

We first solve the following LP problem to obtain $z_{\alpha,\max}^{k,p}$ and $x_{\alpha,\max}^{k,p}$ $(k = L, U)$ with regard to the problems (4.128) and (4.129):

Table 4.7 α_i-cuts of $\tilde{c}_1$ and $\tilde{c}_2$

α_i	0	0.25	0.5	0.75	1
$[\tilde{c}_1]_\alpha$	[0.5, 10]	[1, 7]	[2, 5]	[3.2, 4.5]	7
$[\tilde{c}_2]_\alpha$	[1.4, 11]	[5, 10]	[6, 9]	[6.5, 7.5]	4

$$z_{\alpha,\max}^{k,p} = \max Z_\alpha^{k,p}(x^p) = \tilde{c}_{1,\alpha}^k x_1^p + \tilde{c}_{2,\alpha}^k x_2^p$$

s.t.

$$x_1^p + 4x_2^p \leq 100, x_1^p + 3x_2^p \leq 76, x_1^p + 2x_2^p \leq 53,$$
$$3x_1^p + 5x_2^p \leq 138, 3x_1^p + 4x_2^p \leq 120, 7x_1^p + 8x_2^p \leq 260 \qquad (4.140)$$
$$x_1^p + x_2^p \leq 36, 3x_1^p + 2x_2^p \leq 103, 2x_1 + x_2^p \leq 68,$$
$$3 - (2x_1^p + 1x_2^p) \geq (\alpha - 1)(2 + (x_1^p + 2x_2^p))$$
$$x_1^p, x_2^p \geq 0.$$

The results are reported in Table 4.8.

Now, we solve the following LP problem to obtain $z_{\alpha,\max}^{k,n}$ and $x_{\alpha,\max}^{k,n}$ $(k = L, U)$:

$$z_{\alpha,\max}^{k,n} = \max Z_\alpha^{k,n}(x^n) = \tilde{c}_{1,\alpha}^k x_1^n + \tilde{c}_{2,\alpha}^k x_2^n$$

s.t.

$$x_1^n + 4x_2^n \leq 100, x_1^n + 3x_2^n \leq 76, x_1^n + 2x_2^n \leq 53,$$
$$3x_1^n + 5x_2^n \leq 138, 3x_1^n + 4x_2^n \leq 120, 7x_1^n + 8x_2^n \leq 260 \qquad (4.141)$$
$$x_1^n + x_2^n \leq 36, 3x_1^n + 2x_2^n \leq 103, 2x_1^n + x_2^n \leq 68,$$
$$2 - (3x_1^n + 2x_2^n) \geq (1 - \alpha)(1 + (2x_1^n + x_2^n))$$
$$x_1^n, x_2^n \geq 0.$$

The results are reported in Table 4.9.

Now, we can compute the minimum objective values by setting $z_{\alpha,\min}^{L,p} = z_\alpha^{L,p}(x_{\alpha,\max}^{U,p})$, $z_{\alpha,\min}^{U,p} = z_\alpha^{U,p}(x_{\alpha,\max}^{L,p})$, $z_{\alpha,\min}^{L,n} = z_\alpha^{L,n}(x_{\alpha,\max}^{U,n})$ and $z_{\alpha,\min}^{U,n} = z_\alpha^U(x_{\alpha,\max}^{L,n})$. The results are given in Table 4.10.

Table 4.8 The results from the possibility viewpoint

α	0	0.25	0.5	0.75	1
$x_{\alpha,\max}^{L,p}$	(7, 23)	(0, 25)	(2.6667, 24.3333)	(0, 7)	(0, 3)
$z_{\alpha,\max}^{L,p}$	35.7	125	151.333	45.5	21
$x_{\alpha,\max}^{U,p}$	(20, 15)	(11.8065, 20.5161)	(2.6667, 24.3333)	(0, 7)	(0, 3)
$z_{\alpha,\max}^{U,p}$	365	287.8065	232.3333	52.5	21

Table 4.9 The results from the necessity viewpoint

α	0	0.25	0.5	0.75	1
$x_{\alpha,\max}^{L,n}$	(0, 0.3333)	(0, 0.454545)	(0, 0.6)	(0, 0.454545)	(0.66667, 0)
$z_{\alpha,\max}^{L,n}$	0.466667	2.272727	3.6	5.055556	4.66667
$x_{\alpha,\max}^{U,n}$	(0, 0.3333)	(0, 0.454545)	(0, 0.6)	(0, 0.454545)	(0.66667, 0)
$z_{\alpha,\max}^{U,n}$	3.66667	4.545455	35.4	5.833333	4.66667

Table 4.10 The lower limits of the objective functions

α	0	0.25	0.5	0.75	1
$z_{\alpha,\min}^{L,p}$	31	114.3871	151.3332	45.5	21
$z_{\alpha,\min}^{U,p}$	323	250	232.3332	52.5	21
$z_{\alpha,\min}^{L,n}$	0.466667	2.272727	3.6	5.055556	4.66667
$z_{\alpha,\min}^{U,n}$	3.66667	4.545455	35.4	5.833333	4.66667

Table 4.11 The results from two different viewpoints

α	0	0.25	0.5	0.75
x^{p*}	(13.9060, 19.25639)	(6.688747, 23.10375)	(2.6662, 24.333)	(0,7)
w^{p*}	0.6195652	0.736877	1	1
x^{n*}	(0, 0.33333)	(0, 0.4545454)	(0, 0.6)	(0.777778)
w^{n*}	1	1	1	1

Finally, the models (4.134) and (4.135) are evolved for this example and Table 4.11 shows the optimal solution for various α-cuts. For instance, in the possibility case, the following proposed model under $\alpha = 0.25$ is formulated:

$$\max \omega^p$$
$$s.t. \quad x_1^p + 5x_2^p \geq 114.3871 + \omega^p(125 - 114.3871),$$
$$7x_1^p + 10x_2^p \geq 250 + \omega^p(287.8065 - 250),$$
$$x_1^p + 4x_2^p \leq 100, x_1^p + 3x_2^p \leq 76, x_1^p + 2x_2^p \leq 53,$$
$$3x_1^p + 5x_2^p \leq 138, 3x_1^p + 4x_2^p \leq 120, 7x_1^p + 8x_2^p \leq 260 \qquad (4.142)$$
$$x_1^p + x_2^p \leq 36, 3x_1^p + 2x_2^p \leq 103, 2x_1 + x_2^p \leq 68,$$
$$3 - (2x_1^p + 1x_2^U) \geq (0.25 - 1)\left(2 + (x_1^p + 2x_2^p)\right),$$
$$\omega^p \leq 1, x_1^p, x_2^p \geq 0.$$

Analogously, the proposed model for the necessity case with $\alpha = 0.25$ is formulated as follows:

$$\max \omega^n$$
$$s.t. \quad x_1^n + 5x_2^n \geq 2.272727 + \omega^n(2.272727 - 2.272727),$$
$$7x_1^n + 10x_2^n \geq 4.545455 + \omega^n(4.545455 - 4.545455),$$
$$x_1^n + 4x_2^n \leq 100, x_1^n + 3x_2^n \leq 76, x_1^n + 2x_2^n \leq 53,$$
$$3x_1^n + 5x_2^n \leq 138, 3x_1^n + 4x_2^n \leq 120, 7x_1^n + 8x_2^n \leq 260 \qquad (4.143)$$
$$x_1^n + x_2^n \leq 36, 3x_1^n + 2x_2^n \leq 103, 2x_1^n + x_2^n \leq 68,$$
$$2 - (3x_1^n + 2x_2^n) \geq (1 - 0.25)\left(1 + (2x_1^n + x_2^n)\right);$$
$$\omega^n \leq 1, x_1^n, x_2^n \geq 0.$$

The arithmetic average of the objective value can be calculated as follows with regard to (4.136):

$$z^* = \frac{1}{4r} \sum_{i=1}^{4} \left(z_{\alpha_i}^{L,n}(x^{n*}) + z_{\alpha_i}^{U,n}(x^{n*}) + z_{\alpha_i}^{L,p}(x^{p*}) + z_{\alpha_i}^{U,p}(x^{p*}) \right) = 79.0796. \quad (4.144)$$

4.7 Type-5 LPP with Fuzzy Parameters

The typical form of LPPs in which the decision variables, the coefficients of the decision variables in the constraints and the right-hand side of the constraints are fuzzy numbers, can be formulated as follows:

$$\min \tilde{z} = \sum_{j=1}^{n} c_j \tilde{x}_j$$

$$s.t. \quad \sum_{j=1}^{n} a_{ij} \tilde{x}_j \geq \tilde{b}_i, \; i = 1, 2, \ldots, m, \quad (4.145)$$

$$\tilde{x}_j \geq \tilde{0}, \quad j = 1, 2, \ldots, n.$$

It is interesting to mention that the problem (4.145) is the same with the FVLP problem (3.51) and three simplex based solution approaches have been proposed for solving it in Chapter 3. In what follows we explore a non-simplex based approach proposed by Saati et al. [21] for solving Type-5 LPP with fuzzy parameters (4.145).

For the purpose of simplicity and without loss of generality, Saati et al. [21] assumed that all fuzzy numbers are trapezoidal fuzzy numbers. They developed a new solution method for solving the problem (4.145) without the use of fuzzy arithmetic. To do this, they defined some new definitions for non-negative trapezoidal fuzzy numbers and fuzzy variables.

Definition 4.8 $\tilde{a} = (a_1, a_2, a_3, a_4)$ is a non-negative trapezoidal fuzzy number if $a_1 + a_4 \geq 0$ and $a_2 + a_3 \geq 0$. Obviously, $\tilde{a}$ is a non-positive trapezoidal fuzzy number if $a_1 + a_4 \leq 0$ and $a_2 + a_3 \leq 0$.

Definition 4.9 Let $\tilde{a} = (a_1, a_2, a_3, a_4)$ be a non-negative trapezoidal fuzzy number. Therefore $\tilde{a}$ corresponds to the following relations:

$$a_4 \geq a_3, a_3 \geq a_2, a_2 \geq a_1, a_1 + a_4 \geq 0, a_2 + a_3 \geq 0$$

Definition 4.10 A trapezoidal fuzzy variable $\tilde{x} = (x_1, x_2, x_3, x_4)$ is a non-negative trapezoidal fuzzy variable if we impose the following conditions:

$$x_4 \geq x_3, x_3 \geq x_2, x_2 \geq x_1, x_1 + x_4 \geq 0, x_2 + x_3 \geq 0$$

Definition 4.11 Let $\sum_{j=1}^{n} a_{ij}\tilde{x}_j \geq \tilde{b}_i$ be the constraint i of the model (4.145). The constraint corresponds with the following four conditions:

$$\sum_{j=1}^{n} a_{ij}x_{1,j} \geq b_{1,i}, \sum_{j=1}^{n} a_{ij}x_{2,j} \geq b_{2,i}, \sum_{j=1}^{n} a_{ij}x_{3,j} \geq b_{3,i}, \sum_{j=1}^{n} a_{ij}x_{4,j} \geq b_{4,i}$$

Note that Definition 4.8 is given for identification of the positive fuzzy numbers, while Definition 4.11 enables us to solve the linear inequalities with fuzzy variables and fuzzy right-hand side variables so that the solution is able to satisfy the four conditions. Using this approach, Saati et al. [21] proposed a solution technique to obtain the fuzzy optimal solution and the crisp optimal solution by solving one LP.

In order to obtain the crisp optimal solution on top of the fuzzy solutions, they define a new variable, namely x, which is smaller than or equal to x_3 and bigger than or equal to x_2 (i.e., $x_2 \leq x \leq x_3$).

Based on Definitions 4.10 and 4.11, Saati et al. [21] formulated the following crisp LP problem in order to obtain the crisp and optimal solutions of the problem (4.145):

$$\min z = \sum_{j=1}^{n} c_j x_j$$

$$s.t. \quad \sum_{j=1}^{n} a_{ij}x_{1,j} \geq b_{1,i}, \ i = 1, 2, \ldots, m,$$

$$\sum_{j=1}^{n} a_{ij}x_{2,j} \geq b_{2,i}, \ i = 1, 2, \ldots, m,$$

$$\sum_{j=1}^{n} a_{ij}x_{3,j} \geq b_{3,i}, \ i = 1, 2, \ldots, m,$$

(4.146)

$$\sum_{j=1}^{n} a_{ij}x_{4,j} \geq b_{4,i}, \ i = 1, 2, \ldots, m,$$

$$x_{2,j} \leq x_j \leq x_{3,j}, \ j = 1, 2, \ldots, n,$$

$$x_{4,j} - x_{3,j} \geq 0, \ j = 1, 2, \ldots, n,$$

$$x_{3,j} - x_{2,j} \geq 0, \ j = 1, 2, \ldots, n,$$

$$x_{2,j} - x_{1,j} \geq 0, \ j = 1, 2, \ldots, n,$$

$$x_{1,j} + x_{4,j} \geq 0, \ j = 1, 2, \ldots, n,$$

$$x_{2,j} + x_{3,j} \geq 0, \ j = 1, 2, \ldots, n,$$

Note that the constraints corresponding to non-negativity of the trapezoidal fuzzy variable $\tilde{x}_j = (x_{1,j}, x_{2,j}, x_{3,j}, x_{4,j})$ in the model (4.146) that is the set of constraints can be shown as:

$$
\begin{bmatrix}
0 & -1 & 0 & 0 & 1 \\
0 & 0 & 1 & 0 & -1 \\
0 & 0 & -1 & 1 & 0 \\
0 & -1 & 1 & 0 & 0 \\
-1 & 1 & 0 & 0 & 0 \\
1 & 0 & 0 & 1 & 0 \\
0 & 1 & 1 & 0 & 0
\end{bmatrix}
\begin{bmatrix}
x_{1,j} \\
x_{2,j} \\
x_{3,j} \\
x_{4,j} \\
x_j
\end{bmatrix}
\geq
\begin{bmatrix}
0 \\
0 \\
0 \\
0 \\
0 \\
0 \\
0
\end{bmatrix}
$$

In consequence, $\tilde{x}_j = (x_{1,j}, x_{2,j}, x_{3,j}, x_{4,j})$ is the component j of the fuzzy optimal solution and x_j is the component j of the crisp optimal solution for the Type-5 LPP with fuzzy parameters (4.145). Thus, the fuzzy and crisp objective function values can be computed, respectively, as $z = \sum_{j=1}^{n} c_j x_j$ and $\tilde{z} = \sum_{j=1}^{n} c_j \tilde{x}_j$.

Saati et al. [21] also proved that the optimal solution of the following problem with positive a_{ij} $(i = 1, 2, \ldots, m, j = 1, 2, \ldots, n)$ is equal to the crisp optimal solution of the problem (4.146):

$$
\min z = \sum_{j=1}^{n} c_j x_j
$$

$$
s.t. \quad \sum_{j=1}^{n} a_{ij} x_j \geq b_{2,i}, \quad i = 1, 2, \ldots, m,
$$

$$
x_j \geq 0, \quad j = 1, 2, \ldots, n,
$$

(4.147)

It is interesting to mention that the model (4.146) involves $4m + 7n$ constraints and $5n$ decision variables whilst the model (4.147) has m constraints and n decision variables.

Saati et al. [21] used the numeric example proposed by Mahdavi-Amiri and Nasseri [34] to demonstrate the advantage of the proposed model as well as comparing their results with Mahdavi-Amiri and Nasseri's [34] results.

Example 4.11 [21] Consider the following Type-5 LPP with fuzzy parameters:

$$
\max \tilde{z} = 6\tilde{x}_1 + 10\tilde{x}_2
$$

$$
s.t. \quad 2\tilde{x}_1 + 5\tilde{x}_2 \geq (3, 5, 8, 13)
$$

$$
3\tilde{x}_1 + 4\tilde{x}_2 \geq (4, 6, 10, 16)
$$

$$
\tilde{x}_1, \tilde{x}_2 \geq \tilde{0}
$$

(4.148)

The optimal solutions and the objective function as trapezoidal fuzzy numbers for Mahdavi-Amiri and Nasseri [34] are as follows:

$$\tilde{x}_1 = (-4.571, -0.286, 4.286, 9.715),$$
$$\tilde{x}_2 = (-3.286, -0.714, 1.714, 7.428), \qquad (4.149)$$
$$\tilde{z} = (-60.288, -8.857, 42.857, 02.571)$$

Figure 4.3 shows the fuzzy objective function value given in (4.149).

According to Saati et al. approach [21], the following crisp LP is formulated in order to obtain the fuzzy and crisp optimal solutions of the problem (4.158) with regard to the problem (4.146):

$$\min z = 6x_1 + 10x_2$$
$$s.t. \ 2x_{1,1} + 5x_{1,2} \geq 3, 2x_{2,1} + 5x_{2,2} \geq 5,$$
$$2x_{3,1} + 5x_{3,2} \geq 8, 2x_{4,1} + 5x_{4,2} \geq 13,$$
$$3x_{1,1} + 4x_{1,2} \geq 4, 3x_{2,1} + 4x_{2,2} \geq 6,$$
$$3x_{3,1} + 4x_{3,2} \geq 10, 3x_{4,1} + 4x_{4,2} \geq 16,$$
$$x_{2,1} \leq x_1 \leq x_{3,1}, x_{2,2} \leq x_2 \leq x_{3,2}, \qquad (4.150)$$
$$x_{4,1} - x_{3,1} \geq 0, x_{4,2} - x_{3,2} \geq 0,$$
$$x_{3,1} - x_{2,1} \geq 0, x_{3,2} - x_{2,2} \geq 0,$$
$$x_{2,1} - x_{1,1} \geq 0, x_{2,2} - x_{1,2} \geq 0,$$
$$x_{1,1} + x_{4,1} \geq 0, x_{1,2} + x_{4,2} \geq 0,$$
$$x_{2,1} + x_{3,1} \geq 0, x_{2,2} + x_{3,2} \geq 0.$$

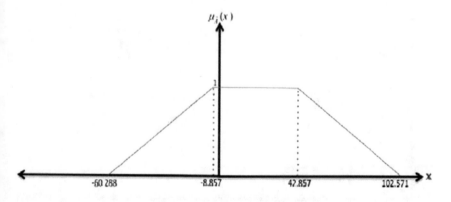

Fig. 4.3 The fuzzy optimal objective function value for Mahdavi-Amiri and Nasseri's model

The fuzzy and crisp optimal solutions of the model (4.150) are as follows:

$$\tilde{x}_1 = (x_{1,1}, x_{2,1}, x_{3,1}, x_{4,1}) = (1.143, 1.429, 2.929, 5.429)$$
$$\tilde{x}_1 = (x_{1,2}, x_{2,2}, x_{3,2}, x_{4,2}) = (0.143, 0.429, 0.429, 0.429)$$
$$\tilde{z} = 6\tilde{x}_1 + 10\tilde{x}_2 = (8.288, 12.864, 21.864, 36.864) \qquad (4.151)$$
$$x = (x_1, x_2) = (1.429, 0.429)$$
$$z = 6x_1 + 10x_2 = 12.857$$

The fuzzy optimal value of the objective function is depicted in Fig. 4.4.

It is interesting to mention that the method proposed by Saati et al.'s approach [21] gives the crisp and fuzzy optimal solutions (consequently crisp and fuzzy optimal objective function values) by solving the LP problem simultaneously.

It should also be noted that by considering the problem (4.147) for solving Example 4.11, the following model is formulated:

$$\min z = 6x_1 + 10x_2$$
$$s.t. \quad 2x_1 + 5x_2 \geq 5,$$
$$3x_1 + 4x_2 \geq 6, \qquad (4.152)$$
$$x_1, x_2 \geq 0.$$

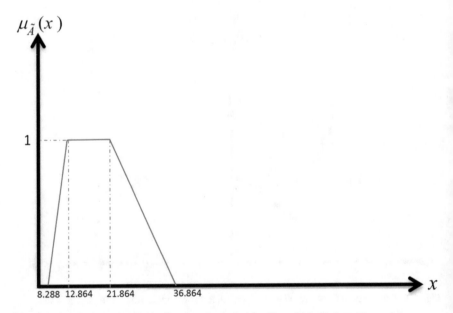

Fig. 4.4 The fuzzy optimal objective function value for Hatami-Marbini et al.'s model

After solving the LP problem (4.152), the optimal solutions and the optimal value are $x = (x_1, x_2) = (1.429, 0.429)$ and $z = 12.857$ $z = 12.857$, which are matched with those given in (4.151) obtained by solving the problem (4.150).

Comparison of the results of Saati et al.'s model [21] with Mahdavi-Amiri and Nasseri's model [34] shows that the fuzzy objective function of the former is considerably better with respect to two reasons. First, the coefficients of the objective function are positive, and the fuzzy decision variables are non-negative. Therefore, the optimal value of the objective function must be non-negative, while a portion of the respective fuzzy number in Mahdavi-Amiri and Nasseri's model [34] is placed in the negative area (see Fig. 4.3). However, the fuzzy optimal value of the objective function proposed by Saati et al. [21] lies in the positive territory as shown in Fig. 4.4. This finding confirms that Saati et al.'s model produces a more rational solution compared with the model proposed by Mahdavi-Amiri and Nasseri [34]. Second, the model in this example is the minimization problem, thereby; its aim is to decrease the value of the objective function as much as possible, subject to the constraints. As shown in Figs. 4.3 and 4.4, the fuzzy objective function of the proposed model proposed by Saati et al. [21] is much smaller than Mahdavi-Amiri and Nasseri's model [34]. The distance between the lower and upper bounds of the fuzzy objective function for Mahdavi-Amiri and Nasseri's model [34] is 162.859, while the associated distance in Saati et al.'s model is 28.576.

4.8 Type-6 LPP with Fuzzy Parameters

The typical form of LPPs in which the decision variables, the coefficients of the decision variables in the objective function, the coefficients of the decision variables in the constraints and the right-hand-side of the constraints are symmetric trapezoidal fuzzy numbers, can be formulated as follows [35]:

$$\max \tilde{z} = \sum_{j=1}^{n} \tilde{c}_j \tilde{x}_j$$

$$s.t. \quad \sum_{j=1}^{n} a_{ij} \tilde{x}_j \preceq \tilde{b}_i, \ i = 1, 2, \ldots, m, \tag{4.153}$$

$$\tilde{x}_j \succeq \tilde{0}, \quad j = 1, 2, \ldots, n.$$

It is interesting to mention that the problem (4.145) is the same with the LPP with fuzzy parameters of type 3 (3.111) discussed in Chap. 3. Three simplex based techniques were explored in Chap. 3 for solving the problem (4.153). In what follows we explore a non-simplex based approach proposed by Kumar and Kaur [22] for solving Type-6 LPP with fuzzy parameters (4.153).

Let $\tilde{x}_j = (x_{2,j} - \alpha_j, x_{2,j}, x_{3,j}, x_{3,j} + \alpha_j)$, $\tilde{c}_j = (c_{2,j} - \beta_j, c_{2,j}, c_{3,j}, c_{3,j} + \beta_j)$ and $\tilde{b}_i = (b_{2,i} - \delta_i, b_{2,i}, b_{3,i}, b_{3,i} + \delta_i)$. Kumar and Kaur [22] formulated the following crisp LP problem in order to obtain the fuzzy optimal solution of the problem (4.153):

$$\max z = \sum_{j=1}^{n} \frac{c_{2,j}+c_{3,j}}{4}x_{2,j} + \sum_{j=1}^{n} \frac{c_{2,j}+c_{3,j}}{4}x_{3,j}$$

$$s.t. \quad \sum_{j=1}^{n} a_{ij}x_{2,j} + \sum_{j=1}^{n} a_{ij}x_{3,j} \le \frac{b_{2,i}+b_{3,i}}{2}, \quad i = 1,2,\ldots,m,$$

$$x_{2,j} + x_{3,j} \ge 0, \qquad j = 1,2,\ldots,n,$$

$$x_{2,j} \le x_{3,j}, \qquad\quad j = 1,2,\ldots,n,$$

$$\alpha_j \ge 0, \qquad\qquad j = 1,2,\ldots,n. \tag{4.154}$$

Let us compare the number of constraints and variables in the Models (4.153) with the number of constraints and variables in the Model (4.154). The Model (4.154) has $m + 2n$ constraints and $4n$ variables (without considering slacks), while the Model (4.153) has m constraints and n variables. This shows that the Model (4.154) has $2n$ constraints and $3n$ variables more than the Model (4.153), and hence utilizing the Model (4.153) is computationally more efficient than using the Model (4.154). As a result, the number of functional constraints is increased and the increased number of constraints directly affects the computational time of the simplex method.

Kumar and Kaur [22] solved the following numeric example proposed by Ganesan and Veeramani [35] with their proposed approach as well as comparing their results with Ganesan and Veeramani's results.

Example 4.12 [22] Consider the following Type-6 LPP with fuzzy parameters:

$$\max \tilde{z} = (11,13,15,17)\tilde{x}_1 + (9,12,14,17)\tilde{x}_2 + (13,15,17,19)\tilde{x}_3$$
$$s.t. \quad 12\tilde{x}_1 + 13\tilde{x}_2 + 12\tilde{x}_3 \preceq (69,475,505,511)$$
$$14\tilde{x}_1 + 13\tilde{x}_3 \preceq (52,460,480,488) \tag{4.155}$$
$$12\tilde{x}_1 + 15\tilde{x}_3 \preceq (460,465,495,500)$$
$$\tilde{x}_1,\tilde{x}_2,x_3 \succeq \tilde{0}$$

The optimal solutions and the objective function as symmetric trapezoidal fuzzy numbers for Ganesan and Veeramani [35] and Ebrahimnejad and Tavana [36] are as follows:

$$\tilde{x}_1 \approx (0,0,0,0),$$
$$\tilde{x}_2 \approx \left(\frac{231}{169},\frac{405}{169},\frac{1045}{169},\frac{1219}{169}\right),$$
$$\tilde{x}_3 \approx \left(\frac{452}{13},\frac{460}{13},\frac{1480}{13},\frac{1488}{13}\right), \tag{4.156}$$
$$\tilde{z} \approx \left(\frac{74416}{169},\frac{94235}{169},\frac{120265}{169},\frac{140084}{169}\right)$$

Additionally, according to Kumar and Kaur's approach, the fuzzy optimal solution of the problem (4.155) can be obtained by solving the following crisp LP problem with regard to the problem (4.154):

$$\max z = 7x_{2,1} + 7x_{3,1} + \frac{13}{2}x_{2,2} + \frac{13}{2}x_{3,2} + 8x_{2,3} + 8x_{3,3}$$

$$s.t. \quad 12x_{2,1} + 12x_{3,1} + 13x_{2,2} + 13x_{3,2} + 12x_{2,3} + 12x_{3,3} \leq 490$$

$$14x_{2,1} + 14x_{3,1} + 12x_{2,3} + 12x_{3,3} \leq 470$$

$$12x_{2,1} + 12x_{3,1} + 15x_{2,2} + 15x_{3,2} \leq 480$$

$$x_{2,1} + x_{3,1} \geq 0$$

$$x_{2,2} + x_{3,22} \geq 0 \tag{4.157}$$

$$x_{2,3} + x_{3,3} \geq 0$$

$$x_{2,1} \leq x_{3,1}$$

$$x_{2,2} \leq x_{3,2}$$

$$x_{2,3} \leq x_{3,3}$$

$$\alpha_1, \alpha_2, \alpha_3 \geq 0.$$

After solving the crisp problem (4.157), the following fuzzy optimal solution is obtained:

$$\tilde{x}_1 \approx (0, 0, 0, 0),$$

$$\tilde{x}_2 \approx \left(\frac{730}{169}, \frac{730}{169}, \frac{730}{169}, \frac{730}{169}\right),$$

$$\tilde{x}_3 \approx \left(\frac{470}{13}, \frac{470}{13}, \frac{470}{13}, \frac{470}{13}\right), \tag{4.158}$$

$$\tilde{z} \approx \left(\frac{86000}{169}, \frac{100410}{169}, \frac{114090}{169}, \frac{128500}{169}\right)$$

Although the fuzzy optimal solution obtained by all three methods (Ganesan and Veeramani's method [35], Ebrahimnejad and Tavana's method [36], Kumar and Kaur's method [22]) are equivalent (based on Definition 3.17), in the fuzzy optimal solution obtained by Kumar and Kaur's method [22], the left and the right spread of each fuzzy decision variable are zero, and the starting and end points of the cores are equal. It is somewhat unusual that the right-hand-side of the constraint is fuzzy while the obtained decision variables are crisp. In addition, the crisp solution derived from Kumar and Kaur's method lacks the flexibility required for real-life implementation. Finally, there is no need to define the decision variables as fuzzy variables in Kumar and Kaur's method since the final solution is ultimately crisp. In summary, Ebrahimnejad and Tavana's method (discussed in Chap. 3) produces a fuzzy optimal solution that is equivalent to the competing methods of Ganesan and Veeramani [35] and Kumar and Kaur [22] but is simpler and computationally more efficient than these other two methods.

4.9   Type-7 LPP with Fuzzy Parameters

The general form of LPPs in which the decision variables, the coefficients of the decision variables in the objective function, the coefficients of the constraint matrix and the right-hand side of the constraints, are fuzzy numbers, can be formulated as follows:

$$
\begin{aligned}
\max \tilde{z} &= \sum_{j=1}^{n} \tilde{c}_j \otimes \tilde{x}_j \\
s.t. \quad &\sum_{j=1}^{n} \tilde{a}_{ij} \otimes \tilde{x}_j \preceq \tilde{b}_i, \; i = 1, 2, \ldots, m, \\
&\tilde{x}_j \geq 0, \quad j = 1, 2, \ldots, n.
\end{aligned}
\tag{4.159}
$$

In this section, we will explore several approaches for solving the Type-7 LPP with fuzzy parameters (4.159).

4.9.1   Hashemi et al.'s Approach

Hashemi et al.'s [23] proposed a two-phase approach to find the optimal solutions of the problem (4.159) based on the comparison of the mean and standard deviation of fuzzy numbers. In the first phase, they maximized the possibilistic mean value of the fuzzy objective function and obtained a set of feasible solutions. In the second phase, they minimized the standard deviation of the original fuzzy objective function, by considering all basic feasible solutions obtained at the end of the first phase.

Hashemi et al. [23] assumed that all the decision parameters as well as the variables of the problem (4.159) are symmetric LR fuzzy numbers and applied new fuzzy arithmetic operations on symmetric fuzzy numbers from Nasrabadi and Nasrabadi [37] to develop their approach. Therefore, we first define these new fuzzy arithmetic operations on symmetric fuzzy numbers.

Definition 4.12 A symmetric fuzzy number $\tilde{a}$ with the reference function L, denoted by $\tilde{a} = (a - \alpha, a, a + \alpha)_L$, is defined as

$$
\mu_{\tilde{a}}(x) = L\left(\frac{x - a}{\alpha}\right), \; \alpha \geq 0
\tag{4.160}
$$

Definition 4.13 Let $\tilde{a}_1 = (a_1 - \alpha_1, a_1, a_1 + \alpha_1)_L$ and $\tilde{a}_2 = (a_2 - \alpha_2, a_2, a_2 + \alpha_2)_L$ be two symmetric fuzzy numbers with the reference function L. The fuzzy arithmetic operations on symmetric fuzzy numbers are defined as follows:

$$\tilde{a}_1 + \tilde{a}_2 = ((a_1 + a_2) - (\alpha_1 + \alpha_2), a_1 + a_2, (a_1 + a_2) + (\alpha_1 + \alpha_2))_L$$

$$\tilde{a}_1 - \tilde{a}_2 = ((a_1 - a_2) - (\alpha_1 + \alpha_2), a_1 - a_2, (a_1 - a_2) + (\alpha_1 + \alpha_2))_L$$

$$k\tilde{a}_1 = (ka_1 - |k|\alpha_1, ka_1, a_1 + |k|\alpha_1)_L, k \in R$$

$$\tilde{a}_1 \otimes \tilde{a}_2 = (a_1 a_2 - \alpha_1 \alpha_2, a_1 a_2, a_1 a_2 + \alpha_1 \alpha_2)_L$$

In order to define the fuzzy inequality of the problem (4.159), Hashemi et al. [23] proposed a new ranking function in which each fuzzy number is mapped into a point on the real plane $\mathbb{R} \times \mathbb{R}$ based on its mean value and standard deviation.

Definition 4.14 The possibilistic mean value and variance of the fuzzy number $\tilde{a}$, denoted by $M(\tilde{a})$ and $V(\tilde{a})$, respectively, are defined as:

$$M(\tilde{a}) = \int_0^1 \lambda \left[(\tilde{a})_\lambda^L + (\tilde{a})_\lambda^U \right] d\lambda \tag{4.161}$$

$$Var(\tilde{a}) = \frac{1}{2} \int_0^1 \lambda \left[(\tilde{a})_\lambda^U - (\tilde{a})_\lambda^L \right]^2 d\lambda \tag{4.162}$$

Also, the standard deviation of $\tilde{a}$ is defined by $SD(\tilde{a}) = \sqrt{Var(\tilde{a})}$.

The possibilistic mean value, and variance and standard deviation of the symmetric fuzzy number $\tilde{a} = (a - \alpha, a, a + \alpha)_L$ are given as follows:

$$M(\tilde{a}) = a, \ Var(\tilde{a}) = \frac{\alpha^2}{6}, \ SD(\tilde{a}) = \frac{\alpha}{\sqrt{6}} \tag{4.163}$$

Definition 4.15 A nonzero vector X is called lexicographically positive and denoted by $X >_{LG} 0$ if the first nonzero component of X is positive. Also, a lexicographically nonnegative vector is either a zero vector or else a lexicographically positive vector, denoted by $X \geq_{LG} 0$.

Definition 4.16 Consider the following function:

$$\Re : F(\mathbb{R}) \to \mathbb{R} \times \mathbb{R}$$

$$\tilde{a} \mapsto \begin{bmatrix} M(\tilde{a}) \\ -SD(\tilde{a}) \end{bmatrix}$$

The comparison on fuzzy numbers $\tilde{a}_1$ and $\tilde{a}_2$ is defined by the following lexicography ordering:

$$\tilde{a}_1 \succeq \tilde{a}_2 \Leftrightarrow \Re(\tilde{a}_1) - \Re(\tilde{a}_2) = \begin{bmatrix} M(\tilde{a}_1) - M(\tilde{a}_2) \\ -SD(\tilde{a}_1) + SD(\tilde{a}_2) \end{bmatrix} \geq_{LG} 0$$

$$\tilde{a}_1 \succ \tilde{a}_2 \Leftrightarrow \Re(\tilde{a}_1) - \Re(\tilde{a}_2) = \begin{bmatrix} M(\tilde{a}_1) - M(\tilde{a}_2) \\ -SD(\tilde{a}_1) + SD(\tilde{a}_2) \end{bmatrix} >_{LG} 0$$

$$\tilde{a}_1 \approx \tilde{a}_2 \Leftrightarrow \Re(\tilde{a}_1) - \Re(\tilde{a}_2) = \begin{bmatrix} M(\tilde{a}_1) - M(\tilde{a}_2) \\ -SD(\tilde{a}_1) + SD(\tilde{a}_2) \end{bmatrix} = \begin{bmatrix} 0 \\ 0 \end{bmatrix}$$

Theorem 4.9 *For any two symmetric fuzzy numbers* $\tilde{a}_1 = (a_1 - \alpha_1, a_1, a_1 + \alpha_1)_L$ *and* $\tilde{a}_2 = (a_2 - \alpha_2, a_2, a_2 + \alpha_2)_L$ *the following relations hold*:

i. $\Re(\tilde{a}_1 + \tilde{a}_2) = \Re(\tilde{a}_1) + \Re(\tilde{a}_2)$
ii. $\Re(\tilde{a}_1).\Re(\tilde{a}_2) = \Re(\tilde{a}_1 \otimes \tilde{a}_2) \geq_{LG} 0$

Proof According to Definition 4.13 and the relation (4.163), we have

$$\Re(\tilde{a}_1 + \tilde{a}_2) = \begin{bmatrix} a_1 + a_2 \\ -\dfrac{\alpha_1 + \alpha_2}{6} \end{bmatrix} = \begin{bmatrix} a_1 \\ -\dfrac{\alpha_1}{6} \end{bmatrix} + \begin{bmatrix} a_2 \\ -\dfrac{\alpha_2}{6} \end{bmatrix} = \Re(\tilde{a}_1) + \Re(\tilde{a}_2).$$

Also,

$$\Re(\tilde{a}_1).\Re(\tilde{a}_2) - \Re(\tilde{a}_1 \otimes \tilde{a}_2) = \begin{bmatrix} a_1 \\ -\dfrac{\alpha_1}{6} \end{bmatrix} . \begin{bmatrix} a_2 \\ -\dfrac{\alpha_2}{6} \end{bmatrix} - \begin{bmatrix} a_1 a_2 \\ -\dfrac{\alpha_1 \alpha_2}{6} \end{bmatrix}$$

$$= \begin{bmatrix} a_1 a_2 \\ \dfrac{\alpha_1 \alpha_2}{6} \end{bmatrix} - \begin{bmatrix} a_1 a_2 \\ -\dfrac{\alpha_1 \alpha_2}{6} \end{bmatrix} = \begin{bmatrix} 0 \\ \dfrac{\alpha_1 \alpha_2}{3} \end{bmatrix} \geq_{LG} 0$$

These complete the proof.   □

Let us assume that all fuzzy numbers are symmetric ones. Thus, $\tilde{a}_{ij}, \tilde{c}_j, \tilde{b}_i$ and $\tilde{x}_j$ are represented as $\tilde{a}_{ij} = (a_{ij} - \alpha_{ij}, a_{ij}, a_{ij} - \alpha_{ij})_L$, $\tilde{c}_j = (c_j - \omega_j, c_j, c_j + \omega_j)$, $\tilde{b}_i = (b_i - \beta_i, b_i, b_i + \beta_i)$ and $\tilde{x}_j = (x_j - r_j, x_j, x_j - r_j)_L$, respectively. The following Theorem shows that the constraints of the problem (4.159) can be transformed into linear constraints.

Theorem 4.10 *The constraints of the problem* (4.159) *are equivalent to the following linear system*:

$$\sum_{j=1}^{n} a_{ij} x_j + s_{i1} = b_i, \quad i = 1, 2, \ldots, m,$$

$$\sum_{j=1}^{n} \alpha_{ij} r_j + s_{i2} = \beta_i, \quad i = 1, 2, \ldots, m,$$

$$
\begin{aligned}
Ks_{i1} - s_{i2} \geq 0, && i = 1, 2, \ldots, m, \\
Kx_j - r_j \geq 0, && j = 1, 2, \ldots, m, \\
x_j, r_j \geq 0, && j = 1, 2, \ldots, n, \\
s_{i1} \geq 0, \ s_{i2} \text{ free in sign}, && i = 1, 2, \ldots, m.
\end{aligned}
\tag{4.164}
$$

where K is a very large value and is positive.

Proof Let S_i be the set of all $\tilde{x} = (\tilde{x}_1, \tilde{x}_2, \ldots, \tilde{x}_n)$ which satisfy the constraint i of the problem (4.159). Thus, $\tilde{x} \in S_i$ if and only if:

$$
\begin{bmatrix}
b_i - \displaystyle\sum_{j=1}^{n} a_{ij}x_j \\[2mm]
- \beta_i + \displaystyle\sum_{j=1}^{n} \alpha_{ij}r_j
\end{bmatrix}
\geq_{LG} 0
\tag{4.165}
$$

Clearly, Eq. (4.165) holds if either $\displaystyle\sum_{j=1}^{n} a_{ij}x_j < b_i$ or $\displaystyle\sum_{j=1}^{n} a_{ij}x_j = b_i$ and $\displaystyle\sum_{j=1}^{n} \alpha_{ij}r_j \geq \beta_i$. In other words, Eq. (4.165) holds if and only if:

$$
\begin{aligned}
\sum_{j=1}^{n} a_{ij}x_j + s_{i1} = b_i, && i = 1, 2, \ldots, m, \\
\sum_{j=1}^{n} \alpha_{ij}r_j + s_{i2} = \beta_i, && i = 1, 2, \ldots, m, \\
s_{i2} \leq 0 \text{ if } s_{i1} = 0, && i = 1, 2, \ldots, m.
\end{aligned}
\tag{4.166}
$$

The set of constraints (4.166) is equivalent to the following system:

$$
\begin{aligned}
\sum_{j=1}^{n} a_{ij}x_j + s_{i1} = b_i, && i = 1, 2, \ldots, m, \\
\sum_{j=1}^{n} \alpha_{ij}r_j + s_{i2} = \beta_i, && i = 1, 2, \ldots, m, \\
Ks_{i1} - s_{i2} \geq 0, && i = 1, 2, \ldots, m, \\
s_{i1} \geq 0, \ s_{i2} \text{ free in sign}, && i = 1, 2, \ldots, m.
\end{aligned}
\tag{4.167}
$$

Similarly, $\tilde{x}_j = (x_j - r_j, x_j, x_j - r_j)_L \succeq (0, 0, 0)_L$ if and only if

$$
Kx_j - r_j \geq 0, \ x_j, r_j \geq 0.
\tag{4.168}
$$

According to (4.167) and (4.168), the constraints of the problem (4.159) are equivalent to the linear system (4.164). $\qquad\square$

In practice, the value of K must be determined by decision makers. Note that the bigger the value of K is, the more important the mean value is, in comparison with the standard deviation of fuzzy numbers.

In order to obtain an optimal solution of the problem (4.159), in the first phase, the set of all feasible solutions that maximize the mean value of the fuzzy objective function subject to the constraints (4.164), is determined Then, in the next phase, the standard deviation of the original fuzzy objective function subject to the set obtained in the first phase; is minimized. This procedure is called the two-phase approach and is described as follows.

Phase I: Solve the following LP problem:

$$
\max z = \sum_{j=1}^{n} c_j x_j
$$

$$
\text{s.t.} \quad \sum_{j=1}^{n} a_{ij} x_j \le b_i, \ i = 1, 2, \ldots, m, \tag{4.169}
$$

$$
x_j \ge 0, \qquad j = 1, 2, \ldots, n.
$$

Clearly, if the above LP problem is infeasible, then the original problem (4.159) is also infeasible. Otherwise, let the alternative optimal solutions be represented by

$$
X_t^* = (x_{t1}^*, x_{t2}^*, \ldots, x_{tm}^*) \ (t = 1, 2, \ldots, T). \ \text{Set} \ I_t = \left\{ i \left| \sum_{j=1}^{n} a_{ij} x_j^* = b_i \right. \right\} \ \text{for} \ 1 \le t \le T.
$$

Then, solve the following LP problem:

$$
\max z = \sum_{j=1}^{n} \omega_j r_{tj}
$$

$$
\text{s.t.} \quad \sum_{j=1}^{n} \alpha_{ij} r_{tj} \ge \beta_i, \ i \in I_t \tag{4.170}
$$

$$
r_{tj} \ge 0, \ j = 1, 2, \ldots ., n, \ r_{tj} = 0 \ \text{if} \ x_{tj}^* = 0
$$

Let $R_t^* = (r_{t1}^*, r_{t2}^*, \ldots, r_{tm}^*)$ be an optimal solution of the LP problem (4.170). So, $(X_t^*, R_t^*) = (x_{t1}^*, x_{t2}^*, \ldots, x_{tm}^*, r_{t1}^*, r_{t2}^*, \ldots, r_{tm}^*) \ (t = 1, 2, \ldots, T)$ are the solutions of the problem (4.159) that maximize the possibilistic mean value of the fuzzy objective function. It is obvious that any convex combination of these solutions is also a feasible solution that maximizes the possibilistic mean value of the fuzzy objective function.

Phase II: Solve the following LP problem:

$$
\min z = \sum_{j=1}^{n} \omega_j \sum_{t=1}^{T} \lambda_t r_{tj}^* = \sum_{t=1}^{T} \lambda_t \sum_{j=1}^{n} \omega_j r_{tj}^*
$$

$$
\text{s.t.} \quad \sum_{t=1}^{T} \lambda_t = 1, \tag{4.171}
$$

$$
\lambda_t \ge 0, \quad t = 1, 2, \ldots, T.
$$

Finding an optimal solution of the problem (4.171) is a very easy task. We simply find the minimum $\sum_{j=1}^{n} \omega_j r_{tj}^*$ over index t, say $\sum_{j=1}^{n} \omega_j r_{pj}^*$, then set $\lambda_p = 1$, $\lambda_j = 0 \ (j \neq p)$.

Example 4.13 [23] Consider the following Type-7 LPP with fuzzy parameters:

$$\max z = (1.9, 2, 0.1)_L \otimes \tilde{x}_1 + (4.7, 5, 5.3)_L \otimes \tilde{x}_2$$
$$s.t. \quad (3.7, 4, 4.3)_L \otimes \tilde{x}_1 + (9.4, 10, 10.6)_L \otimes \tilde{x}_2 \preceq (24, 25, 26)_L,$$
$$(2.8, 3, 3.2)_L \otimes \tilde{x}_1 + (3.8, 4, 4.2)_L \otimes \tilde{x}_2 \preceq (14.2, 15, 15.8)_L,$$
$$\tilde{x}_1 = (x_1 - r_1, x_1, x_1 - r_1) \succeq (0, 0, 0)_L, \tilde{x}_2 = (x_2 - r_2, x_2, x_2 - r_2) \succeq (0, 0, 0)_L.$$

$$(4.172)$$

In phase I, the set of all feasible solutions that maximize the mean value of the objective function is obtained through the following LP problem with regard to the problem (4.169):

$$\max z = 2x_1 + 5x_2$$
$$s.t. \ 4x_1 + 10x_2 \leq 25,$$
$$3x_1 + 4x_2 \leq 15,$$
$$x_1, \ x_2 \geq 0.$$

$$(4.173)$$

The optimal solutions of the problem (4.173) are $X_1^* = (3.5714, 1.0714)$ and $X_2^* = (0, 2.5)$. In this case we have $I_1 = \{1, 2\}$ and $I_2 = \{1\}$.

Thus, we solve the following LP problem for $X_1^* = (3.5714, 1.0714)$ with regard to the problem (4.170):

$$Min \, z = 0.1r_{11} + 0.3r_{12}$$
$$s.t. \ \ 0.3r_{11} + 0.6r_{12} \geq 1,$$
$$0.2r_{11} + 0.2r_{12} \geq 0.8,$$
$$r_{11}, \ r_{12} \geq 0.$$

$$(4.174)$$

The optimal solution of the problem (4.174) is $R_1^* = (4, 0)$.

Similarly, we solve the following LP problem for $X_2^* = (0, 2.5)$ with regard to the problem (4.170):

$$Min \, z = 0.1r_{11} + 0.3r_{12}$$
$$s.t. \ 0.3r_{11} + 0.6r_{12} \geq 1,$$
$$r_{11} = 0, \ r_{12} \geq 0.$$

$$(4.175)$$

The optimal solution of the problem (4.175) is $R_2^* = (0, 0.1666)$.

Therefore, we obtain $\tilde{X}_1^*$ and $\tilde{X}_2^*$ as the solutions that maximize the possibilistic mean value of the fuzzy objective function of the problem (4.172):

$$\tilde{X}_1^* = \big((-0.4286, 3.5714, 7.5714)_L, (1.0714, 1.0714, 1.0714)_L\big)$$
$$\tilde{X}_2^* = \big((0, 0, 0)_L, (0.834, 2.5, 4.166)_L\big) \tag{4.176}$$

In phase II, the following LP problem is solved with regard to the problem (4.171) and the solution (4.176):

$$Min\, z = 0.4\lambda_1 + 0.5\lambda_2$$
$$s.t.\ \lambda_1 + \lambda_2 = 1, \tag{4.177}$$
$$\lambda_1 + \lambda_2 \geq 0.$$

The optimal solution of the problem (4.177) is $\lambda_1 = 1, \lambda_2 = 0$. Thus, $\tilde{X}_1^* = \big((-0.4286, 3.5714, 7.5714)_L, (1.0714, 1.0714, 1.0714)_L\big)$ is the optimal solution of the problem (4.172) with the fuzzy objective value $\tilde{z}^* = (12.1, 12.5, 12.9)_L$.

4.9.2 Hosseinzadeh Lotfi et al.'s Approach

Hosseinzadeh Lotfi et al. [24] proposed a method for solving the fully FLP problem (4.159) with equality constraints. This method can be used to find the non-negative fuzzy optimal solution of such fully FLP problems with equality constraints in which all the parameters and variables are represented by non-negative triangular fuzzy numbers. According to this method all the coefficients are approximated into their nearest symmetric fuzzy numbers.

Definition 4.17 Let $\tilde{a}$ be a general fuzzy number and $[\tilde{a}^L(r), \tilde{a}^U(r)]$ be its parametric form. The nearest symmetric triangular fuzzy number of $\tilde{a}$ is given by $NS(\tilde{a}) = (a - \sigma, a, a + \sigma)$, where

$$a = \frac{1}{2} \int_0^1 \big(\tilde{a}^L(r) + \tilde{a}^U(r)\big) dr \tag{4.178}$$

$$\sigma = \frac{3}{2} \int_0^1 \big(\tilde{a}^U(r) - \tilde{a}^L(r)\big)(1 - r) dr \tag{4.179}$$

The parametric form of the triangular fuzzy number $\tilde{a} = (a_1, a_2, a_3)$ is given by $[\tilde{a}^L(r), \tilde{a}^U(r)] = [a_1 + (a_2 - a_1)r, a_3 - (a_3 - a_2)r]$. Therefore, according to the Eqs. (4.178) and (4.179), its nearest symmetric triangular fuzzy number is given by $NS(\tilde{a}) = (a - \sigma, a, a + \sigma)$, where

$$a = \frac{a_1 + 2a_2 + a_3}{4} \tag{4.180}$$

$$\sigma = \frac{a_3 - a_1}{2} \tag{4.181}$$

Assuming all the parameters and variables are non-negative triangular fuzzy numbers, the problem (4.159) with equality constraints can be rewritten as follows:

$$\max \tilde{z} \approx \sum_{j=1}^{n} (c_{1,j}, c_{2,j}, c_{3,j}) \otimes (x_{1,j}, x_{2,j}, x_{3,j})$$

$$s.t. \sum_{j=1}^{n} (a_{1,ij}, a_{2,ij}, a_{3,ij}) \otimes (x_{1,j}, x_{2,j}, x_{3,j}) \approx (b_{1,i}, b_{2,i}, b_{3,i}), \ i = 1, 2, \ldots, m,$$

$(x_{1,j}, x_{2,j}, x_{3,j})$ is a non negative trinagular fuzzy number, $j = 1, 2, \ldots, n.$

$$\tag{4.182}$$

By considering the parametric forms of the fuzzy parameters and variables, the problem (4.182) can be converted into the following problem:

$$\max \sum_{j=1}^{n} \left[\tilde{c}_j^L(r), \tilde{c}_j^U(r) \right] \left[\tilde{x}_j^L(r), \tilde{x}_j^U(r) \right]$$

$$s.t. \sum_{j=1}^{n} \left[\tilde{a}_{ij}^L(r), \tilde{a}_{ij}^U(r) \right] \left[\tilde{x}_j^L(r), \tilde{x}_j^U(r) \right] = \left[\tilde{b}_i^L(r), \tilde{b}_i^U(r) \right], \ i = 1, 2, \ldots, m, \tag{4.183}$$

$$x_{1,j} \geq 0, x_{2,j} \geq x_{1,j}, x_{3,j} \geq x_{2,j} \, j = 1, 2, \ldots, n.$$

where

$$\begin{aligned}
\tilde{c}_j^L(r) &= c_{1,j} + (c_{2,j} - c_{1,j})r, \ \tilde{c}_j^U(r) = c_{3,j} - (c_{3,j} - c_{2,j})r \\
\tilde{x}_j^L(r) &= x_{1,j} + (x_{2,j} - x_{1,j})r, \ \tilde{x}_j^U(r) = x_{3,j} - (x_{3,j} - x_{2,j})r \\
\tilde{a}_{ij}^L(r) &= a_{1,ij} + (a_{2,ij} - a_{1,ij})r, \ \tilde{a}_{ij}^U(r) = a_{3,ij} - (a_{3,ij} - a_{2,ij})r \\
\tilde{b}_i^L(r) &= b_{1,i} + (b_{2,i} - b_{1,i})r, \ \tilde{b}_i^U(r) = b_{3,i} - (b_{3,i} - b_{2,i})r
\end{aligned} \tag{4.184}$$

Using the arithmetic operations of intervals, the problem (4.183) can be reformulated as follows:

$$\max \sum_{j=1}^{n} \left[\tilde{c}_j^L(r)\tilde{x}_j^L(r), \tilde{c}_j^U(r)\tilde{x}_j^U(r) \right]$$

$$s.t. \sum_{j=1}^{n} \left[\tilde{a}_{ij}^L(r)\tilde{x}_j^L(r), \tilde{a}_{ij}^U(r)\tilde{x}_j^U(r) \right] = \left[\tilde{b}_i^L(r), \tilde{b}_i^U(r) \right], \ i = 1, 2, \ldots, m, \tag{4.185}$$

$$x_{1,j} \geq 0, x_{2,j} \geq x_{1,j}, x_{3,j} \geq x_{2,j} \ j = 1, 2, \ldots, n.$$

By approximating all the coefficients of the problem (4.184) into their nearest symmetric fuzzy numbers with regard to the Eqs. (4.180) and (4.181), the problem (4.185) is reformulated as follows:

$$\max \left(\sum_{j=1}^{n} \left((cx)_j - \sigma_{(cx)_j} \right), \sum_{j=1}^{n} (cx)_j, \sum_{j=1}^{n} \left((cx)_j + \sigma_{(cx)_j} \right) \right)$$

$$s.t. \left(\sum_{j=1}^{n} \left((a_i x)_j - \sigma_{(a_i x)_j} \right), \sum_{j=1}^{n} (a_i x)_j, \sum_{j=1}^{n} \left((a_i x)_j + \sigma_{(a_i x)_j} \right) \right) = (b_i - \sigma_{b_i}, b_i, b_i + \sigma_{b_i})$$

$$x_{1,j} \geq 0, x_{2,j} \geq x_{1,j}, x_{3,j} \geq x_{2,j} \ j = 1, 2, \ldots, n. \tag{4.186}$$

where

$$(cx)_j = c_{2j}x_{2,j} + \frac{1}{4}c_{2j}(x_{3,j} - x_{2,j}) + \frac{1}{4}(c_{3,j} - c_{2,j})x_{2,j} + \frac{1}{6}(c_{3,j} - c_{2,j})(x_{3,j} - x_{2,j})$$

$$- \frac{1}{4}c_{2,j}(x_{2,j} - x_{1,j}) - \frac{1}{4}(c_{2,j} - c_{1,j})x_{2,j} + \frac{1}{6}(c_{2,j} - c_{1,j})(x_{2,j} - x_{1,j}),$$

$$\sigma_{(cx)_j} = \frac{1}{2}c_{2,j}(x_{3,j} - x_{2,j}) + \frac{1}{2}(c_{3,j} - c_{2,j})x_{2,j} + \frac{3}{8}(c_{3,j} - c_{2,j})(x_{3,j} - x_{2,j})$$

$$+ \frac{1}{2}c_{2,j}(x_{2,j} - x_{1,j}) + \frac{1}{2}(c_{2,j} - c_{1,j})x_{2,j} - \frac{3}{8}(c_{2,j} - c_{1,j})(x_{2,j} - x_{1,j}),$$

$$(a_i x)_j = a_{2,ij}x_{2,j} + \frac{1}{4}a_{ij}(x_{3,j} - x_{2,j}) + \frac{1}{4}(a_{3,ij} - a_{2,ij})x_{2,j} + \frac{1}{6}(a_{3,ij} - a_{2,ij})(x_{3,j} - x_{2,j})$$

$$- \frac{1}{4}a_{ij}(x_{2,j} - x_{1,j}) - \frac{1}{4}(a_{2,ij} - a_{1,ij})x_{2,j} + \frac{1}{6}(a_{2,ij} - a_{1,ij})(x_{2,j} - x_{1,j}),$$

$$\sigma_{(a_i x)_j} = \frac{1}{2}a_{2,ij}(x_{3,j} - x_{2,j}) + \frac{1}{2}(a_{3,ij} - a_{2,ij})x_{2,j} + \frac{3}{8}(a_{3,ij} - a_{2,ij})(x_{3,j} - x_{2,j})$$

$$+ \frac{1}{2}a_{2,ij}(x_{2,j} - x_{1,j}) + \frac{1}{2}(a_{2,ij} - a_{1,ij})x_{2,j} - \frac{3}{8}(a_{2,ij} - a_{1,ij})(x_{2,j} - x_{1,j})$$

$$b_i = \frac{b_{1,i} + 2b_{2,i} + b_{3,i}}{4}, \sigma_{b_i} = \frac{b_{3,i} - b_{1,i}}{2}. \tag{4.187}$$

Hosseinzadeh Lotfi et al. [24] used the following definition for maximization of the problem (4.187) involving symmetric triangular fuzzy numbers.

Definition   4.18 Let $NS(\tilde{a}_1) = (a_1 - \sigma_1, a_1, a_1 + \sigma_1)$ and $NS(\tilde{a}_2) = (a_2 - \sigma_2, a_2, a_2 + \sigma_2)$ be two symmetric triangular fuzzy numbers. We say $\tilde{a}_1 \prec \tilde{a}_2$ if either $a_1 < a_2$ or $a_1 = a_2$ and $\sigma_1 > \sigma_2$. In the case of equality, we have $\tilde{a}_1 = \tilde{a}_2$ if $a_1 = a_2$ and $\sigma_1 = \sigma_2$. Finally, $\tilde{a}_1 \preceq \tilde{a}_2$ if either $a_1 < a_2$ or $a_1 = a_2$ and $\sigma_1 \geq \sigma_2$.

Now according to Definition 4.18 for ranking the fuzzy numbers, for maximizing the objective function of the problem (4.186), in the first step we must solve a maximization problem for the core and in the second step a minimization problem for the spread.

According to the above discussion, in the first step the following LP problem is solved to maximize the core:

$$\max z = \sum_{j=1}^{n} (cx)_j$$

$$s.t. \quad \sum_{j=1}^{n} (a_i x)_j = b_i, \ i = 1, 2, \ldots, m,$$

$$\sum_{j=1}^{n} \sigma_{(a_i x)_j} = \sigma_{b_i}, \ i = 1, 2, \ldots, m,$$

$$x_{1,j} \geq 0, x_{2,j} \geq x_{1,j}, x_{3,j} \geq x_{2,j} \, j = 1, 2, \ldots, n.$$

(4.188)

where $(cx)_j, \sigma_{(a_i x)_j}, b_i$ and σ_{b_i} are given by Eq. (4.187).

If the crisp LP problem (4.188) has a unique optimal solution $x_{1,j}^*, x_{2,j}^*, x_{3,j}^* \ (j = 1, 2, \ldots, n)$, then $\tilde{x}_j^* = (x_{1,j}^*, x_{2,j}^*, x_{3,j}^*) \ (j = 1, 2, \ldots, n)$ will be the fuzzy optimal solution of (4.182). If it has alternative optimal solutions then in the second step the following problem is solved to minimize the spread, where z^* is the optimal objective function value of the problem (4.188):

$$\max \sigma = \sum_{j=1}^{n} \sigma_{(cx)_j}$$

$$s.t. \quad \sum_{j=1}^{n} (cx)_j = z^*$$

$$\sum_{j=1}^{n} (a_i x)_j = b_i, \quad i = 1, 2, \ldots, m,$$

$$\sum_{j=1}^{n} \sigma_{(a_i x)_j} = \sigma_{b_i}, \quad i = 1, 2, \ldots, m,$$

$$x_{1,j} \geq 0, x_{2,j} \geq x_{1,j}, x_{3,j} \geq x_{2,j} \, j = 1, 2, \ldots, n.$$

(4.189)

Example 4.14 [24] Consider the following Type-7 LPP with fuzzy parameters:

$$\max \tilde{z} = (0, 1, 4) \otimes \tilde{x}_1 + (2, 4, 5) \otimes \tilde{x}_2$$
$$s.t. \quad (2, 3, 7) \otimes \tilde{x}_1 + (2, 4, 5) \otimes \tilde{x}_2 = (6, 18, 46),$$
$$(0, 2, 4) \otimes \tilde{x}_1 + (3, 5, 8) \otimes \tilde{x}_2 = (6, 19, 52),$$
$$\tilde{x}_1 = (x_{1,1}, x_{2,1}, x_{3,1}) \succeq (0, 0, 0), \tilde{x}_2 = (x_{1,2}, x_{2,2}, x_{3,2}) \succeq (0, 0, 0).$$

(4.190)

In the first step, to maximize the core of the objective function, the following LP problem is solved with regard to the problem (4.188):

$$\max z = 0.083x_{1,1} + 0.666x_{2,1} + 0.75x_{3,1} + 0.666x_{1,2} + 1.916x_{2,2} + 1.166x_{3,2}$$
$$s.t. \quad 0.583x_{1,1} + 1.750_{2,1} + 1.416x_{3,1} + 0.666x_{1,2} + 1.916x_{2,2} + 1.166x_{3,2} = 22,$$
$$0.166x_{1,1} + x_{2,1} + 0.833x_{3,1} + 0.916x_{1,2} + 2.583x_{2,2} + 1.750x_{3,2} = 24,$$
$$-1.125x_{1,1} + 0.625x_{2,1} + 3.000x_{3,1} - 1.250x_{1,2} + 0.375x_{2,2} + 2.375x_{3,2} = 20,$$
$$-0.250x_{1,1} + 0.500x_{2,1} + 1.750x_{3,1} - 1.750x_{1,2} + 0.625x_{2,2} + 3.625x_{3,2} = 23,$$
$$x_{1,1} \geq 0, x_{2,1} - x_{1,1} \geq 0, x_{3,1} - x_{2,1} \geq 0,$$
$$x_{1,2} \geq 0, x_{2,2} - x_{1,2} \geq 0, x_{3,2} - x_{2,2} \geq 0.$$

(4.191)

The optimal value of the problem is $z^* = 17.48811$. Thus, in the second step the following problem is solved in order to minimize the spread of the fuzzy objective function with regard to the problem (4.189):

$$\max z = -0.125x_{1,1} + 0.5x_{2,1} + 1.625x_{3,1} - 1.25x_{1,2} + 0.375x_{2,2} + 2.375x_{3,2}$$
$$s.t. \quad 0.583x_{1,1} + 1.750_{2,1} + 1.416x_{3,1} + 0.666x_{1,2} + 1.916x_{2,2} + 1.166x_{3,2} = 22,$$
$$0.166x_{1,1} + x_{2,1} + 0.833x_{3,1} + 0.916x_{1,2} + 2.583x_{2,2} + 1.750x_{3,2} = 24,$$
$$-1.125x_{1,1} + 0.625x_{2,1} + 3.000x_{3,1} - 1.250x_{1,2} + 0.375x_{2,2} + 2.375x_{3,2} = 20,$$
$$-0.250x_{1,1} + 0.500x_{2,1} + 1.750x_{3,1} - 1.750x_{1,2} + 0.625x_{2,2} + 3.625x_{3,2} = 23,$$
$$0.083x_{1,1} + 0.666x_{2,1} + 0.75x_{3,1} + 0.666x_{1,2} + 1.916x_{2,2} + 1.166x_{3,2} = 17.48811$$
$$x_{1,1} \geq 0, x_{2,1} - x_{1,1} \geq 0, x_{3,1} - x_{2,1} \geq 0,$$
$$x_{1,2} \geq 0, x_{2,2} - x_{1,2} \geq 0, x_{3,2} - x_{2,2} \geq 0.$$

(4.192)

The optimal solution of the problem (4.192) is as follows:

$$x_{1,1}^* = 1.456, x_{2,1}^* = 1.456, x_{3,1}^* = 3.311,$$
$$x_{1,2}^* = 0.879, x_{2,2}^* = 4.324, x_{3,1}^* = 4.324.$$

(4.193)

Therefore, the fuzzy optimal solution of the problem (4.190) is given as follows:

$$\tilde{x}_1^* = (x_{1,1}^*, x_{2,1}^*, x_{3,1}^*) = (1.456, 1.456, 3.311),$$
$$\tilde{x}_2^* = (x_{1,2}^*, x_{2,2}^*, x_{3,2}^*) = (0.879, 4.324, 4.324).$$

(4.194)

The main shortcoming of Hosseinzadeh Lotfi et al.'s method is that the obtained solutions are approximated and do not satisfy the constraints exactly. For example, the obtained fuzzy optimal solution given in (4.193) does not satisfy exactly the constraints of the problem (4.190). Moreover, this method cannot be used for solving those fuzzy FLP problems with equality constraints in which the decision variables are represented by non-negative triangular or trapezoidal fuzzy number and some or all of the coefficients are either represented by unrestricted triangular or trapezoidal fuzzy numbers. Thus, in the next subsection, we explore the method by proposed Kumar et al. [26] to overcome the above-mentioned shortcomings.

4.9.3 Kumar et al.'s Approach

Kumar et al. [26] proposed a method for solving the fully FLP problem (4.159) with equality constraints in which the decision variables are represented by non-negative triangular or trapezoidal fuzzy number and some or all of the coefficients are either represented by unrestricted triangular or trapezoidal fuzzy numbers. Consider the following fully FLP problem with equality constraints:

$$\max \tilde{z} \approx \sum_{j=1}^{n} \tilde{c}_j \otimes \tilde{x}_j$$

$$s.t. \quad \sum_{j=1}^{n} \tilde{a}_{ij} \otimes \tilde{x}_j \approx \tilde{b}_i, \quad i = 1, 2, \ldots, m,$$

$$\tilde{x}_j \succeq \tilde{0}, \quad j = 1, 2, \ldots, n.$$

(4.195)

Let us assume that all fuzzy numbers are trapezoidal. Hence, the problem (4.195) can be reformulated as follows:

$$\max \tilde{z} \approx \sum_{j=1}^{n} (c_{1,j}, c_{2,j}, c_{3,j}, c_{4,j}) \otimes (x_{1,j}, x_{2,j}, x_{3,j}, x_{4,j})$$

$$s.t. \sum_{j=1}^{n} (a_{1,ij}, a_{2,ij}, a_{3,ij}, a_{4,ij}) \otimes (x_{1,j}, x_{2,j}, x_{3,j}, x_{4,j}) \approx (b_{1,i}, b_{2,i}, b_{3,i}, b_{4,i}), \quad i = 1, 2, \ldots, m,$$

$$(x_{1,j}, x_{2,j}, x_{3,j}, x_{4,j}) \text{ is a non negative trapezoidal fuzzy number, } j = 1, 2, \ldots, n.$$

(4.196)

Assuming $(a_{1,ij}, a_{2,ij}, a_{3,ij}, a_{4,ij}) \otimes (x_{1,j}, x_{2,j}, x_{3,j}, x_{4,j}) = (m_{1,ij}, m_{2,ij}, m_{3,ij}, , m_{4,ij})$
and using the ranking function given in Remark 1.15 for the objective function, the
FLP (4.196) can be rewritten as follows:

$$\max \Re(\tilde{z}) = \Re \left(\sum_{j=1}^{n} (c_{1,j}, c_{2,j}, c_{3,j}, c_{4,j}) \otimes (x_{1,j}, x_{2,j}, x_{3,j}, x_{4,j}) \right)$$

$$s.t \sum_{j=1}^{n} (m_{1,ij}, m_{2,ij}, m_{3,ij}, m_{4,ij}) \approx (b_{1,i}, b_{2,i}, b_{3,i}, b_{4,i}), \ i = 1, 2, \ldots, m,$$

$(x_{1,j}, x_{2,j}, x_{3,j}, x_{4,j})$ is a non negative trapezoidal fuzzy number, $j = 1, 2, \ldots, n$.

$$(4.197)$$

Using Theorem 1.10 and Definitions 1.31 and 1.32, the FLP problem (4.197)
becomes:

$$\max \Re(\tilde{z}) = \Re \left(\sum_{j=1}^{n} (c_{1,j}, c_{2,j}, c_{3,j}, c_{4,j}) \otimes (x_{1,j}, x_{2,j}, x_{3,j}, x_{4,j}) \right)$$

$$s.t. \sum_{j=1}^{n} m_{1,ij} = b_{1,i}, \ i = 1, 2, \ldots, m,$$

$$\sum_{j=1}^{n} m_{2,ij} = b_{2,i}, \ i = 1, 2, \ldots, m,$$

$$\sum_{j=1}^{n} m_{3,ij} = b_{3,i}, \ i = 1, 2, \ldots, m, \qquad (4.198)$$

$$\sum_{j=1}^{n} m_{4,ij} = b_{4,i}, \ i = 1, 2, \ldots, m,$$

$$x_{1,j} \geq 0, x_{2,j} - x_{1,j} \geq 0, x_{3,j} - x_{2,j} \geq 0, x_{4,j} - x_{3,j} \geq 0, \ j = 1, 2, \ldots, n.$$

The optimal solution of the crisp LP problem (4.198) can be considered as the
optimal solution of the problem (4.195).

Example 4.15 [26] Consider the following Type-7 LPP with fuzzy parameters:

$$\max z = (1, 6, 9, 12) \otimes (x_{1,1}, x_{2,1}, x_{3,1}, x_{4,1}) + (2, 3, 8, 9) \otimes (x_{1,2}, x_{2,2}, x_{3,2}, x_{4,2})$$

$$s.t. \ (2, 3, 4, 5) (x_{1,1}, x_{2,1}, x_{3,1}, x_{4,1}) + (1, 2, 3, 4) (x_{1,2}, x_{2,2}, x_{3,2}, x_{4,2}) = (6, 16, 30, 48),$$

$$(-1, 1, 2, 3) (x_{1,1}, x_{2,1}, x_{3,1}, x_{4,1}) + (1, 3, 4, 6) (x_{1,2}, x_{2,2}, x_{3,2}, x_{4,2}) = (0, 17, 30, 54)$$

$$(x_{1,1}, x_{2,1}, x_{3,1}, x_{4,1}) \text{ and } (x_{1,2}, x_{2,2}, x_{3,2}, x_{4,2}) \text{ are non negative fuzzy numbers.}$$

$$(4.199)$$

This problem is converted into the following crisp LP problem with regard to the
problem (4.198):

$$\max \Re(\tilde{z}) = \frac{1}{4}\left[(x_{1,1} + 2x_{1,2} + 6x_{2,1} + 3x_{2,2} + 9x_{3,1} + 8x_{3,2} + 12x_{4,1} + 9x_{4,2})\right]$$

$$\begin{aligned}
s.t. \quad & 2x_{1,1} + x_{1,2} = 6, \\
& -x_{4,1} + x_{1,2} = 0, \\
& 3x_{2,1} + 2x_{2,2} = 16, \\
& x_{2,1} + 3x_{2,2} = 17, \\
& 4x_{3,1} + 3x_{3,2} = 30, \\
& 2x_{3,1} + 4x_{3,2} = 30, \\
& 5x_{4,1} + 4x_{4,2} = 48, \\
& 3x_{4,1} + 6x_{4,2} = 54, \\
& x_{1,1} \geq 0, x_{2,1} - x_{1,1} \geq 0, x_{3,1} - x_{2,1} \geq 0, x_{4,1} - x_{3,1} \geq 0, \\
& x_{1,2} \geq 0, x_{2,2} - x_{1,2} \geq 0, x_{3,2} - x_{2,2} \geq 0, x_{4,2} - x_{3,2} \geq 0.
\end{aligned}$$

$$\text{(4.200)}$$

The optimal solution of the crisp LP problem (4.200) is as follows:

$$x_{1,1}^* = 1, x_{2,1}^* = 2, x_{3,1}^* = 3, x_{4,1}^* = 4, x_{1,2}^* = 4, x_{2,2}^* = 5, x_{3,2}^* = 6, x_{4,2}^* = 7 \quad \text{(4.201)}$$

Hence, the fuzzy optimal solution of the fully FLP problem (4.199) is

$$\begin{aligned}
\tilde{x}^* = (\tilde{x}_1^*, \tilde{x}_2^*) &= \left(\left(x_{1,1}^*, x_{2,1}^*, x_{3,1}^*, x_{4,1}^*\right), \left(x_{1,2}^*, x_{2,2}^*, x_{3,2}^*, x_{4,2}^*\right)\right) \\
&= ((1,2,3,4),(4,5,6,7))
\end{aligned}$$

$$\text{(4.202)}$$

Putting the fuzzy optimal solution (4.202) in the objective function of the problem (4.199) gives $\tilde{z}^* = (9, 27, 75, 111)$.

In contrast to the approaches proposed in [24], Kumar et al.'s method [26] gives the fuzzy optimal solution. In contrast to the simplex based approach [34, 35, 38–41], Kumar et al.'s method [26] gives non-negative fuzzy optimal solutions. In contrast to the method proposed by Hashemi et al. [23], not only can this method be used for solving FLP problems involving symmetric triangular fuzzy numbers, but it can also be used for solving FLP problems involving asymmetric triangular and trapezoidal fuzzy numbers. Finally, in contrast to the method proposed by Hosseinzaeh Lotfi et al. [24], according to this method, the constraints of the fuzzy problem are satisfied exactly by the fuzzy optimal solution.

Example 4.16 Again consider the Type-7 LPP with fuzzy parameters (4.190) given in Example 4.14.

This problem is converted into the following crisp LP problem with regard to the problem (4.197):

$$\max \Re(\tilde{z}) = \frac{1}{4}\left[(2x_{2,1} + 4x_{3,1} + 2x_{1,2} + 8x_{2,2} + 5x_{3,2})\right]$$

$$
\begin{aligned}
s.t. \quad & 2x_{1,1} + 2x_{1,2} = 6, \\
& 3x_{2,1} + 4x_{2,2} = 18, \\
& 7x_{3,1} + 5x_{3,2} = 46, \\
& 3x_{1,2} = 6, \\
& 2x_{2,1} + 5x_{2,2} = 19, \\
& 4x_{3,1} + 8x_{3,2} = 52, \\
& x_{1,1} \geq 0, x_{2,1} - x_{1,1} \geq 0, x_{3,1} - x_{2,1} \geq 0, \\
& x_{1,2} \geq 0, x_{2,2} - x_{1,2} \geq 0, x_{3,2} - x_{2,2} \geq 0.
\end{aligned}
\tag{4.203}
$$

The optimal solution of the crisp LP problem (4.203) is as follows:

$$x_{1,1}^* = 1, x_{2,1}^* = 2, x_{3,1}^* = 3, x_{1,2}^* = 2, x_{2,2}^* = 3, x_{3,2}^* = 5 \tag{4.204}$$

Hence, the fuzzy optimal solution of the fully FLP problem (4.190) is

$$\tilde{x}^* = (\tilde{x}_1^*, \tilde{x}_2^*) = \left(\left(x_{1,1}^*, x_{2,1}^*, x_{3,1}^*\right), \left(x_{1,2}^*, x_{2,2}^*, x_{3,2}^*\right)\right) = ((1, 2, 3), (2, 3, 5)) \tag{4.205}$$

As we see the fuzzy optimal solution (4.205) obtained by using Kumar et al.'s method, exactly satisfies the constraints of the FLP problem (4.190).

Remark 4.1 Kumar and Kaur [42] proposed a similar approach for solving a general form of Type-7 LPP with fuzzy parameters (4.159) with both fuzzy equality and inequality constraints. In order to define the fuzzy inequality constraints, they have used the linear ranking function given in Remark 1.15 and for the fuzzy equality constraints have used the same method discussed in this subsection. To illustrate this approach we solve the following fully FLP problem.

Example 4.18 Consider the following Type-7 LPP with fuzzy parameters:

$$\max \tilde{z} = (1, 6, 9, 12) \otimes (x_{1,1}, x_{2,1}, x_{3,1}, x_{4,1}) + (2, 3, 8, 9) \otimes (x_{1,2}, x_{2,2}, x_{3,2}, x_{4,2})$$

$$
\begin{aligned}
s.t. \quad & (2, 3, 4, 5) \otimes (x_{1,1}, x_{2,1}, x_{3,1}, x_{4,1}) + (1, 2, 3, 4) \otimes (x_{1,2}, x_{2,2}, x_{3,2}, x_{4,2}) = (6, 16, 30, 48), \\
& (0, 1, 2, 3) \otimes (x_{1,1}, x_{2,1}, x_{3,1}, x_{4,1}) + (1, 2, 3, 4) \otimes (x_{1,2}, x_{2,2}, x_{3,2}, x_{4,2}) \succeq (4, 7, 8, 11), \\
& (1, 2, 3, 4) \otimes (x_{1,1}, x_{2,1}, x_{3,1}, x_{4,1}) + (2, 3, 4, 5) \otimes (x_{1,2}, x_{2,2}, x_{3,2}, x_{4,2}) \preceq (5, 30, 40, 45), \\
& (x_{1,1}, x_{2,1}, x_{3,1}, x_{4,1}) \text{ and } (x_{1,2}, x_{2,2}, x_{3,2}, x_{4,2}) \text{ are non negative fuzzy numbers.}
\end{aligned}
$$

$$\tag{4.206}$$

This problem is reformulated as follows with regard to Remark 4.1:

$$\max \Re(\tilde{z}) = \Re\big((1,6,9,12) \otimes (x_{1,1},x_{2,1},x_{3,1},x_{4,1}) + (2,3,8,9) \otimes (x_{1,2},x_{2,2},x_{3,2},x_{4,2})\big)$$
$$s.t. \ (2,3,4,5) \otimes (x_{1,1},x_{2,1},x_{3,1},x_{4,1}) + (1,2,3,4) \otimes (x_{1,2},x_{2,2},x_{3,2},x_{4,2}) = (6,16,30,48),$$
$$\Re\big((0,1,2,3) \otimes (x_{1,1},x_{2,1},x_{3,1},x_{4,1}) + (1,2,3,4) \otimes (x_{1,2},x_{2,2},x_{3,2},x_{4,2})\big) \ge \Re(4,7,8,11),$$
$$\Re\big((1,2,3,4) \otimes (x_{1,1},x_{2,1},x_{3,1},x_{4,1}) + (2,3,4,5) \otimes (x_{1,2},x_{2,2},x_{3,2},x_{4,2})\big) \le \Re(5,30,40,45),$$
$$(x_{1,1},x_{2,1},x_{3,1},x_{4,1}) \text{ and } (x_{1,2},x_{2,2},x_{3,2},x_{4,2}) \text{ are non negative fuzzy numbers.}$$

$$(4.207)$$

This problem is converted into the following crisp LP problem with regard to the linear ranking function given in Remark 1.15 and fuzzy arithmetic operations:

$$\max \Re(\tilde{z}) = \frac{1}{4}\big[(x_{1,1} + 2x_{1,2} + 6x_{2,1} + 3x_{2,2} + 9x_{3,1} + 8x_{3,2} + 12x_{4,1} + 9x_{4,2})\big]$$

$$s.t. \quad 2x_{1,1} + x_{1,2} = 6,$$
$$3x_{2,1} + 2x_{2,2} = 16,$$
$$4x_{3,1} + 3x_{3,2} = 30,$$
$$5x_{4,1} + 4x_{4,2} = 48,$$
$$x_{1,2} + x_{2,1} + 2x_{2,2} + 2x_{3,1} + 3x_{3,2} + 3x_{4,1} + 4x_{4,2} \ge 30,$$
$$x_{1,1} + 2x_{1,2} + 2x_{2,1} + 3x_{2,2} + 3x_{3,1} + 4x_{3,2} + 4x_{4,1} + 5x_{4,2} \le 120,$$
$$x_{1,1} \ge 0, x_{2,1} - x_{1,1} \ge 0, x_{3,1} - x_{2,1} \ge 0, x_{4,1} - x_{3,1} \ge 0,$$
$$x_{1,2} \ge 0, x_{2,2} - x_{1,2} \ge 0, x_{3,2} - x_{2,2} \ge 0, x_{4,2} - x_{3,2} \ge 0.$$

$$(4.208)$$

The optimal solution of the crisp LP problem (4.208) is as follows:

$$x_{1,1}^* = 0, x_{2,1}^* = 1.333333, x_{3,1}^* = 1.333333, x_{4,1}^* = 3.022222,$$
$$x_{1,2}^* = 6, x_{2,2}^* = 6, x_{3,2}^* = 8.222222, x_{4,2}^* = 8.222222.$$

$$(4.209)$$

Hence, the fuzzy optimal solution of the fully FLP problem (4.206) is

$$\tilde{x}^* = (\tilde{x}_1^*, \tilde{x}_2^*) = \Big(\big(x_{1,1}^*, x_{2,1}^*, x_{3,1}^*, x_{4,1}^*\big), \big(x_{1,2}^*, x_{2,2}^*, x_{3,2}^*, x_{4,2}^*\big)\Big)$$
$$= \big((0, 1.333333, 1.333333, 3.022222), (6, 6, 8.222222, 8.222222)\big)$$

$$(4.210)$$

Putting the fuzzy optimal solution (4.202) in the objective function of the problem (4.206) gives $\tilde{z}^* = (12, 25.999998, 77.777773, 110.266639)$.

4.9.4   Ezzati et al.'s Approach

Based on a new lexicographic ordering on triangular fuzzy numbers, Ezzati et al. [27] proposed a novel algorithm for solving Type-7 LPP with fuzzy parameters by converting it to its equivalent MOLP problem.

In order to define the concept of maximization of the fuzzy objective function, Ezzati et al. [27] proposed the following definition to compare two arbitrary triangular fuzzy numbers based on the lexicographic method.

Definition 4.19 Let $\tilde{a}_1 = (a_{1,1}, a_{2,1}, a_{3,1})$ and $\tilde{a}_2 = (a_{1,2}, a_{2,2}, a_{3,2})$ be two triangular fuzzy numbers. We say $\tilde{a}_1 \prec \tilde{a}_2$ if and only if $a_{2,1} < a_{2,2}$ or $a_{2,1} = a_{2,2}$ and $a_{3,1} - a_{1,1} > a_{3,2} - a_{1,2}$ or $a_{2,1} = a_{2,2}$, $a_{3,1} - a_{1,1} = a_{3,2} - a_{1,2}$ and $a_{3,1} + a_{1,1} < a_{3,2} + a_{1,2}$. Finally, $\tilde{a}_1 \preceq \tilde{a}_2$ if either $\tilde{a}_1 \prec \tilde{a}_2$ or $\tilde{a}_1 = \tilde{a}_2$.

Consider the following fully FLP problem with equality constraints and assume that all fuzzy numbers are triangular:

$$
\begin{aligned}
\max \tilde{z} &\approx \sum_{j=1}^{n} \tilde{c}_j \otimes \tilde{x}_j \\
s.t. \ \sum_{j=1}^{n} &\tilde{a}_{ij} \otimes \tilde{x}_j = \tilde{b}_i, \ i = 1, 2, \ldots, m,
\end{aligned}
\tag{4.211}
$$

$\tilde{x}_j$ is a non negative triangular fuzzy number, $j = 1, 2, \ldots, n$.

where $\tilde{c}_j = (c_{1,j}, c_{2,j}, c_{3,j})$, $\tilde{a}_{ij} = (a_{1,ij}, a_{2,ij}, a_{3,ij})$, $\tilde{x}_j = (x_{1,j}, x_{2,j}, x_{3,j})$ and $\tilde{b}_i = (b_{1,i}, b_{2,i}, b_{3,i})$.

Let $\tilde{a}_{ij} \otimes \tilde{x}_j = \left((ax)_{1,ij}, (ax)_{2,ij}, (ax)_{3,ij} \right)$ and $\tilde{c}_j \otimes \tilde{x}_j = \left((cx)_{1,j}, (cx)_{2,j}, (cx)_{3,j} \right)$. Thus, with regard to fuzzy arithmetic operations, the problem (2.222) can be shown as follows:

$$
\begin{aligned}
\max \tilde{z} &= \left(\sum_{j=1}^{n} (cx)_{1,j}, \sum_{j=1}^{n} (cx)_{2,j}, \sum_{j=1}^{n} (cx)_{3,j} \right) \\
s.t. \ \left(\sum_{j=1}^{n} (ax)_{1,ij}, \right. & \left. \sum_{j=1}^{n} (ax)_{2,ij}, \sum_{j=1}^{n} (ax)_{3,ij} \right) = (b_{1,i}, b_{2,i}, b_{3,i}), \ i = 1, 2, \ldots, m, \\
x_{1,j} &\geq 0, x_{2,j} - x_{1,j} \geq 0, x_{3,j} - x_{2,j} \geq 0, \ j = 1, 2, \ldots, n.
\end{aligned}
\tag{4.212}
$$

Equivalently, the problem (4.212) may be written as follows:

$$\max \tilde{z} = \left(\sum_{j=1}^{n} (cx)_{1,j}, \sum_{j=1}^{n} (cx)_{2,j}, \sum_{j=1}^{n} (cx)_{3,j} \right)$$

$$s.t. \quad \sum_{j=1}^{n} (ax)_{1,ij} = b_{1,i}, \ i = 1, 2, \ldots, m,$$

$$\sum_{j=1}^{n} (ax)_{2,ij} = b_{2,i}, \ i = 1, 2, \ldots, m,$$ \hfill (4.213)

$$\sum_{j=1}^{n} (ax)_{3,ij} = b_{3,i}, \ i = 1, 2, \ldots, m,$$

$$x_{1,j} \geq 0, x_{2,j} - x_{1,j} \geq 0, x_{3,j} - x_{2,j} \geq 0, \ j = 1, 2, \ldots, n.$$

Regarding Definition 4.19, the problem (4.213) is converted to the MOLP problem with three crisp objective functions, as follows:

$$\max \sum_{j=1}^{n} (cx)_{2,j}$$

$$\min \left(\sum_{j=1}^{n} (cx)_{3,j} - \sum_{j=1}^{n} (cx)_{3,j} \right)$$

$$\max \left(\sum_{j=1}^{n} (cx)_{3,1} + \sum_{j=1}^{n} (cx)_{1,j} \right)$$

$$s.t. \quad \sum_{j=1}^{n} (ax)_{1,ij} = b_{1,i}, \ i = 1, 2, \ldots, m,$$ \hfill (4.214)

$$\sum_{j=1}^{n} (ax)_{2,ij} = b_{2,i}, \ i = 1, 2, \ldots, m,$$

$$\sum_{j=1}^{n} (ax)_{3,ij} = b_{3,i}, \ i = 1, 2, \ldots, m,$$

$$x_{1,j} \geq 0, x_{2,j} - x_{1,j} \geq 0, x_{3,j} - x_{2,j} \geq 0, \ j = 1, 2, \ldots, n.$$

Now, the lexicographic method will be used to obtain an optimal solution of the MOLP problem (4.214). So, we have:

$$\max z_1 = \sum_{j=1}^{n} (cx)_{2,j}$$

$$\text{s.t.} \quad \sum_{j=1}^{n} (ax)_{1,ij} = b_{1,i}, \ i = 1, 2, \ldots, m,$$

$$\sum_{j=1}^{n} (ax)_{2,ij} = b_{2,i}, \ i = 1, 2, \ldots, m, \tag{4.215}$$

$$\sum_{j=1}^{n} (ax)_{3,ij} = b_{3,i}, \ i = 1, 2, \ldots, m,$$

$$x_{1,j} \geq 0, x_{2,j} - x_{1,j} \geq 0, x_{3,j} - x_{2,j} \geq 0, \ j = 1, 2, \ldots, n.$$

If the problem (4.215) has a unique optimal solution, then it is an optimal solution of (4.211). Otherwise, we solve the following problem over the optimal solutions of problem (4.215) where z_1^* is the optimal objective function value of the problem (4.215):

$$\min z_2 = \left(\sum_{j=1}^{n} (cx)_{3,j} - \sum_{j=1}^{n} (cx)_{1,j} \right)$$

$$\text{s.t.} \quad \sum_{j=1}^{n} (cx)_{2,j} = z_1^*$$

$$\sum_{j=1}^{n} (ax)_{1,ij} = b_{1,i}, \ i = 1, 2, \ldots, m,$$

$$\sum_{j=1}^{n} (ax)_{2,ij} = b_{2,i}, \ i = 1, 2, \ldots, m, \tag{4.216}$$

$$\sum_{j=1}^{n} (ax)_{3,ij} = b_{3,i}, \ i = 1, 2, \ldots, m,$$

$$x_{1,j} \geq 0, x_{2,j} - x_{1,j} \geq 0, x_{3,j} - x_{2,j} \geq 0, \ j = 1, 2, \ldots, n.$$

If the problem (4.216) has a unique optimal solution, then it is an optimal solution of (4.211). Otherwise, we solve the following problem over the optimal solutions of the problem (4.216) where z_2^* is the optimal objective function value of the problem (4.216):

$$\max z_3 = \left(\sum_{j=1}^{n} (cx)_{1,j} + \sum_{j=1}^{n} (cx)_{3,j} \right)$$

$$s.t. \quad \left(\sum_{j=1}^{n} (cx)_{3,j} - \sum_{j=1}^{n} (cx)_{1,j} \right) = z_2^*$$

$$\sum_{j=1}^{n} (cx)_{2,j} = z_1^*$$

$$\sum_{j=1}^{n} (ax)_{1,ij} = b_{1,i}, \ i = 1, 2, \ldots, m, \qquad (4.217)$$

$$\sum_{j=1}^{n} (ax)_{2,ij} = b_{2,i}, \ i = 1, 2, \ldots, m,$$

$$\sum_{j=1}^{n} (ax)_{3,ij} = b_{3,i}, \ i = 1, 2, \ldots, m,$$

$$x_{1,j} \geq 0, x_{2,j} - x_{1,j} \geq 0, x_{3,j} - x_{2,j} \geq 0, \ j = 1, 2, \ldots, n.$$

If the crisp LP problem (4.217) has an optimal solution as $x_{1,j}^*, x_{2,j}^*, x_{3,j}^*$ $(j = 1, 2, \ldots, n)$, then $\tilde{x}_j^* = (x_{1,j}^*, x_{2,j}^*, x_{3,j}^*)$ $(j = 1, 2, \ldots, n)$ will be the fuzzy optimal solution of (4.211).

Theorem 4.11 *If $\tilde{x}_j^* = (x_{1,j}^*, x_{2,j}^*, x_{3,j}^*)$ $(j = 1, 2, \ldots, n)$ is an optimal solution of the problems (4.215)–(4.217) (and hence is a lexicographic optimal solution of the problem (4.214)), then it is also an exact optimal solution of the problem (4.211).*

Proof By contradiction, let $\tilde{x}_j^* = (x_{1,j}^*, x_{2,j}^*, x_{3,j}^*)$ be an optimal solution of (4.215)–(4.217), but it is not the exact optimal solution of the problem (4.211). Therefore, a feasible solution of the problem (4.211), exists, namely $\tilde{x}_j^\circ = (x_{1,j}^\circ, x_{2,j}^\circ, x_{3,j}^\circ)$, such that

$$\tilde{c}_j \otimes \tilde{x}_j^* = \left((cx)_{1,j}^*, (cx)_{2,j}^*, (cx)_{3,j}^* \right) \prec \tilde{c}_j \otimes \tilde{x}_j^\circ = \left((cx)_{1,j}^\circ, (cx)_{2,j}^\circ, (cx)_{3,j}^\circ \right)$$

So, with respect to Definition 4.19, we have three conditions, as follows:

Case 1: Let $(cx)_{2,j}^* < (cx)_{2,j}^\circ$. Since $\tilde{x}_j^\circ = (x_{1,j}^\circ, x_{2,j}^\circ, x_{3,j}^\circ)$ is a feasible solution of the problem (4.211), thus it is a feasible solution of the problem (4.216) in which the objective value in $(x_{1,j}^\circ, x_{2,j}^\circ, x_{3,j}^\circ)$ is greater than the objective value in $(x_{1,j}^*, x_{2,j}^*, x_{3,j}^*)$. This is, however, a contradiction.

Case 2: Let $(cx)_{2,j}^* = (cx)_{2,j}^\circ$ and $(cx)_{3,j}^\circ - (cx)_{1,j}^\circ < (cx)_{3,j}^* - (cx)_{1,j}^*$. Hence, $(x_{1,j}^\circ, x_{2,j}^\circ, x_{3,j}^\circ)$ is a feasible solution of the problem (4.216) in which the objective value in $(x_{1,j}^\circ, x_{2,j}^\circ, x_{3,j}^\circ)$ is less than the objective value in $(x_{1,j}^*, x_{2,j}^*, x_{3,j}^*)$. This is, however, a contradiction.

Case 3: Let $(cx)^*_{2,j} = (cx)^\circ_{2,j}$, $(cx)^\circ_{3,j} - (cx)^\circ_{1,j} = (cx)^*_{3,j} - (cx)^*_{1,j}$ and $(cx)^\circ_{3,j} + (cx)^\circ_{1,j} > (cx)^*_{3,j} + (cx)^*_{1,j}$. Hence, $(x^\circ_{1,j}, x^\circ_{2,j}, x^\circ_{3,j})$ is a feasible solution of the problem (4.217) in which the objective value in $(x^\circ_{1,j}, x^\circ_{2,j}, x^\circ_{3,j})$ is greater than the objective value in $(x^*_{1,j}, x^*_{2,j}, x^*_{3,j})$. This is, however, a contradiction.

Therefore, $\tilde{x}^*_j = (x^*_{1,j}, x^*_{2,j}, x^*_{3,j})$ is also an exact optimal solution of the problem (4.211).   □

Example 4.19 Consider the following Type-7 LPP with fuzzy parameters:

$$
\begin{aligned}
\max \tilde{z} = {}& (10, 15, 17) \otimes (x_{1,1}, x_{2,1}, x_{3,1}) + (10, 16, 20) \otimes (x_{1,2}, x_{2,2}, x_{3,2}) \\
& + (10, 14, 17) \otimes (x_{1,3}, x_{2,3}, x_{3,3}) + (10, 12, 14) \otimes (x_{1,4}, x_{2,4}, x_{3,4}) \\
\text{s.t.} \quad & (8, 10, 13) \otimes (x_{1,1}, x_{2,1}, x_{3,1}) + (10, 11, 13) \otimes (x_{1,2}, x_{2,2}, x_{3,2}) \\
& + (9, 12, 13) \otimes (x_{1,3}, x_{2,3}, x_{3,3}) + (11, 15, 17) \otimes (x_{1,4}, x_{2,4}, x_{3,4}) = (271.75, 411.75, 573.75), \\
& (12, 14, 16) \otimes (x_{1,1}, x_{2,1}, x_{3,1}) + (14, 18, 19) \otimes (x_{1,2}, x_{2,2}, x_{3,2}) \\
& + (14, 17, 20) \otimes (x_{1,3}, x_{2,3}, x_{3,3}) + (13, 14, 18) \otimes (x_{1,4}, x_{2,4}, x_{3,4}) = (385.5, 539.5, 759.5), \\
& (x_{1,1}, x_{2,1}, x_{3,1}), (x_{1,2}, x_{2,2}, x_{3,2}), (x_{1,3}, x_{2,3}, x_{3,3}), (x_{1,4}, x_{2,4}, x_{3,4}) \succeq (0, 0, 0).
\end{aligned}
$$

$$(4.218)$$

Equivalently, the problem (4.218) may be written as follows with regard to the problem (4.213):

$$
\begin{aligned}
\max \tilde{z} = {}& (10x_{1,1} + 10x_{1,2} + 10x_{1,3} + 10x_{1,4}, 15x_{2,1} + 16x_{2,2} + 14x_{2,3} + 12x_{2,4}, \\
& 17x_{3,1} + 20x_{3,2} + 17x_{3,3} + 14x_{3,4}) \\
\text{s.t.} \quad & 8x_{1,1} + 10x_{1,2} + 9x_{1,3} + 11x_{1,4} = 271.75, \\
& 10x_{2,1} + 11x_{2,2} + 12x_{2,3} + 15x_{2,4} = 411.75, \\
& 13x_{3,1} + 13x_{3,2} + 13x_{3,3} + 17x_{3,4} = 573.5, \\
& 12x_{1,1} + 14x_{1,2} + 14x_{1,3} + 13x_{1,4} = 385.5, \\
& 14x_{2,1} + 18x_{2,2} + 17x_{2,3} + 14x_{2,4} = 539.5, \\
& 16x_{3,1} + 19x_{3,2} + 20x_{3,3} + 18x_{3,4} = 759.5. \\
& x_{1,1} \geq 0, x_{2,1} - x_{1,1} \geq 0, x_{3,1} - x_{2,1} \geq 0, \\
& x_{1,2} \geq 0, x_{2,2} - x_{1,2} \geq 0, x_{3,2} - x_{2,2} \geq 0, \\
& x_{1,3} \geq 0, x_{2,3} - x_{1,3} \geq 0, x_{3,3} - x_{2,3} \geq 0, \\
& x_{1,4} \geq 0, x_{2,4} - x_{1,4} \geq 0, x_{3,4} - x_{2,4} \geq 0.
\end{aligned}
$$

$$(4.219)$$

Thus, the problem (4.219) is reformulated as the following MOLP problem with regard to the problem (4.214):

$$\max z_1 = 15x_{2,1} + 16x_{2,2} + 14x_{2,3} + 12x_{2,4}$$
$$\min z_2 = (17x_{3,1} + 20x_{3,2} + 17x_{3,3} + 14x_{3,4}) - (10x_{1,1} + 10x_{1,2} + 10x_{1,3} + 10x_{1,4})$$
$$\max z_3 = (10x_{1,1} + 10x_{1,2} + 10x_{1,3} + 10x_{1,4}) + (17x_{3,1} + 20x_{3,2} + 17x_{3,3} + 14x_{3,4})$$
$$s.t. \quad 8x_{1,1} + 10x_{1,2} + 9x_{1,3} + 11x_{1,4} = 271.75,$$
$$10x_{2,1} + 11x_{2,2} + 12x_{2,3} + 15x_{2,4} = 411.75,$$
$$13x_{3,1} + 13x_{3,2} + 13x_{3,3} + 17x_{3,4} = 573.5$$
$$12x_{1,1} + 14x_{1,2} + 14x_{1,3} + 13x_{1,4} = 385.5,$$
$$14x_{2,1} + 18x_{2,2} + 17x_{2,3} + 14x_{2,4} = 539.5,$$
$$16x_{3,1} + 19x_{3,2} + 20x_{3,3} + 18x_{3,4} = 759.5.$$
$$x_{1,1} \geq 0, x_{2,1} - x_{1,1} \geq 0, x_{3,1} - x_{2,1} \geq 0,$$
$$x_{1,2} \geq 0, x_{2,2} - x_{1,2} \geq 0, x_{3,2} - x_{2,2} \geq 0,$$
$$x_{1,3} \geq 0, x_{2,3} - x_{1,3} \geq 0, x_{3,3} - x_{2,3} \geq 0,$$
$$x_{1,4} \geq 0, x_{2,4} - x_{1,4} \geq 0, x_{3,4} - x_{2,4} \geq 0.$$

$$(4.220)$$

Using the lexicographic method and regarding problems (4.215), (4.216) and (4.217), the optimal solution of the MOLP problem (4.220) is given as follows:

$$x_{1,1}^* = x_{2,1}^* = x_{3,1}^* = 17.27,$$
$$x_{1,2}^* = x_{2,2}^* = x_{3,2}^* = 2.16,$$
$$x_{1,3}^* = 4.64, x_{2,3}^* = 9.97, x_{3,3}^* = 16.36,$$
$$x_{1,4}^* = x_{2,4}^* = x_{3,4}^* = 6.36.$$

$$(4.221)$$

Thus, the fuzzy optimal solution of the problem (4.218) is given as follows:

$$\tilde{x}_1^* = (x_{1,1}^*, x_{2,1}^*, x_{3,1}^*) = (17.27, 17.27, 17.27),$$
$$\tilde{x}_2^* = (x_{1,2}^*, x_{2,2}^*, x_{3,2}^*) = (2.16, 2.16, 2.16),$$
$$\tilde{x}_3^* = (x_{1,3}^*, x_{2,3}^*, x_{3,3}^*) = (4.64, 9.97, 16.36),$$
$$x_3^* = (x_{1,4}^*, x_{2,4}^*, x_{3,4}^*) = (6.36, 6.36, 6.36).$$

$$(4.222)$$

Putting the fuzzy optimal solution (4.222) in the objective function of the problem (4.218) gives $\tilde{z}^* = \tilde{c} \otimes \tilde{x}^* = ((cx)_1^*, (cx)_2^*, (cx)_3^*) = (304.58, 509.79, 704.37)$.

It should be noted that using Kumar et al.'s method [26] the optimal solution and the optimal objective function value of the problem (4.218) are given as follows:

$$\tilde{x}_1^* = (x_{1,1}^*, x_{2,1}^*, x_{3,1}^*) = (15.28, 15.28, 15.28),$$

$$\tilde{x}_2^* = (x_{1,2}^*, x_{2,2}^*, x_{3,2}^*) = (2.40, 2.40, 9.10),$$

$$\tilde{x}_3^* = (x_{1,3}^*, x_{2,3}^*, x_{3,3}^*) = (611.25, 11.25), \qquad (4.223)$$

$$x_3^* = (x_{1,4}^*, x_{2,4}^*, x_{3,4}^*) = (6.49, 6.49, 6.49),$$

$$\tilde{z}_{\text{Kumar et al.}}^* = (301.83, 503.24, 724, 15).$$

By comparing the results of Ezzati et al.'s method [27] with Kumar et al.'s method [26], based on Definition 4.19, it can be concluded that the result of Ezzati et al.'s method is more reliable, because $509.79 = (cx)_{2, \text{ Ezzati et al.}}^* > (cx)_{2, \text{ Kumar et al.}}^* = 503.24$ and so $\tilde{z}_{\text{Ezzati et al.}}^* \succ \tilde{z}_{\text{Kumar et al.}}^*$.

It should be noted that Ezzati et al. [27] claimed that the fuzzy optimal solution of fully FLP problems with inequality constraints can also be obtained through the same method by transforming it into fully FLP problems with equality constraints. Bhardwaj and Kumar [43] proved that fully FLP problems with inequality constraints cannot be transformed them into fully FLP problems with equality constraints and hence, the method proposed by Ezzati et al. [27] to find the fuzzy optimal solution of fully FLP problems with equality constraints; cannot be used for finding the fuzzy optimal solution of fully FLP problems with inequality constraints.

4.9.5 Das et al.'s Approach

Das et al. [28] designed a new strategy to solve Type-7 LPP with fuzzy parameters based on MOLP problems and the lexicographic method assuming that all fuzzy parameters and variables are non-negative trapezoidal fuzzy numbers.

In order to define the concept of maximization of the fuzzy objective function, Das et al. [28] proposed the following definition to compare two arbitrary trapezoidal fuzzy numbers based on the lexicographic method.

Definition 4.20 Let $\tilde{a}_1 = (a_{1,1}, a_{2,1}, a_{3,1}, a_{4,1})$ and $\tilde{a}_2 = (a_{1,2}, a_{2,2}, a_{3,2}, a_{4,2})$ be two triangular fuzzy numbers. We say $\tilde{a}_1 \prec \tilde{a}_2$ if and only if $a_{2,1} - a_{1,1} > a_{2,2} - a_{1,2}$, or $a_{2,1} - a_{1,1} = a_{2,2} - a_{1,2}$ and $a_{2,1} < a_{2,2}$, or $a_{2,1} - a_{1,1} = a_{2,2} - a_{1,2}$, $a_{2,1} = a_{2,2}$ and $\frac{a_{2,1} + a_{3,1}}{2} < \frac{a_{2,2} + a_{3,2}}{2}$, or $a_{2,1} - a_{1,1} = a_{2,2} - a_{1,2}$, $a_{2,1} = a_{2,2}$, $\frac{a_{2,1} + a_{3,1}}{2} = \frac{a_{2,2} + a_{3,2}}{2}$ and $a_{4,1} - a_{3,1} < a_{4,2} - a_{3,2}$. Finally, $\tilde{a}_1 \preceq \tilde{a}_2$ if either $\tilde{a}_1 \prec \tilde{a}_2$ or $\tilde{a}_1 = \tilde{a}_2$.

Consider the following fully FLP problem with equality constraints and assume that all fuzzy numbers are non-negative trapezoidal fuzzy numbers:

$$\max \tilde{z} \approx \sum_{j=1}^{n} \tilde{c}_j \otimes \tilde{x}_j$$

$$s.t. \ \sum_{j=1}^{n} \tilde{a}_{ij} \otimes \tilde{x}_j = \tilde{b}_i, \ i = 1, 2, \ldots, m, \tag{4.224}$$

$\tilde{x}_j$ is a non negative trapezoidal fuzzy number, $j = 1, 2, \ldots, n$.

where $\tilde{c}_j = (c_{1,j}, c_{2,j}, c_{3,j}, c_{4,j})$, $\tilde{a}_{ij} = (a_{1,ij}, a_{2,ij}, a_{3,ij}, a_{4,ij})$, $\tilde{x}_j = (x_{1,j}, x_{2,j}, x_{3,j}, x_{4,j})$ and $\tilde{b}_i = (b_{1,i}, b_{2,i}, b_{3,i}, b_{4,i})$.

Let $\tilde{a}_{ij} \otimes \tilde{x}_j = \left((ax)_{1,ij}, (ax)_{2,ij}, (ax)_{3,ij}, (ax)_{4,ij}\right)$ and $\tilde{c}_j \otimes \tilde{x}_j = \left((cx)_{1,j}, (cx)_{2,j}, (cx)_{3,j}, (cx)_{4,j}\right)$. Thus, with regard to fuzzy arithmetic operations, the problem (2.235) can be shown as follows:

$$\max \tilde{z} = \left(\sum_{j=1}^{n} (cx)_{1,j}, \sum_{j=1}^{n} (cx)_{2,j}, \sum_{j=1}^{n} (cx)_{3,j}, \sum_{j=1}^{n} (cx)_{4,j}\right)$$

$$s.t. \ \left(\sum_{j=1}^{n} (ax)_{1,ij}, \sum_{j=1}^{n} (ax)_{2,ij}, \sum_{j=1}^{n} (ax)_{3,ij}, \sum_{j=1}^{n} (ax)_{4,ij}\right) = (b_{1,i}, b_{2,i}, b_{3,i}, b_{4,i}), \ i = 1, 2, \ldots, m,$$

$$x_{1,j} \geq 0, x_{2,j} - x_{1,j} \geq 0, x_{3,j} - x_{2,j} \geq 0, x_{4,j} - x_{3,j} \geq 0, \ j = 1, 2, \ldots, n.$$
$$\tag{4.225}$$

Equivalently, the problem (4.225) may be written as follows:

$$\max \tilde{z} = \left(\sum_{j=1}^{n} (cx)_{1,j}, \sum_{j=1}^{n} (cx)_{2,j}, \sum_{j=1}^{n} (cx)_{3,j}, \sum_{j=1}^{n} (cx)_{4,j}\right)$$

$$s.t. \ \sum_{j=1}^{n} (ax)_{1,ij} = b_{1,i}, \ i = 1, 2, \ldots, m,$$

$$\sum_{j=1}^{n} (ax)_{2,ij} = b_{2,i}, \ i = 1, 2, \ldots, m,$$

$$\sum_{j=1}^{n} (ax)_{3,ij} = b_{3,i}, \ i = 1, 2, \ldots, m, \tag{4.226}$$

$$\sum_{j=1}^{n} (ax)_{4,ij} = b_{4,i}, \ i = 1, 2, \ldots, m,$$

$$x_{1,j} \geq 0, x_{2,j} - x_{1,j} \geq 0, x_{3,j} - x_{2,j} \geq 0, x_{4,j} - x_{3,j} \geq 0, \ j = 1, 2, \ldots, n.$$

Regarding Definition 4.20, the problem (4.226) is converted to the MOLP problem with four crisp objective functions as follows:

$$\min z_1 = \left(\sum_{j=1}^{n} (cx)_{2,j} - \sum_{j=1}^{n} (cx)_{1,j} \right)$$

$$\max z_2 = \sum_{j=1}^{n} (cx)_{2,j}$$

$$\max z_3 = \frac{1}{2} \left(\sum_{j=1}^{n} (cx)_{2,j} + \sum_{j=1}^{n} (cx)_{3,j} \right)$$

$$\max z_4 = \left(\sum_{j=1}^{n} (cx)_{4,1} - \sum_{j=1}^{n} (cx)_{3,j} \right)$$

$$s.t. \quad \sum_{j=1}^{n} (ax)_{1,ij} = b_{1,i}, \ i = 1, 2, \ldots, m,$$

$$\sum_{j=1}^{n} (ax)_{2,ij} = b_{2,i}, \ i = 1, 2, \ldots, m,$$

$$\sum_{j=1}^{n} (ax)_{3,ij} = b_{3,i}, \ i = 1, 2, \ldots, m,$$

$$\sum_{j=1}^{n} (ax)_{4,ij} = b_{4,i}, \ i = 1, 2, \ldots, m,$$

$$x_{1,j} \geq 0, x_{2,j} - x_{1,j} \geq 0, x_{3,j} - x_{2,j} \geq 0, x_{4,j} - x_{3,j} \geq 0, \ j = 1, 2, \ldots, n.$$

(4.227)

Now, the lexicographic method will be used to obtain an optimal solution of the MOLP problem (4.227). Thus, we have:

$$\min z_1 = \left(\sum_{j=1}^{n} (cx)_{2,j} - \sum_{j=1}^{n} (cx)_{1,j} \right)$$

$$s.t. \quad \sum_{j=1}^{n} (ax)_{1,ij} = b_{1,i}, \ i = 1, 2, \ldots, m,$$

$$\sum_{j=1}^{n} (ax)_{2,ij} = b_{2,i}, \ i = 1, 2, \ldots, m,$$

$$\sum_{j=1}^{n} (ax)_{3,ij} = b_{3,i}, \ i = 1, 2, \ldots, m,$$

$$\sum_{j=1}^{n} (ax)_{4,ij} = b_{4,i}, \ i = 1, 2, \ldots, m,$$

$$x_{1,j} \geq 0, x_{2,j} - x_{1,j} \geq 0, x_{3,j} - x_{2,j} \geq 0, x_{4,j} - x_{3,j} \geq 0, \ j = 1, 2, \ldots, n.$$

(4.228)

If the problem (4.228) has a unique optimal solution, then it is an optimal solution of (4.224). Otherwise, we solve the following problem over the optimal solutions of the problem (4.228) where z_1^* is the optimal objective function value of the problem (4.228):

$$\max z_2 = \sum_{j=1}^{n} (cx)_{2,j}$$

$$s.t. \quad \sum_{j=1}^{n} (cx)_{2,j} - \sum_{j=1}^{n} (cx)_{1,j} = z_1^*$$

$$\sum_{j=1}^{n} (ax)_{1,ij} = b_{1,i}, \ i = 1, 2, \ldots, m,$$

$$\sum_{j=1}^{n} (ax)_{2,ij} = b_{2,i}, \ i = 1, 2, \ldots, m,$$

$$\sum_{j=1}^{n} (ax)_{3,ij} = b_{3,i}, \ i = 1, 2, \ldots, m,$$

$$\sum_{j=1}^{n} (ax)_{4,ij} = b_{4,i}, \ i = 1, 2, \ldots, m,$$

$$x_{1,j} \geq 0, x_{2,j} - x_{1,j} \geq 0, x_{3,j} - x_{2,j} \geq 0, x_{4,j} - x_{3,j} \geq 0, \ j = 1, 2, \ldots, n.$$

$$(4.229)$$

If the problem (4.229) has a unique optimal solution, then it is an optimal solution of (4.224). Otherwise, we solve the following problem over the optimal solutions of the problem (4.229) where z_2^* is the optimal objective function value of the problem (4.229):

$$\max z_3 = \frac{1}{2} \left(\sum_{j=1}^{n} (cx)_{2,j} + \sum_{j=1}^{n} (cx)_{2,j} \right)$$

$$s.t. \quad \sum_{j=1}^{n} (cx)_{2,j} - \sum_{j=1}^{n} (cx)_{1,j} = z_1^*$$

$$\sum_{j=1}^{n} (cx)_{2,j} = z_2^*$$

$$\sum_{j=1}^{n} (ax)_{1,ij} = b_{1,i}, i = 1, 2, \ldots, m,$$

$$\sum_{j=1}^{n} (ax)_{2,ij} = b_{2,i}, i = 1, 2, \ldots, m,$$

$$\sum_{j=1}^{n} (ax)_{3,ij} = b_{3,i}, i = 1, 2, \ldots, m,$$

$$\sum_{j=1}^{n} (ax)_{4,ij} = b_{4,i}, \ i = 1, 2, \ldots, m,$$

(4.230)

$$x_{1,j} \geq 0, x_{2,j} - x_{1,j} \geq 0, x_{3,j} - x_{2,j} \geq 0, x_{4,j} - x_{3,j} \geq 0, \ j = 1, 2, \ldots, n.$$

If the problem (4.230) has a unique optimal solution, then it is an optimal solution of (4.224). Otherwise, we solve the following problem over the optimal solutions of the problem (4.230) where z_3^* is the optimal objective function value of the problem (4.230):

$$\max z_4 = \left(\sum_{j=1}^{n} (cx)_{4,j} - \sum_{j=1}^{n} (cx)_{3,j} \right)$$

$$s.t. \quad \sum_{j=1}^{n} (cx)_{2,j} - \sum_{j=1}^{n} (cx)_{1,j} = z_1^*$$

$$\sum_{j=1}^{n} (cx)_{2,j} = z_2^*$$

$$\frac{1}{2} \left(\sum_{j=1}^{n} (cx)_{2,j} + \sum_{j=1}^{n} (cx)_{2,j} \right) = z_3^*$$

$$\sum_{j=1}^{n} (ax)_{1,ij} = b_{1,i}, \ i = 1, 2, \ldots, m,$$

$$\sum_{j=1}^{n} (ax)_{2,ij} = b_{2,i}, \ i = 1, 2, \ldots, m,$$

$$\sum_{j=1}^{n} (ax)_{3,ij} = b_{3,i}, \ i = 1, 2, \ldots, m,$$

$$\sum_{j=1}^{n} (ax)_{4,ij} = b_{4,i}, \ i = 1, 2, \ldots, m,$$

$$x_{1,j} \geq 0, x_{2,j} - x_{1,j} \geq 0, x_{3,j} - x_{2,j} \geq 0, x_{4,j} - x_{3,j} \geq 0, \ j = 1, 2, \ldots, n.$$

(4.231)

If the crisp LP problem (4.231) has an optimal solution as $x_{1,j}^*, x_{2,j}^*, x_{3,j}^*, x_{4,j}^*$ $(j = 1, 2, \ldots, n)$, then $\tilde{x}_j^* = (x_{1,j}^*, x_{2,j}^*, x_{3,j}^*, x_{4,j}^*)$ $(j = 1, 2, \ldots, n)$ will be the fuzzy optimal solution of (4.224).

Theorem 4.12 i$\tilde{x}_j^* = (x_{1,j}^*, x_{2,j}^*, x_{3,j}^*, x_{4,j}^*)$ $(j = 1, 2, \ldots, n)$ *is an optimal solution of the problems* (4.228)–(4.231) *(and hence is a lexicographic optimal solution of the problem* (4.227)), *then it is also an exact optimal solution of the problem* (4.224).

Proof By contradiction, let $\tilde{x}_j^* = (x_{1,j}^*, x_{2,j}^*, x_{3,j}^*, x_{4,j}^*)$ be an optimal solution of (4.228)–(4.231), but it is not the exact optimal solution of the problem (4.224). Therefore, a feasible solution of the problem (4.224) exists, namely $\tilde{x}_j^\circ = (x_{1,j}^\circ, x_{2,j}^\circ, x_{3,j}^\circ, x_{4,j}^\circ)$, such that

$$\tilde{c}_j \otimes \tilde{x}_j^* = \left((cx)_{1,j}^*, (cx)_{2,j}^*, (cx)_{3,j}^*, (cx)_{4,j}^*\right) \prec \tilde{c}_j \otimes \tilde{x}_j^\circ$$
$$= \left((cx)_{1,j}^\circ, (cx)_{2,j}^\circ, (cx)_{3,j}^\circ, (cx)_{4,j}^\circ\right)$$

So, with respect to Definition 4.20, we have four conditions as follows:

Case 1: Let $(cx)_{2,j}^* - (cx)_{1,j}^* > (cx)_{2,j}^\circ - (cx)_{1,j}^\circ$. Since $\tilde{x}_j^\circ = (x_{1,j}^\circ, x_{2,j}^\circ, x_{3,j}^\circ, x_{4,j}^\circ)$ is a feasible solution of the problem (4.224), it is a feasible solution of the problem (4.228) in which the objective value in $(x_{1,j}^\circ, x_{2,j}^\circ, x_{3,j}^\circ, x_{4,j}^\circ)$ is less than the objective value in $(x_{1,j}^*, x_{2,j}^*, x_{3,j}^*, x_{4,j}^*)$. This is, however, a contradiction.

Case 2: Let $(cx)_{2,j}^* - (cx)_{1,j}^* = (cx)_{2,j}^\circ - (cx)_{1,j}^\circ$ and $(cx)_{2,j}^* < (cx)_{2,j}^\circ$. Hence, $(x_{1,j}^\circ, x_{2,j}^\circ, x_{3,j}^\circ, x_{4,j}^\circ)$ is a feasible solution of the problem (4.229) in which the objective value in $(x_{1,j}^\circ, x_{2,j}^\circ, x_{3,j}^\circ, x_{4,j}^\circ)$ is greater than the objective value in $(x_{1,j}^*, x_{2,j}^*, x_{3,j}^*, x_{4,j}^*)$. This is, however, a contradiction.

Case 3: Let $(cx)_{2,j}^* - (cx)_{1,j}^* = (cx)_{2,j}^\circ - (cx)_{1,j}^\circ$, $(cx)_{2,j}^* = (cx)_{2,j}^\circ$ and $\frac{1}{2}\left((cx)_{2,j}^\circ + (cx)_{3,j}^\circ\right) > \frac{1}{2}\left((cx)_{2,j}^* + (cx)_{3,j}^*\right)$. Hence, $(x_{1,j}^\circ, x_{2,j}^\circ, x_{3,j}^\circ, x_{4,j}^\circ)$ is a feasible solution of the problem (4.230) in which the objective value in $(x_{1,j}^\circ, x_{2,j}^\circ, x_{3,j}^\circ, x_{4,j}^\circ)$ is greater than the objective value in $(x_{1,j}^\circ, x_{2,j}^\circ, x_{3,j}^\circ, x_{4,j}^\circ)$. This is, however, a contradiction.

Case 4: Let $(cx)_{2,j}^* - (cx)_{1,j}^* = (cx)_{2,j}^\circ - (cx)_{1,j}^\circ$, $(cx)_{2,j}^* = (cx)_{2,j}^\circ$, $\frac{1}{2}\left((cx)_{2,j}^* + (cx)_{3,j}^\circ\right) = \frac{1}{2}\left((cx)_{2,j}^* + (cx)_{3,j}^*\right)$ and $\left((cx)_{4,j}^\circ - (cx)_{3,j}^\circ\right) > \left((cx)_{4,j}^* - (cx)_{3,j}^*\right)$. Hence, $(x_{1,j}^\circ, x_{2,j}^\circ, x_{3,j}^\circ, x_{4,j}^\circ)$ is a feasible solution of the problem (4.231) in which the objective value in $(x_{1,j}^\circ, x_{2,j}^\circ, x_{3,j}^\circ, x_{4,j}^\circ)$ is greater than the objective value in $(x_{1,j}^\circ, x_{2,j}^\circ, x_{3,j}^\circ, x_{4,j}^\circ)$. This is, however, a contradiction.

Therefore, $(x_{1,j}^\circ, x_{2,j}^\circ, x_{3,j}^\circ, x_{4,j}^\circ)$ is also an exact optimal solution of the problem (4.224).   □

Example 4.20 Reconsider the Type-7 LPP with fuzzy parameters (4.199) given in Example 4.15.

Regarding the problem (4.227), this problem is converted to the following MOLP problem with four crisp objective functions:

$$\min z_1 = 6x_{2,1} + 3x_{2,2} - x_{1,1} - 2x_{1,2}$$
$$\max z_2 = 6x_{2,1} + 3x_{2,2}$$
$$\max z_3 = \frac{1}{2}\left[6x_{2,1} + 3x_{2,2} + 9x_{3,1} + 8x_{3,2}\right]$$

$$\max z_4 = 12x_{4,1} + 9x_{4,2} - 9x_{3,1} - 8x_{3,2}$$

$$\begin{aligned}
s.t. \quad & 2x_{1,1} + x_{1,2} = 6, \\
& -x_{4,1} + x_{1,2} = 0, \\
& 3x_{2,1} + 2x_{2,2} = 16, \\
& x_{2,1} + 3x_{2,2} = 17, \\
& 4x_{3,1} + 3x_{3,2} = 30, \\
& 2x_{3,1} + 4x_{3,2} = 30, \\
& 5x_{4,1} + 4x_{4,2} = 48, \\
& 3x_{4,1} + 6x_{4,2} = 54, \\
& x_{1,1} \geq 0, x_{2,1} - x_{1,1} \geq 0, x_{3,1} - x_{2,1} \geq 0, x_{4,1} - x_{3,1} \geq 0, \\
& x_{1,2} \geq 0, x_{2,2} - x_{1,2} \geq 0, x_{3,2} - x_{2,2} \geq 0, x_{4,2} - x_{3,2} \geq 0.
\end{aligned}$$

(4.232)

Now, the lexicographic method utilized by Das et al. [28] is used to obtain an optimal solution of the MOLP problem (4.232). In the first step the following crisp LP problem is solved with regard to the problem (4.228):

$$\min z_1 = 6x_{2,1} + 3x_{2,2} - x_{1,1} - 2x_{1,2}$$

$$\begin{aligned}
s.t. \quad & 2x_{1,1} + x_{1,2} = 6, \\
& -x_{4,1} + x_{1,2} = 0, \\
& 3x_{2,1} + 2x_{2,2} = 16, \\
& x_{2,1} + 3x_{2,2} = 17, \\
& 4x_{3,1} + 3x_{3,2} = 30, \\
& 2x_{3,1} + 4x_{3,2} = 30, \\
& 5x_{4,1} + 4x_{4,2} = 48, \\
& 3x_{4,1} + 6x_{4,2} = 54, \\
& x_{1,1} \geq 0, x_{2,1} - x_{1,1} \geq 0, x_{3,1} - x_{2,1} \geq 0, x_{4,1} - x_{3,1} \geq 0, \\
& x_{1,2} \geq 0, x_{2,2} - x_{1,2} \geq 0, x_{3,2} - x_{2,2} \geq 0, x_{4,2} - x_{3,2} \geq 0.
\end{aligned}$$

(4.233)

The optimal objective function value of the problem (4.233) is $z_1^* = 18$. Thus, in the second step the following crisp LP problem is solved with regard to the problem (4.229):

$$\max z_2 = 6x_{2,1} + 3x_{2,2}$$

$$\begin{aligned}
s.t. \quad & 6x_{2,1} + 3x_{2,2} - x_{1,1} - 2x_{1,2} = 18 \\
& 2x_{1,1} + x_{1,2} = 6, \\
& -x_{4,1} + x_{1,2} = 0, \\
& 3x_{2,1} + 2x_{2,2} = 16, \\
& x_{2,1} + 3x_{2,2} = 17,
\end{aligned}$$

$$4x_{3,1} + 3x_{3,2} = 30,$$
$$2x_{3,1} + 4x_{3,2} = 30,$$
$$5x_{4,1} + 4x_{4,2} = 48,$$
$$3x_{4,1} + 6x_{4,2} = 54,$$
$$x_{1,1} \geq 0, x_{2,1} - x_{1,1} \geq 0, x_{3,1} - x_{2,1} \geq 0, x_{4,1} - x_{3,1} \geq 0,$$
$$x_{1,2} \geq 0, x_{2,2} - x_{1,2} \geq 0, x_{3,2} - x_{2,2} \geq 0, x_{4,2} - x_{3,2} \geq 0.$$

(4.234)

The optimal objective function value of the problem (4.234) is $z_2^* = 27$. Thus, in the third step the following crisp LP problem is solved with regard to the problem (4.230):

$$\max z_3 = \frac{1}{2} \left[6x_{2,1} + 3x_{2,2} + 9x_{3,1} + 8x_{3,2} \right]$$

s.t. $$6x_{2,1} + 3x_{2,2} = 27$$
$$6x_{2,1} + 3x_{2,2} - x_{1,1} - 2x_{1,2} = 18$$
$$2x_{1,1} + x_{1,2} = 6,$$
$$- x_{4,1} + x_{1,2} = 0,$$
$$3x_{2,1} + 2x_{2,2} = 16,$$
$$x_{2,1} + 3x_{2,2} = 17,$$
$$4x_{3,1} + 3x_{3,2} = 30,$$
$$2x_{3,1} + 4x_{3,2} = 30,$$
$$5x_{4,1} + 4x_{4,2} = 48,$$
$$3x_{4,1} + 6x_{4,2} = 54,$$
$$x_{1,1} \geq 0, x_{2,1} - x_{1,1} \geq 0, x_{3,1} - x_{2,1} \geq 0, x_{4,1} - x_{3,1} \geq 0,$$
$$x_{1,2} \geq 0, x_{2,2} - x_{1,2} \geq 0, x_{3,2} - x_{2,2} \geq 0, x_{4,2} - x_{3,2} \geq 0.$$

(4.235)

The optimal objective function value of the problem (4.235) is $z_3^* = 51$. Thus, in the final step the following crisp LP problem is solved with regard to the problem (4.231):

$$\max z_4 = 12x_{4,1} + 9x_{4,2} - 9x_{3,1} - 8x_{3,2}$$

s.t. $$\frac{1}{2} \left[6x_{2,1} + 3x_{2,2} + 9x_{3,1} + 8x_{3,2} \right] = 51,$$
$$6x_{2,1} + 3x_{2,2} = 27,$$
$$6x_{2,1} + 3x_{2,2} - x_{1,1} - 2x_{1,2} = 18,$$
$$2x_{1,1} + x_{1,2} = 6,$$
$$- x_{4,1} + x_{1,2} = 0,$$
$$3x_{2,1} + 2x_{2,2} = 16,$$

$$x_{2,1} + 3x_{2,2} = 17,$$
$$4x_{3,1} + 3x_{3,2} = 30,$$
$$2x_{3,1} + 4x_{3,2} = 30,$$
$$5x_{4,1} + 4x_{4,2} = 48, \tag{4.236}$$
$$3x_{4,1} + 6x_{4,2} = 54,$$
$$x_{1,1} \geq 0, x_{2,1} - x_{1,1} \geq 0, x_{3,1} - x_{2,1} \geq 0, x_{4,1} - x_{3,1} \geq 0,$$
$$x_{1,2} \geq 0, x_{2,2} - x_{1,2} \geq 0, x_{3,2} - x_{2,2} \geq 0, x_{4,2} - x_{3,2} \geq 0.$$

The optimal solution of the crisp LP problem (4.236) is as follows:

$$x_{1,1}^* = 1, x_{2,1}^* = 2, x_{3,1}^* = 3, x_{4,1}^* = 4, x_{1,2}^* = 4, x_{2,2}^* = 5, x_{3,2}^* = 6, x_{4,2}^* = 7 \tag{4.237}$$

Hence, the fuzzy optimal solution of the fully FLP problem (4.199) is

$$\tilde{x}^* = (\tilde{x}_1^*, \tilde{x}_2^*) = \left(\left(x_{1,1}^*, x_{2,1}^*, x_{3,1}^*, x_{4,1}^* \right), \left(x_{1,2}^*, x_{2,2}^*, x_{3,2}^*, x_{4,2}^* \right) \right)$$
$$= ((1, 2, 3, 4), (4, 5, 6, 7)) \tag{4.238}$$

Putting the fuzzy optimal solution (4.238) in the objective function of the problem (4.199) gives $\tilde{z}^* = (9, 27, 75, 111)$.

Remark 4.2 Das et al. [28] explored that the fully FLP problems with inequality constraints can be solved with a similar approach. In order to define the fuzzy inequality constraints, they have used the following partial order:

$$\tilde{a}_1 = (a_{1,1}, a_{2,1}, a_{3,1}, a_{4,1}) \preceq \tilde{a}_2 = (a_{1,2}, a_{2,2}, a_{3,2}, a_{4,2})$$
$$\Leftrightarrow a_{2,1} - a_{1,1} \leq a_{2,2} - a_{1,2}, a_{2,1} \leq a_{2,2}, a_{3,1} \leq a_{3,2}, a_{4,1} - a_{3,1} \leq a_{4,2} - a_{3,2} \tag{4.239}$$

Das et al. [28], based on the order given in (2.250), extended the proposed approach for solving FLP problems with equality constraints to the FLP problems with inequality constraints. Hence, in the same method, two different approaches are used for the inequality of two fuzzy numbers, which is erroneous. In fact, they have used the order given in Definition 4.20 for the objective function and the order given in Remark 4.2 for the inequality constraints. Moreover, they have used a wrong procedure for maximization of the fuzzy objective function by converting the fuzzy objective function into the following four crisp objective functions:

$$\min z_1 = \left(\sum_{j=1}^{n} (cx)_{3,j} - \sum_{j=1}^{n} (cx)_{2,j} \right)$$

$$\max z_2 = \sum_{j=1}^{n} (cx)_{3,j}$$

$$\max z_3 = \frac{1}{2} \left(\sum_{j=1}^{n} (cx)_{2,j} + \left(\sum_{j=1}^{n} (cx)_{3,j} - \sum_{j=1}^{n} (cx)_{1,j} \right) \right) \qquad (4.240)$$

$$\max z_4 = \left(\left[\sum_{j=1}^{n} (cx)_{4,1} - \sum_{j=1}^{n} (cx)_{3,j} \right] - \left[\sum_{j=1}^{n} (cx)_{2,1} - \sum_{j=1}^{n} (cx)_{1,j} \right] \right)$$

The above four objective functions are obviously inconsistent with those given in the problem (4.227). On the basis of the above discussion, it can be concluded that the algorithm, proposed by Das et al. [28] cannot be used for finding the fuzzy optimal solution of fully FLP problems with inequality constraints. However, if the order given in Definition 4.20 is used for both the objective function and inequality constraints, we can overcome the above-mentioned shortcomings of the Das et al.'s algorithm. To illustrate this approach, we solve the following fully FLP problem.

Example 4.21 Consider the Type-7 LPP with fuzzy parameters.

$$\max \tilde{z} = (7, 10, 14, 17) \otimes \tilde{x}_1 + (8, 13, 15, 20) \otimes \tilde{x}_2$$
$$s.t. \quad (11, 13, 15, 17) \otimes \tilde{x}_1 + (7, 11, 13, 17) \otimes \tilde{x}_2 \preceq (94, 100, 102, 108),$$
$$(12, 14, 16, 18) \otimes \tilde{x}_1 + (8, 12, 14, 18) \otimes \tilde{x}_2 \preceq (104, 112, 114, 122),$$
$$\tilde{x}_1 = (x_{1,1}, x_{2,1}, x_{3,1}, x_{4,1}), \tilde{x}_2 = (x_{1,2}, x_{2,2}, x_{3,2}, x_{4,2}) \succeq \tilde{0}.$$

$$(4.241)$$

Equivalently, this problem is reformulated as follows with regard to fuzzy arithmetic operations:

$$\max \tilde{z} = (7x_{1,1} + 8x_{1,2}, 10x_{2,1} + 13x_{2,2}, 14x_{3,1} + 15x_{3,2}, 17x_{4,1} + 20x_{4,2})$$
$$s.t. \quad (11x_{1,1} + 7x_{1,2}, 13x_{2,1} + 11x_{2,2}, 15x_{3,1} + 13x_{3,2}, 17x_{4,1} + 17x_{4,2}) \preceq (94, 100, 102, 108),$$
$$(12x_{1,1} + 8x_{1,2}, 14x_{2,1} + 12x_{2,2}, 16x_{3,1} + 14x_{3,2}, 18x_{4,1} + 18x_{4,2}) \preceq (104, 112, 114, 122),$$
$$x_{1,1} \geq 0, x_{2,1} - x_{1,1} \geq 0, x_{3,1} - x_{2,1} \geq 0, x_{4,1} - x_{3,1} \geq 0,$$
$$x_{1,2} \geq 0, x_{2,2} - x_{1,2} \geq 0, x_{3,2} - x_{2,2} \geq 0, x_{4,2} - x_{3,2} \geq 0.$$

$$(4.242)$$

This problem is converted into the following MOLP problem with regard to Definition 4.20:

$$\min z_1 = 10x_{2,1} + 13x_{2,2} - 7x_{1,1} - 8x_{1,2}$$

$$\max z_2 = 10x_{2,1} + 13x_{2,2}$$

$$\max z_3 = \frac{1}{2}\left[10x_{2,1} + 13x_{2,2} + 14x_{3,1} + 15x_{3,2}\right]$$

$$\max z_4 = 17x_{4,1} + 20x_{4,2} - 14x_{3,1} - 15x_{3,2}$$

$$
\begin{aligned}
\text{s.t.} \quad & 13x_{2,1} + 11x_{2,2} - 11x_{1,1} - 7x_{1,2} \ge 6, \\
& 13x_{2,1} + 11x_{2,2} \le 100, \\
& \frac{1}{2}\left(13x_{2,1} + 11x_{2,2} + 15x_{3,1} + 13x_{3,2}\right) \le \frac{1}{2}(202), \\
& 17x_{4,1} + 17x_{4,2} - 15x_{3,1} - 13x_{3,2} \le 6, \\
& 14x_{2,1} + 12x_{2,2} - 12x_{1,1} - 8x_{1,2} \ge 8, \\
& 14x_{2,1} + 12x_{2,2} \le 112, \\
& \frac{1}{2}\left(14x_{2,1} + 12x_{2,2} + 16x_{3,1} + 14x_{3,2}\right) \le \frac{1}{2}(226), \\
& 18x_{4,1} + 18x_{4,2} - 16x_{3,1} - 14x_{3,2} \le 8, \\
& x_{1,1} \ge 0, x_{2,1} - x_{1,1} \ge 0, x_{3,1} - x_{2,1} \ge 0, x_{4,1} - x_{3,1} \ge 0, \\
& x_{1,2} \ge 0, x_{2,2} - x_{1,2} \ge 0, x_{3,2} - x_{2,2} \ge 0, x_{4,2} - x_{3,2} \ge 0.
\end{aligned}
$$

$$(4.243)$$

The optimal solution of the MOLP problem (4.243) with regard to the lexicographic method is given as follows lexicographic:

$$
\begin{aligned}
& x_{1,1}^* = 0, x_{2,1}^* = 0.5714286, x_{3,1}^* = 3, x_{4,1}^* = 3, \\
& x_{1,2}^* = 0, x_{2,2}^* = 0, x_{3,2}^* = 0, x_{4,2}^* = 0.
\end{aligned}
$$

$$(4.244)$$

Hence, the fuzzy optimal solution of the fully FLP problem (4.241) is as follows:

$$
\begin{aligned}
\tilde{x}^* = (\tilde{x}_1^*, \tilde{x}_2^*) &= \left(\left(x_{1,1}^*, x_{2,1}^*, x_{3,1}^*, x_{4,1}^*\right), \left(x_{1,2}^*, x_{2,2}^*, x_{3,2}^*, x_{4,2}^*\right)\right) \\
&= ((0, 0.5714286, 3, 3), (0, 0, 0, 0))
\end{aligned}
$$

$$(4.245)$$

Putting the fuzzy optimal solution (4.245) in the objective function of the problem (4.241) gives $\tilde{z}^* = (0, 5.71428, 42, 51)$.

Note that the fuzzy optimal solution of the problem (4.241) using Das et al. [28] is $\tilde{x}^* = (\tilde{x}_1^*, \tilde{x}_2^*) = ((0, 3, 3, 6), (4.38, 4.38, 4.38, 4.38))$ which obviously does not satisfy the fuzzy constraints of the problem (4.241).

4.10 Conclusions

In this chapter we classified the LPPs with fuzzy parameters that can be solved based on non-simplex algorithms into seven different groups and discussed various approaches for each group. Two main concepts that have key roles in solving LPPs with fuzzy parameters according to non-simplex algorithms were the concept of fuzzy optimization (maximization or minimization) of the fuzzy objective function and the concept of fuzzy equality or inequality of the fuzzy constraints. Regarding the variety of techniques that exist for defining the mentioned concepts, various approaches were explored for solving LPPs with fuzzy parameters in this chapter. The limitations and advantages of each approach were also pointed out. Moreover, the solution approaches were illustrated with several numerical examples.

References

1. Ramik, J., Rimanek, J.: Inequality relation between fuzzy numbers and its use in fuzzy optimization. Fuzzy Sets Syst. **16**(2), 123–138 (1985)
2. Tanaka, H., Ichihashi, H., Asai, K.: A formulation of fuzzy linear programming problems based on comparison of fuzzy numbers. Control Cybern. **13**, 186–194 (1984)
3. Dubois, D.: Linear programming with fuzzy data, in analysis of fuzzy information. Application in Engineering and Sciences, vol. III, pp. 241–263. CRC Press, Boca Raton (1987)
4. Li, H., Gong, Z.: Fuzzy linear programming with possibility and necessity relation. In: Fuzzy Information and Engineering. Advances in Intelligent and Soft Computing, pp. 305–311. Springer, Berlin (2010)
5. Lai, Y.J., Hwang, C.L.: A new approach to some possibilistic linear programming problem. Fuzzy Sets Syst. **49**(2), 121–133 (1992)
6. Wan, S.P., Dong, J.Y.: Possibility linear programming with trapezoidal fuzzy numbers. Appl. Math. Model. **38**, 1660–1672 (2014)
7. Rommelfanger, H., Hanuschech, R., Wolf, J.: Linear programming with fuzzy objective. Fuzzy Sets Syst. **29**(1), 311–348 (1989)
8. Zimmerman, H.J.: Fuzzy programming and linear programming with several objective functions. Fuzzy Sets Syst. **1**(1), 45–55 (1978)
9. Delgado, M., Verdegay, J.L., Vila, M.A.: Imprecise costs in mathematical programming problems. Control Cybern. **16**(2), 113–121 (1987)
10. Maeda, T.: Fuzzy linear programming problems as bi-criteria optimization problems. Appl. Math. Comput. **120**(1–3), 109–121 (2001)
11. Zhang, G., Wu, Y.H., Remias, M., Lu, J.: Formulation of fuzzy linear programming problems as four-objective constrained optimization problems. Appl. Math. Comput. **139**(2–3), 383–399 (2003)
12. Li, G., Guo, R.: Comments on "Formulation of fuzzy linear programming problems as four-objective constrained optimization problems". Appl. Math. Comput. **186**(1), 941–944 (2007)
13. Cai, Q., Hao, Z., Pan, S.: The solution of linear programming with LR-fuzzy numbers in objective function. In: Fuzzy Information and Engineering (ICFIE), pp. 988–999. Springer, Berlin (2007)
14. Lai, Y.J., Hwang, C.L.: Fuzzy Mathematical Programming. Springer, Berlin (1992)
15. Buckley, J.J.: Possibilistic linear programming with triangular fuzzy numbers. Fuzzy Sets Syst. **26**(1), 135–138 (1988)
16. Buckley, J.J.: Solving possibilistic linear programming. Fuzzy Sets Syst. **31**(3), 329–341 (1989)
17. Negi, D.S., Lee, E.S.: Possibility programming by the comparison of fuzzy numbers. Comput. Math Appl. **25**(9), 43–50 (1993)

18. Fullér, R.: On stability in fuzzy linear programming problems. Fuzzy Sets Syst. **30**(3), 339–344 (1989)
19. Hatami-Marbini, A., Agrell, P.J., Tavana, M., Emrouznejad, A.: A stepwise fuzzy linear programming model with possibility and necessity relations. J. Intell. Fuzzy Syst. **25**(1), 81–93 (2013)
20. Buckley, J.J., Feuring, T.: Evolutionary algorithm solution to fuzzy problems: fuzzy linear programming. Fuzzy Sets Syst. **109**(1), 35–53 (2000)
21. Saati, S., Tavana, M., Hatami-Marbini, A., Hajiakhondi, E.: A fuzzy linear programming model with fuzzy parameters and decision variables. Int. J. Inf. Decis. Sci. **7**(4), 312–333 (2015)
22. Kumar, A., Kaur, J.: A new method for solving fuzzy linear programs with trapezoidal fuzzy numbers. J. Fuzzy Set Valued Anal. **Art. ID 00102**, 12 (2011)
23. Hashemi, S.M., Modarres, M., Nasrabadi, E., Nasrabadi, M.M.: Fully fuzzified linear programming, solution and duality. Intell. Fuzzy Syst. **17**(1), 253–261 (2006)
24. Hosseinzadeh Lotfi, F., Allahviranloo, T., Alimardani Jondabeh, M., Alizadeh, L.: Solving a full fuzzy linear programming using lexicography method and fuzzy approximate solution. Appl. Math. Model. **33**(7), 3151–3156 (2009)
25. Kumar, A., Kuar, A.: Optimal solution of fuzzy transportation problems based on fuzzy linear programming formulation. J. Adv. Res. Appl. Math. **2**(4), 70–84 (2010)
26. Kumar, A., Kaur, J., Singh, P.: A new method for solving fully fuzzy linear programming problems. Appl. Math. Modell. **35**, 817–823 (2011)
27. Ezzati, R., Khorram, E., Enayati, R.: A new algorithm to solve fully fuzzy linear programming problems using the MOLP problem. Appl. Math. Model. **39**, 3183–3193 (2015)
28. Das, S.K., Mandal, T., Edalatpanah, S.A.: A mathematical model for solving fully fuzzy linear programming problem with trapezoidal fuzzy numbers. Appl. Intell. **46**(3), 509–519 (2017)
29. Cheng, H., Huang, W., Cai, J.: Solving a fully fuzzy linear programming problem through compromise programming. J. Appl. Math. **Art. ID 726296**, 10 (2013)
30. Hosseinzadeh, A., Edalatpanah, S.A.: A new approach for solving fully fuzzy linear programming by using the lexicography method. Adv. Fuzzy Syst. **Art. ID 1538496**, 6 (2016)
31. Ishibuchi, H., Tanaka, H.: Multiobjective programming in optimization of the interval objective function. Eur. J. Oper. Res. **48**, 219–225 (1990)
32. Chanas, S.: The use of parametric programming in fuzzy linear programming problems. Fuzzy Sets Syst. **11**, 243–251 (1983)
33. Dubois, D., Prade, H.: Fuzzy Sets and Systems: Theory and Applications. Academic, New York (1980)
34. Mahdavi-Amiri, N., Nasseri, S.H.: Duality results and a dual simplex method for linear programming problems with trapezoidal fuzzy variables. Fuzzy Sets Syst. **158**(17), 1961–1978 (2007)
35. Ganesan, K., Veeramani, P.: Fuzzy linear programming with trapezoidal fuzzy numbers. Ann. Oper. Res. **143**(1), 305–315 (2006)
36. Ebrahimnejad, A., Tavana, M.: A novel method for solving linear programming problems with symmetric trapezoidal fuzzy numbers. Appl. Math. Model. **38**(17–18), 4388–4395 (2014)
37. Nasrabadi, M.M., Nasrabadi, E.: A mathematical programming approach to fuzzy linear regression analysis. Appl. Math. Comput. **155**, 873–881 (2004)
38. Ebrahimnejad, A., Nasseri, S.H., Hosseinzadeh Lotfi, F.: Bounded linear programs with trapezoidal fuzzy numbers. Int. J. Uncertainty Fuzziness Knowl. Based Syst. **18**(3), 269–286 (2010)
39. Ebrahimnejad, A., Nasseri, S.H., HosseinzadehLotfi, F., Soltanifar, M.: A primal-dual method for linear programming problems with fuzzy variables. Eur. J. Ind. Eng. **4**(2), 189–209 (2010)
40. Mahdavi-Amiri, N., Nasseri, S.H.: Duality in fuzzy number linear programming by use of a certain linear ranking function. Appl. Math. Comput. **180**(1), 206–216 (2006)

41. Maleki, H.R., Tata, M., Mashinchi, M.: Linear programming with fuzzy variables. Fuzzy Sets Syst. **109**(1), 21–33 (2000)
42. Kumar, A., Kaur, J.: Fuzzy optimal solution of fully fuzzy linear programming problems using ranking function. J. Intell. Fuzzy Syst. **26**, 337–344 (2014)
43. Bhardwaj, B., Kumar, A.: A note on the paper "A new algorithm to solve fully fuzzy linear programming problems using the MOLP problem". J Appl. Math. Model. **39**(19), 5982–5985 (2015)

Chapter 5
Fuzzy Transportation Problem

5.1 Introduction

The transportation problem (TP) is an important network-structured LP problem that arises in several contexts and can be applied to a wide variety of situations, such as scheduling, production, investment, deciding plant location and inventory control. The central concept in the TP is to determine the minimum total transportation cost of a commodity for satisfying the demand at destinations using the available supply at the origins.

The TP with equality constraints is formulated mathematically as follows:

$$\min z = \sum_{i=1}^{m} \sum_{j=1}^{n} c_{ij} x_{ij}$$

$$s.t. \quad \sum_{j=1}^{n} x_{ij} = a_i, \quad i = 1, 2, \ldots, m,$$

$$\sum_{i=1}^{m} x_{ij} = b_j, \quad j = 1, 2, \ldots, n, \qquad (5.1)$$

$$x_{ij} \geq 0, \qquad i = 1, 2, \ldots, m, j = 1, 2, \ldots, n.$$

where, m is the total number of origins, n is the total number of destinations, a_i is the supply of the commodity at the ith origin, b_j is the demand for the commodity at the jth destination, c_{ij} is the cost of transportation for a unit quantity of the commodity from the ith origin to the jth destination, and x_{ij} is the quantity of the commodity that should be transported from the ith origin to the jth destination to minimize the total transportation cost $z = \sum_{i=1}^{m} \sum_{j=1}^{n} c_{ij} x_{ij}$.

Remark 5.1 Model (5.1) is feasible if, and only if, the total supply is equal to the total demand, i.e. $\sum_{i=1}^{m} a_i = \sum_{j=1}^{n} b_j$.

© Springer International Publishing AG, part of Springer Nature 2018
A. Ebrahimnejad and J. L. Verdegay, *Fuzzy Sets-Based Methods and Techniques for Modern Analytics*, Studies in Fuzziness and Soft Computing 364,
https://doi.org/10.1007/978-3-319-73903-8_5

Remark 5.2 If $\sum_{i=1}^{m} a_i = \sum_{j=1}^{n} b_j$ then the TP (5.1) is said to be a balanced TP, otherwise it is called an unbalanced TP.

In a similar way, the mathematical description of the conventional TP with inequality constraints is:

$$\min z = \sum_{i=1}^{m} \sum_{j=1}^{n} c_{ij} x_{ij}$$

$$s.t. \quad \sum_{j=1}^{n} x_{ij} \le a_i, \quad i = 1, 2, \ldots, m, \tag{5.2}$$

$$\sum_{i=1}^{m} x_{ij} \ge b_j, \quad j = 1, 2, \ldots, n,$$

$$x_{ij} \ge 0, \quad i = 1, 2, \ldots, m, j = 1, 2, \ldots, n.$$

Remark 5.3 Model (5.2) is feasible if, and only if, the total supply is greater or equal to the total demand, i.e. $\sum_{i=1}^{m} a_i \ge \sum_{j=1}^{n} b_j$.

Generally, transportation problems are solved with the assumption that the transportation costs, supply and demand are specified precisely. However, in many cases, the decision maker does not possess exact information about the coefficients for the transportation problem. If the information is vague, that is, if it lacks precision, the corresponding coefficients or elements defining the problem can be formulated using fuzzy sets, giving rise to fuzzy transportation problems (FTPs).

On the one hand, because the transportation problem is a special type of LP problem, a straightforward approach is to apply existing FLP techniques to the FTP. On the other hand, because of the special structure of the TP, a host of specialized approaches have been proposed for the FTP. As in the case of generalized FLP problems, the categories mentioned in Chaps. 2, 3, and 4 can be considered for classifying the FTPs. Here, we focus on four main FTPs, classified as follows:

- FTPs of type-I: Those FTPs in which fuzziness exists in the transportation cost belong to this group.
- FTPs of type-II: Those FTPs in which fuzziness exists in the supply and demand as well as decision variables belong to this group.
- FTPs of type-III: Those FTPs in which fuzziness exists in both the transportation cost and the supply and demand belong to this group.
- FTPs of type-IV: Those FTPs in which fuzziness exists in all of the model parameters (costs, supply, and demand) as well as decision variables belong to this group.

A comprehensive literature review on solving FTPs is presented in the next section.

5.2   A Brief Literature Review on FTPs

An overview of the research on various FTPs and a discussion of the different models for solving them are presented as follows.

In 1978, Zimmermann [1] used FLP approaches to solve the linear vector maximum problem. He showed that solutions obtained by the FLP were always efficient and developed a method for solving the FTP. Ohegeartaigh [2] proposed an algorithm for solving transportation problems with fuzzy constraints. He also examined the relationship between the algebraic structure of the optimum solution of the deterministic problem and its fuzzy equivalent. Chanas et al. [3] investigated TPs with a fuzzy supply and demand and solved these via the parametric programming technique in terms of the Bellman-Zadeh criterion. Their method provided a solution that simultaneously satisfied the constraints and the goal to the maximal degree. In addition, Chanas et al. [4] formulated classical, interval and FTPs and focused on the relationship between interval transportation and fuzzy transportation. For each case, they provided auxiliary classical models, whereby optimal solutions were obtained for the original problems. In the FTP considered here, fuzziness was only present in the constraint set. Chanas and Kuchta [5] proposed the concept of the optimal solution of the TP with fuzzy coefficients and an algorithm for determining all of the efficient solutions. Using this algorithm, the FTP is converted into a bi-criteria TP with precise objective functions. Jimenez and Verdegay [6, 7] investigated a fuzzy solid TP (FSTP) in which the supply, demand and conveyance capacities were represented by trapezoidal fuzzy numbers, and they applied a parametric approach for obtaining the fuzzy solution. Liu and Kao [8] developed a procedure based on the extension principle for determining the membership function of the fuzzy total transportation cost for TPs with equality and inequality constraints on the unit transportation costs, supply quantities, and demand quantities. Liu [9] proposed a new approach for solving FSTPs using the method proposed by Liu and Kao [8]. This method was able to derive the fuzzy objective value of the FSTP when the cost coefficients, supply and demand quantities and conveyance capacities were fuzzy numbers. Chiang [10] used interval-valued fuzzy numbers instead of normal fuzzy numbers to represent the availability and demand and proposed a method for obtaining the optimal solution of single-objective TPs by representing the availability and demand as interval-valued triangular fuzzy numbers. Gupta and Kumar [11] generalized the method proposed by Chaing [10] for obtaining the solution of a linear multi-objective TP by representing all the parameters as interval-valued trapezoidal fuzzy numbers. Liang et al. [12] developed a novel interactive possibilistic linear programming approach for solving transportation planning decision problems with imprecise goal and constraints. The proposed approach minimized the total transportation costs in terms of imprecise cost coefficients, available supply and forecast demand. The proposed strategy considered simultaneously minimizing the most possible values, maximizing the possibility of obtaining lower values, and minimizing the risk of obtaining higher values for the imprecise objective function.

Gani and Razak [13] presented a two-stage cost-minimizing FTP in which the supply and demand were trapezoidal fuzzy numbers. They used a parametric approach for obtaining a fuzzy solution with the aim of minimizing the sum of the transportation costs in two stages. According to the solution definition of a fuzzy equation based on goal programming and the algorithm for calculating the fuzzy number difference, Li et al. [14] developed an exact TP for a fuzzy environment and discussed a balanced TP with fuzzy shipping cost coefficients. Chen et al. [15] extended the classical transportation problem by considering the preference of arcs in a transportation path, the flexibility of the supply and demand, and the randomness of the unit transportation cost of each path. Lin [16] used a genetic algorithm to solve the TP with fuzzy supply and demand coefficients. Here, ranking fuzzy numbers with signed-distance measurements were applied for the evaluation and selection of the algorithm.

Dinagar and Palanivel [17] investigated the FTP with the aid of trapezoidal fuzzy numbers and proposed a fuzzy modified-distribution method to obtain the optimal solution in terms of fuzzy numbers. Pandian and Natarajan [18] introduced a new algorithm, the fuzzy zero-point method, for determining the fuzzy optimal solution for FTPs in which the transportation cost, supply and demand were represented by trapezoidal fuzzy numbers. Chakraborty and Chakraborty [19] proposed a method for the minimization of the transportation cost and transportation time for a fuzzy supply, demand, and transportation cost. A method involving a novel representation of trapezoidal fuzzy numbers was proposed by Kumar and Kaur [20] for finding the optimal solution of FTPs on the basis of the FLP problem. In addition, a method based on the ranking function was proposed for solving the FTP assuming that transportation cost, supply, and demand of the commodity are represented by generalized trapezoidal fuzzy numbers [21]. Moreover, a method was proposed for obtaining the exact fuzzy optimal solution of an unbalanced FTP by representing all the parameters as LR flat fuzzy numbers [22]. Senthilkumar and Vengataasalam [23] discussed the solution of a fuzzy FTP with fuzzy quantities and solved the problem in two stages. In the first stage, the FTP was reduced to a crisp system by using the lower and upper bounds of fuzzy quantities. In the second stage, the crisp TPs were solved by the usual simplex method. Ojha et al. [24] applied the modified sub-gradient method for optimization and showed its effectiveness in a fuzzy multi-item balanced TP where the unit transportation costs were imprecise. Shanmugasundari and Ganesan [25] proposed an algorithm for solving FTPs in which the supply, demand, and transportation costs were uncertain. They developed fuzzy versions of Vogel's method and the MODI method for obtaining the fuzzy IBFS and fuzzy optimal solution, respectively, without converting the problem into a classical TP. Kaur and Kumar [26] proposed new methods for determining the initial basic feasible solution (IBFS) of transportation problems by using a ranking function, in which the transportation costs were represented as generalized trapezoidal fuzzy numbers instead of normal fuzzy trapezoidal numbers. They then generalized the fuzzy modified-distribution method to determine the optimal solution of the FTP using the IBFS. Ebrahimnejad [27] proved that once the ranking function is selected, the FTP considered by Kaur and Kumar [26] is converted into a deterministic one, which is easily

solved by standard transportation algorithms. Ebrahimnejad [28, 29] proposed a
two-step method for solving FTPs where all of the parameters were represented by
non-negative triangular fuzzy numbers. Moreover, Ebrahimnejad [30] proposed an
FLP approach for solving the TP involving interval-valued trapezoidal fuzzy numbers by comparing the interval-valued fuzzy numbers using the signed distance
ranking. Sudhagar and Ganesan [31] proposed an algorithm for obtaining the optimal
solution of FTPs, where the supply, demand, and cost coefficients were fuzzy
numbers. Ebrahimnejad [32] used an example to show that this method does not
always lead to a fuzzy optimal solution. Ebrahimnejad [33] proposed a novel algorithm, based on a new lexicographic ordering on triangular fuzzy numbers, which
proposed to solve the FTP by converting it into an MOLP problem with
three-objective functions. Ebrahimnejad [34] proposed a new method for solving
FTPs in which the transportation costs and supply and demand were represented by
non-negative LR flat fuzzy numbers. In the proposed method, the FTP is converted
into four TPs, which are solved using standard transportation simplex algorithms.
Ebrahimnejad and Verdegay [35] formulated a TP in which costs are triangular
intuitionistic fuzzy numbers and availabilities and demands are taken as exact
numerical values. They proposed an efficient computational solution approach for
solving intuitionistic FTPs based on classical transportation algorithms. Moreover,
Ebrahimnejad and Verdegay [36] investigated a TP having uncertainty as well as
hesitation in prediction of the transportation cost, availabilities and demands of
products. They then proposed a novel solution approach for solving fully intuitionistic fuzzy TP based on classical LP algorithms.

5.3   FTPs of Type-I

The FTPs of type-I, in which a decision maker considers the transportation costs as
fuzzy numbers to deal efficiently with the uncertainty arising in the prediction of the
transportation cost, but (s)he is sure about the supply and demand values of the
commodity, can be formulated as follows:

$$\min z = \sum_{i=1}^{m} \sum_{j=1}^{n} \tilde{c}_{ij} x_{ij}$$

$$s.t. \quad \sum_{j=1}^{n} x_{ij} = a_i, \quad i = 1, 2, \ldots, m,$$

$$\sum_{i=1}^{m} x_{ij} = b_j, \quad j = 1, 2, \ldots, n, \tag{5.3}$$

$$x_{ij} \geq 0, \qquad i = 1, 2, \ldots, m, j = 1, 2, \ldots, n.$$

where transportation costs are represented by generalized trapezoidal fuzzy
numbers.

Let $\tilde{u}_i$ and $\tilde{v}_i$ be the fuzzy dual variables associated with the ith row and jth column constraints, respectively, then the fuzzy dual of the FTP of type-I (5.3) will be as follows [26]:

$$\max \quad \sum_{i=1}^{m} a_i \tilde{u}_i + \sum_{j=1}^{n} b_j \tilde{v}_j \tag{5.4}$$
$$s.t. \ \tilde{u}_i + \tilde{v}_j \preceq \tilde{c}_{ij} \quad i = 1, 2, \ldots, m, j = 1, 2, \ldots, n.$$

Remark 5.4 The fuzzy dual of the FTP of type-I (5.3) defined as the FLP problem (5.4) has been obtained according to the definition of the dual problem for the LP problems with fuzzy coefficient costs discussed in Chap. 4.

In this section, we shall explore two approaches for solving the FTP of type-I (5.3).

5.3.1 Kaur and Kumar's Approach

Kaur and Kumar [26] proposed a new method based on ranking function for solving the FTP of type-I (5.3). They modified some existing methods for finding the initial basic feasible solution (IBFS) and fuzzy optimal solution using the ranking function given in Remark 1.15.

Based on the linear function given in Remark 1.15, Kaur and Kumar [26] introduced three methods, namely the generalized fuzzy North-West corner method (GFNWCM), the generalized fuzzy least-cost method (GFLCM) and the generalized fuzzy Vogel's approximation method (GFFAM) to find the initial BFS for the FTP of type-I (5.3). They then proposed the generalized fuzzy modified distribution method (GFMODIM) to find the fuzzy optimal solution for the FTP of type-I (5.3) with the help of BFS. They used the following theorem to provide optimality criteria for the obtained BFS.

Theorem 5.1 *Let the FTP of type-I (5.3) have a BFS with B as a basis matrix. If $\tilde{u}_i + \tilde{v}_j \preceq \tilde{c}_{ij}$ for all non-basic variables, the current BFS is optimal.*

The two main steps of the fuzzy transportation algorithm proposed by Kaur and Kumar [26] for solving the FTP of type-I (5.3) are summarized as follows:

Step 1: Find an IBFS using the GFNWCM, GFLCM or GFVAM.

Step 2: Find the optimal solution of the FTP of type-I (5.3) using the GFMODIM.

The steps of the GFNWCM for finding the IBFS to carry out Step 1 are as follows:

(1) Select the north-west corner (NWC) cell of the fuzzy transportation table $m \times n$ and allocate as much as possible to the variable corresponding to the determined cell.

(2) Adjust the supply and demand and delete the satisfied row or column.
(3) Repeat the process for the reduced table until the table is reduced to 1×1.

The steps for finding the IBFS using the GFLCM to carry out Step 1 are as follows:

(1) Determine the smallest fuzzy cost of the fuzzy transportation table $m \times n$ using the linear function given in Remark 1.15 and allocate as much as possible to the variable corresponding to the determined cell.
(2) Adjust the supply and demand and delete the satisfied row or column.
(3) Repeat the process for the reduced table until the table is reduced to 1×1.

The steps for finding the IBFS using the GFVAM to carry out Step 1 are as follows:

(1) Compute the fuzzy penalty for each row and each column of the fuzzy transportation table of order $m \times n$.
(2) Select the highest fuzzy penalty using the linear function given in Remark 1.15.
(3) Determine the cell with the smallest cost in the selected row or column using the linear function given in Remark 1.15.
(4) Allocate as much as possible to the variable corresponding to the determined cell, adjust the supply and demand and delete the satisfied row or column.
(5) Repeat the process for the reduced table until the table is reduced to 1×1.

Remark 5.5 The penalty measure for each row (column) is determined by subtracting the smallest unit fuzzy cost element in the row (column) from the next smallest unit fuzzy cost element in the same row (column).

Let us suppose that an IBFS obtained using the GFNWCM, GFLCM or GFVAM with basis B is at hand. The main steps of the GFMODIM to carry out Step 2 are summarized as follows:

(1) For each cell (i,j), define fuzzy dual variables $\tilde{u}_i$ and $\tilde{v}_j$ associated with the ith row and jth column, respectively.
(2) Solve the intuitionistic fuzzy system $\tilde{u}_i + \tilde{v}_j = \tilde{c}_{ij}$ for each basic cell (i,j).
(3) Compute $\tilde{d}_{ij} = \tilde{c}_{ij} - (\tilde{u}_i + \tilde{v}_j)$ for each non-basic cell (i,j). If $\tilde{d}_{ij} \succeq \tilde{0}$ for each non-basic cell (i,j), then stop, the current BFS is optimal. Otherwise select an entering cell (a non-basic cell with the most negative rank $\tilde{d}_{ij}$ based on the linear function given in Remark 1.15).
(4) Determine an existing cell; obtain the new BFS using the standard transportation methods and repeat Step 1.

Example 5.1   [26] Table 5.1 gives the supply (a_i) of the commodity available at three sources $S_i, i = 1, 2, 3$ and their demand (b_j) at three destinations $D_j, j = 1, 2, 3$, and the approximate cost for transporting one unit quantity of said commodity from each source to each destination is represented by a generalized trapezoidal fuzzy number. Determine the fuzzy optimal transportation of the commodity such that the total fuzzy transportation cost is minimized.

Table 5.1 Tabular representation of the chosen FTP

	D_1	D_2	D_3	a_i
S_1	(1, 4, 9, 19; 0.5)	(1, 2, 5, 9; 0.4)	(2, 5, 8, 18; 0.5)	10
S_2	(8, 9, 12, 26; 0.5)	(3, 5, 8, 12; 0.2)	(7, 9, 13, 28; 0.4)	14
S_3	(11, 12, 20, 27; 0.5)	(0, 5, 10, 15; 0.8)	(4, 5, 8, 11; 0.6)	15
b_j	15	14	10	

Each of the three suggested methods (GFNWCM, GFLCM or GFVAM) can be used to obtain an IBFS to the FTP given in Table 5.1. Then, starting from the obtained IBFS, we can find the optimal solution using the GFMODIM.

We first use the GFNWCM to find the IBFS and then apply the GFMODIM on the IBFS, obtained by using the GFNWCM, to find the fuzzy optimal solution.

The NWC cell of Table 5.1 corresponds to S_1 and D_1. The maximum possible allocation at this position is minimum $\{10, 15\} = 10$. Allocate $x_{11} = 10$ in the NWC cell of Table 5.1. After allocation in the NWC cell the remaining demand at destination D_1 is changed to $15 - 10 = 5$. Since now no commodity is available at S_1 for supplying to any other destination ignore the row corresponding to S_1. Due to ignoring a row, Table 5.1 of order 3×3 reduces into a table of order 2×3. The reduced table, after allocating x_{11} and ignoring the row of S_1, is shown in Table 5.2.

Now as explained in the proposed algorithm, repeating the same process until the fuzzy transportation problem is reduced into a table of order 1×1, the obtained IBFS of the given problem is

$$
\begin{aligned}
x_{11} &= 10, \; x_{12} = 0, \; x_{13} = 0, \\
x_{21} &= 5, \; x_{22} = 9, \; x_{23} = 0, \\
x_{31} &= 0, \; x_{32} = 5, \; x_{33} = 10.
\end{aligned}
\tag{5.5}
$$

Now the GFMODIM is applied on the IBFS (5.5), to find the fuzzy optimal solution. According to Step (1) and Step (2) of the GFMODIM proposed by Kaur and Kumar [26] the following fuzzy system should be solved to test the optimality of the IBFS (5.5):

$$
\begin{aligned}
\tilde{u}_1 + \tilde{v}_1 &= (1, 4, 9, 19; 0.5), \\
\tilde{u}_2 + \tilde{v}_1 &= (8, 9, 12, 26; 0.5), \\
\tilde{u}_2 + \tilde{v}_2 &= (3, 5, 8, 12; 0.2), \\
\tilde{u}_3 + \tilde{v}_2 &= (0, 5, 10, 15; 0.8), \\
\tilde{u}_3 + \tilde{v}_3 &= (4, 5, 8, 11; 0.6).
\end{aligned}
\tag{5.6}
$$

Table 5.2 Tabular representation of the problem after the first iteration of the GFNWCM

	D_1	D_2	D_3	a_i
S_2	(8, 9, 12, 26; 0.5)	(3, 5, 8, 12; 0.2)	(7, 9, 13, 28; 0.4)	14
S_3	(11, 12, 20, 27; 0.5)	(0, 5, 10, 15; 0.8)	(4, 5, 8, 11; 0.6)	15
b_j	5	14	10	

With $\tilde{u}_2 = (0,0,0,0;1)$, the fuzzy system (5.6) gives

$$\tilde{v}_1 = (8,\ 9,\ 12,\ 26;\ 0.5) - (0,0,0,0;1) = (8,\ 9,\ 12,\ 26;\ 0.5),$$
$$\tilde{u}_1 = (1,\ 4,\ 9,\ 19;\ 0.5) - (8,\ 9,\ 12,\ 26;\ 0.5) = (-25,\ -8,\ 0,\ 11;\ 0.5),$$
$$\tilde{v}_2 = (3,\ 5,\ 8,\ 12;\ 0.2) - (0,0,0,0;1) = (3,\ 5,\ 8,\ 12;\ 0.2), \qquad (5.7)$$
$$\tilde{u}_3 = (0,\ 5,\ 10,\ 15;\ 0.8) - (3,\ 5,\ 8,\ 12;\ 0.2) = (-12,\ -3,\ 5,\ 12;\ 0.2),$$
$$\tilde{v}_3 = (4,\ 5,\ 8,\ 11;\ 0.6) - (-12,\ -3,\ 5,\ 12;\ 0.2) = (-8,\ 0,\ 11,\ 23;\ 0.2).$$

According to Step (3) of the GFMODIM proposed by Kaur and Kumar [26], it is required to compute $\tilde{d}_{ij} = \tilde{c}_{ij} - (\tilde{u}_i + \tilde{v}_j)$ for each non-basic cell (i, j). Thus, we have:

$$\tilde{d}_{12} = \tilde{c}_{12} - (\tilde{u}_1 + \tilde{v}_2) = (-22, -6, 8, 31; 0.2),$$
$$\tilde{d}_{13} = \tilde{c}_{13} - (\tilde{u}_1 + \tilde{v}_3) = (-32, -6, 0, 51; 0.2),$$
$$\tilde{d}_{23} = \tilde{c}_{23} - (\tilde{u}_2 + \tilde{v}_3) = (-16, 2, 13, 36; 0.2), \qquad (5.8)$$
$$\tilde{d}_{31} = \tilde{c}_{31} - (\tilde{u}_3 + \tilde{v}_1) = (-27, -5, 14, 31; 0.2).$$

Since $\Re(\tilde{d}_{ij}) \geq 0$ for all non-basic cell (i,j), the BFS (5.5) is the optimal solution of the given FTP. The minimum fuzzy transportation cost is achieved as follows:

$$\sum_{i=1}^{3}\sum_{j=1}^{3} \tilde{c}_{ij}x_{ij} = 10(1,\ 4,\ 9,\ 19;\ 0.5) \oplus 5(8,\ 9,\ 12,\ 26;\ 0.5) \oplus 9(3,\ 5,\ 8,\ 12;\ 0.2)$$

$$\oplus\ 5(0,\ 5,\ 10,\ 15;\ 0.8) \oplus 10(4,\ 5,\ 8,\ 11;\ 0.6)$$
$$= (10,\ 40,\ 90,\ 190;\ 0.5) \oplus (40,\ 45,\ 60,\ 130;\ 0.5)$$
$$\oplus\ (27,\ 45,\ 72,\ 108;\ 0.2) \oplus (0,\ 25,\ 50,\ 75;\ 0.8)$$
$$\oplus\ (40,\ 50,\ 80,\ 110;\ 0.6) = (117,\ 205,\ 352,\ 613;\ 0.2)$$

$$(5.9)$$

Now, we use the GFLCM to find the IBFS and then apply the GFMODIM on the IBFS, obtained by using the GFLCM, to find the fuzzy optimal solution of the same FTP.

The lowest fuzzy cost in Table 5.1 corresponds to S_1 and D_2. The maximum possible allocation at this position is minimum $\{10, 14\} = 10$. Allocate $x_{12} = 10$ in the cell corresponding to S_1 and D_2 of Table 5.1. After allocation in this cell the remaining demand at destination D_2 is changed to $14 - 10 = 4$. Since now no commodity is available at S_1 for supplying to any other destination ignore the row corresponding to S_1. Due to ignoring a row, Table 5.1 of order 3×3 reduces into a table of order 2×3. The reduced table, after allocating x_{12} and ignoring row of S_1, is shown in Table 5.3.

Now as explained in the proposed algorithm, repeating the same process until the fuzzy transportation problem is reduced into a table of order 1×1, the obtained IBFS of the given problem is

$$x_{11} = 0, x_{12} = 10, x_{13} = 0,$$
$$x_{21} = 10, x_{22} = 4, x_{23} = 0, \tag{5.10}$$
$$x_{31} = 5, x_{32} = 0, x_{33} = 10.$$

Now the GFMODIM is applied on the IBFS (5.10), to find the fuzzy optimal solution. According to Step (1) and Step (2) of the GFMODIM proposed by Kaur and Kumar [26] the following fuzzy system should be solved to test the optimality of the IBFS (5.10):

$$\tilde{u}_1 + \tilde{v}_2 = (1, 2, 5, 9; 0.4),$$
$$\tilde{u}_2 + \tilde{v}_1 = (8, 9, 12, 26; 0.5),$$
$$\tilde{u}_2 + \tilde{v}_2 = (3, 5, 8, 12; 0.2), \tag{5.11}$$
$$\tilde{u}_3 + \tilde{v}_1 = (11, 12, 20, 27; 0.5),$$
$$\tilde{u}_3 + \tilde{v}_3 = (4, 5, 8, 11; 0.6).$$

With $\tilde{u}_2 = (0, 0, 0, 0; 1)$, the fuzzy system (5.11) gives

$$\tilde{v}_1 = (8, 9, 12, 26; 0.5) - (0, 0, 0, 0; 1) = (8, 9, 12, 26; 0.5),$$
$$\tilde{v}_2 = (3, 5, 8, 12; 0.2) - (0, 0, 0, 0; 1) = (3, 5, 8, 12; 0.2),$$
$$\tilde{u}_1 = (1, 2, 5, 9; 0.4) - (3, 5, 8, 12; 0.2) = (-11, -6, 0, 6; 0.2), \tag{5.12}$$
$$\tilde{u}_3 = (11, 12, 20, 27; 0.5) - (8, 9, 12, 26; 0.5) = (-15, 0, 11, 19; 0.2),$$
$$\tilde{v}_3 = (4, 5, 8, 11; 0.6) - (-15, 0, 11, 19; 0.2) = (-15, -6, 8, 16; 0.2).$$

Table 5.3 Tabular representation of the problem after the first iteration of the GFLCM

	D_1	D_2	D_3	a_i
S_2	(8, 9, 12, 26; 0.5)	(3, 5, 8, 12; 0.2)	(7, 9, 13, 28; 0.4)	14
S_3	(11, 12, 20, 27; 0.5)	(0, 5, 10, 15; 0.8)	(4, 5, 8, 11; 0.6)	15
b_j	15	4	10	

According to Step (3) of the GFMODIM proposed by Kaur and Kumar [26], it is required to compute $\tilde{d}_{ij} = \tilde{c}_{ij} - (\tilde{u}_i + \tilde{v}_j)$ for each non-basic cell (i, j). Thus, we have:

$$
\begin{aligned}
\tilde{d}_{11} &= \tilde{c}_{11} - (\tilde{u}_1 + \tilde{v}_1) = (-31, -8, 6, 22; 0.2), \\
\tilde{d}_{13} &= \tilde{c}_{13} - (\tilde{u}_1 + \tilde{v}_3) = (-20, -3, 20, 44; 0.2), \\
\tilde{d}_{23} &= \tilde{c}_{23} - (\tilde{u}_2 + \tilde{v}_3) = (-9, 1, 19, 43; 0.2), \\
\tilde{d}_{32} &= \tilde{c}_{32} - (\tilde{u}_3 + \tilde{v}_2) = (-31, -14, 5, 27; 0.2).
\end{aligned}
\tag{5.13}
$$

Since the rank of $\tilde{d}_{11} = (-31, -8, 6, 22; 0.2)$ is most negative, then the fuzzy solution (5.10) is not fuzzy optimal. In this case, x_{11} is selected as the entering variable. According to the classical transportation algorithm, x_{12} is selected as the leaving variable and the new BFS is found as follows:

$$
\begin{aligned}
x_{11} &= 10, \ x_{12} = 0, \ x_{13} = 0, \\
x_{21} &= 0, \ x_{22} = 14, \ x_{23} = 0, \\
x_{31} &= 5, \ x_{32} = 0, \ x_{33} = 10.
\end{aligned}
\tag{5.14}
$$

Now as explained in the proposed algorithm, repeating the same process until $\Re(\tilde{d}_{ij}) \geq 0$ for each non-basic cell (i, j), the same fuzzy optimal solution given in (5.5) is obtained.

Finally, we use the GFVAM to find the IBFS and then apply the GFMODIM on the IBFS, obtained by using the GFVAG, to find the fuzzy optimal solution of the same FTP.

According to Step (1) of the GFVAM proposed by Kaur and Kumar [26], it is required to compute the fuzzy penalty regarding Remark 5.5 for each row and each column of Table 5.1. Take the first row and choose its smallest fuzzy cost and subtract this from the next smallest fuzzy cost. This is the penalty corresponding to the first row i.e., $R\tilde{F}P_1 = (2, 5, 8, 18; 0.5) - (1, 2, 5, 9; 0.4) = (-7, 0, 6, 17; 0.5)$. In this way, the fuzzy raw penalty (FRP) for each row and fuzzy column penalty (FCP) for each column are computed. The results are given as follows:

$$
\begin{aligned}
F\tilde{R}P_1 &= (2, 5, 8, 18; 0.5) - (1, 2, 5, 9; 0.4) = (-7, 0, 6, 17; 0.5), \\
F\tilde{R}P_2 &= (8, 9, 12, 26; 0.5) - (3, 5, 8, 12; 0.2) = (-4, 1, 7, 23; 0.2), \\
F\tilde{R}P_3 &= (0, 5, 10, 15; 0.8) - (4, 5, 8, 11; 0.6) = (-11, -3, 5, 11; 0.6), \\
F\tilde{C}P_3 &= (8, 9, 12, 26; 0.5) - (1, 4, 9, 19; 0.5) = (-11, 0, 8, 25; 0.5), \\
F\tilde{C}P_2 &= (3, 5, 8, 12; 0.2) - (1, 2, 5, 9; 0.4) = (-6, 0, 6, 11; 0.2), \\
F\tilde{C}P_3 &= (2, 5, 8, 18; 0.5) - (4, 5, 8, 11; 0.6) = (-9, -3, 3, 11; 0.5).
\end{aligned}
\tag{5.15}
$$

The highest fuzzy penalty $(-4, 1, 7, 23; 0.2)$ is that corresponding to the second row. The minimum fuzzy cost corresponding to this row is $(3, 5, 8, 12; 0.2)$. The maximum possible allocation at this position is minimum $\{14, 14\} = 14$. Allocate $x_{22} = 14$ in the cell corresponding to S_2 and D_2 of Table 5.1. After allocation in this cell the remaining demand at destination D_2 is changed to $14 - 14 = 0$. Since now no commodity is available at S_2 for supplying to any other destination ignore the row corresponding to S_2. Due to ignoring a row, Table 5.2 of order 3×3 reduces into a table of order 2×3. The reduced table, after allocating x_{22} and ignoring the row of S_2, is shown in Table 5.4.

Now as explained in the proposed algorithm, repeating the same process until the fuzzy transportation problem is reduced into a table of order 1×1, the obtained IBFS of the given problem is

$$x_{11} = 10, \ x_{12} = 0, \ x_{13} = 0,$$
$$x_{21} = 0, \ x_{22} = 14, \ x_{23} = 0, \qquad (5.16)$$
$$x_{31} = 5, \ x_{32} = 0, \ x_{33} = 10.$$

Now the GFMODIM is applied on the IBFS (5.16), to find the fuzzy optimal solution. According to Step (1) and Step (2) of the GFMODIM proposed by Kaur and Kumar [26] the following fuzzy system should be solved to test the optimality of the IBFS (5.16):

$$\tilde{u}_1 + \tilde{v}_1 = (1, 4, 9, 19; 0.5),$$
$$\tilde{u}_2 + \tilde{v}_2 = (3, 5, 8, 12; 0.2),$$
$$\tilde{u}_3 + \tilde{v}_1 = (11, 12, 20, 27; 0.5), \qquad (5.17)$$
$$\tilde{u}_3 + \tilde{v}_2 = (0, 5, 10, 15; 0.8),$$
$$\tilde{u}_3 + \tilde{v}_3 = (8, 9, 12, 26; 0.5).$$

With $\tilde{u}_3 = (0, 0, 0, 0; 1)$, the fuzzy system (5.17) gives

$$\tilde{v}_1 = (1, 4, 9, 19; 0.5),$$
$$\tilde{v}_2 = (3, 5, 8, 12; 0.2),$$
$$\tilde{v}_3 = (8, 9, 12, 26; 0.5), \qquad (5.18)$$
$$\tilde{u}_1 = (-26, -16, -3, 8; 0.5),$$
$$\tilde{u}_2 = (-12, -5, 3, 12; 0.2).$$

Table 5.4 Tabular representation of the problem after the first iteration of the GFVAM

	D_1	D_2	D_3	a_i
S_1	(1, 4, 9, 19; 0.5)	(1, 2, 5, 9; 0.4)	(2, 5, 8, 18; 0.5)	10
S_3	(11, 12, 20, 27; 0.5)	(0, 5, 10, 15; 0.8)	(4, 5, 8, 11; 0.6)	15
b_j	15	0	10	

According to Step (3) of the GFMODIM proposed by Kaur and Kumar [26], it is required to compute $\tilde{d}_{ij} = \tilde{c}_{ij} - (\tilde{u}_i + \tilde{v}_j)$ for each non-basic cell (i, j). Thus, we have:

$$
\begin{aligned}
\tilde{d}_{12} &= \tilde{c}_{12} - (\tilde{u}_1 + \tilde{v}_2) = (-22, -5, 16, 35; 0.4), \\
\tilde{d}_{13} &= \tilde{c}_{13} - (\tilde{u}_1 + \tilde{v}_3) = (-17, 0, 19, 40; 0.5), \\
\tilde{d}_{21} &= \tilde{c}_{21} - (\tilde{u}_2 + \tilde{v}_1) = (-31, -14, 5, 27; 0.2), \\
\tilde{d}_{23} &= \tilde{c}_{23} - (\tilde{u}_2 + \tilde{v}_3) = (-16, -2, 13, 36; 0.2).
\end{aligned}
\tag{5.19}
$$

Since the rank of $\tilde{d}_{21} = (-31, -14, 5, 27; 0.2)$ is most negative, then the fuzzy solution (5.19) is not fuzzy optimal. In this case, x_{21} is selected as the entering variable. According to the classical transportation algorithm, x_{13} is selected as the leaving variable and the new BFS is found as follows:

$$
\begin{aligned}
x_{11} &= 10, \ x_{12} = 0, \ x_{13} = 0, \\
x_{21} &= 5, \ x_{22} = 9, \ x_{23} = 0, \\
x_{31} &= 0, \ x_{32} = 5, \ x_{33} = 10.
\end{aligned}
\tag{5.20}
$$

Now as explained in the proposed algorithm, repeating the same process until $\Re(\tilde{d}_{ij}) \geq 0$ for each non-basic cell (i, j), the same fuzzy optimal solution given in (5.5) is obtained.

5.3.2   Ebrahimnejad's Approach

Ebrahimnejad [27] proved that once the ranking function is chosen, the FTP of type-I (5.3) is converted into a crisp one, which is easily solved by the standard transportation algorithms. Thus, the method proposed by Ebrahimnejad [27] is simpler and computationally more efficient than the method proposed by Kaur and Kumar [26]. Since the technique proposed by Ebrahimnejad [27] is based on a classical approach it is very easy to understand and to apply in real life TPs for the decision makers.

According to the ranking function given in Remark 1.15, it is possible to define a rank for each generalized trapezoidal fuzzy number. It follows that if $\tilde{A}_i = (a_i, b_i, c_i, d_i; w_i), (i = 1, 2, \ldots, k)$ be k generalized trapezoidal fuzzy numbers, then $\Re(\tilde{A}_i) = \frac{w(a_i + b_i + c_i + d_i)}{4}$, where $w = \min\{w_i, i = 1, 2, \ldots, k\}$. Thus, the FTP of type-I (5.2) is converted into an equivalent crisp TP. To do this, substitute the rank of each generalized trapezoidal fuzzy number instead of the corresponding generalized trapezoidal fuzzy number in the FTP under consideration. This leads to an equivalent crisp TP which can be solved by the standard transportation algorithms.

Then all arithmetic operations are done on the crisp numbers. As a result, the computational effort with this approach is significantly decreased.

The main contribution of this technique is the reduction of the computational complexity of the method proposed by Kaur and Kumar [26]. In particular, it is shown that the method proposed by Ebrahimnejad [27] needs fewer elementary operations such as additions, multiplications, and comparisons as compared to Kaur and Kumar's method [26].

According to Kaur and Kumar's method [26], to carry out Step (1) using the GFCLM it is required to determine the smallest fuzzy transportation cost in fuzzy transportation tableau (FTT) of order $m \times n$ until the tableau is reduced into a table of order 1×1. To do this, it is required to compare the fuzzy transportation costs using the ranking function given in Remark 1.15 until the FTT is reduced into an FTT of order 1×1. While based on Ebrahimnejad's method the comparison of fuzzy costs is done just once. Moreover, to find the IBFS using the GFVAM it is required to compute a fuzzy penalty for each row of the FTT of order $m \times n$ by subtracting the smallest entry from the next smallest entry in each step of this method. In a similar way, a fuzzy penalty is computed for each column of the FTT of order $m \times n$ in all iterations. After that, the highest fuzzy penalty is determined based on the ranking function given in Remark 1.15. These steps are repeated until the FTT is reduced into an FTT of order 1×1. As we see when doing Steps (1)–(3) using GFVAM, a lot of fuzzy arithmetic operations such as additions, subtractions and comparison on generalized trapezoidal fuzzy numbers are required. While based on Ebrahimnejad's method the comparison of fuzzy costs is done once and all arithmetic operations are done on real numbers.

In addition, to carry out Step 2 using the GFMODIM proposed by Kaur and Kumar [26] requires solving the fuzzy system $\tilde{u}_i + \tilde{v}_j = \tilde{c}_{ij}$ with $m + n - 1$ fuzzy equations corresponding to basic variables. After solving this fuzzy system, the fuzzy value $\tilde{d}_{ij}$ for each non-basic variable is obtained based on $\tilde{d}_{ij} = \tilde{c}_{ij} - (\tilde{u}_i + \tilde{v}_j)$. Finally, the entering column is determined according to the most negative rank of $\tilde{d}_{ij}$. This step requires to a lot of fuzzy additions and subtractions on generalized trapezoidal fuzzy numbers. While based on Ebrahimnejad's method, the entering column is found without solving any fuzzy system and without any fuzzy arithmetic operations. These results confirm that Ebrahimnejad's method is simpler and computationally more efficient than the method proposed by Kaur and Kumar [26].

Now, in order to illustrate the method proposed by Ebrahimnejad [27], the FTP given in Example 5.1 is considered.

Example 5.2 [27] Consider the FTP of type-I given in Table 5.1. We first put the rank order of each fuzzy transportation cost (given in Table 5.1) instead of the corresponding fuzzy number to obtain the classical TP. The results are given in Table 5.5.

The crisp TP given in Table 5.5 can be solved by the standard transportation algorithms. We first apply three famous methods, namely the north-west corner method, the least cost method and Vogel's approximation method for finding the IBFS and then we use the modified distribution method on the IBFS to obtain the

	D_1	D_2	D_3	a_i
S_1	1.65	0.85	1.65	10
S_2	2.75	1.4	2.85	14
S_3	3.5	1.5	1.4	15
b_j	15	14	10	

Table 5.5 Tabular representation of classical transportation problem

optimal solution. Finally, we compare the result obtained with that found by Kaur and Kumar [26].

If we use the north-west corner method to obtain the IBFS, we obtain the following solution, which is matched with the IBFS (5.5) obtained based on the GFNWCM proposed by Kaur and Kumar [26]:

$$x_{11} = 10, \ x_{12} = 0, \ x_{13} = 0,$$
$$x_{21} = 5, \ x_{22} = 9, \ x_{23} = 0, \qquad (5.21)$$
$$x_{31} = 0, \ x_{32} = 5, \ x_{33} = 10.$$

Now applying the modified distribution method on the IBFS given in Eq. (5.21) shows that this initial solution is also the optimal solution of the FTP given in Table 5.1. It needs to be pointed out that Kaur and Kumar [26] obtained the same optimal solution by using the GGMODIM on the IBFS given in Eq. (5.21). However, in their method all arithmetic operations are performed on the generalized trapezoidal fuzzy numbers, while in Ebrahimnejad's method all arithmetic operations are done on real numbers.

In addition, if we apply the least cost method to obtain the IBFS, we obtain the following solution which is matched with the IBFS (5.10) obtained based on the GFLCM proposed by Kaur and Kumar [26]:

$$x_{11} = 0, \ x_{12} = 10, \ x_{13} = 0,$$
$$x_{21} = 10, \ x_{22} = 4, \ x_{23} = 0, \qquad (5.22)$$
$$x_{31} = 5, \ x_{32} = 0, \ x_{33} = 10.$$

In this case, using the modified distribution method on the obtained IBFS given in Eq. (5.22), after three iterations the same optimal solution given in (5.5) is obtained. This means that both approaches give the same optimal solution for the FTP given in Table 5.1.

Finally, if we use the Vogel's approximation method to find the IBFS, we reach the solution (5.16) which is matched with the IBFS obtained based on the GFVAM proposed by Kaur and Kumar [26].

In this case, if we apply the modified distribution method on the IBFS given in Eq. (5.16), we find the same optimal solution given in (5.5) after two iterations. Kaur and Kumar [26] obtained the same optimal solution using the GFMODIM on the IBFS given in Eq. (5.16).

As shown here, the optimal solution of the classical method is equivalent to the optimal solution derived from Kaur and Kumar's method [26]. However, Ebrahimnejad's method is by far simpler and computationally more efficient than the fuzzy method proposed by Kaur and Kumar [26].

In sum the main advantages of the proposed method in this subsection are explored as follows:

- The proposed technique does not use the goal and parametric approaches which are difficult to apply in real life situations.
- By applying the proposed approach for finding the fuzzy optimal solution, there is no need for much knowledge of the FLP technique, Zimmerman approach and crisp LP which are difficult to learn for a new decision maker.
- The proposed method to solve the FTP of type-I (5.3) is based on traditional transportation algorithms. Thus, the existing and easily available software can be used for the same. However, the existing method [26] to solve the FTP should be implemented into a programming language.
- To solve the FTP of type-I (5.3) by using the existing method [26] requires using arithmetic operations of generalized fuzzy numbers. While if the proposed technique is used for the same then there is a need to use arithmetic operations of real numbers. This proves that it is much easier to apply Ebrahimnejad's method, as compared to the existing method [26].
- Moreover, it is possible to assume a generic ranking index for comparing the fuzzy numbers involved in the FTP of type-I (5.3), in such a way that each time that the decision maker wants to solve the FTP problem under consideration (s) he can choose (or propose) the ranking index that best suits the FTP problem.

5.4 FTPs of Type-II

The general form of FTPs with fuzzy supply and fuzzy demand as well as decision variables can be formulated as follows:

$$
\min \tilde{z} = \sum_{i=1}^{m} \sum_{j=1}^{n} c_{ij} \tilde{x}_{ij}
$$

$$
s.t. \quad \sum_{j=1}^{n} \tilde{x}_{ij} = \tilde{a}_i, \quad i = 1, 2, \ldots, m,
$$

$$
\sum_{i=1}^{m} \tilde{x}_{ij} = \tilde{b}_j, \quad j = 1, 2, \ldots, n, \tag{5.23}
$$

$$
\tilde{x}_{ij} \succeq \tilde{0}, \qquad i = 1, 2, \ldots, m, j = 1, 2, \ldots, n.
$$

Senthilkumar and Vengataasalam [23] proposed a method for solving the FTP of type-II (5.23) with equality constraints by expressing the supply and demand as triangular fuzzy numbers in parametric forms.

Assume that the parametric forms of fuzzy numbers $\tilde{a}_i$ and $\tilde{b}_j$ are presented by $\left[\tilde{a}_i^L(r), \tilde{a}_i^U(r)\right]$ and $\left[\tilde{b}_j^L(r), \tilde{b}_j^U(r)\right]$, respectively. In addition, the fuzzy decision variable $\tilde{x}_{ij}$ is represented as the parametric form $\left[\tilde{x}_{ij}^L(r), \tilde{x}_{ij}^U(r)\right]$. In this case, Senthilkumar and Vengataasalam [23] proposed the following four crisp problems in order to find the parametric form of the fuzzy optimal solution of the FTP of type-II (5.23):

$$
\begin{aligned}
\min \quad & \sum_{i=1}^{m}\sum_{j=1}^{n} c_{ij}\tilde{x}_{ij}^L(0) \\
s.t. \quad & \sum_{j=1}^{n} \tilde{x}_{ij}^L(0) = \tilde{a}_i^L(0), \quad i = 1, 2, \ldots, m, \\
& \sum_{i=1}^{m} \tilde{x}_{ij}^L(0) = \tilde{b}_j^L(0), \quad j = 1, 2, \ldots, n, \\
& \tilde{x}_{ij}^L(0) \geq 0, \qquad\quad i = 1, 2, \ldots, m, j = 1, 2, \ldots, n.
\end{aligned}
\tag{5.24}
$$

$$
\begin{aligned}
\min \quad & \sum_{i=1}^{m}\sum_{j=1}^{n} c_{ij}\tilde{x}_{ij}^L(1) \\
s.t. \quad & \sum_{j=1}^{n} \tilde{x}_{ij}^L(1) = \tilde{a}_i^L(1), \quad i = 1, 2, \ldots, m, \\
& \sum_{i=1}^{m} \tilde{x}_{ij}^L(1) = \tilde{b}_j^L(1), \quad j = 1, 2, \ldots, n, \\
& \tilde{x}_{ij}^L(1) \geq 0, \qquad\quad i = 1, 2, \ldots, m, j = 1, 2, \ldots, n.
\end{aligned}
\tag{5.25}
$$

$$
\begin{aligned}
\min \quad & \sum_{i=1}^{m}\sum_{j=1}^{n} c_{ij}\tilde{x}_{ij}^U(0) \\
s.t. \quad & \sum_{j=1}^{n} \tilde{x}_{ij}^U(0) = \tilde{a}_i^U(0), \quad i = 1, 2, \ldots, m, \\
& \sum_{i=1}^{m} \tilde{x}_{ij}^U(0) = \tilde{b}_j^U(0), \quad j = 1, 2, \ldots, n, \\
& \tilde{x}_{ij}^U(0) \geq 0, \qquad\quad i = 1, 2, \ldots, m, \quad j = 1, 2, \ldots, n.
\end{aligned}
\tag{5.26}
$$

$$
\begin{aligned}
\min \quad & \sum_{i=1}^{m}\sum_{j=1}^{n} c_{ij}\tilde{x}_{ij}^U(1) \\
s.t. \quad & \sum_{j=1}^{n} \tilde{x}_{ij}^U(1) = \tilde{a}_i^U(1), \quad i = 1, 2, \ldots, m, \\
& \sum_{i=1}^{m} \tilde{x}_{ij}^U(1) = \tilde{b}_j^U(1), \quad j = 1, 2, \ldots, n, \\
& \tilde{x}_{ij}^U(1) \geq 0, \qquad\quad i = 1, 2, \ldots, m, \quad j = 1, 2, \ldots, n.
\end{aligned}
\tag{5.27}
$$

Assume that $\tilde{x}_{ij}^{L*}(0)$, $\tilde{x}_{ij}^{L*}(1)$, $\tilde{x}_{ij}^{U*}(0)$ and $\tilde{x}_{ij}^{U*}(1)$ are the optimal solutions of the crisp TPs (5.24), (5.25), (5.26) and (5.27), respectively. In this case, the fuzzy optimal solution of the FTP of type (5.24) is given as follows:

$$\tilde{x}_{ij}^* = \left[\tilde{x}_{ij}^{L*}, \tilde{x}_{ij}^{U*} \right]$$
$$= \left[\left(\tilde{x}_{ij}^{L*}(1) - \tilde{x}_{ij}^{L*}(0) \right) r + \tilde{x}_{ij}^{L*}(0), \left(\tilde{x}_{ij}^{U*}(1) - \tilde{x}_{ij}^{U*}(0) \right) r + \tilde{x}_{ij}^{U*}(0) \right] \qquad (5.28)$$

Remark 5.6 In the case of triangular fuzzy numbers, the problems (5.25) and (5.27) will be the same.

Example 5.3 [23] Consider the following FTP of type-II:

$$
\begin{aligned}
\min \tilde{z} &= 8\tilde{x}_{11} + 6\tilde{x}_{12} + 12\tilde{x}_{13} + 5\tilde{x}_{21} + 7\tilde{x}_{22} + 10\tilde{x}_{23} \\
\text{s.t.} \quad & \tilde{x}_{11} + \tilde{x}_{12} + \tilde{x}_{13} = (13, 15, 17), \\
& \tilde{x}_{21} + \tilde{x}_{22} + \tilde{x}_{23} = (8, 11, 11) \\
& \tilde{x}_{11} + \tilde{x}_{21} = (3, 5, 7), \\
& \tilde{x}_{12} + \tilde{x}_{22} = (7, 9, 9), \\
& \tilde{x}_{13} + \tilde{x}_{23} = (11, 12, 12), \\
& \tilde{x}_{11}, \tilde{x}_{12}, \tilde{x}_{13}, \tilde{x}_{21}, \tilde{x}_{22}, \tilde{x}_{23}, \succeq \tilde{0}.
\end{aligned}
\qquad (5.29)
$$

The parametric form of the FTP of type-II (5.29) is given as follows:

$$
\begin{aligned}
\min \tilde{z} &= 8\tilde{x}_{11} + 6\tilde{x}_{12} + 12\tilde{x}_{13} + 5\tilde{x}_{21} + 7\tilde{x}_{22} + 10\tilde{x}_{23} \\
\text{s.t.} \quad & \tilde{x}_{11} + \tilde{x}_{12} + \tilde{x}_{13} = (13 + 2r, 17 - 2r), \\
& \tilde{x}_{21} + \tilde{x}_{22} + \tilde{x}_{23} = (8 + 3r, 11) \\
& \tilde{x}_{11} + \tilde{x}_{21} = (3 + 2r, 7 - 2r), \\
& \tilde{x}_{12} + \tilde{x}_{22} = (7 + 2r, 9), \\
& \tilde{x}_{13} + \tilde{x}_{23} = (11 + r, 12), \\
& \tilde{x}_{11}, \tilde{x}_{12}, \tilde{x}_{13}, \tilde{x}_{21}, \tilde{x}_{22}, \tilde{x}_{23}, \succeq \tilde{0}.
\end{aligned}
\qquad (5.30)
$$

The problem (5.30) is divided into the following four crisp problems with regard to the problems (5.24), (5.25), (5.26) and (5.27):

$$\min 8\tilde{x}_{11}^{L}(0) + 6\tilde{x}_{12}^{L}(0) + 12\tilde{x}_{13}^{L}(0) + 5\tilde{x}_{21}^{L}(0) + 7\tilde{x}_{22}^{L}(0) + 10\tilde{x}_{23}^{L}(0)$$

$$s.t. \quad \tilde{x}_{11}^{L}(0) + \tilde{x}_{12}^{L}(0) + \tilde{x}_{13}^{L}(0) = 13$$

$$\tilde{x}_{21}^{L}(0) + \tilde{x}_{22}^{L}(0) + \tilde{x}_{23}^{L}(0) = 8$$

$$\tilde{x}_{11}^{L}(0) + \tilde{x}_{21}^{L}(0) = 3,$$

$$\tilde{x}_{12}^{L}(0) + \tilde{x}_{22}^{L}(0) = 7,$$

$$\tilde{x}_{13}^{L}(0) + \tilde{x}_{23}^{L}(0) = 11,$$

$$\tilde{x}_{11}^{L}(0), \tilde{x}_{12}^{L}(0), \tilde{x}_{13}^{L}(0), \tilde{x}_{21}^{L}(0), \tilde{x}_{22}^{L}(0), \tilde{x}_{23}^{L}(0) \geq 0.$$

$$(5.31)$$

$$\min 8\tilde{x}_{11}^{L}(1) + 6\tilde{x}_{12}^{L}(1) + 12\tilde{x}_{13}^{L}(1) + 5\tilde{x}_{21}^{L}(1) + 7\tilde{x}_{22}^{L}(1) + 10\tilde{x}_{23}^{L}(1)$$

$$s.t. \quad \tilde{x}_{11}^{L}(1) + \tilde{x}_{12}^{L}(1) + \tilde{x}_{13}^{L}(1) = 15$$

$$\tilde{x}_{21}^{L}(1) + \tilde{x}_{22}^{L}(1) + \tilde{x}_{23}^{L}(1) = 11$$

$$\tilde{x}_{11}^{L}(1) + \tilde{x}_{21}^{L}(1) = 5,$$

$$\tilde{x}_{12}^{L}(0) + \tilde{x}_{22}^{L}(1) = 9,$$

$$\tilde{x}_{13}^{L}(0) + \tilde{x}_{23}^{L}(1) = 12,$$

$$\tilde{x}_{11}^{L}(1), \tilde{x}_{12}^{L}(1), \tilde{x}_{13}^{L}(1), \tilde{x}_{21}^{L}(1), \tilde{x}_{22}^{L}(1), \tilde{x}_{23}^{L}(1) \geq 0.$$

$$(5.32)$$

$$\min 8\tilde{x}_{11}^{U}(0) + 6\tilde{x}_{12}^{U}(0) + 12\tilde{x}_{13}^{U}(0) + 5\tilde{x}_{21}^{U}(0) + 7\tilde{x}_{22}^{U}(0) + 10\tilde{x}_{23}^{U}(0)$$

$$s.t. \quad \tilde{x}_{11}^{U}(0) + \tilde{x}_{12}^{U}(0) + \tilde{x}_{13}^{U}(0) = 17$$

$$\tilde{x}_{21}^{U}(0) + \tilde{x}_{22}^{U}(0) + \tilde{x}_{23}^{U}(0) = 11$$

$$\tilde{x}_{11}^{U}(0) + \tilde{x}_{21}^{U}(0) = 7,$$

$$\tilde{x}_{12}^{U}(0) + \tilde{x}_{22}^{U}(0) = 9,$$

$$\tilde{x}_{13}^{U}(0) + \tilde{x}_{23}^{U}(0) = 12,$$

$$\tilde{x}_{11}^{U}(0), \tilde{x}_{12}^{U}(0), \tilde{x}_{13}^{U}(0), \tilde{x}_{21}^{U}(0), \tilde{x}_{22}^{U}(0), \tilde{x}_{23}^{U}(0) \geq 0.$$

$$(5.33)$$

$$\min 8\tilde{x}_{11}^{U}(1) + 6\tilde{x}_{12}^{U}(1) + 12\tilde{x}_{13}^{U}(1) + 5\tilde{x}_{21}^{U}(1) + 7\tilde{x}_{22}^{U}(1) + 10\tilde{x}_{23}^{U}(1)$$

$$s.t. \quad \tilde{x}_{11}^{U}(1) + \tilde{x}_{12}^{U}(1) + \tilde{x}_{13}^{U}(1) = 15$$

$$\tilde{x}_{21}^{U}(1) + \tilde{x}_{22}^{U}(1) + \tilde{x}_{23}^{U}(1) = 11$$

$$\tilde{x}_{11}^{U}(1) + \tilde{x}_{21}^{U}(1) = 5,$$

$$\tilde{x}_{12}^{U}(1) + \tilde{x}_{22}^{U}(1) = 9,$$

$$\tilde{x}_{13}^{U}(1) + \tilde{x}_{23}^{U}(1) = 12,$$

$$\tilde{x}_{11}^{U}(1), \tilde{x}_{12}^{U}(1), \tilde{x}_{13}^{U}(1), \tilde{x}_{21}^{U}(1), \tilde{x}_{22}^{U}(1), \tilde{x}_{23}^{U}(1) \geq 0.$$

$$(5.34)$$

Solving the crisp TP (5.31) using the standard algorithm for solving crisp TP gives the following optimal solution:

$$\tilde{x}_{11}^{L*}(0) = 0, \tilde{x}_{12}^{L*}(0) = 7, \tilde{x}_{13}^{L*}(0) = 6, \tilde{x}_{21}^{L*}(0) = 3, \tilde{x}_{22}^{L*}(0) = 0, \tilde{x}_{23}^{L*}(0) = 5 \quad (5.35)$$

Similarly, we obtain the following optimal solutions by solving TPs (5.32), (5.33) and (5.34):

$$\tilde{x}_{11}^{L*}(1) = 0, \tilde{x}_{12}^{L*}(1) = 9, \tilde{x}_{13}^{L*}(1) = 6, \tilde{x}_{21}^{L*}(1) = 5, \tilde{x}_{22}^{L*}(1) = 0, \tilde{x}_{23}^{L*}(1) = 6 \quad (5.36)$$

$$\tilde{x}_{11}^{U*}(0) = 0, \tilde{x}_{12}^{U*}(0) = 9, \tilde{x}_{13}^{U*}(0) = 8, \tilde{x}_{21}^{U*}(0) = 7, \tilde{x}_{22}^{U*}(0) = 0, \tilde{x}_{23}^{U*}(0) = 4 \quad (5.37)$$

$$\tilde{x}_{11}^{U*}(1) = 0, \tilde{x}_{12}^{U*}(1) = 9, \tilde{x}_{13}^{U*}(1) = 6, \tilde{x}_{21}^{U*}(1) = 5, \tilde{x}_{22}^{U*}(1) = 0, \tilde{x}_{23}^{U*}(1) = 6 \quad (5.38)$$

Hence, the fuzzy optimal solution of the FTP of type-II (5.29) is given as follows with regret to Eq. (5.28):

$$\begin{aligned}
\tilde{x}_{11}^* &= [0,0], \tilde{x}_{12}^* = [7+2r,9], \tilde{x}_{13}^* = [6, 8-2r], \\
\tilde{x}_{21}^* &= [3+2r, 7-2r], \tilde{x}_{22}^* = [0,0], \tilde{x}_{23}^* = [5+r, 4+2r].
\end{aligned} \quad (5.39)$$

The optimal solution (5.39) can be represented graphically as Fig. 5.1.

Note that the upper bound of the parametric form of a fuzzy number should be non-increasing over [0, 1], while the upper bound of $\tilde{x}_{23}^*$ in the optimal solution (5.39), i.e. $4+2r$ is increasing over [0, 1]. This means that the optimal solution of the FTP based on this approach is not necessarily in the form of triangular fuzzy numbers even if the data of the FTP under consideration are represented in terms of triangular fuzzy numbers.

Fig. 5.1 The optimal solution of the FTP of type-II (5.30)

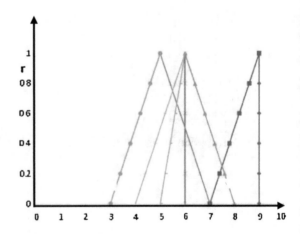

5.5 FTPs of Type-III

Liu and Kao [8] developed a procedure based on the extension principle to derive the fuzzy objective value of the FTP, in that the transportation cost and the supply and demand quantities are fuzzy numbers.

They investigated two different kinds of the FTP of type-III: one with inequality constraints and the other with equality constraints. In this section, we will explore their proposed approach for solving these two different types of FTPs.

5.5.1 FTP of Type-III with Inequality Constraints

The general form of FTP of type-III with inequality constraints can be formulated as follows:

$$
\min \tilde{z} = \sum_{i=1}^{m} \sum_{j=1}^{n} \tilde{c}_{ij} x_{ij}
$$

$$
s.t. \quad \sum_{j=1}^{n} x_{ij} \leq \tilde{a}_i, \quad i = 1, 2, \ldots, m,
$$

$$
\sum_{i=1}^{m} x_{ij} \geq \tilde{b}_j, \quad j = 1, 2, \ldots, n, \tag{5.40}
$$

$$
x_{ij} \geq 0, \quad i = 1, 2, \ldots, m, j = 1, 2, \ldots, n
$$

Obviously, when the transportation costs or the supply and demand quantities are fuzzy numbers, the total transportation cost will be fuzzy as well. Liu and Kao [8] developed a solution procedure to calculate the fuzzy objective value of the FTP of type-III, where at least one of the parameters is a fuzzy number.

Assume that the membership functions of the fuzzy transportation cost $\tilde{c}_{ij}$, the fuzzy supply $\tilde{a}_i$ and the fuzzy demand $\tilde{b}_j$ are represented by $\mu_{\tilde{c}_{ij}}$, $\mu_{\tilde{a}_i}$ and $\mu_{\tilde{b}_j}$, respectively. Thus we have:

$$
\tilde{c}_{ij} = \left\{ \left(c_{ij}, \mu_{\tilde{c}_{ij}}(c_{ij}) \right) \middle| c_{ij} \in S(\tilde{c}_{ij}) \right\}
$$

$$
\tilde{a}_i = \left\{ (a_i, \mu_{\tilde{a}_i}(a_i)) \middle| a_i \in S(\tilde{a}_i) \right\} \tag{5.41}
$$

$$
\tilde{b}_j = \left\{ \left(b_j, \mu_{\tilde{b}_j}(b_j) \right) \middle| b_j \in S(\tilde{b}_j) \right\}
$$

where $S(\tilde{c}_{ij})$, $S(\tilde{a}_i)$ and $S(\tilde{b}_j)$ are the supports of $\tilde{c}_{ij}$, $\tilde{a}_i$ and $\tilde{b}_j$, respectively.

Denote the α-cuts of $\tilde{c}_{ij}$, $\tilde{a}_i$ and $\tilde{b}_j$ as follows:

$$
\begin{aligned}
[\tilde{c}_{ij}]_\alpha &= \left[(\tilde{c}_{ij})_\alpha^L, (\tilde{c}_{ij})_\alpha^U \right] \\
&= \left[\min_{c_{ij}} \left\{ c_{ij} \in S(\tilde{c}_{ij}) \middle| \mu_{\tilde{c}_{ij}}(c_{ij}) \geq \alpha \right\}, \max_{c_{ij}} \left\{ c_{ij} \in S(\tilde{c}_{ij}) \middle| \mu_{\tilde{c}_{ij}}(c_{ij}) \geq \alpha \right\} \right]
\end{aligned}
\tag{5.42}
$$

$$
\begin{aligned}
[\tilde{a}_i]_\alpha &= \left[(\tilde{a}_i)_\alpha^L, (\tilde{a}_i)_\alpha^U \right] \\
&= \left[\min_{a_i} \left\{ a_i \in S(\tilde{a}_i) \middle| \mu_{\tilde{a}_i}(a_i) \geq \alpha \right\}, \max_{a_i} \left\{ a_i \in S(\tilde{a}_i) \middle| \mu_{\tilde{a}_i}(a_i) \geq \alpha \right\} \right]
\end{aligned}
\tag{5.43}
$$

$$
\begin{aligned}
[\tilde{b}_j]_\alpha &= \left[(\tilde{b}_j)_\alpha^L, (\tilde{b}_j)_\alpha^U \right] \\
&= \left[\min_{b_j} \left\{ b_j \in S(\tilde{b}_j) \middle| \mu_{\tilde{b}_j}(b_j) \geq \alpha \right\}, \max_{b_j} \left\{ b_j \in S(\tilde{b}_j) \middle| \mu_{\tilde{b}_j}(b_j) \geq \alpha \right\} \right]
\end{aligned}
\tag{5.44}
$$

Liu and Kao [8] applied the Zadeh's extension principle in order to derive the membership function of the total transportation cost $\tilde{z}$.

Based on the extension principle, the membership function $\mu_{\tilde{z}}$ can be defined as:

$$
\mu_{\tilde{z}}(z) = \sup_{c,a,b} \min \left\{ \mu_{\tilde{c}_{ij}}(c_{ij}), \mu_{\tilde{a}_i}(a_i), \mu_{\tilde{b}_j}(b_j), \forall i,j \middle| z = z(c,a,b) \right\}
\tag{5.45}
$$

where $z(c,a,b)$ is defined in the Model (5.2). If the α-cuts of $\tilde{z}$ at all α values degenerate to the same point, then the total transportation cost is a crisp number. Otherwise, it is a fuzzy number. In the Eq. (5.45), several membership functions are involved. To derive $\mu_{\tilde{z}}$ in closed form is hardly possible. According to (5.45), $\mu_{\tilde{z}}$ is the minimum of $\mu_{\tilde{c}_{ij}}(c_{ij})$, $\mu_{\tilde{a}_i}(a_i)$ and $\mu_{\tilde{b}_j}(b_j), \forall i,j$. It is required $\mu_{\tilde{c}_{ij}}(c_{ij}) \geq \alpha$, $\mu_{\tilde{a}_i}(a_i) \geq \alpha$, $\mu_{\tilde{b}_j}(b_j) \geq \alpha$ and at least one $\mu_{\tilde{c}_{ij}}(c_{ij})$, $\mu_{\tilde{a}_i}(a_i)$ or $\mu_{\tilde{b}_j}(b_j), \forall i,j$ equal to α such that $z = z(c,a,b)$ to satisfy $\mu_{\tilde{z}}(z) = \alpha$. To find the membership function $\tilde{z}$, it suffices to find the left shape function and the right shape function of $\tilde{z}$, which is equivalent to finding the lower bound $(\tilde{z})_\alpha^L$ and upper bound $(\tilde{z})_\alpha^U$ of the α-cut of $\tilde{z}$. Since $(\tilde{z})_\alpha^L$ is the minimum of $z(c,a,b)$ and $(\tilde{z})_\alpha^U$ is the maximum of $z(c,a,b)$, they can be expressed as:

$$
(\tilde{z})_\alpha^L = \min \left\{ z(c,a,b) \middle| (\tilde{c}_{ij})_\alpha^L \leq c_{ij} \leq (\tilde{c}_{ij})_\alpha^U, (\tilde{a}_i)_\alpha^L \leq a_i \leq (\tilde{a}_i)_\alpha^U, (\tilde{b}_j)_\alpha^L \leq b_j \leq (\tilde{b}_j)_\alpha^U, \forall i,j \right\}
\tag{5.46}
$$

$$
(\tilde{z})_\alpha^U = \max \left\{ z(c,a,b) \middle| (\tilde{c}_{ij})_\alpha^L \leq c_{ij} \leq (\tilde{c}_{ij})_\alpha^U, (\tilde{a}_i)_\alpha^L \leq a_i \leq (\tilde{a}_i)_\alpha^U, (\tilde{b}_j)_\alpha^L \leq b_j \leq (\tilde{b}_j)_\alpha^U, \forall i,j \right\}
\tag{5.47}
$$

which can be reformulated as the following pair of two-level mathematical programs:

$$(\tilde{z})^L_\alpha = \min_{\substack{(\tilde{c}_{ij})^L_\alpha \le c_{ij} \le (\tilde{c}_{ij})^U_\alpha \\ (\tilde{a}_i)^L_\alpha \le a_i \le (\tilde{a}_i)^U_\alpha \\ (\tilde{b}_j)^L_\alpha \le b_j \le (\tilde{b}_j)^U_\alpha \\ \forall i,j}} \begin{cases} \min & \sum_{i=1}^m \sum_{j=1}^n c_{ij} x_{ij} \\ s.t. & \sum_{j=1}^n x_{ij} \le a_i, \quad i = 1, 2, \ldots, m, \\ & \sum_{i=1}^m x_{ij} \ge b_j, \quad j = 1, 2, \ldots, n, \\ & x_{ij} \ge 0, \qquad i = 1, 2, \ldots, m, j = 1, 2, \ldots, n. \end{cases}$$

(5.48)

$$(\tilde{z})^U_\alpha = \max_{\substack{(\tilde{c}_{ij})^L_\alpha \le c_{ij} \le (\tilde{c}_{ij})^U_\alpha \\ (\tilde{a}_i)^L_\alpha \le a_i \le (\tilde{a}_i)^U_\alpha \\ (\tilde{b}_j)^L_\alpha \le b_j \le (\tilde{b}_j)^U_\alpha \\ \forall i,j}} \begin{cases} \min & \sum_{i=1}^m \sum_{j=1}^n c_{ij} x_{ij} \\ s.t. & \sum_{j=1}^n x_{ij} \le a_i, \quad i = 1, 2, \ldots, m, \\ & \sum_{i=1}^m x_{ij} \ge b_j, \quad j = 1, 2, \ldots, n, \\ & x_{ij} \ge 0, \qquad i = 1, 2, \ldots, m, j = 1, 2, \ldots, n. \end{cases}$$

(5.49)

At least one c_{ij}, a_i, or b_j must hit the boundary of their α-cuts to satisfy $\mu_{\tilde{z}}(z) = \alpha$.

To ensure the TP of the second level to be feasible, it is necessary that the constraint $\sum_{i=1}^m a_i \ge \sum_{j=1}^n b_j$ be imposed in the first level. Hence, the Models (5.48) and (5.49) become:

$$(\tilde{z})^L_\alpha = \min_{\substack{(\tilde{c}_{ij})^L_\alpha \le c_{ij} \le (\tilde{c}_{ij})^U_\alpha \\ (\tilde{a}_i)^L_\alpha \le a_i \le (\tilde{a}_i)^U_\alpha \\ (\tilde{b}_j)^L_\alpha \le b_j \le (\tilde{b}_j)^U_\alpha, \\ \forall i,j \\ \sum_{i=1}^m a_i \ge \sum_{j=1}^n b_j,}} \begin{cases} \min & \sum_{i=1}^m \sum_{j=1}^n c_{ij} x_{ij} \\ s.t. & \sum_{j=1}^n x_{ij} \le a_i, \quad i = 1, 2, \ldots, m, \\ & \sum_{i=1}^m x_{ij} \ge b_j, \quad j = 1, 2, \ldots, n, \\ & x_{ij} \ge 0, \qquad i = 1, 2, \ldots, m, j = 1, 2, \ldots, n. \end{cases}$$

(5.50)

$$(\tilde{z})_{\alpha}^{U} = \begin{array}{c} \max \\ (\tilde{c}_{ij})_{\alpha}^{L} \leq c_{ij} \leq (\tilde{c}_{ij})_{\alpha}^{U} \\ (\tilde{a}_{i})_{\alpha}^{L} \leq a_{i} \leq (\tilde{a}_{i})_{\alpha}^{U} \\ (\tilde{b}_{j})_{\alpha}^{L} \leq b_{j} \leq (\tilde{b}_{j})_{\alpha}^{U}, \\ \forall i,j \\ \sum_{i=1}^{m} a_{i} \geq \sum_{j=1}^{n} b_{j}, \end{array} \left\{ \begin{array}{ll} \min & \sum_{i=1}^{m} \sum_{j=1}^{n} c_{ij} x_{ij} \\ \\ s.t. & \sum_{j=1}^{n} x_{ij} \leq a_{i}, \quad i = 1, 2, \ldots, m, \\ \\ & \sum_{i=1}^{m} x_{ij} \geq b_{j}, \quad j = 1, 2, \ldots, n, \\ \\ & x_{ij} \geq 0, \qquad i = 1, 2, \ldots, m, j = 1, 2, \ldots, n. \end{array} \right.$$

$$(5.51)$$

The Models (5.50) and (5.51) will be infeasible for any α level if $\sum_{i=1}^{m} (\tilde{a}_{i})_{\alpha=0}^{U} \geq \sum_{j=1}^{n} (\tilde{b}_{j})_{\alpha=0}^{L}$. In other words, an FTP is feasible if the upper bound of the total fuzzy supply is greater than or equal to the lower bound of the total fuzzy demand.

To derive the lower bound of the objective value in the Model (5.50), we can directly set c_{ij} to its lower bound $(\tilde{c}_{ij})_{\alpha}^{L}, \forall i,j$, to find the minimum objective value. Hence, the Model (5.50) can be reformulated as:

$$(\tilde{z})_{\alpha}^{L} = \begin{array}{c} \min \\ (\tilde{a}_{i})_{\alpha}^{L} \leq a_{i} \leq (\tilde{a}_{i})_{\alpha}^{U} \\ (\tilde{b}_{j})_{\alpha}^{L} \leq b_{j} \leq (\tilde{b}_{j})_{\alpha}^{U}, \\ \forall i,j \\ \sum_{i=1}^{m} a_{i} \geq \sum_{j=1}^{n} b_{j}, \end{array} \left\{ \begin{array}{ll} \min & \sum_{i=1}^{m} \sum_{j=1}^{n} (\tilde{c}_{ij})_{\alpha}^{L} x_{ij} \\ \\ s.t. & \sum_{j=1}^{n} x_{ij} \leq a_{i}, \qquad i = 1, 2, \ldots, m, \\ \\ & \sum_{i=1}^{m} x_{ij} \geq b_{j}, \qquad j = 1, 2, \ldots, n, \\ \\ & x_{ij} \geq 0, \qquad i = 1, 2, \ldots, m, j = 1, 2, \ldots, n. \end{array} \right.$$

$$(5.52)$$

Since the Model (5.52) is to find the minimum of all the minimum objective values, one can insert the constraints of level 1 into level 2 and simplify the two-level mathematical program to the conventional one-level program as follows:

$$(\tilde{z})_{\alpha}^{L} = \min \sum_{i=1}^{m} \sum_{j=1}^{n} (\tilde{c}_{ij})_{\alpha}^{L} x_{ij}$$

$$s.t. \quad \sum_{j=1}^{n} x_{ij} \leq a_{i}, \qquad i = 1, 2, \ldots, m,$$

$$\sum_{i=1}^{m} x_{ij} \geq b_{j}, \qquad j = 1, 2, \ldots, n,$$

$$\sum_{i=1}^{m} a_{i} \geq \sum_{j=1}^{n} b_{j},$$

$$(\tilde{a}_{i})_{\alpha}^{L} \leq a_{i} \leq (\tilde{a}_{i})_{\alpha}^{U}, \quad i = 1, 2, \ldots, m,$$

$$(\tilde{b}_{j})_{\alpha}^{L} \leq b_{j} \leq (\tilde{b}_{j})_{\alpha}^{U}, \quad j = 1, 2, \ldots, n,$$

$$x_{ij} \geq 0, \qquad i = 1, 2, \ldots, m, j = 1, 2, \ldots, n.$$

$$(5.53)$$

This model is an LP which can be solved easily. In this model, since all c_{ij} have been set to the lower bounds of their α-cuts, that is, $\mu_{\tilde{c}_{ij}}(c_{ij}) = \alpha$, this assures $\mu_{\tilde{z}}(z) = \alpha$ as required by (5.45).

To solve the Model (5.51), the dual of the level 2 problem is formulated to become a maximization problem to be consistent with the maximization operation of level 1. It is well-known from the duality theorem of LP that the primal model and the dual model have the same objective value. Thus, the Model (5.51) becomes:

$$(\tilde{z})_\alpha^U = \begin{matrix} \max \\ \left(\tilde{c}_{ij}\right)_\alpha^L \le c_{ij} \le \left(\tilde{c}_{ij}\right)_\alpha^U \\ \left(\tilde{a}_i\right)_\alpha^L \le a_i \le \left(\tilde{a}_i\right)_\alpha^U \\ \left(\tilde{b}_j\right)_\alpha^L \le b_j \le \left(\tilde{b}_j\right)_\alpha^U, \\ \forall i,j \\ \sum_{i=1}^{m} a_i \ge \sum_{j=1}^{n} b_j, \end{matrix} \quad \begin{cases} \max - \sum_{i=1}^{m} a_i u_i + \sum_{j=1}^{n} b_j v_j \\ s.t. \quad -u_i + v_j \le c_{ij}, \quad i = 1,2,\ldots,m, j = 1,2,\ldots,n, \\ \qquad u_i, v_j \ge 0, \quad i = 1,2,\ldots,m, j = 1,2,\ldots,n. \end{cases}$$

$$(5.54)$$

To derive the upper bound of the objective value in the Model (5.54), one can directly set c_{ij} to its upper bound $\left(\tilde{c}_{ij}\right)_\alpha^U, \forall i,j$, because this gives the largest feasible space. Hence, the Model (5.54) can be reformulated as:

$$(\tilde{z})_\alpha^U = \begin{matrix} \max \\ \left(\tilde{a}_i\right)_\alpha^L \le a_i \le \left(\tilde{a}_i\right)_\alpha^U \\ \left(\tilde{b}_j\right)_\alpha^L \le b_j \le \left(\tilde{b}_j\right)_\alpha^U, \\ \forall i,j \\ \sum_{i=1}^{m} a_i \ge \sum_{j=1}^{n} b_j, \end{matrix} \quad \begin{cases} \max - \sum_{i=1}^{m} a_i u_i + \sum_{j=1}^{n} b_j v_j \\ s.t. \quad -u_i + v_j \le \left(\tilde{c}_{ij}\right)_\alpha^U, \quad i = 1,2,\ldots,m, j = 1,2,\ldots,n, \\ \qquad u_i, v_j \ge 0, \quad i = 1,2,\ldots,m, j = 1,2,\ldots,n. \end{cases}$$

$$(5.55)$$

Now, since both level 1 and level 2 perform the same maximization operation, their constraints can be combined to form the following one-level mathematical program:

$$(\tilde{z})_\alpha^U = \max - \sum_{i=1}^{m} a_i u_i + \sum_{j=1}^{n} b_j v_j$$

$$s.t. \quad -u_i + v_j \le \left(\tilde{c}_{ij}\right)_\alpha^U, \quad i = 1,2,\ldots,m, j = 1,2,\ldots,n,$$

$$\sum_{i=1}^{m} a_i \ge \sum_{j=1}^{n} b_j, \tag{5.56}$$

$$\left(\tilde{a}_i\right)_\alpha^L \le a_i \le \left(\tilde{a}_i\right)_\alpha^U, \quad i = 1,2,\ldots,m,$$

$$\left(\tilde{b}_j\right)_\alpha^L \le b_j \le \left(\tilde{b}_j\right)_\alpha^U, \quad j = 1,2,\ldots,n,$$

$$u_i, v_j \ge 0, \quad i = 1,2,\ldots,m, j = 1,2,\ldots,n.$$

This model is a linearly constrained non-linear program. There are several effective and efficient methods for solving this problem [37]. Similar to the Model (5.53) since all c_{ij} have been set to the upper bounds of their α-cuts, that is, $\mu_{\tilde{c}_{ij}}(c_{ij}) = \alpha$, this ensures that $\mu_{\tilde{z}}(z) = \alpha$ as required by (5.45).

Remark 5.7 For two possibility levels α_1 and α_2 such that $0 < \alpha_1 < \alpha_2 \leq 1$, the feasible spaces defined by α_1 in the Models (5.53) and (5.56) are smaller than those defined by α_2. Consequently, $(\tilde{z})^L_{\alpha_1} \geq (\tilde{z})^L_{\alpha_2}$ and $(\tilde{z})^U_{\alpha_1} \leq (\tilde{z})^U_{\alpha_2}$; in other words, the left shape function is non-decreasing and the right shape function is non-increasing. This property, based on the definition of convex fuzzy set, assures the convexity of $\tilde{z}$.

If both $(\tilde{z})^L_\alpha$ and $(\tilde{z})^U_\alpha$ are invertible with respect to α, then a left shape function $\left[(\tilde{z})^L_\alpha\right]^{-1}$ and a right shape function $\left[(\tilde{z})^U_\alpha\right]^{-1}$ can be obtained. Hence, the membership function $\tilde{z}$ is constructed as:

$$
\mu_{\tilde{z}}(z) = \begin{cases} \left[(\tilde{z})^L_\alpha\right]^{-1}, & (\tilde{z})^L_{\alpha=0} \leq z \leq (\tilde{z})^L_{\alpha=1}, \\ 1, & (\tilde{z})^L_{\alpha=1} \leq z \leq (\tilde{z})^U_{\alpha=1}, \\ \left[(\tilde{z})^U_\alpha\right]^{-1}, & (\tilde{z})^L_{\alpha=1} \leq z \leq (\tilde{z})^U_{\alpha=1}. \end{cases} \tag{5.57}
$$

In most cases, the values of $(\tilde{z})^L_\alpha$ and $(\tilde{z})^U_\alpha$ may not be solved analytically. However, the numerical solutions for $(\tilde{z})^L_\alpha$ and $(\tilde{z})^U_\alpha$ at different possibility levels α can be collected to approximate the shapes of $\left[(\tilde{z})^L_\alpha\right]^{-1}$ and $\left[(\tilde{z})^U_\alpha\right]^{-1}$.

Example 5.4 [8] Consider the following FTP problem of type-III:

$$
\begin{aligned}
\min \tilde{z} = {} & 10x_{11} + 50x_{12} + 80x_{13} + (60, 70, 80, 90)x_{21} + 60x_{22} + 20x_{23} \\
s.t. \quad & x_{11} + x_{12} + x_{13} \leq (70, 90, 100), \\
& x_{21} + x_{22} + x_{23} \leq (40, 60, 70, 80), \\
& x_{11} + x_{21} \geq (30, 40, 50, 70), \\
& x_{12} + x_{22} \geq (20, 30, 40, 50), \\
& x_{13} + x_{23} \geq (40, 50, 80), \\
& x_{11}, x_{12}, x_{13}, x_{21}, x_{22}, x_{23} \geq 0.
\end{aligned} \tag{5.58}
$$

The total supply is $\tilde{a}_1 + \tilde{a}_2 = (110, 150, 160, 180)$ and the total demand is $\tilde{b}_1 + \tilde{b}_2 + \tilde{b}_3 = (90, 120, 140, 200)$, as shown in Fig. 5.2. The spread of the total fuzzy supply overlaps with the spread of the total fuzzy demand. In other words, the upper bound of the total fuzzy supply is greater than the lower bound of the total fuzzy demand, implying that the problem is feasible.

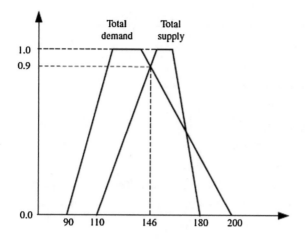

Fig. 5.2 Total supply and demand of Example 5.4

According to the Models (5.53) and (5.56), the lower and upper bounds of $\tilde{z}$ at possibility level α can be solved as:

$$(\tilde{z})_\alpha^L = \min 10x_{11} + 50x_{12} + 80x_{13} + (60 + 10\alpha)x_{21} + 60x_{22} + 20x_{23}$$

$s.t.$ $x_{11} + x_{12} + x_{13} \leq a_1,$

$x_{21} + x_{22} + x_{23} \leq a_2,$

$x_{11} + x_{21} \geq b_1,$

$x_{12} + x_{22} \geq b_2,$

$x_{13} + x_{23} \geq b_3,$ (5.59)

$a_1 + a_2 \geq b_1 + b_2 + b_3,$

$70 + 20\alpha \leq a_1 \leq 100 - 10\alpha, 40 + 20\alpha \leq a_2 \leq 80 - 10\alpha,$

$30 + 10\alpha \leq b_1 \leq 70 - 20\alpha, 20 + 10\alpha \leq b_2 \leq 50 - 10\alpha,$

$40 + 10\alpha \leq b_3 \leq 80 - 30\alpha,$

$x_{11}, x_{12}, x_{13}, x_{21}, x_{22}, x_{23} \geq 0.$

$$(\tilde{z})_\alpha^U = \max - a_1 u_1 - a_2 u_2 + b_1 v_1 + b_2 v_2 + b_3 v_3$$

$$\begin{aligned}
s.t. \quad & -u_1 + v_1 \leq 10, \\
& -u_1 + v_2 \leq 50, \\
& -u_1 + v_3 \leq 80, \\
& -u_2 + v_1 \leq 90 - 10\alpha, \\
& -u_2 + v_2 \leq 60, \\
& -u_2 + v_3 \leq 20, \\
& a_1 + a_2 \geq d_1 + d_2 + d_3, \\
& 70 + 20\alpha \leq a_1 \leq 100 - 10\alpha, 40 + 20\alpha \leq a_2 \leq 80 - 10\alpha, \\
& 30 + 10\alpha \leq d_1 \leq 70 - 20\alpha, 20 + 10\alpha \leq d_2 \leq 50 - 10\alpha, \\
& 40 + 10\alpha \leq d_3 \leq 80 - 30\alpha, \\
& u_1, u_2, v_1, v_2, v_3 \geq 0.
\end{aligned} \tag{5.60}$$

The α-cuts of the total transportation cost are given in Table 5.6 for different values of $\alpha \in [0, 1]$. The α-cut of $\tilde{z}$ represents the possibility that the transportation cost will appear in the associated range. Specifically, the α-cut for $\alpha = 1$ shows what the total transportation cost is most likely to be and the α-cut for $\alpha = 0$ shows the range in which the total transportation cost could appear.

In this example, while the total transportation cost is fuzzy, its most likely value falls between 2900 and 3500, and it is impossible for its value to fall outside the range of 2100 and 5800 (see Fig. 5.3).

Table 5.6 The α-cuts of the total transportation cost for Example 5.4

α	0.0	0.1	0.2	0.3	0.4	0.5	0.6	0.7	0.8	0.9	1.0
$(\tilde{z})_\alpha^U$	2100	2180	2260	2340	2420	2500	2580	2660	2740	2820	2900
$(\tilde{z})_\alpha^U$	5800	5600	5400	5200	5000	4800	4440	4080	3860	3680	3500

Fig. 5.3 The membership function of the objective function for Example 5.4

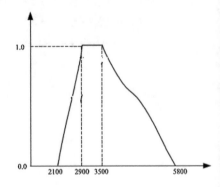

For the $\alpha = 0$ cut of $\tilde{z}$, the lower bound of $z^* = 2100$ occurs at $x_{11}^* = 30, x_{12}^* = 20, x_{23}^* = 40, x_{13}^* = x_{21}^* = x_{22}^* = 0$ with $b_1 = 30$, $b_2 = 20$, $b_3 = 40$ and $a_1 = 70$, $a_2 = 40$. The upper bound of $z^* = 5800$ occurs at $x_{11}^* = 30, x_{12}^* = 30, x_{13}^* = 40, x_{23}^* = 40, x_{21}^* = x_{22}^* = 0$ with $b_1 = 30$, $b_2 = 30$, $b_3 = 80$ and $a_1 = 100, a_2 = 40$. At the other extreme end of $\alpha = 1$, the lower bound of $z^* = 2900$ occurs at $x_{11}^* = 40, x_{12}^* = 30, x_{23}^* = 50, x_{13}^* = x_{21}^* = x_{22}^* = 0$ with $b_1 = 40$, $b_2 = 30$, $b_3 = 50$ and $a_1 = 70, a_2 = 50$. The upper bound of $z^* = 3500$ occurs at $x_{11}^* = 50, x_{12}^* = 40, x_{23}^* = 50, x_{13}^* = x_{21}^* = x_{22}^* = 0$ with $b_1 = 50$, $b_2 = 40$, $b_3 = 50$ and $a_1 = 90, a_2 = 50$. Notably, the total transportation cost associated with the largest total quantity being shipped need not be the highest. In this example, the largest possible amount to be shipped is 180, which is the largest total supply. The total transportation cost for this amount is $z^* = 4600$. However, as noted previously, the highest cost is $z^* = 5800$, occurring at the total amount of 140.

5.5.2 FTP of Type-III with Equality Constraints

The general form of the FTP of type-III with equality constraints can be formulated as follows:

$$\min \tilde{z} = \sum_{i=1}^{m} \sum_{j=1}^{n} \tilde{c}_{ij} x_{ij}$$

$$s.t. \quad \sum_{j=1}^{n} x_{ij} = \tilde{a}_i, \quad i = 1, 2, \ldots, m,$$

$$\sum_{i=1}^{m} x_{ij} = \tilde{b}_j, \quad j = 1, 2, \ldots, n, \qquad (5.61)$$

$$x_{ij} \geq 0, \quad i = 1, 2, \ldots, m, j = 1, 2, \ldots, n.$$

Similar to the discussion of the inequality-constraint case, the lower and upper bounds of $\tilde{z}$ at possibility level α can be solved from the following pair of two-level mathematical programs:

$$(\tilde{z})_{\alpha}^{L} = \min_{\substack{(\tilde{c}_{ij})_{\alpha}^{L} \leq c_{ij} \leq (\tilde{c}_{ij})_{\alpha}^{U} \\ (\tilde{a}_i)_{\alpha}^{L} \leq a_i \leq (\tilde{a}_i)_{\alpha}^{U} \\ (\tilde{b}_j)_{\alpha}^{L} \leq b_j \leq (\tilde{b}_j)_{\alpha}^{U}, \\ \forall i,j \\ \sum_{i=1}^{m} a_i = \sum_{j=1}^{n} b_j,}} \left\{ \begin{array}{l} \min \sum_{i=1}^{m} \sum_{j=1}^{n} c_{ij} x_{ij} \\ \\ s.t. \quad \sum_{j=1}^{n} x_{ij} = a_i, \quad i = 1, 2, \ldots, m, \\ \\ \quad \sum_{i=1}^{m} x_{ij} = b_j, \quad j = 1, 2, \ldots, n, \\ \\ \quad x_{ij} \geq 0, \quad i = 1, 2, \ldots, m, j = 1, 2, \ldots, n. \end{array} \right.$$

$$(5.62)$$

$$(\tilde{z})_\alpha^U = \begin{array}{c} \max \\ (\tilde{c}_{ij})_\alpha^L \le c_{ij} \le (\tilde{c}_{ij})_\alpha^U \\ (\tilde{a}_i)_\alpha^L \le a_i \le (\tilde{a}_i)_\alpha^U \\ (\tilde{b}_j)_\alpha^L \le b_j \le (\tilde{b}_j)_\alpha^U, \\ \forall i,j \\ \sum_{i=1}^m a_i = \sum_{j=1}^n b_j, \end{array} \left\{ \begin{array}{ll} \min & \displaystyle\sum_{i=1}^m \sum_{j=1}^n c_{ij} x_{ij} \\ s.t. & \displaystyle\sum_{j=1}^n x_{ij} = a_i, \quad i = 1,2,\ldots,m, \\ & \displaystyle\sum_{i=1}^m x_{ij} = b_j, \quad j = 1,2,\ldots,n, \\ & x_{ij} \ge 0, \quad\quad i = 1,2,\ldots,m, j = 1,2,\ldots,n. \end{array} \right.$$

$$(5.63)$$

Similar to the discussion of the inequality-constraint case, the corresponding pair of one-level mathematical models are:

$$(\tilde{z})_\alpha^L = \min \sum_{i=1}^m \sum_{j=1}^n (\tilde{c}_{ij})_\alpha^L x_{ij}$$

$$\begin{array}{ll} s.t. & \displaystyle\sum_{j=1}^n x_{ij} = a_i, & i = 1,2,\ldots,m, \\[2mm] & \displaystyle\sum_{i=1}^m x_{ij} = b_j, & j = 1,2,\ldots,n, \\[2mm] & \displaystyle\sum_{i=1}^m a_i = \sum_{j=1}^n b_j, \\[2mm] & (\tilde{a}_i)_\alpha^L \le a_i \le (\tilde{a}_i)_\alpha^U, & i = 1,2,\ldots,m, \\[1mm] & (\tilde{b}_j)_\alpha^L \le b_j \le (\tilde{b}_j)_\alpha^U, & j = 1,2,\ldots,n, \\[1mm] & x_{ij} \ge 0, & i = 1,2,\ldots,m, j = 1,2,\ldots,n. \end{array}$$

$$(5.64)$$

$$(\tilde{z})_\alpha^U = \max \sum_{i=1}^m a_i u_i + \sum_{j=1}^n b_j v_j$$

$$\begin{array}{ll} s.t. & u_i + v_j \le (\tilde{c}_{ij})_\alpha^U, & i = 1,2,\ldots,m, j = 1,2,\ldots,n, \\[2mm] & \displaystyle\sum_{i=1}^m a_i = \sum_{j=1}^n b_j, \\[2mm] & (\tilde{a}_i)_\alpha^L \le a_i \le (\tilde{a}_i)_\alpha^U, & i = 1,2,\ldots,m, \\[1mm] & (\tilde{b}_j)_\alpha^L \le b_j \le (\tilde{b}_j)_\alpha^U, & j = 1,2,\ldots,n, \\[1mm] & u_i, v_j \text{ free in sign}, & i = 1,2,\ldots,m, j = 1,2,\ldots,n. \end{array}$$

$$(5.65)$$

The lower and upper bounds of the total transportation cost at α level can be obtained by solving the problems (5.64) and (5.65). The α-level sets $\left[(\tilde{z})_\alpha^L, (\tilde{z})_\alpha^U\right]$ at different possibility levels constitute the membership function $\mu_{\tilde{z}}(z)$.

Example 5.5 [8] Consider the following FTP problem of type-III with equality constraints:

$$\min \tilde{z} = 10x_{11} + 50x_{12} + 80x_{13} + (60, 70, 80, 90)x_{21} + 60x_{22} + 20x_{23}$$
$$s.t. \quad x_{11} + x_{12} + x_{13} = (70, 90, 100),$$
$$x_{21} + x_{22} + x_{23} = (40, 60, 70, 80),$$
$$x_{11} + x_{21} = (30, 40, 50, 70), \tag{5.66}$$
$$x_{12} + x_{22} = (20, 30, 40, 50),$$
$$x_{13} + x_{23} = (40, 50, 80),$$
$$x_{11}, x_{12}, x_{13}, x_{21}, x_{22}, x_{23} \geq 0.$$

According to the Models (5.65) and (5.66), the lower and upper bounds of $\tilde{z}$ at possibility level α can be obtained by solving the following problems:

$$(\tilde{z})_\alpha^L = \min 10x_{11} + 50x_{12} + 80x_{13} + (60 + 10\alpha)x_{21} + 60x_{22} + 20x_{23}$$
$$s.t. \quad x_{11} + x_{12} + x_{13} = a_1,$$
$$x_{21} + x_{22} + x_{23} = a_2,$$
$$x_{11} + x_{21} = b_1,$$
$$x_{12} + x_{22} = b_2,$$
$$x_{13} + x_{23} = b_3, \tag{5.67}$$
$$a_1 + a_2 = b_1 + b_2 + b_3,$$
$$70 + 20\alpha \leq a_1 \leq 100 - 10\alpha, 40 + 20\alpha \leq a_2 \leq 80 - 10\alpha,$$
$$30 + 10\alpha \leq b_1 \leq 70 - 20\alpha, 20 + 10\alpha \leq b_2 \leq 50 - 10\alpha,$$
$$40 + 10\alpha \leq b_3 \leq 80 - 30\alpha,$$
$$x_{11}, x_{12}, x_{13}, x_{21}, x_{22}, x_{23} \geq 0.$$

$$(\tilde{z})_\alpha^U = \max a_1 u_1 + a_2 u_2 + b_1 v_1 + b_2 v_2 + b_3 v_3$$
$$s.t. \quad u_1 + v_1 \leq 10,$$
$$u_1 + v_2 \leq 50,$$
$$u_1 + v_3 \leq 80,$$
$$u_2 + v_1 \leq 90 - 10\alpha,$$
$$u_2 + v_2 \leq 60,$$
$$u_2 + v_3 \leq 20, \tag{5.68}$$
$$a_1 + a_2 = d_1 + d_2 + d_3,$$
$$70 + 20\alpha \leq a_1 \leq 100 - 10\alpha, 40 + 20\alpha \leq a_2 \leq 80 - 10\alpha,$$
$$30 + 10\alpha \leq d_1 \leq 70 - 20\alpha, 20 + 10\alpha \leq d_2 \leq 50 - 10\alpha,$$
$$40 + 10\alpha \leq d_3 \leq 80 - 30\alpha,$$
$$u_1, u_2, v_1, v_2, v_3 \text{ free in sign.}$$

Table 5.7 The α-cuts of the total transportation cost for Example 5.4

α	0.0	0.1	0.2	0.3	0.4	0.5	0.6	0.7	0.8	0.9	1.0
$(\tilde{z})_\alpha^U$	2300	2400	2500	2600	2700	2800	2900	3040	3260	3680	Infeasible
$(\tilde{z})_\alpha^U$	5800	5600	5400	5200	5000	4800	4440	4080	3860	3680	Infeasible

Fig. 5.4 The membership function of the objective function for Example 5.5

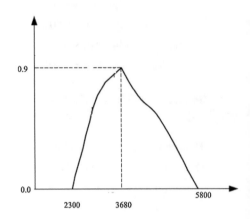

The bounds of the total transportation cost for different values of $\alpha \in [0,1]$ are reported in Table 5.7. The membership function $\tilde{z}$ of this example is illustrated in Fig. 5.4. For α greater than 0.9, the problem is infeasible. In other words, the maximum degree to which the constraints could be satisfied is equal to 0.9. It is also worthwhile to note that the membership function of the objective value of this example is contained in that of Example 5.4. The reason is simply because equality constraints are more restrictive than inequality constraints.

At $\alpha = 0$ $\tilde{z}$, the lower bound of the objective function is $z^* = 2100$ occurs at $x_{11}^* = 50, x_{12}^* = 20, x_{23}^* = 40, x_{13}^* = x_{21}^* = x_{22}^* = 0$ with $b_1 = 50$, $b_2 = 20$, $b_3 = 40$ and $a_1 = 70, a_2 = 40$. The upper bound of $z^* = 5800$ occurs at $x_{11}^* = 30, x_{12}^* = 30, x_{13}^* = 40, x_{23}^* = 40, x_{21}^* = x_{22}^* = 0$ with $b_1 = 30$, $b_2 = 30$, $b_3 = 80$ and $a_1 = 100$, $a_2 = 40$. At $\alpha = 0.9$, $\alpha-$ cut is a single point 3660. The associated optimal solution is $x_{11}^* = 52, x_{12}^* = 36, x_{22}^* = 5, x_{23}^* = 53, x_{13}^* = x_{21}^* = 0$ with $b_1 = 52$, $b_2 = 41$, $b_3 = 53$ and $a_1 = 88, a_2 = 58$. Note that this point corresponds to the intersection of the left shape function of the total supply and the right shape function of the total demand, which also has a membership degree of 0.9.

5.6   FTPs of Type-IV

The general form of FTPs with fuzzy transportation costs, fuzzy supply and demand as well as decision variables can be formulated as follows:

$$\min \tilde{z} = \sum_{i=1}^{m} \sum_{j=1}^{n} \tilde{c}_{ij} \tilde{x}_{ij}$$

$$s.t. \quad \sum_{j=1}^{n} \tilde{x}_{ij} = \tilde{a}_i, \quad i = 1, 2, \ldots, m,$$

$$\sum_{i=1}^{m} \tilde{x}_{ij} = \tilde{b}_j, \quad j = 1, 2, \ldots, n, \qquad (5.69)$$

$$\tilde{x}_{ij} \succeq \tilde{0}, \qquad i = 1, 2, \ldots, m, j = 1, 2, \ldots, n.$$

In the Model (5.69), we assume non-negative LR flat fuzzy numbers to quantify the ambiguity in the decision parameters. Thus, $\tilde{c}_{ij}$, $\tilde{a}_i$, $\tilde{b}_j$ and $\tilde{x}_{ij}$ are represented by the non-negative LR flat fuzzy numbers $\left(c_{ij,1}, c_{ij,2}, c_{ij,2}, c_{ij,4}\right)_{LR}$, $\left(a_{i,1}, a_{i,2}, a_{i,3}, a_{i,4}\right)_{LR}$, $\left(b_{j,1}, b_{j,2}, b_{j,3}, b_{j,4}\right)_{LR}$ and $\left(x_{ij,1}, x_{ij,2}, x_{ij,3}, x_{ij,4}\right)_{LR}$, respectively.

Remark 5.8 If $\sum_{i=1}^{m} \tilde{a}_i = \sum_{j=1}^{n} \tilde{b}_j$ then the FTP of type-IV (5.69) is said to be a balanced FTP, otherwise it is called an unbalanced FTP.

Kumar and Kaur [22, 38] proposed an approach to convert the unbalanced kind of FTP of type-IV (5.69) into a balanced one. Assume $\sum_{i=1}^{m} \tilde{a}_i = (a_1, a_2, a_3, a_4)_{LR}$ and $\sum_{j=1}^{n} \tilde{b}_j = (b_1, b_2, b_3, b_4)_{LR}$. The steps of the proposed approach for converting the unbalanced FTP into a balanced problem are as follows:

Step 1: If $a_1 \leq b_1, a_2 \leq b_2, a_3 \leq b_3$, and $a_4 \leq b_4$ then introduce a dummy source with fuzzy availability $(b_1 - a_1, b_2 - a_2, b_3 - a_3, b_4 - a_4)_{LR}$. Assume the fuzzy transportation cost for one unit quantity of the product from the introduced dummy source to all destinations as zero LR flat fuzzy number; otherwise, go to Step 2.

Step 2: If $a_1 \geq b_1, a_2 \geq b_2, a_3 \geq b_3$, and $a_4 \leq b_4$ then introduce a dummy destination with fuzzy demand $(a_1 - b_1, a_2 - b_2, a_3 - b_3, a_4 - b_4)_{LR}$. Assume the fuzzy transportation cost for one unit quantity of the product from all sources to the introduced dummy destination as zero LR flat fuzzy number; otherwise, go to Step 3.

Step 3: If neither condition given in Step 1 nor Step 2 is satisfied then introduce (1) a dummy source with fuzzy availability $(d_1, d_2, d_3, d_4)_{LR}$ where $d_1 = \max\{0, b_1 - a_1\}$, $d_2 = d_1 + \max\{0, (b_2 - b_1) - (a_2 - a_1)\}$, $d_3 = d_2 + \max\{0, (b_3 - b_2) - (a_3 - a_2)\}$ and $d_4 = d_3 + \max\{0, (b_4 - b_3) - (a_4 - a_3)\}$, and (2) a dummy destination with fuzzy demand $(e_1, e_2, e_3, e_4)_{LR}$ where $e_1 = \max\{0, a_1 - b_1\}$, $e_2 = e_1 + \max\{0, (a_2 - a_1) - (b_2 - b_1)\}$, $e_3 = e_2 + \max\{0, (a_3 - a_2) - (b_3 - b_2)\}$ and $e_4 = e_3 + \max\{0, (a_4 - a_3) - (b_4 - b_3)\}$. Assume the fuzzy transportation cost for one unit quantity of the product from the introduced dummy source to all destinations and from all sources to the introduced dummy destination as zero LR flat fuzzy number.

As we discussed in the past decade, researchers have discussed various properties of the FTP and proposed an assortment of models. Here, the shortcomings of the widely-used methods for solving the FTP are presented. The main defects of the existing methods can be classified as follows:

(1) The existing methods [5, 26] can be applied only to solve FTPs where the transportation costs are represented as fuzzy numbers. These methods cannot be used for solving the FTP (5.69) where the supply and demand are fuzzy numbers as well.

(2) The existing methods [5, 13] can be used for solving FTPs where the supply and demand are fuzzy, and the transportation costs are represented as real numbers. These methods cannot be used for solving the FTP (5.69), where all the parameters are fuzzy.

(3) Liu and Kao [8] proposed a method for numerically driving the membership function of the fuzzy total transportation cost using fuzzy unit shipping costs, supply quantities, and the demand quantities. Here, the membership degree of a specific transportation cost is approximated using the α-cuts. This method does not provide the mathematical form of the membership function so that the membership degree can be calculated directly. Additionally, it requires the solution of a non-linear program.

(4) In the existing methods [2, 3, 8], the optimal solutions of the FTP are real numbers. In the case of fuzzy data this represents a compromise.

(5) In the existing methods [17, 18, 21, 25], there is a negative part in the fuzzy optimal quantities of the commodity to be transformed from some sources to some destinations and in the total fuzzy transportation cost; thus the quantity of the commodity and the transportation cost may be negative. However, a negative quantity of the commodity and negative transportation costs have no physical meaning, especially, when the transportation costs are non-negative fuzzy numbers. That is, when the fuzzy coefficients of the objective function, the fuzzy values of the supply and demand and the fuzzy decision variable are non-negative fuzzy numbers, the fuzzy optimal value of the objective function must be non-negative, whereas some of the respective fuzzy numbers in the existing methods [17, 18, 21, 25] are negative.

Now, we explore two approaches for solving the FTP problem of type-IV (5.69) in which all fuzzy parameters are represented by non-negative LR flat fuzzy numbers. These two approaches overcome the above-mentioned shortcomings.

5.6.1   Kumar and Kaur's Approach

Kumar and Kaur [22] proposed a new method based on FLP formulation to find the fuzzy optimal solution of the FTP of type-IV (5.69). To do this they defined the concepts of fuzzy optimal solution and alternative fuzzy optimal solution to the FTP of type-IV (5.69).

Definition 5.1 The fuzzy optimal solution of the balanced kind of FTP of type-IV (5.69) is a fuzzy number $\tilde{x}_{ij}$, which satisfies the following characteristics:

(i) $\tilde{x}_{ij}$ is a non-negative LR flat fuzzy number,

(ii) $\sum_{j=1}^{n} \tilde{x}_{ij} = \tilde{a}_i, i = 1, 2, \ldots, m, \sum_{i=1}^{m} \tilde{x}_{ij} = \tilde{b}_j, \quad j = 1, 2, \ldots, n,$

(iii) If there is any non-negative fuzzy LR flat fuzzy number $\tilde{x}'_{ij}$ such that $\sum_{j=1}^{n} \tilde{x}'_{ij} = \tilde{a}_i, i = 1, 2, \ldots, m, \sum_{i=1}^{m} \tilde{x}'_{ij} = \tilde{b}_j, j = 1, 2, \ldots, n,$ then $\sum_{i=1}^{m} \sum_{j=1}^{n} \Re(\tilde{c}_{ij}\tilde{x}_{ij}) \leq \sum_{i=1}^{m} \sum_{j=1}^{n} \Re\left(\tilde{c}_{ij}\tilde{x}'_{ij}\right).$

Remark 5.9 The index $\Re(\tilde{z})$ is calculated for the LR flat fuzzy number $\tilde{z} = (z_1, z_2, z_3, z_4)_{LR}$ from its $\alpha-$ cut, $[\tilde{z}]_\alpha = \{x \in R; \mu_{\tilde{z}}(x) \geq \alpha\}$, according to the following formula:

$$\Re(\tilde{z}) = \frac{1}{2} \int_0^1 (\inf[\tilde{z}]_\alpha + \sup[\tilde{z}]_\alpha) d\alpha$$

Definition 5.2 Let $\tilde{x}_{ij}$ be a fuzzy optimal solution of the FTP of type-IV (5.69) and one or more $\tilde{x}'_{ij}$ exist such that

(i) $\tilde{x}'_{ij}$ is a non-negative LR flat fuzzy number,

(ii) $\sum_{j=1}^{n} \tilde{x}'_{ij} = \tilde{a}_i, i = 1, 2, \ldots, m, \sum_{i=1}^{m} \tilde{x}'_{ij} = \tilde{b}_j, j = 1, 2, \ldots, n,$

(iii) $\sum_{i=1}^{m} \sum_{j=1}^{n} \Re(\tilde{c}_{ij}\tilde{x}_{ij}) = \sum_{i=1}^{m} \sum_{j=1}^{n} \Re\left(\tilde{c}_{ij}\tilde{x}'_{ij}\right),$

then $\tilde{x}'_{ij}$ is said to be an alternative fuzzy optimal solution to the FTP of type-IV (5.69).

The steps of the proposed method by Kumar and Kaur [22] for solving the FTP of type-IV (5.69) are as follows:

Step 1: Find the total fuzzy supply $\sum_{i=1}^{m} \tilde{a}_i$ and the total fuzzy demand $\sum_{j=1}^{n} \tilde{b}_j$. Examine whether the problem is balanced or unbalanced, i.e., $\sum_{i=1}^{m} \tilde{a}_i = \sum_{j=1}^{n} \tilde{b}_j$ or $\sum_{i=1}^{m} \tilde{a}_i \neq \sum_{j=1}^{n} \tilde{b}_j$. If the problem is balanced, go to Step 2; otherwise convert the unbalanced problem into the balanced one as discussed in this section and then proceed to Step 2.

Step 2: Let $\sum_{i=1}^{m} \sum_{j=1}^{n} \tilde{c}_{ij}\tilde{x}_{ij} = (z_1, z_2, z_3, z_4)_{LR}$ and formulate the FTP of type-IV (5.69) into the following fuzzy problem based on Definition 5.1:

$$\min z = \Re(z_1, z_2, z_3, z_4)_{LR}$$

$$s.t. \quad \sum_{j=1}^{n} \tilde{x}_{ij} = \tilde{a}_i, \quad i = 1, 2, \ldots, m,$$

$$\sum_{i=1}^{m} \tilde{x}_{ij} = \tilde{b}_j, \quad j = 1, 2, \ldots, n, \tag{5.70}$$

$$\tilde{x}_{ij} \succeq \tilde{0}, \quad i = 1, 2, \ldots, m, j = 1, 2, \ldots, n.$$

The index $\Re(\tilde{z})$ has been given in Remark 5.9.

Step 3: Equivalently, the fuzzy problem (5.70) may be reformulated as follows:

$$\min z = \Re(z_1, z_2, z_3, z_4)_{LR}$$

$$s.t. \quad \sum_{j=1}^{n} \left(x_{ij,1}, x_{ij,2}, x_{ij,3}, x_{ij,4}\right)_{LR} = \left(a_{i,1}, a_{i,2}, a_{i,3}, a_{i,4}\right)_{LR}, \quad i = 1, 2, \ldots, m,$$

$$\sum_{i=1}^{m} \left(x_{ij,1}, x_{ij,2}, x_{ij,3}, x_{ij,4}\right)_{LR} = \left(b_{j,1}, b_{j,2}, b_{j,3}, b_{j,4}\right)_{LR}, \quad j = 1, 2, \ldots, n, \tag{5.71}$$

$$\left(x_{ij,1}, x_{ij,2}, x_{ij,3}, x_{ij,4}\right)_{LR} \succeq \tilde{0}, \qquad\qquad i = 1, 2, \ldots, m,$$
$$\qquad\qquad\qquad\qquad\qquad\qquad\qquad\qquad\qquad j = 1, 2, \ldots, n.$$

Step 4: The fuzzy problem (5.71) is converted to the following model with regard to the arithmetic operations on fuzzy numbers:

$$\min z = \Re(z_1, z_2, z_3, z_4)_{LR}$$

$$s.t. \quad \left(\sum_{j=1}^{n} x_{ij,1}, \sum_{j=1}^{n} x_{ij,2}, \sum_{j=1}^{n} x_{ij,3}, \sum_{j=1}^{n} x_{ij,4}\right)_{LR} = \left(a_{i,1}, a_{i,2}, a_{i,3}, a_{i,4}\right)_{LR}, \quad i = 1, 2, \ldots, m,$$

$$\left(\sum_{i=1}^{m} x_{ij,1}, \sum_{i=1}^{m} x_{ij,2}, \sum_{i=1}^{m} x_{ij,3}, \sum_{i=1}^{m} x_{ij,4}\right)_{LR} = \left(b_{j,1}, b_{j,2}, b_{j,3}, b_{j,4}\right)_{LR}, \quad j = 1, 2, \ldots, n,$$

$$\left(x_{ij,1}, x_{ij,2}, x_{ij,3}, x_{ij,4}\right)_{LR} \succeq \tilde{0}, \qquad\qquad i = 1, 2, \ldots, m,$$
$$\qquad\qquad\qquad\qquad\qquad\qquad\qquad\qquad\qquad j = 1, 2, \ldots, n.$$
$$\tag{5.72}$$

Step 5: The fuzzy problem (5.72) is converted to the following LP problem:

$$\min \quad z = \Re(z_1, z_2, z_3, z_4)_{LR}$$

$$s.t. \quad \sum_{j=1}^{n} x_{ij,1} = a_{i,1}, \sum_{j=1}^{n} x_{ij,2} = a_{i,2}, \sum_{j=1}^{n} x_{ij,3} = a_{i,3}, \sum_{j=1}^{n} x_{ij,4} = a_{i,4}, \quad i = 1, 2, \ldots, m,$$

$$\sum_{i=1}^{m} x_{ij,1} = b_{j,1}, \sum_{i=1}^{m} x_{ij,2} = b_{j,2}, \sum_{i=1}^{m} x_{ij,3} = b_{j,3}, \sum_{i=1}^{m} x_{ij,4} = b_{j,4}, \quad j = 1, 2, \ldots, n,$$

$$x_{ij,1} \geq 0, x_{ij,2} - x_{ij,1} \geq 0, x_{ij,3} - x_{ij,2} \geq 0, x_{ij,4} - x_{ij,3} \geq 0, \qquad i = 1, 2, \ldots, m,$$
$$\qquad\qquad\qquad\qquad\qquad\qquad\qquad\qquad\qquad\qquad\qquad j = 1, 2, \ldots, n.$$
$$\tag{5.73}$$

Step 6: Solve the crisp LP problem (5.73) to find the optimal solution $x_{ij,1}^*, x_{ij,2}^*, x_{ij,3}^*$ and $x_{ij,4}^*$.

Step 7: Find the fuzzy optimal solution $\tilde{x}_{ij}^*$ by putting the values of $x_{ij,1}^*, x_{ij,2}^*, x_{ij,3}^*$ and $x_{ij,4}^*$ in $\tilde{x}_{ij}^* = \left(x_{ij,1}^*, x_{ij,2}^*, x_{ij,3}^*, x_{ij,4}^*\right)_{LR}$.

Step 8: Find the minimum total fuzzy transportation cost by putting the values of $\tilde{x}_{ij}^*$ in $\sum_{i=1}^{m} \sum_{j=1}^{n} \tilde{c}_{ij} \tilde{x}_{ij}$.

Example 5.6 Data are collected from a trader that supplies a product (TMT), which is made from raw materials INGOT and BILLET, to different centers after obtaining the product from different plants. The trader supplies the product from three plants, Fortune Metals (Mandi Gobindgarh), Kamdhenu Saria (Bhiwadi), and Goel Group (Raipur) to four different centers Ludhiana, Delhi, Himachal Pradesh, and Leh Ladakh. On the basis of the trader's perception the approximate transportation cost per ton (in thousands of rupees), the approximate availability of the product (in tons) at different plants, and the approximate demand for the product (in tons) at different centers are represented by *LR* flat fuzzy numbers, as shown in Table 1. Here, $L(x) = R(x) = \max\{0, 1 - x\}$ (Table 5.8).

The trader wishes to determine the approximate quantity of the product that should be transported from each plant to each center so that the total approximate transportation cost is minimized. This problem can be formulated as the following FTP:

$$\min\ (19, 20, 21, 22)_{LR}\tilde{x}_{11} + (59, 62, 63, 65)_{LR}\tilde{x}_{12} + (90, 95, 97, 99)_{LR}\tilde{x}_{13}$$
$$+ (150, 160, 165, 170)_{LR}\tilde{x}_{14} + (97, 99, 103, 105)_{LR}\tilde{x}_{21} + (15, 17, 19, 21)_{LR}\tilde{x}_{22}$$
$$+ (110, 112, 115, 119)_{LR}\tilde{x}_{23} + (190, 210, 220, 240)_{LR}\tilde{x}_{24} + (260, 262, 264, 270)\tilde{x}_{31}$$
$$+ (240, 247, 249, 255)_{LR}\tilde{x}_{32} + (272, 274, 279, 290)_{LR}\tilde{x}_{33} + (320, 326, 332, 340)_{LR}\tilde{x}_{34}$$

$$s.t.\quad \tilde{x}_{11} + \tilde{x}_{12} + \tilde{x}_{13} + \tilde{x}_{14} = (3500, 3555, 3580, 4000)_{LR},$$
$$\tilde{x}_{21} + \tilde{x}_{22} + \tilde{x}_{23} + \tilde{x}_{24} = (3125, 3175, 3190, 3200)_{LR},$$
$$\tilde{x}_{31} + \tilde{x}_{32} + \tilde{x}_{33} + \tilde{x}_{34} = (2475, 2995, 3275, 3400)_{LR},$$
$$\tilde{x}_{11} + \tilde{x}_{21} + \tilde{x}_{31} = (2050, 2500, 2700, 3050)_{LR},$$
$$\tilde{x}_{12} + \tilde{x}_{22} + \tilde{x}_{32} = (3000, 3050, 3100, 3200)_{LR},$$
$$\tilde{x}_{13} + \tilde{x}_{23} + \tilde{x}_{33} = (2100, 2150, 2190, 2250)_{LR},$$
$$\tilde{x}_{14} + \tilde{x}_{24} + \tilde{x}_{34} = (1950, 2025, 2055, 2100)_{LR},$$
$$\tilde{x}_{11}, \tilde{x}_{12}, \tilde{x}_{13}, \tilde{x}_{14}, \tilde{x}_{21}, \tilde{x}_{22}, \tilde{x}_{23}, \tilde{x}_{24}, \tilde{x}_{31}, \tilde{x}_{32}, \tilde{x}_{33}, \tilde{x}_{34} \in \ell\Re(R)^+.$$

$$(5.74)$$

The total fuzzy supply $= (9110, 9725, 10045, 10600)_{LR} =$ the total fuzzy demand, and thus this is a balanced FTP. According to steps 2, 3, 4 and 5 of the proposed algorithm by Kumar and Kaur [22], we should solve the following LP model:

Table 5.8 Data of Example 5.6

Source	Destination				Supply (in tons)
	Ludhiana	Delhi	Kullu	Leh Ladakh	
Gobindgarh	$(19, 20, 21, 22)_{LR}$	$(59, 62, 63, 65)_{LR}$	$(90, 95, 97, 99)_{LR}$	$(150, 160, 165, 170)_{LR}$	$(3500, 3555, 3580, 4000)_{LR}$
Bhiwadi	$(97, 99, 103, 105)_{LR}$	$(15, 17, 19, 21)_{LR}$	$(110, 112, 115, 119)_{LR}$	$(190, 210, 220, 240)_{LR}$	$(3125, 3175, 3190, 3200)_{LR}$
Raipur	$(260, 262, 264, 270)_{LR}$	$(240, 247, 249, 255)_{LR}$	$(272, 274, 279, 290)_{LR}$	$(320, 326, 332, 340)_{LR}$	$(2475, 2995, 3275, 3400)_{LR}$
Demand (in tons)	$(2050, 2500, 2700, 3050)_{LR}$	$(3000, 3050, 3100, 3200)_{LR}$	$(2100, 2150, 2190, 2250)_{LR}$	$(1950, 2025, 2055, 2100)_{LR}$	

$$\min \frac{1}{4}(19x_{11,1} + 59x_{12,1} + 90x_{13,1} + 150x_{14,1} + 97x_{21,1}$$

$$+ 15x_{22,1} + 110x_{23,1} + 190x_{24,1} + 260x_{31,1} + 240x_{32,1}$$

$$+ 272x_{33,1} + 320x_{34,1} + 20x_{11,2} + 62x_{12,2} + 95x_{13,2}$$

$$+ 160x_{14,2} + 99x_{21,2} + 17x_{22,2} + 112x_{23,2} + 210x_{24,2}$$

$$+ 262x_{31,2} + 247x_{32,2} + 274x_{33,2} + 326x_{34,2} + 21x_{11,3}$$

$$+ 63x_{12,3} + 97x_{13,3} + 165x_{14,3} + 103x_{21,3} + 19x_{22,3}$$

$$+ 115x_{23,3} + 220x_{24,3} + 264x_{31,3} + 249x_{32,3} + 279x_{33,3}$$

$$+ 332x_{34,3} + 22x_{11,4} + 65x_{12,4} + 99x_{13,4} + 170x_{14,4}$$

$$+ 105x_{21,4} + 21x_{22,4} + 119x_{23,4} + 240x_{24,4}$$

$$+ 270x_{31,4} + 255x_{32,4} + 290x_{33,4} + 340x_{34,4})$$

$$s.t. x_{11,1} + x_{12,1} + x_{13,1} + x_{14,1} = 3500_{21,1} + x_{22,1} + x_{23,1} + x_{24,1} = 3125,$$

$$x_{31,1} + x_{32,1} + x_{33,1} + x_{34,1} = 2475, x_{11,1} + x_{21,1} + x_{31,1} = 2050,$$

$$x_{12,1} + x_{22,1} + x_{32,1} = 3000, x_{13,1} + x_{23,1} + x_{33,1} = 2100,$$

$$x_{14,1} + x_{24,1} + x_{34,1} = 1950,$$

$$x_{11,2} + x_{12,2} + x_{13,2} + x_{14,2} = 3550, x_{21,2} + x_{22,2} + x_{23,2} + x_{24,2} = 3175,$$

$$x_{31,2} + x_{32,2} + x_{33,2} + x_{34,2} = 2995, x_{11,2} + x_{21,2} + x_{31,2} = 2500,$$

$$x_{12,2} + x_{22,2} + x_{32,2} = 3050, x_{13,2} + x_{23,2} + x_{33,2} = 3150,$$

$$x_{14,2} + x_{24,2} + x_{34,2} = 2025,$$

$$x_{11,3} + x_{12,3} + x_{13,3} + x_{14,3} = 3580, x_{21,3} + x_{22,3} + x_{23,3} + x_{24,3} = 3190,$$

$$x_{31,3} + x_{32,3} + x_{33,3} + x_{34,3} = 3275, x_{11,3} + x_{21,3} + x_{31,3} = 2700,$$

$$x_{12,3} + x_{22,3} + x_{32,3} = 3100, x_{13,3} + x_{23,3} + x_{33,3} = 2190,$$

$$x_{14,3} + x_{24,3} + x_{34,3} = 2055,$$

$$x_{11,4} + x_{12,4} + x_{13,4} + x_{14,4} = 4000, x_{21,4} + x_{22,4} + x_{23,4} + x_{24,4} = 3200,$$

$$x_{31,4} + x_{32,4} + x_{33,4} + x_{34,4} = 3400, x_{11,4} + x_{21,4} + x_{31,4} = 3050,$$

$$x_{12,4} + x_{22,4} + x_{32,4} = 3200, x_{13,4} + x_{23,4} + x_{33,4} = 2250,$$

$$x_{14,4} + x_{24,4} + x_{34,4} = 2100,$$

$$x_{11,4} \geq x_{11,3}, x_{11,3} \geq x_{11,2}, x_{11,2} \geq x_{11,1}, x_{12,4} \geq x_{12,3}, x_{12,3} \geq x_{12,2}, x_{12,2} \geq x_{12,1},$$

$$x_{13,4} \geq x_{13,3}, x_{13,3} \geq x_{13,2}, x_{13,2} \geq x_{13,1}, x_{14,4} \geq x_{14,3}, x_{14,3} \geq x_{14,2}, x_{14,2} \geq x_{14,1},$$

$$x_{21,4} \geq x_{21,3}, x_{21,3} \geq x_{21,2}, x_{21,2} \geq x_{21,1}, x_{22,4} \geq x_{22,3}, x_{22,3} \geq x_{22,2}, x_{22,2} \geq x_{22,1},$$

$$x_{23,4} \geq x_{23,3}, x_{23,3} \geq x_{23,2}, x_{23,2} \geq x_{23,1}, x_{24,4} \geq x_{24,3}, x_{24,3} \geq x_{24,2}, x_{24,2} \geq x_{24,1},$$

$$x_{31,4} \geq x_{31,3}, x_{31,3} \geq x_{31,2}, x_{31,2} \geq x_{31,1}, x_{32,4} \geq x_{32,3}, x_{32,3} \geq x_{32,2}, x_{32,2} \geq x_{32,1},$$

$$x_{33,4} \geq x_{33,3}, x_{33,3} \geq x_{33,2}, x_{33,2} \geq x_{33,1}, x_{34,4} \geq x_{34,3}, x_{34,3} \geq x_{34,2}, x_{34,2} \geq x_{34,1},$$

$$x_{11,1}, x_{12,1}, x_{13,1}, x_{14,1}, x_{21,1}, x_{22,1}, x_{23,1}, x_{24,1}, x_{31,1}, x_{32,1}, x_{33,1}, x_{34,1} \geq 0.$$

$$(5.75)$$

The optimal solution of the transportation problem (5.75) is as follows:

$$
\begin{aligned}
&x_{11,1}^* = 2050, x_{12,1}^* = 0, x_{13,1}^* = 1450, x_{14,1}^* = 0, \\
&x_{21,1}^* = 0, x_{22,1}^* = 3000, x_{23,1}^* = 125, x_{24,1}^* = 0, \\
&x_{31,1}^* = 0, x_{32,1}^* = 0, x_{33,1}^* = 525, x_{34,1}^* = 1950, \\
&x_{11,2}^* = 2105, x_{12,2}^* = 0, x_{13,2}^* = 1450, x_{14,2}^* = 0, \\
&x_{21,2}^* = 0, x_{22,2}^* = 3050, x_{23,2}^* = 125, x_{24,2}^* = 0, \\
&x_{31,2}^* = 395, x_{32,2}^* = 0, x_{33,2}^* = 575, x_{34,2}^* = 2025, \\
&x_{11,3}^* = 2130, x_{12,3}^* = 0, x_{13,3}^* = 1450, x_{14,3}^* = 0, \\
&x_{21,3}^* = 0, x_{22,3}^* = 3060, x_{23,3}^* = 125, x_{24,3}^* = 0, \\
&x_{31,3}^* = 570, x_{32,3}^* = 35, x_{33,3}^* = 615, x_{34,3}^* = 2055, \\
&x_{11,4}^* = 2480, x_{12,4}^* = 10, x_{13,4}^* = 1510, x_{14,4}^* = 0, \\
&x_{21,4}^* = 0, x_{22,4}^* = 3075, x_{23,4}^* = 125, x_{24,4}^* = 0, \\
&x_{31,4}^* = 570, x_{32,4}^* = 115, x_{33,4}^* = 615, x_{34,4}^* = 2100,
\end{aligned}
\tag{5.76}
$$

Thus, according to Step 6, the fuzzy optimal solution of the FTP (5.74) is given as follows:

$$
\begin{aligned}
&\tilde{x}_{11}^* = (2050, 2105, 2130, 2480)_{LR}, \ \tilde{x}_{12}^* = (0,0,0,10)_{LR}, \\
&\tilde{x}_{13}^* = (1450, 1450, 1450, 1510)_{LR}, \ \tilde{x}_{14}^* = (0,0,0,0)_{LR}, \\
&\tilde{x}_{21}^* = (0,0,0,0)_{LR}, \ \tilde{x}_{22}^* = (3000, 3050, 3065, 3075)_{LR}, \\
&\tilde{x}_{23}^* = (125, 125, 125, 125)_{LR}, \ \tilde{x}_{14}^* = (0,0,0,0)_{LR}, \\
&\tilde{x}_{31}^* = (0, 395, 570, 570)_{LR}, \ \tilde{x}_{32}^* = (0,0,35,115)_{LR}, \\
&\tilde{x}_{33}^* = (525, 575, 615, 615)_{LR}, \ \tilde{x}_{34}^* = (1950, 2025, 2055, 2100)_{LR}.
\end{aligned}
\tag{5.77}
$$

Now, according to Step 7, the minimum transportation cost is given as follows:

$$
\tilde{z}^* = \sum_{i=1}^{3} \sum_{j=1}^{4} \tilde{c}_{ij} \otimes \tilde{x}_{ij}^* = (999500, 1166890, 1271030, 1359725)_{LR}
\tag{5.78}
$$

5.6.2   Ebrahimnejad's Approach

Ebrahimnejad [34] proposed a new method for solving the FTP of type-IV (5.69) where all parameters are represented by non-negative LR flat fuzzy numbers. According to this approach, the FTP is converted into four transportation problems, which are solved using standard transportation simplex algorithms.

The contributions of the approach proposed by Ebrahimnejad [34] are summarized as follows: (1) In the FTP under consideration, all of the parameters, such as the transportation costs, supplies and demands are considered as fuzzy numbers. (2) According to this approach, the FTP is converted into four classical transportation problems. The integration of the optimal solution of the four sub-problems provides the optimal solution of the primary FTP. (3) In contrast to most existing approaches, which provide a precise solution, this method provides a fuzzy optimal solution. (4) In contrast to existing methods that include negative parts in the obtained fuzzy optimal solution and fuzzy optimal cost, this method provides a non-negative fuzzy optimal solution and optimal cost. (5) Similarly to the competing methods in the literature, this method is applicable for all types of LR flat fuzzy numbers. (6) The computation complexity is greatly reduced compared with commonly-used existing methods in the literature.

The FTP of type-IV (5.69) can be written as follows:

$$\min \sum_{i=1}^{m} \sum_{j=1}^{n} \left(c_{ij,1}, c_{ij,2}, c_{ij,3}, c_{ij,4}\right)_{LR} \left(x_{ij,1}, x_{ij,2}, x_{ij,3}, x_{ij,4}\right)_{LR}$$

$$s.t. \quad \sum_{j=1}^{n} \left(x_{ij,1}, x_{ij,2}, x_{ij,3}, x_{ij,4}\right)_{LR} = \left(a_{i,1}, a_{i,2}, a_{i,3}, a_{i,4}\right)_{LR}, \quad i = 1, 2, \ldots, m,$$

$$\sum_{i=1}^{m} \left(x_{ij,1}, x_{ij,2}, x_{ij,3}, x_{ij,4}\right)_{LR} = \left(b_{j,1}, b_{j,2}, b_{j,3}, b_{j,4}\right)_{LR}, \quad j = 1, 2, \ldots, n,$$

$$\left(x_{ij,1}, x_{ij,2}, x_{ij,3}, x_{ij,4}\right)_{LR} \in \ell\Re(R)^{+}$$

$$(5.79)$$

Equivalently, with regard to Theorems 1.9, 1.12 and 1.13, and Definitions 1.31 and 1.32, the FTP (5.79) may be rewritten as follows:

$$\min \left(\sum_{i=1}^{m} \sum_{j=1}^{n} c_{ij,1} x_{ij,1}, \sum_{i=1}^{m} \sum_{j=1}^{n} c_{ij,2} x_{ij,2}, \sum_{i=1}^{m} \sum_{j=1}^{n} c_{ij,3} x_{ij,3}, \sum_{i=1}^{m} \sum_{j=1}^{n} c_{ij,4} x_{ij,4} \right)_{LR}$$

$$s.t. \quad \sum_{i=1}^{m} x_{ij,1} = a_{i,1}, \quad i = 1, 2, \ldots, m, \qquad\qquad (1)$$

$$\sum_{i=1}^{m} x_{ij,2} = a_{i,2}, \quad i = 1, 2, \ldots, m, \qquad\qquad (2)$$

$$\sum_{i=1}^{m} x_{ij,3} = a_{i,3}, \quad i = 1, 2, \ldots, m, \qquad\qquad (3)$$

$$\sum_{i=1}^{m} x_{ij,4} = a_{i,4}, \quad i = 1, 2, \ldots, m, \qquad\qquad (4)$$

$$\sum_{j=1}^{n} x_{ij,1} = b_{j,1}, \quad j = 1, 2, \ldots, n, \tag{5}$$

$$\sum_{j=1}^{n} x_{ij,2} = b_{j,2}, \quad j = 1, 2, \ldots, n, \tag{6}$$

$$\sum_{j=1}^{n} x_{ij,3} = b_{j,3}, \quad j = 1, 2, \ldots, n, \tag{7}$$

$$\sum_{j=1}^{n} x_{ij,4} = b_{j,4}, \quad j = 1, 2, \ldots, n, \tag{8}$$

$$x_{ij,4} - x_{ij,3} \geq 0, \quad i = 1, 2, \ldots, m, j = 1, 2, \ldots, n, \tag{9}$$
$$x_{ij,3} - x_{ij,2} \geq 0, \quad i = 1, 2, \ldots, m, j = 1, 2, \ldots, n, \tag{10}$$
$$x_{ij,2} - x_{ij,1} \geq 0, \quad i = 1, 2, \ldots, m, j = 1, 2, \ldots, n, \tag{11}$$
$$x_{ij,1} \geq 0, \quad i = 1, 2, \ldots, m, j = 1, 2, \ldots, n, \tag{12}$$

(5.80)

The objective function of the Model (5.80) is obtained by the approximate multiplication of two non-negative *LR* flat fuzzy numbers. Constraints (1)–(8) in the Model (5.80) are simply obtained using the equality of two *LR* flat fuzzy numbers. To impose the non-negative $\tilde{x}_{ij}$ while preserving its form as a non-negative *LR* flat fuzzy number, we have $x_{ij,4} \geq x_{ij,3} \geq x_{ij,2} \geq x_{ij,1} \geq 0$, which corresponds to Constraints (9)–(12) in the Model (5.80).

We recall that Kumar and Kaur [22] used the classical LP problem (5.73) for obtaining the optimal solution of the FTP (5.69). We observe that they minimized $\sum_{i=1}^{m} \sum_{j=1}^{n} \Re(\tilde{c}_{ij}\tilde{x}_{ij})$ instead of $\sum_{i=1}^{m} \sum_{j=1}^{n} \tilde{c}_{ij}\tilde{x}_{ij}$. Therefore, assuming $\tilde{x}^*$ to be an optimal solution of the problem (5.73) with the optimal objective function value $\sum_{i=1}^{m} \sum_{j=1}^{n} \Re\left(\tilde{c}_{ij}\tilde{x}_{ij}^*\right)$ and $\tilde{x}'$ to be a feasible solution of the problem (5.73) with the objective function value $\sum_{i=1}^{m} \sum_{j=1}^{n} \Re\left(\tilde{c}_{ij}\tilde{x}_{ij}'\right)$ such that $\sum_{i=1}^{m} \sum_{j=1}^{n} \Re\left(\tilde{c}_{ij}\tilde{x}_{ij}^*\right) = \sum_{i=1}^{m} \sum_{j=1}^{n} \Re\left(\tilde{c}_{ij}\tilde{x}_{ij}'\right)$, both of these are optimal solutions of the problem (5.73) and also optimal solutions of the FTP (5.69). Thus, according to their approach $\sum_{i=1}^{m} \sum_{j=1}^{n} \tilde{c}_{ij}\tilde{x}_{ij}^* = \sum_{i=1}^{m} \sum_{j=1}^{n} \tilde{c}_{ij}\tilde{x}_{ij}'$, if, and only if, $\sum_{i=1}^{m} \sum_{j=1}^{n} \Re\left(\tilde{c}_{ij}\tilde{x}_{ij}^*\right) = \sum_{i=1}^{m} \sum_{j=1}^{n} \Re\left(\tilde{c}_{ij}\tilde{x}_{ij}'\right)$, which is obviously incorrect. In addition, for the equality constraints of the FTP (5.39), they used Definition 1.31. Hence, in the same method, two different approaches are used for the equality of two fuzzy numbers, which is erroneous. Moreover, as discussed in the following, this approach has a high computational complexity when used for solving the FTP (5.80).

In summary, the existing methods for solving FTPs fall short in five important aspects: (1) not considering fully FTPs; (2) providing non-positive fuzzy optimal solutions and non-positive fuzzy transportation costs; (3) not providing fuzzy optimal solutions; (4) not providing the mathematical form of the membership function of the fuzzy optimal solution and fuzzy transportation cost; and (5) having a high computational complexity.

The model proposed by Ebrahimnejad [34] overcomes the first defect by considering all of the parameters as fuzzy numbers. Constraints (9)–(12) in the Model (5.80) ensure that the optimal solution of the FTP (5.80) is a non-negative fuzzy number. Hence, it is possible to overcome the second and third defects by solving the FTP (5.80). The solution process given in the following ensures that the obtained fuzzy optimal solution and total transportation cost maintain the form of non-negative LR flat fuzzy numbers. Therefore, in contrast to the model proposed by Liu and Kao [8], the mathematical form of the membership function of the fuzzy optimal solution and fuzzy transportation cost is obtained. Finally, a comparison of the complexity between the model proposed by Ebrahimnejad [34] and that proposed by Kumar and Kaur [22] shows that the former is simpler and more computationally efficient.

Now we shall explore the approach proposed by Ebrahimnejad [34] for solving the FTP of type-IV (5.69).

The Model (5.80) is an FTP with one fuzzy variable in the objective function. This model can be solved by converting it into four bounded transportation problems. The main contribution of this method is to reduce the complexity of the method proposed by Kumar and Kaur [22] for solving the FTP (5.80). In addition, a non-negative fuzzy solution can be obtained by using Ebrahimnejad's method as with the method of Kumar and Kaur [22]. We first recall the following order suggested by Ramik and Řimanek [39] and utilized by Okada and Soper [40] for the fuzzy shortest path problem:

Theorem 5.2 *Let* $\tilde{a} = (a_1, a_2, a_3, a_4)_{LR}$ *and* $\tilde{b} = (b_1, b_2, b_3, b_4)_{LR}$ *be two LR flat fuzzy numbers. Then,* $\tilde{a} \preceq \tilde{b}$ *if, and only if,* $a_1 \leq b_1, \ a_2 \leq b_2, \ a_3 \leq b_3, \ a_4 \leq b_4.$

Constraints (9)–(11) in the Model (5.80) are used to preserve the form of the optimal solutions as non-negative LR flat fuzzy numbers. Without these constraints, the feasible space of the FTP (5.80) is separable in terms of $x_{ij,1}, x_{ij,2}, x_{ij,3}$ and $x_{ij,4}$. Thus, Ebrahimnejad [34] proposed a new approach for solving the FTP (5.80) by first removing Constraints (9)–(11) from the feasible space and then decomposing the FTP (5.80) into four bounded transportation problems. He proved that the integration of the optimal solution of the four sub-problems not only provides the optimal solution of the FTP (5.80), but also satisfies Constraints (9)–(11).

Here, we explore his approach for solving the FTP (5.80) with respect to the order given in Theorem 5.2. The steps of the proposed method are as follows:

Step 1: Solve the following transportation problem using the classical transportation simplex algorithm:

$$z_1^* = \min \sum_{i=1}^{m} \sum_{j=1}^{n} c_{ij,1} x_{ij,1}$$

$$s.t. \quad \sum_{i=1}^{m} x_{ij,1} = a_{i,1}, \quad i = 1, 2, \ldots, m, \qquad (1)$$

$$\sum_{j=1}^{n} x_{ij,1} = b_{j,1}, \quad j = 1, 2, \ldots, n, \qquad (2)$$

$$x_{ij,1} \geq 0, \qquad i = 1, 2, \ldots, m, j = 1, 2, \ldots, n, \quad (3)$$

(5.81)

The optimal value of the objective function of the Model (5.81), z_1^*, is the first component of the fuzzy optimal total transportation cost.

Step 2: Solve the following bounded transportation problem using the bounded transportation simplex algorithm [41, 42] assuming that $x_1^* = (x_{ij,1}^*)_{nm \times 1}$ is the optimal solution of the problem (5.81):

$$z_2^* = \min \sum_{i=1}^{m} \sum_{j=1}^{n} c_{ij,2} x_{ij,2}$$

$$s.t. \quad \sum_{i=1}^{m} x_{ij,2} = a_{i,2}, \quad i = 1, 2, \ldots, m, \qquad (1)$$

$$\sum_{j=1}^{n} x_{ij,2} = b_{j,2}, \quad j = 1, 2, \ldots, n, \qquad (2)$$

$$x_{ij,2} \geq x_{ij,1}^*, \qquad i = 1, 2, \ldots, m, j = 1, 2, \ldots, n, \quad (3)$$

(5.82)

The optimal value of the objective function of the Model (5.82), z_2^*, is the second component of the fuzzy optimal total transportation cost.

Proposition 5.1 *The optimal value of the objective function of the Model* (5.81) *is less than or equal to that of the objective function of the Model* (5.82).

Proof Let $x_2^* = (x_{ij,2}^*)_{nm \times 1}$ be the optimal solution of the Model (5.82). On the one hand, Constraint (3) in the Model (5.82) stipulates that $x_{ij,1}^* \leq x_{ij,2}^*$. On the other hand, owing to the form of the fuzzy number $(c_{ij,1}, c_{ij,2}, c_{ij,2}, c_{ij,4})_{LR}$ we have $c_{ij,1} \leq c_{ij,2}$. Therefore, $\sum_{i=1}^{m} \sum_{j=1}^{n} c_{ij,1} x_{ij,1}^* \leq \sum_{i=1}^{m} \sum_{j=1}^{n} c_{ij,2} x_{ij,2}^*$, which means that $z_1^* \leq z_2^*$. $\square$

Step 3: Solve the following bounded transportation problem using the bounded transportation simplex algorithm assuming that $x_2^* = (x_{ij,2}^*)_{nm \times 1}$ is the optimal solution of the problem (5.82):

$$z_3^* = \min \sum_{i=1}^{m} \sum_{j=1}^{n} c_{ij,3} x_{ij,3}$$

$$\text{s.t.} \quad \sum_{i=1}^{m} x_{ij,3} = a_{i,3}, \quad i = 1, 2, \ldots, m, \qquad (1)$$

$$\sum_{j=1}^{n} x_{ij,3} = b_{j,3}, \quad j = 1, 2, \ldots, n, \qquad (2) \qquad (5.83)$$

$$x_{ij,3} \geq x_{ij,2}^*, \qquad i = 1, 2, \ldots, m, j = 1, 2, \ldots, n, \quad (3)$$

The optimal value of the objective function of the Model (5.83), z_3^*, is the third component of the fuzzy optimal total transportation cost. Moreover, the following can be stated (proof omitted).

Proposition 5.2 [34] The optimal value of the objective function of the Model (5.82) is less than or equal to that of the objective function of the Model (5.83).

Step 4: Solve the following bounded transportation problem using the bounded transportation simplex algorithm assuming that $x_3^* = (x_{ij,3}^*)_{nm \times 1}$ is the optimal solution of the problem (5.83):

$$z_4^* = \min \sum_{i=1}^{m} \sum_{j=1}^{n} c_{ij,4} x_{ij,4}$$

$$\text{s.t.} \quad \sum_{i=1}^{m} x_{ij,4} = a_{i,4}, \quad i = 1, 2, \ldots, m, \qquad (1)$$

$$\sum_{j=1}^{n} x_{ij,4} = b_{j,4}, \quad j = 1, 2, \ldots, n, \qquad (2) \qquad (5.84)$$

$$x_{ij,4} \geq x_{ij,3}^*, \qquad i = 1, 2, \ldots, m, j = 1, 2, \ldots, n, \quad (3)$$

The optimal value of the objective function of model (5.84), z_4^*, is the fourth component of the fuzzy optimal total transportation cost. Moreover, the following can be stated (proof omitted).

Proposition 5.3 [34] *The optimal value of the objective function of the Model (5.83) is less than or equal to that of the objective function of the Model (5.84).*

Step 5: Determine the fuzzy optimal solution $\tilde{x}_{ij}^*$ by substituting the values of $x_{ij,1}^*$, $x_{ij,2}^*$, $x_{ij,3}^*$ and $x_{ij,4}^*$ in $\tilde{x}_{ij}^* = (x_{ij,1}^*, x_{ij,2}^*, x_{ij,3}^*, x_{ij,4}^*)_{LR}$.

Step 6: Determine the total minimum fuzzy transportation cost by substituting the values of $\tilde{x}_{ij}^*$ in $\sum_{i=1}^{m} \sum_{j=1}^{n} \tilde{c}_{ij} \tilde{x}_{ij}$.

Theorem 5.3 [34] *The fuzzy optimal total transportation cost, $\tilde{z}^* = (z_1^*, z_2^*, z_3^*, z_4^*)$, maintains the form of a non-negative LR flat fuzzy number.*

Proof This is straightforward because of Propositions 5.1–5.3.   □

Theorem 5.4 [34] *The integration of the optimal solutions of the four sub-problems (5.81)–(5.84) provides the optimal solution of the FTP (5.80).*

Proof Let $\tilde{\bar{x}}_{ij}^* = (\bar{x}_{ij,1}^*, \bar{x}_{ij,2}^*, \bar{x}_{ij,3}^*, \bar{x}_{ij,4}^*)_{LR}$ be an arbitrary fuzzy feasible solution of the FTP (3). It follows that

(i) $\sum_{i=1}^{m} \bar{x}_{ij,1}^* = a_{i,1} (i = 1, 2, \ldots, m)$, $\sum_{j=1}^{n} \bar{x}_{ij,1}^* = b_{j,1} (j = 1, 2, \ldots, n)$,
   $\bar{x}_{ij,1}^* \geq 0 (i = 1, 2, \ldots, m, j = 1, 2, \ldots, n)$

(ii) $\sum_{i=1}^{m} \bar{x}_{ij,2}^* = s_{i,2} (i = 1, 2, \ldots, m)$, $\sum_{j=1}^{n} \bar{x}_{ij,2}^* = d_{j,2} (j = 1, 2, \ldots, n)$,
   $\bar{x}_{ij,2}^* \geq \bar{x}_{ij,1}^* (i = 1, 2, \ldots, m, j = 1, 2, \ldots, n)$

(iii) $\sum_{i=1}^{m} \bar{x}_{ij,3}^* = s_{i,3} (i = 1, 2, \ldots, m)$, $\sum_{j=1}^{n} \bar{x}_{ij,3}^* = d_{j,3} (j = 1, 2, \ldots, n)$,
   $\bar{x}_{ij,3}^* \geq \bar{x}_{ij,2}^* (i = 1, 2, \ldots, m, j = 1, 2, \ldots, n)$

(iv) $\sum_{i=1}^{m} \bar{x}_{ij,4}^* = s_{i,4} (i = 1, 2, \ldots, m)$, $\sum_{j=1}^{n} \bar{x}_{ij,4}^* = d_{j,4} (j = 1, 2, \ldots, n)$,
   $\bar{x}_{ij,4}^* \geq \bar{x}_{ij,3}^* (i = 1, 2, \ldots, m, j = 1, 2, \ldots, n)$

The conditions (i), (ii), (iii) and (iv) imply that $\bar{x}_{ij,1}^*, \bar{x}_{ij,2}^*, \bar{x}_{ij,3}^*$ and $\bar{x}_{ij,4}^*$ are feasible solutions of the problems (5.81)–(5.84), respectively. Moreover, owing to the optimality of $x_{ij,1}^*, x_{ij,2}^*, x_{ij,3}^*$ and $x_{ij,4}^*$, for the problems (5.81)–(5.84), respectively, we conclude that $\sum_{i=1}^{m} \sum_{j=1}^{n} c_{ij,1} x_{ij,1}^* \leq \sum_{i=1}^{m} \sum_{j=1}^{n} c_{ij,1} \bar{x}_{ij,1}^*, \sum_{i=1}^{m} \sum_{j=1}^{n} c_{ij,2} x_{ij,2}^* \leq \sum_{i=1}^{m} \sum_{j=1}^{n} c_{ij,2} \bar{x}_{ij,2}^*, \sum_{i=1}^{m} \sum_{j=1}^{n} c_{ij,3} x_{ij,3}^* \leq \sum_{i=1}^{m} \sum_{j=1}^{n} c_{ij,3} \bar{x}_{ij,3}^*$ and $\sum_{i=1}^{m} \sum_{j=1}^{n} c_{ij,4} x_{ij,4}^* \leq \sum_{i=1}^{m} \sum_{j=1}^{n} c_{ij,4} \bar{x}_{ij,4}^*$, respectively. Regarding Theorem 5.2, this means that

$$\left(\sum_{i=1}^{m} \sum_{j=1}^{n} c_{ij,1} x_{ij,1}^*, \sum_{i=1}^{m} \sum_{j=1}^{n} c_{ij,2} x_{ij,2}^*, \sum_{i=1}^{m} \sum_{j=1}^{n} c_{ij,3} x_{ij,3}^*, \sum_{i=1}^{m} \sum_{j=1}^{n} c_{ij,4} x_{ij,4}^* \right)_{LR}$$
$$\preceq \left(\sum_{i=1}^{m} \sum_{j=1}^{n} c_{ij,1} \bar{x}_{ij,1}^*, \sum_{i=1}^{m} \sum_{j=1}^{n} c_{ij,2} \bar{x}_{ij,2}^*, \sum_{i=1}^{m} \sum_{j=1}^{n} c_{ij,3} \bar{x}_{ij,3}^*, \sum_{i=1}^{m} \sum_{j=1}^{n} c_{ij,4} \bar{x}_{ij,4}^* \right)_{LR}$$

Therefore, $\sum_{i=1}^{m} \sum_{j=1}^{n} \tilde{c}_{ij} \tilde{x}_{ij}^* \preceq \sum_{i=1}^{m} \sum_{j=1}^{n} \tilde{c}_{ij} \tilde{\bar{x}}_{ij}^*$; hence $\tilde{x}_{ij}^* = (x_{ij,1}^*, x_{ij,2}^*, x_{ij,3}^*, x_{ij,4}^*)_{LR}$ is the fuzzy optimal solution of the FTP (5.80).   □

Constraints (3), in the Models (5.81)–(5.84) ensure that the obtained fuzzy optimal solution maintains the form of a non-negative LR flat fuzzy number.

The advantages of the method proposed by Ebrahimnejad [34] over the existing methods for solving FTPs are now discussed.

(1) The proposed approach can be applied for solving fully fuzzy transportation problems where all the parameters are represented as fuzzy numbers.

(2) The optimal solutions are non-negative fuzzy numbers i.e., there are no negative parts in the fuzzy quantities of the commodity or the total fuzzy transportation cost.

(3) In contrast to the existing method [27], the proposed method provides fuzzy optimal solutions that indicate possible outcomes with a certain degree of membership to the decision maker. This is especially useful for strategic decisions in cases where more uncertainty exists.

(4) Because the fuzzy optimal solution of the FTP (5.80) is obtained using classical transportation algorithms, different methods for determining the IBFS of Problems (5.81)–(5.84) such as the north-west-corner method, least-cost method and fuzzy Vogel's approximation method yield the same total transportation cost.

(5) The main advantage of the proposed method is that utilizing problems (5.81)–(5.84) for solving the FTP (5.80) is highly economical compared with the problem (5.73) from a computational viewpoint, regarding the number of constraints and variables. There is a direct relationship between the computational complexity of LP problems and their number of constraints and variables. Because the memory size needed for maintaining the basis (or its inverse) in the simplex algorithm is given by the square of the number of constraints, reducing the number of constraints in LP models is crucial for increasing the computational efficiency. Thus, decreasing the number of constraints and variables in LP models leads to a reduction in the complexity of the LP models solved by the simplex algorithm and almost all of the interior-point methods, such as Khachian's ellipsoid algorithm and Karmarkar's projective algorithm [43]. It has been shown that the LP problem

$$\max \, (\min) \; c^T x$$
$$s.t. \; Ax = b \tag{5.85}$$
$$x \geq 0$$

where A is $m \times n$, can be solved using Khachian's ellipsoid algorithm and Karmarkar's projective algorithm within an effort of $O[n^4 L]$ and $O\left[(m \times n)^6 L\right]$.

Here L is the number of binary bits required to record all the data of the problem and is known as the input length of the LP problem [43]. Thus, there is a direct relationship between the number of constraints and variables in LP problems and the computational complexity of the algorithms. Now, we compare the number of constraints and variables between the problems (5.81)–(5.84) and the problem (5.73). The problem (5.81) has $(m + n)$ constraints (without considering the non-negative constraints $x_{ij,1} \geq 0$) and mn variables (without considering slacks), whereas the problem (5.73) has $4(m+n)+3mn$ constraints and $4mn$ variables. There is a similar relationship in the number of constraints and variables between the problems (5.82)–(5.84) and the problem (5.73). Thus, the problem (5.73) has a complexity of $\left[(4mn)^4 L_1\right]$ and $\left[((4mn)(4(m+n)+3mn))^6 L_1\right]$ in Khachichian's and Karmarkar's algorithms, respectively, whereas the problems (5.81)–(5.84) have the total of complexities $4\left[(mn)^4 L_2\right]$ and $4\left[((mn)(m+n))^6 L_2\right]$, respectively. This shows that the new algorithm is highly economical compared with the problem (5.73) from a computational viewpoint. Importantly, Constraints (3) in the Models (5.82)–(5.84) show that the variables $x_{ij,2}$, $x_{ij,3}$ and $x_{ij,4}$ are bounded. The bounded simplex method handles these constraints implicitly, similarly to the simplex method for handling the constraints $x_{ij,1} \geq 0$. This means that these bounded constraints do not increase the number of constraints directly. Hence, from a computational viewpoint, it is recommended to use the problems (5.81)–(5.83) rather than the problem (5.73) for solving the FTP (5.69) or (5.80).

(6) In contrast to existing methods [20, 26, 27], the approach proposed by Ebrahimnejad [34] does not utilize fuzzy ranking functions for modeling the objective and constraint functions.

(7) Because the classical transportation simplex algorithms are used for solving the FTP, the method proposed by Ebrahimnejad [34] is easy to apply for finding the fuzzy optimal solution of the FTP (5.69) in real-world applications, compared with the existing methods.

(8) It is highlighted by [45, 46] that the computational efforts required to solve an FTP problem can be reduced; if *LR* flat fuzzy numbers are used to convey the subjective evaluations of decision makers. Using such fuzzy numbers allows us to compare the proposed method with most of the existing approaches in the literature. As the proposed scheme considers the extreme points of *LR* flat numbers, the approximated multiplication does not impact the results, i.e., exact formulas yield the same results as the presented approach.

Now, two examples are solved using the Ebrahimnejad's method [34] and the results are discussed. The first application example leads to a balanced FTP, and the second example leads to an unbalanced FTP.

Example 5.7 [34] Reconsider Example 5.6 and the balanced FTP of type-IV given in (5.74). The fuzzy optimal solution of the FTP (5.74) can be obtained using the method proposed by Ebrahimnejad (2016), as follows:

Step 1: The following classical transportation problem is solved using the standard transportation simplex algorithm:

$$\min 19x_{11,1} + 59x_{12,1} + 90x_{13,1} + 150x_{14,1}$$
$$+ 97x_{21,1} + 15x_{22,1} + 110x_{23,1} + 190x_{24,1}$$
$$+ 260x_{31,1} + 240x_{32,1} + 272x_{33,1} + 320x_{34,1}$$
$$s.t. \quad x_{11,1} + x_{12,1} + x_{13,1} + x_{14,1} = 3500,$$
$$x_{21,1} + x_{22,1} + x_{23,1} + x_{24,1} = 3125,$$
$$x_{31,1} + x_{32,1} + x_{33,1} + x_{34,1} = 2475, \quad (5.86)$$
$$x_{11,1} + x_{21,1} + x_{31,1} = 2050,$$
$$x_{12,1} + x_{22,1} + x_{32,1} = 3000,$$
$$x_{13,1} + x_{23,1} + x_{33,1} = 2100,$$
$$x_{14,1} + x_{24,1} + x_{34,1} = 1950,$$
$$x_{ij,1} \geq 0, \quad i = 1, 2, 3, j = 1, 2, 3, 4.$$

The optimal solution of the transportation problem (5.86) is as follows:

$$x_{11,1}^* = 2050, \ x_{12,1}^* = 0, \ x_{13,1}^* = 1450, \ x_{14,1}^* = 0,$$
$$x_{21,1}^* = 0, \ x_{22,1}^* = 3000, \ x_{23,1}^* = 125, \ x_{24,1}^* = 0,$$
$$x_{31,1}^* = 0, \ x_{32,1}^* = 0, \ x_{33,1}^* = 525, \ x_{34,1}^* = 1950, \quad (5.87)$$
$$\sum_{i=1}^{3} \sum_{j=1}^{4} c_{ij,1} x_{ij,1}^* = 995000.$$

Step 2: Now, the following bounded transportation problem is solved using the bounded transportation simplex algorithm [42]:

$$\min 20x_{11,2} + 62x_{12,2} + 95x_{13,2} + 160x_{14,2}$$
$$+ 99x_{21,2} + 17x_{22,2} + 112x_{23,2} + 210x_{24,2}$$
$$+ 262x_{31,2} + 247x_{32,2} + 274x_{33,2} + 326x_{34,2}$$
$$s.t. \quad x_{11,2} + x_{12,2} + x_{13,2} + x_{14,2} = 3550,$$
$$x_{21,2} + x_{22,2} + x_{23,2} + x_{24,2} = 3175,$$
$$x_{31,2} + x_{32,2} + x_{33,2} + x_{34,2} = 2995,$$
$$x_{11,2} + x_{21,2} + x_{31,2} = 2500, \quad (5.88)$$
$$x_{12,2} + x_{22,2} + x_{32,2} = 3050,$$
$$x_{13,2} + x_{23,2} + x_{33,2} = 3150,$$
$$x_{14,2} + x_{24,2} + x_{34,2} = 2025,$$
$$x_{11,2} \geq 2050, x_{12,2} \geq 0, x_{13,2} \geq 1450, x_{14,2} \geq 0,$$
$$x_{21,2} \geq 0, x_{22,2} \geq 3000, x_{23,2} \geq 125, x_{24,2} \geq 0,$$
$$x_{31,2} \geq 0, x_{32,2} \geq 0, x_{33,2} \geq 525, x_{34,2} \geq 1950.$$

The optimal solution of the bounded transportation problem (5.88) is obtained as follows:

$$x^*_{11,2} = 2105, \ x^*_{12,2} = 0, \ x^*_{13,2} = 1450, \ x^*_{14,2} = 0,$$

$$x^*_{21,2} = 0, \ x^*_{22,2} = 3050, \ x^*_{23,2} = 125, \ x^*_{24,2} = 0,$$

$$x^*_{31,2} = 395, \ x^*_{32,2} = 0, \ x^*_{33,2} = 575, \ x^*_{34,2} = 2025, \tag{5.89}$$

$$\sum_{i=1}^{3} \sum_{j=1}^{4} c_{ij,2} x^*_{ij,2} = 1166890.$$

Step 3: Using the optimal solution (5.89), the following transportation problem is solved:

$$\min 21x_{11,3} + 63x_{12,3} + 97x_{13,3} + 165x_{14,3}$$
$$+ 103x_{21,3} + 19x_{22,3} + 115x_{23,3} + 220x_{24,3}$$
$$+ 264x_{31,3} + 249x_{32,3} + 279x_{33,3} + 332x_{34,3}$$

$$s.t. \quad x_{11,3} + x_{12,3} + x_{13,3} + x_{14,3} = 3580,$$
$$x_{21,3} + x_{22,3} + x_{23,3} + x_{24,3} = 3190,$$
$$x_{31,3} + x_{32,3} + x_{33,3} + x_{34,3} = 3275,$$
$$x_{11,3} + x_{21,3} + x_{31,3} = 2700, \tag{5.90}$$
$$x_{12,3} + x_{22,3} + x_{32,3} = 3100,$$
$$x_{13,3} + x_{23,3} + x_{33,3} = 2190,$$
$$x_{14,3} + x_{24,3} + x_{34,3} = 2055,$$
$$x_{11,3} \geq 2105, x_{12,3} \geq 0, x_{13,3} \geq 1450, x_{14,3} \geq 0,$$
$$x_{21,3} \geq 0, x_{22,3} \geq 3050, x_{23,3} \geq 125, x_{24,3} \geq 0,$$
$$x_{31,3} \geq 394, x_{32,3} \geq 0, x_{33,3} \geq 575, x_{34,3} \geq 2025.$$

The optimal solution of the bounded transportation problem (5.90) is given as follows:

$$x^*_{11,3} = 2130, \ x^*_{12,3} = 0, \ x^*_{13,3} = 1450, \ x^*_{14,3} = 0,$$

$$x^*_{21,3} = 0, \ x^*_{22,3} = 3060, \ x^*_{23,3} = 125, \ x^*_{24,3} = 0,$$

$$x^*_{31,3} = 570, \ x^*_{32,3} = 35, \ x^*_{33,3} = 615, \ x^*_{34,3} = 2055, \tag{5.91}$$

$$\sum_{i=1}^{3} \sum_{j=1}^{4} c_{ij,3} x^*_{ij,3} = 1271030.$$

Step 4: Now, according to the optimal solution (5.91), the following bounded transportation problem is solved:

$$\min 22x_{11,4} + 65x_{12,4} + 99x_{13,4} + 170x_{14,4}$$
$$+ 105x_{21,4} + 21x_{22,4} + 119x_{23,4} + 240x_{24,4}$$
$$+ 270x_{31,4} + 255x_{32,4} + 290x_{33,4} + 340x_{34,4}$$
$$s.t. \quad x_{11,4} + x_{12,4} + x_{13,4} + x_{14,4} = 4000,$$
$$x_{21,4} + x_{22,4} + x_{23,4} + x_{24,4} = 3200,$$
$$x_{31,4} + x_{32,4} + x_{33,4} + x_{34,4} = 3400,$$
$$x_{11,4} + x_{21,4} + x_{31,4} = 3050, \qquad (5.92)$$
$$x_{12,4} + x_{22,4} + x_{32,4} = 3200,$$
$$x_{13,4} + x_{23,4} + x_{33,4} = 2250,$$
$$x_{14,4} + x_{24,4} + x_{34,4} = 2100,$$
$$x_{11,4} \geq 2130, x_{12,4} \geq 0, x_{13,4} \geq 1450, x_{14,4} \geq 0,$$
$$x_{21,4} \geq 0, x_{22,4} \geq 3065, x_{23,4} \geq 125, x_{24,4} \geq 0,$$
$$x_{31,4} \geq 570, x_{32,4} \geq 35, x_{33,4} \geq 615, x_{34,3} \geq 2055.$$

The optimal solution of the bounded transportation problem (5.92) is given as follows:

$$x_{11,4}^* = 2480, \ x_{12,4}^* = 10, \ x_{13,4}^* = 1510, \ x_{14,4}^* = 0,$$
$$x_{21,4}^* = 0, \ x_{22,4}^* = 3075, \ x_{23,4}^* = 125, \ x_{24,4}^* = 0,$$
$$x_{31,4}^* = 570, \ x_{32,4}^* = 115, \ x_{33,4}^* = 615, \ x_{34,4}^* = 2100, \qquad (5.93)$$
$$\sum_{i=1}^{3} \sum_{j=1}^{4} c_{ij,4} x_{ij,4}^* = 1359725.$$

Step 5: By substituting the values of $x_{ij,1}^*$, $x_{ij,2}^*$, $x_{ij,3}^*$ and $x_{ij,4}^*$ in $\tilde{x}_{ij}^* = (x_{ij,1}^*, x_{ij,2}^*, x_{ij,3}^*, x_{ij,4}^*)_{LR}$, the fuzzy optimal solution of the FTP (5.74) is obtained as follows:

$$\tilde{x}_{11}^* = (2050, 2105, 2130, 2480)_{LR}, \ \tilde{x}_{12}^* = (0, 0, 0, 10)_{LR},$$
$$\tilde{x}_{13}^* = (1450, 1450, 1450, 1510)_{LR}, \ \tilde{x}_{14}^* = (0, 0, 0, 0)_{LR},$$
$$\tilde{x}_{21}^* = (0, 0, 0, 0)_{LR}, \ \tilde{x}_{22}^* = (3000, 3050, 3065, 3075)_{LR},$$
$$\tilde{x}_{23}^* = (125, 125, 125, 125)_{LR}, \ \tilde{x}_{14}^* = (0, 0, 0, 0)_{LR}, \qquad (5.94)$$
$$\tilde{x}_{31}^* = (0, 395, 570, 570)_{LR}, \ \tilde{x}_{32}^* = (0, 0, 35, 115)_{LR},$$
$$\tilde{x}_{33}^* = (525, 575, 615, 615)_{LR}, \ \tilde{x}_{34}^* = (1950, 2025, 2055, 2100)_{LR}.$$

Step 6: By substituting the values of $x_{ij,1}^*$, $x_{ij,2}^*$, $x_{ij}^*(i = 1, 2, 3, j = 1, 2, 3, 4)$ in the objective function of the FTP (10), the total fuzzy transportation cost is determined as follows:

$$\sum_{i=1}^{3}\sum_{j=1}^{4} \tilde{c}_{ij}\tilde{x}_{ij} = (999500, 1166890, 1271030, 1359725)_{LR}$$

Now, the results of the FTP (5.74) obtained using the existing and Ebrahimnejad's methods are compared.

(1) There is no negative part in the fuzzy optimal solution or the total fuzzy transportation cost obtained using the proposed method, whereas the existing methods [17, 18, 21, 25] yield fuzzy optimal solutions with negative parts in the fuzzy quantities of some products to be transported from origins to destinations, which has no physical meaning.
(2) To solve the FTP (5.74) with regard to the problem (5.73), the LP problem (5.75) is solved. Although the results obtained using the methods proposed by Kumar and Kaur [22, 38] and the method proposed by Ebrahimnejad [34] are the same (see the results given in (5.77) and (5.94)), there are two important reasons for using the method proposed by Ebrahimnejad [34]:

- The classical LP problem (5.75) applied for solving the FTP (5.74) is not a transportation structured LP problem, whereas the problems (5.86), (5.88), (5.90) and (5.92) for solving the FTP (5.74) are classical transportation problems.
- The classical LP problem (5.75) applied for solving the FTP (5.74) has 64 constraints (without considering the non-negative constraints) and 48 variables (without considering slack variables), whereas the problem (5.86) has only 7 constraints and 12 variables. There is a similar relationship in the numbers of constraints and variables between the problems (5.88), (5.90), (5.92) and the problem (5.75). Hence, utilizing the problems (5.86), (5.88), (5.90), and (5.92) for solving the FTP (5.74) is highly economical compared with the problem (5.75) from a computational viewpoint, regarding the number of constraints and variables.

Hence, from a computation point of view the method proposed by Ebrahimnejad [34] is preferable to the existing methods for solving the FTP.

Remark 5.10 In this example, the unit transportation cost and values of the supply and demand are decided by the trader based on his/her professional knowledge, experience, and the available information. Linguistic values characterized by trapezoidal fuzzy numbers (special case of *LR* flat fuzzy numbers with $L(x) = R(x) = \max \{0, 1 - x\}$) defined on [0, 1] are utilized to describe the uncertainty with the available information. For example, the unit transportation cost between Reipur and Kullu, which is "approximately between 274 and 279" can be represented by a trapezoidal fuzzy number (272, 274, 279, 290). This means that the most probable cost is between 274 and 279. The least probable costs are 272 (optimistic value) and 290 (pessimistic value). The cost is unlikely to be less than 272 or larger than 290, in the opinion of the trader. Between 272 and 274, the degree of the membership function increases monotonically and between 279 and 290, it decreases monotonically.

Remark 5.11 [47] In a fuzzy decision environment, linguistic values may be used to convey the subjective judgments and preferences of decision makers. Numerous methods for constructing membership functions, almost invariably based on the judgments of experts, are described in the literature. The least-square curve fitting method is used for constructing a membership function from samples of membership grades for elements of a given universal. This method requires the selection of a suitable parameterized class of functions. The choice may reflect the opinion of an expert or, alternatively, be based on a theory or previous experience. An example of a class of functions frequently used for representing linguistic terms is *LR* flat membership functions.

Now, we solve an unbalanced FTP and show that the fuzzy optimal solution and minimal total fuzzy cost transformation cost, obtained using the method proposed by Ebrahimnejad [34] and the existing methods [22, 38] are the same, whereas the total number of constraints and variables in the converted transportation problems in the method proposed by Ebrahimnejad [34] is smaller than that in the existing methods. The main purpose of this example is to demonstrate the efficiency of the proposed methods of this section for solving an unbalanced FTP.

Example 5.8 [22, 34] A company has two sources O_1 and O_2, and three destinations D_1, D_2 and D_3. The fuzzy transportation cost for a unit quantity of the commodity from the ith source to the jth destination is

$$\tilde{c}_{ij} = \begin{bmatrix} (10, 20, 30, 40)_{LR} & (50, 60, 70, 90)_{LR} & (80, 90, 110, 120)_{LR} \\ (60, 70, 80, 90)_{LR} & (70, 80, 100, 120)_{LR} & (20, 30, 50, 60)_{LR} \end{bmatrix}$$

The fuzzy supply of the commodity at the first and second origins is $(70, 90, 90, 100)_{LR}$ and $(40, 60, 70, 80)_{LR}$, respectively, and the fuzzy demand for the product at the first, second and third destinations is $(30, 40, 50, 70)_{LR}, (20, 30, 40, 50)_{LR}$ and $(40, 50, 50, 80)_{LR}$, respectively. Here, $L(x) = R(x) = \max\{0, 1 - |x|\}$. The company wishes to determine the fuzzy quantity of the commodity that should be transported from each origin to each destination so that the total fuzzy transportation cost is minimized.

The total fuzzy supply is $\sum_{i=1}^{2} \tilde{s}_i = (110, 150, 160, 180)_{LR}$ and the total fuzzy demand is $\sum_{j=1}^{3} \tilde{d}_j = (90, 120, 140, 200)_{LR}$; therefore, this is an unbalanced FTP. According to the method proposed by Kumar and Kaur [22], discussed at the beginning of this section, the unbalanced FTP can be converted into a balanced FTP, by first introducing an imaginary supply $\tilde{s}_3 = (0, 0, 10, 50)_{LR}$ and imaginary demand $\tilde{d}_4 = (20, 30, 30, 30)_{LR}$ such that $\sum_{i=1}^{2} \tilde{s}_i = (110, 150, 170, 230)_{LR} =$

$\sum_{j=1}^{3} \tilde{d}_j$ and then assigning the fuzzy costs for transporting a unit quantity of the commodity from an imaginary origin to all destinations and from all sources to the imaginary origin as zero LR flat fuzzy numbers. The balanced FTP obtained can be formulated as follows:

$$
\begin{aligned}
\min = {} & (19, 20, 21, 22)_{LR}\tilde{x}_{11} + (59, 62, 63, 65)_{LR}\tilde{x}_{12} + (90, 95, 97, 99)_{LR}\tilde{x}_{13} \\
& + (150, 160, 165, 170)_{LR}\tilde{x}_{14} + (97, 99, 103, 105)_{LR}\tilde{x}_{21} + (15, 17, 19, 21)_{LR}\tilde{x}_{22} \\
& + (110, 112, 115, 119)_{LR}\tilde{x}_{23} + (190, 210, 220, 240)_{LR}\tilde{x}_{24} + (260, 262, 264, 270)_{LR}\tilde{x}_{31} \\
& + (240, 247, 249, 255)_{LR}\tilde{x}_{32} + (272, 274, 279, 290)_{LR}\tilde{x}_{33} + (320, 326, 332, 340)_{LR}\tilde{x}_{34}
\end{aligned}
$$

$s.t.$
$$
\begin{aligned}
& \tilde{x}_{11} + \tilde{x}_{12} + \tilde{x}_{13} + \tilde{x}_{14} = (70, 90, 90, 100)_{LR}, \\
& \tilde{x}_{21} + \tilde{x}_{22} + \tilde{x}_{23} + \tilde{x}_{24} = (40, 60, 70, 80)_{LR}, \\
& \tilde{x}_{31} + \tilde{x}_{32} + \tilde{x}_{33} + \tilde{x}_{34} = (0, 0, 10, 50)_{LR}, \\
& \tilde{x}_{11} + \tilde{x}_{21} + \tilde{x}_{31} = (30, 40, 50, 70)_{LR}, \\
& \tilde{x}_{12} + \tilde{x}_{22} + \tilde{x}_{32} = (20, 30, 40, 50)_{LR}, \\
& \tilde{x}_{13} + \tilde{x}_{23} + \tilde{x}_{33} = (40, 50, 50, 80)_{LR}, \\
& \tilde{x}_{14} + \tilde{x}_{24} + \tilde{x}_{34} = (20, 30, 30, 30)_{LR},
\end{aligned}
$$

$\tilde{x}_{11}, \tilde{x}_{12}, \tilde{x}_{13}, \tilde{x}_{14}, \tilde{x}_{21}, \tilde{x}_{22}, \tilde{x}_{23}, \tilde{x}_{24}, \tilde{x}_{31}, \tilde{x}_{32}, \tilde{x}_{33}, \tilde{x}_{34}$, are non-negative triangular fuzzy numbers.

$$(5.95)$$

Using the Steps 1-6 of Ebrahimnejad's method [34], the fuzzy optimal solution and the minimum total fuzzy transportation cost are obtained as follows:

$$
\begin{aligned}
& \tilde{x}_{11}^{*} = (30, 40, 40, 50)_{LR}, \ \tilde{x}_{12}^{*} = (20, 30, 30, 30)_{LR}, \ \tilde{x}_{13}^{*} = (0, 0, 0, 0)_{LR}, \\
& \tilde{x}_{14}^{*} = (20, 20, 20, 20)_{LR}, \ \tilde{x}_{21}^{*} = (0, 0, 10, 10)_{LR}, \ \tilde{x}_{22}^{*} = (0, 0, 0, 0)_{LR}, \\
& \tilde{x}_{23}^{*} = (40, 50, 50, 60)_{LR}, \ \tilde{x}_{14}^{*} = (0, 10, 10, 10)_{LR}, \tilde{x}_{31}^{*} = (0, 0, 0, 0)_{LR}, \\
& \tilde{x}_{32}^{*} = (0, 0, 10, 20)_{LR}, \ \tilde{x}_{33}^{*} = (0, 0, 0, 20)_{LR}, \ \tilde{x}_{34}^{*} = (0, 0, 0, 0)_{LR}.
\end{aligned}
$$

$$
\sum_{i=1,2,3}^{+} \sum_{j=1,2,3,4}^{+} \tilde{c}_{ij} \otimes \tilde{x}_{ij} = (2100, 4100, 6600, 9200)_{LR}
$$

$$(5.96)$$

The results, shown in (5.96), indicate the Kumar and Kaur's method [22] and Ebrahimnejad's method [34] yield the same results for solving the unbalanced FTP (5.95). However, Ebrahimnejad's method is preferable considering the number of constraints and variables in the converted classical transportation problem.

The minimum total transportation cost is $(2100, 4100, 6600, 9200)_{LR}$, which means that it will always be greater than 2100 and less than 9200. It is likely to be between 4100 and 6600.

5.7 Conclusions

In traditional transportation problems, it is assumed that the decision maker has exact information about the coefficients belonging of the problem. However, in real-life situations, the transportation cost, supply and demand of a commodity may not be known precisely because of uncontrollable factors. To handle such situations, fuzzy set theory is applied in the literature for solving transportation problems. In this chapter we investigated four main FTPs: (1) the FTPs with fuzzy transportation costs; (2) the FTPs with fuzzy supply and demand as well as fuzzy decision variables; (3) the FTPs with fuzzy transportation cost and fuzzy supply and demand; and (4) the fully FTPs in which fuzziness exists in all of the model parameters (costs, supply and demand) as well as decision variables. Then, we explored several approaches for solving the FTPs belonging to each group. In the FTP considered in the fourth group, as the general case of the groups (1)–(3), all the parameters were represented by non-negative *LR* flat fuzzy numbers. Because the proposed methods for solving this kind of FTPs are based on the classical simplex algorithms, they are easy to learn and apply for obtaining fuzzy optimal solutions of balanced and unbalanced FTPs pertaining to real-world applications.

References

1. Zimmerman, H.J.: Fuzzy programming and linear programming with several objective functions. Fuzzy Sets Syst. **1**(1), 45–55 (1978)
2. Oheigeartaigh, M.: A fuzzy transportation algorithm. Fuzzy Sets Syst. **8**(3), 235–243 (1982)
3. Chanas, S., Kolodziejczyk, W., Machaj, A.: A fuzzy approach to the transportation problem. Fuzzy Sets Syst. **13**(3), 211–221 (1984)
4. Chanas, S., Delgado, M., Verdegay, J.L., Vila, M.A.: Interval and fuzzy extensions of classical transportation problems. Transp. Plan. Technol. **17**(2), 203–218 (1993)
5. Chanas, S., Kuchta, D.: A concept of the optimal solution of the transportation problem with fuzzy cost coefficients. Fuzzy Sets Syst. **82**(2), 299–305 (1996)
6. Jimenez, F., Verdegay, J.L.: Uncertain solid transportation problem, Fuzzy Sets Syst. **100**(1–3), 45–57 (1998)
7. Jimenez, F., Verdegay, J.L.: Solving fuzzy solid transportation problems by an evolutionary algorithm based parametric approach. Eur. J. Oper. Res. **117**(3), 485–510 (1999)
8. Liu, S.T., Kao, C.: Solving fuzzy transportation problems based on extension principle. Eur. J. Oper. Res. **153**(3), 661–674 (2004)
9. Liu, S.-T.: Fuzzy total transportation cost measures for fuzzy solid transportation problem. Appl. Math. Comput. **174**(2), 927–941 (2006)
10. Chiang, J.: The optimal solution of the transportation problem with fuzzy demand and fuzzy product. J. Inf. Sci. Eng. **21**, 439–451 (2005)
11. Gupta, A., Kumar, A.: A new method for solving linear multi-objective transportation problems with fuzzy parameters. Appl. Math. Model. **36**, 1421–1430 (2012)
12. Liang, T.F., Chiu, C.S., Cheng, H.W.: Using possibilistic linear programming for fuzzy transportation planning decisions. Hsiuping J. **11**, 93–112 (2005)
13. Gani, A., Razak, K.A.: Two stage fuzzy transportation problem. J. Phys. Sci. **10**, 63–69 (2006)

14. Li, L., Huang, Z., Da, Q., Hu, J.: A new method based on goal programming for solving transportation problem with fuzzy cost. In: International Symposium on Information Processing, pp. 3–8 (2008)
15. Chen, M., Ishii, H., Wu, C.: Transportation problems on a fuzzy network. Int. J. Innov. Comput. Inf. Control **4**, 1105–1109 (2008)
16. Lin, F.T.: Solving the transportation problem with fuzzy coefficients using genetic algorithms. In: IEEE International Conference on Fuzzy Systems, pp. 1468–1473 (2009)
17. Dinagar, D.S., Palanivel, K.: The transportation problem in fuzzy environment. Int. J. Algorithms, Comput. Math. **2**(3), 65–71 (2009)
18. Pandian, P., Natarajan, G.: A new algorithm for finding a fuzzy optimal solution for fuzzy transportation problems. Appl. Math. Sci. **4**(2), 79–90 (2010)
19. Chakraborty, A., Chakraborty, M.: Cost-time minimization in a transportation problem with fuzzy parameters: a case study. J. Transp. Syst. Eng. Inf. Technol. **10**(6), 53–63 (2010)
20. Kumar, A., Kuar, A.: Optimal solution of fuzzy transportation problems based on fuzzy linear programming formulation. J. Adv. Res. Appl. Math. **2**(4), 70–84 (2010)
21. Kumar, A., Kaur, A.: A new method for solving fuzzy transportation problems using ranking function. Appl. Math. Model. **35**(12), 5652–5661 (2011)
22. Kumar, A., Kaur, A.: Application of classical transportation methods to find the fuzzy optimal solution of fuzzy transportation problems. Fuzzy Inf. Eng. **3**(1), 81–99 (2011)
23. Senthilkumar, P., Vengataasalam, S.: A note on the solution of fuzzy transportation problem using fuzzy linear system. J. Fuzzy Set Valued Anal. 1–9 (2013)
24. Ojha, A., Das, B., Mondal, S.K., Maiti, M.: A multi-item transportation problem with fuzzy tolerance. Appl. Soft Comput. **13**, 3703–3712 (2013)
25. Shanmugasundari, M., Ganesan, K.: A novel approach for the fuzzy optimal solution of fuzzy transportation problem. Int. J. Eng. Res. Appl. **3**(1), 1416–1421 (2013)
26. Kaur, A., Kumar, A.: A new approach for solving fuzzy transportation problems using generalized trapezoidal fuzzy numbers. Appl. Soft Comput. **12**(3), 1201–1213 (2012)
27. Ebrahimnejad, A.: A simplified new approach for solving fuzzy transportation problems with generalized trapezoidal fuzzy numbers. Appl. Soft Comput. **19**, 171–176 (2014)
28. Ebrahimnejad, A.: On solving transportation problems with triangular fuzzy numbers: review with some extensions. In: IEEE, 13th Iranian Conference on Fuzzy Systems, pp. 1–4 (2013)
29. Ebrahimnejad, A.: An improved approach for solving fuzzy transportation problem with triangular fuzzy numbers. J. Intel. Fuzzy Syst. **29**, 963–974 (2015)
30. Ebrahimnejad, A.: Fuzzy linear programming approach for solving transportation problems with interval-valued trapezoidal fuzzy numbers. Sadhana **41**(3), 299–316 (2016)
31. Sudhagar, C., Ganesan, K.: A fuzzy approach to transport optimization problem. Optim. Eng. **17**(4), 965–980 (2016)
32. Ebrahimnejad, A.: Note on A fuzzy approach to transport optimization problem. Optim. Eng. **17**(4), 981–985 (2016)
33. Ebrahimnejad, A.: A lexicographic ordering-based approach for solving fuzzy transportation problems with triangular fuzzy numbers. Int. J. Manag. Decis. Mak. **16**(4), 346–374 (2017)
34. Ebrahimnejad, A.: New method for solving fuzzy transportation problems with LR flat fuzzy numbers. Inf. Sci. **357**, 108–124 (2016)
35. Ebrahimnejad, A., Verdegay, J.L.: An efficient computational approach for solving type-2 intuitionistic fuzzy numbers based transportation problems. Int. J. Comput. Intell. Syst. **9**(6), 1154–1173 (2016)
36. Ebrahimnejad, A., Verdegay, J.L.: A new approach for solving fully intuitionistic fuzzy transportation problems. Fuzzy Optim. Decis. Mak. (2017). https://doi.org/10.1007/s10700-017-9280-1
37. Reklaitis, G.V., Ravindran, A., Ragsdell, K.M.: Engineering Optimization. Wiley, New York (1983)
38. Kumar, A., Kaur, A.: Methods for solving unbalanced fuzzy transportation problems. Oper. Res. **12**(3), 287–316 (2012)

39. Ramik, J., Rimanek, J.: Inequality relation between fuzzy numbers and its use in fuzzy optimization. Fuzzy Sets Syst. **16**(2), 123–138 (1985)
40. Okada, S., Soper, T.: A shortest path problem on a network with fuzzy arc lengths. Fuzzy Sets Syst. **109**, 129–140 (2000)
41. Dantzig, G.B., Thapa, M.N.: Linear Programming: 2: Theory and Extensions. Springer, Princeton University Press, Princeton (1963)
42. Murty, G.H.: Linear Programming. Wiley, New York (1983)
43. Bazaraa, M.S., Jarvis, J.J., Sherali, H.D.: Linear Programming and Network Flows, 3rd edn. Wiley-Interscience, Wiley, Hoboken (2005)
44. Ebrahimnejad, A.: Sensitivity analysis in fuzzy number linear programming problems. Math. Comput. Model. **53**(9–10), 1878–1888 (2011)
45. Dubois, D., Prade, H.: Fuzzy Sets and Systems: Theory and Applications. Academic Press, New York (1980)
46. Zimmermann, H.J.: Fuzzy Set Theory and Its Applications. Kluwer Academic Publisher, Dordrecht (2001)
47. Klir, G.J., Yuan, B.: Fuzzy Sets and Fuzzy Logic, Theory and Applications. Prentice-Hall, PTR, Englewood Cliffs (1995)

Printed in the United States
By Bookmasters